VARMINTS

VARMINTS

Mystery Carnivores
of North America

CHAD ARMENT

COACHWHIP PUBLICATIONS

Landisville, Pennsylvania

CONTENTS

ACKNOWLEDGMENTS

Special thanks for varmint-related material in this project goes to Gary Mangiacopra, Rod Dyke, Kevin Stewart, Nick Sucik, Jerry A. Padilla, John Warms, Tal H. Blanco, Ron Schaffner, and Craig Heinselman.

INTRODUCTION

This book is a preliminary cryptozoological investigation into varmint folklore in North America. Varmints, defined for the sake of this volume, are terrestrial carnivorous mammals (canines, felines, bears, and smaller carnivores) that pose a threat to livestock, poultry, domestic animals, and even humans. We're all familiar with urban legends, but rural stories can be just as imaginative. The main difference is that in rural areas, livelihoods depend upon wary vigilance against predators—and there is no doubt that livestock kills occur. But are the animals blamed always the true culprits? And, more to the point, when someone claims to see a predator that isn't recognized by science, is that individual necessarily mistaken?

Cryptozoology, for those who are not familiar with the term, is the methodological investigation of mystery animals—animals for which there is circumstantial evidence (eyewitnesses, tracks, photographs, etc.), but no positive proof and thus no scientific description. Cryptozoology allows a researcher to objectively examine what evidence is so far known, in order to determine methods by which physical evidence can be obtained and analyzed to confirm whether or not a new species has been discovered.

This book is intended as a preliminary investigation: gathering and compiling sighting reports of alleged animals distinct from known species. Using this book as a guide, I believe the local investigator will be able to focus on specific areas, habitats, and stories in order to acquire better evidence (and perhaps even locate lost specimens). In some cases these stories may provide enough details to determine that a misidentification or hoax is likely involved.

Mystery animals are *ethnoknown*—that is, they have acquired a certain degree of recognition by any community (geographical, cultural, or other) that share stories. Otherwise, we would not have heard about these creatures. But, mystery animals are also *folkloric*. They carry cultural baggage: assumptions brought on and influenced by fear, surprise, exaggeration, ridicule, and superstition of the unknown. Even a trustworthy eyewitness may get details wrong or believe they have seen something they really have not. Sightings of strange creatures, especially predators, feed fear—particularly when the animal can take down a cow, horse, or child.

There are many potential explanations that could account for mystery animals: misidentifications, escaped exotics, and hoaxes are all possible. Some mystery animals are just imaginative creations, with no basis in reality. But, there is also the possibility, however remote, that certain sightings relate to species that haven't yet been scientifically described, or which are believed extinct.

Too many people write off North America as unable to produce significant new species. That is a mistake. (See my earlier book, *Cryptozoology: Science & Speculation*, for discussion on that subject.) While there is no call for indiscriminate belief that any and all mystery animals must have their roots in as yet undescribed creatures, there also is no support for wholesale dismissal. Investigation doesn't require reckless belief or disbelief—only action. If there are any unknown predators in the North American landscape, their continued survival (in this day of habitat loss and environmental change) hinges on scientific acknowledgment, and cryptozoology is the methodology with the best chance for such an awareness to emerge.

WHAT ACCOUNTS FOR VARMINTS?

When you look through the state-by-state listings in this book, you will notice that the accounts usually make an effort to distinguish the animal from known species, either from physical description, behavior of the animal, or because the witness "knew" it wasn't a bear (or dog or coyote or cat, etc.). These accounts are the exception, not the rule. There are far more historical cases of predatory animals and livestock kills where the culprit very likely was a known predator, but there wasn't enough of a description to determine which one. (And, of course, there are many cases where the varmint was positively identified as a known species.) I have tried not to include vague cases, except where the varmint was given a colloquial name. Such a name is the first folkloric step in creating the perception of a distinctive animal, separating it from the commonality of known predators and elevating it to ethnoknown status, even if it later turns out to be mundane.

Of course, in many of the recorded cases here, even when there is circumstantial evidence for an unknown varmint, there is usually a "voice of authority" (wildlife officer, biologist, even a newspaper editor) who will caution people not to panic, that the animal is probably just a wandering bear or coyote. (Again, see my earlier book, *Cryptozoology: Science & Speculation* for details on the way newspapers sometimes, often unconsciously, manipulate a news story to fit a folkloric pattern.)

It is true that misidentifications happen. Particularly now, with the proliferation of digital cameras, we can look at a witness' photograph of a "mountain lion" or "black panther" within hours of a sighting and easily see that the witness has misidentified a large domestic cat. Even when this is pointed out, however, some witnesses are unable to move past their own conception of what the animal is. Many people just do not have a good grasp on animal identification. It isn't uncommon for rural folk to claim that people living in cities or suburbs don't know how to identify wildlife—yet, I see misidentifications coming from the rural populace also. (I have noticed, from my interest in herpetology, that this is particularly true with reptile identification. I recall one farmer in Ohio handing me a frozen bag of "water moccasins" that were nothing more than common water snakes.) Identification skills require time and effort, not just seeing the occasional picture in a book or watching a few wildlife documentaries. And, as a brief encounter with a predatory-looking creature is likely to induce shock and possibly fear, it's not unexpected for details to be mistaken. Misidentification will be the largest source of false leads for an investigator. Individual cases, where only a single sighting of a mystery animal from a particular region or time period is known, should be treated with special caution.

Exotic species are also a possible source of varmint sightings. Some cryptozoology researchers try to downplay what they call "circus train wreck" theories, but the fact is, exotic carnivores have been commonly kept in circuses, zoos, traveling shows, and private menageries since the colonial period. Wherever you have wild animals in captivity, you have wild animals that escape. Most will not survive long, or will be quickly recaptured, but there will always be a percentage that survives in the wild.

The range of exotic species kept in captivity today is quite large, though it's kept mostly under the radar of the general public. There are federal regulations for the breeding or exhibition of exotic wildlife, but not simply for their possession, unless they are a protected species. State regulations vary, and an animal in one state may be illegal to possess, while being perfectly legal the next state over. We have seen cases where exotic owners move to another state without realizing the legal consequences, and of course there are many instances where an exotic animal just becomes too much trouble for the owner. Unfortunately, unless the individual is willing to give the animal to a rescue, they sometimes end up dumped in the wild. (I've seen more cases of alligators being found here in the Mid-Atlantic states in the last few years than are reported in the entire "out-of-place gator files" of published Forteana.) We've seen numerous cases of mammalian predators (especially felines) escaping from captivity over the last decade. In some cases, the owner acts responsibly and alerts the authorities; other times, the animal ends up dead with its origin a mystery. There are other exotics that are reported seen, but disappear into the landscape never to be seen again, leaving the public to wonder where they came from, and whether the sightings were real or not.

Hoaxing is another potential source of sighting reports, though perhaps not as common as in other cryptozoological categories. (See my other volumes, *The Historical Bigfoot* and *Boss Snakes*.) Still, hoaxes are known. In some cases they were used to scare people out of certain areas (to protect berry patches or illegal activities), while others just created a public stir (i.e., bored youth using bullroarers). Given the recent Georgia Bigfoot hoax, I think it's appropriate to show that this sort of stunt isn't new. The following is a series of newspaper articles detailing a hoax that sent Los Angeles into a frenzy before it was finally over.

PANTHER HUNT ALARMS L. A.; 150 COPS OUT
Charleston, West Virginia, *Daily Mail*, March 16, 1954

Los Angeles (UP)—A 140-pound black panther was reported loose in a residential area near downtown Los Angeles today, and police said children were thoughtlessly risking their lives to get a look at the big cat.

Some 150 policemen, armed with rifles and shotguns, searched alleys, yards and streets for the cat. At least six officers reported seeing the panther since it was first spotted in an alley late Monday night.

Prowlers Blamed

Other officers, however, expressed doubt that a panther could have been at large in the city three days without being seen. They said the other policemen may have mistaken another large animal for the panther in the darkness Monday night.

The first report came about five hours after a carnival owner reported the panther had escaped from its cage Friday.

Wayne Roberts told officers that a big cat, plus two trained honey-bears, two monkeys, five rare pigeons, a kinkajou and a chimpanzee were set free from a truck, presumably by prowlers who opened their cages.

The chimp and kinkajou were recaptured.

Most residents of the neighborhood watched the search from behind closed windows, and police warned citizens to stay out of the area.

But officers said that many children, unconscious of the danger, ignored police orders and formed their own beating parties, using sticks and clubs.

"They keep coming back as fast as we send them away," one policeman said. "If any of them should stumble across the cat it's curtains."

Black panthers, which are actually a color phase of the common spotted leopard, are among the world's most dangerous game animals.

SEARCH FOR BLACK PANTHER TURNS BRIEFLY TO YUCAIPA

Redlands, California, *Daily Facts*, March 17, 1954

The search for the "black panther" that is thought to have escaped from a carnival truck in Los Angeles or on the road to Phoenix last Friday shifted to Yucaipa for a time yesterday.

The investigation there was just one of many made in Los Angeles and elsewhere after reports had come in of evidence that the panther had been around.

However Police Chief Parker, Los Angeles, said yesterday he is not at all convinced the panther escaped in Los Angeles.

A study of tracks at 328 South 8th street, Yucaipa, indicated that a large animal of the feline family, probably a bobcat, had been prowling in the area, according to Deputy Joseph Glines.

Dep. Sheriff Joseph Glines said he received a call Tuesday morning from Mrs. W. G. Runnerstrom of the 8th street address in Yucaipa. Mrs. Runnerstrom told Glines that about 3 a.m. Tuesday she was awakened by an "ungodly screech, too shrill and loud to be uttered by a cat."

She looked out the window, she said, and in her yard saw an animal "bigger than a cat but not quite as big as a large dog." She said she could not tell the color of the animal.

This morning Mrs. Runnerstrom told the Facts she saw the animal in the moonlight as it wandered around in her backyard only about 20 ft. from her bedroom window.

She said it crossed her back porch and she heard it lapping water from the watering pan that she had by a back yard spigot for her dog.

The dog barked furiously but when she opened the back door to send the dog out to chase the animal away, the dog apparently smelled it and refused to leave the kitchen.

The most noticeable footprint was a large one, larger than the palm of a man's hand. About the same time some neighboring cattle stampeded and ran through the Runnerstrom property.

Mrs. Runnerstrom said the scream she heard was definitely not from a domestic animal, she had not heard a scream like that since she moved away from the vicinity of the zoo at Washington, D. C., where she lived for several years. She didn't couple the Tuesday morning incident with the "Los Angeles panther" because at the time she didn't know about it escaping.

Her son-in-law, James Holden, said that he wouldn't be a bit surprised that it was "the panther" because—something screamed, something was seen, something lapped water, something made the footprints and something stampeded the neighbor's cattle.

Also a small flock of "strange" pigeons was seen in the vicinity. It is possible they escaped at the same time as the panther and were still in the neighborhood.

By coincidence this all took place within a few blocks of Yucaipa Junior High school—home of the Panthers and there are lots of Panthers there.

Glines examined tracks found in the yard and said they might be those of a mountain lion, a bobcat or a kind of lynx cat. He said that a number of these animals will come down from their mountain lairs in the springtime searching for food.

Glines said he doubted the tracks were those of a black panther. He said the tracks of a black panther show three pads and no claws. The tracks in the Yucaipa back yard outline four pads, and some of them show claw marks.

Glines said that although no concerted search is planned, all lawmen in the county's east end are "keeping their eyes open all the time."

Another call, received Monday night, was from a man on 5th street in Yucaipa, who said he saw "what looked like a large dog" in his yard.

Los Angeles (UP) — Frustrated authorities wondered where to look next today for a missing black panther.

Undisguised skepticism was voiced by several police officers who spent hours yesterday racing to various parts of metropolitan Los Angeles investigating reports by frightened citizens.

Wayne Roberts, owner of the 140-pound panther, stuck by his story that the panther escaped from his carnival last Friday and is on the prowl somewhere in Southern California.

"This is no hoax," he said. "The panther is loose. I only wish it were a hoax."

Police said two honeybears, which escaped with the panther while they were being transported from here to Arizona Friday, were captured at Banning, Calif., 85 miles east of Los Angeles and about 25 miles east of Yucaipa.

PANTHER HUNT PUZZLES POLICE

Kalispell, Montana, *Daily Inter Lake*, March 17, 1954

Los Angeles, Calif. (UP)—Frustrated authorities wondered where to look next today for a missing black panther after "huge paw prints" found near Yucaipa, turned out to be those of a cocker spaniel.

Undisguised skepticism was voiced by several police officers who spent hours yesterday racing to various parts of metropolitan Los Angeles investigating report by frightened citizens.

Wayne Roberts, owner of the 140-pound panther, stuck by his story that the panther escaped from his carnival last Friday and is on the prowl somewhere in Southern California.

"This is no hoax," he said. "The panther is loose. I only wish it were a hoax."

Police said two honeybears which escaped with the panther while they were being transported from here to Arizona Friday, were captured at Banning, Calif., 85 miles east of Los Angeles and about 25 miles east of Yucaipa.

The search turned to that area from Los Angeles yesterday when a farmer's wife reported she hat seen the cat when it caused cattle on her ranch to stampede.

Forestry officials and sheriffs deputies who inspected paw prints on the ranch said they were made by a cocker spaniel.

SKEPTICAL, BUT TAKING NO CHANCES

LOS ANGELES POLICE TRY TO TRACK DOWN 'PHANTOM' BLACK PANTHER

Panama City, Florida, *News*, March 17, 1954

Los Angeles, March 16 (UP)—Shot-gun armed police wearily tracked a "phantom" black panther reported wandering in a city residential area but official skepticism grew that the man-killing beast ever got loose m Los Angeles.

Despite reports from citizens who said they saw or heard a prowling 140-pound panther, police and city authorities indicated they doubted the animal was at large.

But no chances were being taken. Additional police crews were armed with shot-guns and high powered rifles and all reports were checked carefully

Wayne Roberts, part owner of a carnival, reported Monday night from Arizona that the panther apparently had been freed from its cage here by prowlers last Friday.

Roberts said prowlers apparently freed the panther along with two trained honeybears, two monkeys, a kinkajou and five rare German pigeons. The monkeys and kinkajou were found in the area in which police centered the search for the panther.

It was not until after the report was made public that police began receiving reports of the animal from alarmed citizens.

"Now everyone is seeing panthers," said one weary desk sergeant. "They're seeing panthers in their dreams."

Police Inspector Hugh Farnum said Roberts story was being checked closely.

"A menagerie of wild animals loose in the heart of the city for five days without being seen just doesn't add up." Farnum said.

Roberts was not available for comment immediately. His landlady said he was out hunting the panther.

Although some parents expressed concern, Board of Education officials said there was no increased absenteeism among children at schools in the neighborhood where the panther hunt centered. Additional police guards were in the neighborhood to reassure worried parents.

The first flurry of reports started when two officers said they thought they sighted the panther Monday night, an hour after the first all-points police bulletin was issued.

Then the police board began lighting up with dozens of additional reports from all over the city. Some said the panther was in backyards. Others placed it in alleys.

Mrs. Sylvia Speriglio, 43, called police excitedly today. She said the panther was slinking away from a lot adjoining her home. Ten squad cars hurried to the Speriglio home. The phantom cat was gone.

"I saw it," insisted Mrs. Speriglio "It's a big long-tailed cat, black as the ace of spades. It wasn't a house cat."

Mrs. Speriglio said her 11-year-old daughter spotted the panther first. The mother sent her 16-year-old son, Milo, armed with a baseball bat out into the lot to investigate. The panther fled, she said.

Police said tracks of some sort were found in the lot. No big game hunters were present to determine if the tracks were that of the hunted ferocious beast.

Shortly after the report from Mrs. Speriglio, five police cars sped to a storm drain where the panther was reported cornered by several neighborhood dogs. Twenty officers warily approached the drain, their shotguns and rifles ready for action.

A group of five entered the huge drain. The "facts" in the case turned out to be a dead foot-long cat, ordinary household variety.

ANIMAL OWNER ADMITS PANTHER STORY A HOAX
Kingsport, Tennessee, *Times*, March 18, 1954

Los Angeles (INS)—The "great Los Angeles panther hunt" ended today, not with a bang but with a fizzle.

After four days of intensive searching for a vicious black panther reported to be on the loose in the city, police called the hunt off when the animal's owner reported it was "all a mistake."

Wayne Roberts, 30, who claimed he was part owner of an animal show, set off the panther scare last Monday. Thousands of children were grabbed from the streets, doors were bolted and armed posses patrolled dark alleys and wooded areas.

But today, Roberts said that the animal had never been loose at all, that it was safely in its cage in Phoenix, Ariz. He said a trailer with a pen for animals in which he thought the panther was being transported was found with an open door in Phoenix.

He claimed he thought it was the same one in which the panther was riding.

Police had suspected a hoax since they questioned Roberts and he proved to be "evasive."

However, Police Chief William Parker said he could not call off the search for the beast until he was sure that the animal was not in the city.

Roberts' sister added the coup de grace to the panther hunt when she disclosed that the big cat was not the ferocious "killer" that Roberts had described.

She declared:

"She was a real old panther and didn't have enough gumption left to do much damage."

Black Panther Scare Jokester Faces Charge
Reno, Nevada, *Evening Gazette*, March 24, 1954

Los Angeles, March 24. (AP)—The man who reported a black panther missing from his animal show has been charged with making a false report.

The black panther scare had hundreds of residents in the Los Angeles area in a dither week before last. The beast supposedly escaped from his cage shortly after Wayne Roberts' show started for Phoenix, Ariz.

"The whole story was a hoax," Police Chief William H. Parker said yesterday. "We were pretty certain of that from the first, but it has taken a lot of checking to substantiate it."

Scores of frightened citizens reported they had seen the black panther. Women who normally walked home from night jobs took taxicabs and children were herded to school by anxious mothers and refused permission to play outdoors after school.

The police department formed a special "panther division" (that's really true), to hunt down the animal.

Chief Parker now says Roberts concocted the whole story and kept it going so he could get a job as publicity man with a carnival.

This is an excellent example of how the power of suggestion can generate false sighting reports. Roberts (whose real name was Lewis E. Smith) eventually served 45 days in prison for charges relating to this fiasco. Not long after this hoax was finally settled, the story was fictionalized and incorporated into the radio series *Dragnet* (episode 252, "The Big Cat," airing June 15, 1954), later to be included in the television series (1956). Roberts tried to sue Jack Webb for $100,000 for invasion of privacy (Zotter 1998), but the case was thrown out.

While many sightings and stories of strange predatory beasts are based on misidentifications and other mistakes, there are some very interesting accounts that suggest the presence of animals yet unknown to science. Predators often have traits that help to keep a species hidden from scientific description: wariness of humans, intelligent and cautious behavior, low population density, nocturnal activity, living alone or in small groups, etc. Many reports come from areas with fewer humans to encounter (or at least fewer humans willing to risk ridicule and send details to scientific authorities). In some cases, people may not realize that the animal is truly unknown. There certainly aren't any academic or government biologists actively pursuing the possibility of unknown carnivores.

I believe we have a good deal yet to discover about potential unknown predators in North America. Certainly, there are still many sighting accounts yet to document, and probably more distinctive mystery animals yet to recognize as ethnoknown. Mystery carnivore research in the U.S. and Canada has focused primarily on eastern cougar sightings, with only sporadic research on "black panthers" and a few other predatory enigmas. There is still much to do.

One additional note: the purpose of this book is to spur interest in carnivorous cryptids. As such, I have decided not to include sightings of the eastern cougar, which technically is not cryptozoological. If those felines still exist, they belong to a well-known living species. The investigative methodology in eastern cougar research is similar to cryptozoology, though serious discussion in the field is too often hijacked by paranoid believers and fanatical skeptics. Some cases in this book could potentially refer to cougar (native or released), but I've not intentionally collected eastern cougar sightings here.

Native Carnivores

American Black Bear
Ursus americanus

This bear is endemic to North America, with 16 subspecies currently recognized in Canada, the United States, and Mexico. Historically, its range included most of North America. Today, it is extirpated from many of the midwest and central states.

Typical adult black bears range from 150-500 pounds for males, and 90-300 pounds for females (NABC 2010). Bears tend to be larger in eastern North America. Black bears vary greatly in body weight over a year's time, due to weight loss during hibernation, but the largest confirmed wild black bear appears to be a male shot in 1998 in Craven County, North Carolina, at 880 pounds. There are unconfirmed accounts of larger bears in the wild. Ernest Thompson Seton noted a 900-pound bear trapped by US Biological Survey officers in Arizona in the 1920s (Etling 2003). Virginia DGIF (2002) noted an old account of a 962-pound black bear supposedly killed in Madison County in the 1880s. In 2001, a bear hit by a car near Winnipeg had an estimated live weight of 886 pounds (ABA 2009). NABC (2010) noted a bear shot in 1994 in St. Louis County, Minnesota, that weighed 876 pounds. Captive bears can also get large. NABC has one black bear, Ted, that originally weighed an estimated 900-1000 pounds before being placed in their sanctuary (and then undergoing some dietary and exercise changes).

Black bears, while omnivorous, are recognized as a potential threat to livestock and domestic animals. Attacks on humans are rare, but known to occur. Recent fatal attacks have been documented in Colorado (UPI 2009), New Mexico (AP 2001), New York (Adcox 2002), Tennessee (Simmons 2000 and Anon. 2006), Utah (2007), British Columbia (Anon. 1997 and CBC News 2007), Manitoba (CTV.ca 2005), Northern Territories (Anon. 2001), and Quebec (CBC News 2000).

©Thomas O'Neil

American Black Bear

National Park Service

American Black Bear Track

Several black bear subspecies display polymorphism, with different and distinctive color morphs. Black bears can be blonde, reddish ("cinnamon"), brown, white, or even a mixture of hues. Historically, this often confused both hunters and biologists into thinking they had encountered a new species of bear. Of particular interest, the silver-tipped "blue" glacier bear and the white (or cream-colored) Kermode bears were initially heralded as new species.

Black-coated bears are by far the most common morph, particularly in the eastern states. In certain western areas, especially the central-to-northern Rockies and in the southwestern desert ranges, non-black bears appear in much higher percentages (Rounds 1987).

True albinos are known, but rare. I have reports of albino black bears from Idaho, Michigan, New Mexico, Pennsylvania, Alberta, and British Columbia. Most recently, one young albino bear was shot in Centre County, Pennsylvania, in 2007 (AP 2007). An adult albino bear was moved from the outskirts of a town in northwestern Montana to Glacier National Park in 2009 to protect it from hunters (Gazette News 2009).

THE GLACIER BEAR

Dall's original description of this bear included: "The general color of the animal resembles that of a Silver fox. The fur is not very long, but remarkably soft and with a rich under fur of a bluish black shade, numbers of the longer hairs being white or having the distal half white and the basal part slaty. The dorsal line from the tip of the nose to the rump, the back of the very short ears, and the outer faces of the limbs are jet black. Numerous long white hairs issue from the ears; black and silver is the prevalent pelage of the sides, neck and rump; the under surface of the belly and the sinuses behind the limbs are grayish white, or even nearly pure white, I am told, in some cases. The sides of the muzzle and the lower anterior part of the cheeks are of a bright tan color, a character I have not seen in any other American bear; and this character is said to be invariable. There is no tint of brown elsewhere in the pelage" (Merriam 1896).

The glacier bear is now considered a color morph of the subspecies *Ursus americanus emmonsii*. This bluish-silver morph is rare, and usually limited to the Yakutat Bay region and the St. Elias range. In 1987, an adult glacier bear wandered south into Juneau and raided garbage cans until it was captured and placed in the Alaska Zoo.

While this morph is fairly distinctive (though with some variation in hue), the name glacier bear is sometimes mistakenly placed on other light-coated bears in the region. One Juneau biologist has stated, "It seems to me that the Juneau area has more color variety for black bears than other parts of Southeast Alaska" (Woodford 2007). He noted one interesting variant: "a black bear with a distinct stripe of longer, reddish-brown hair running the length of its back, like a Mohawk haircut."

The glacier bear morph appears to be in decline as black bears from other regions move into its territory, genetically swamping the population, and decreasing the chance of this recessive trait cropping up. Because it is no longer a truly isolated population, eventually the glacier bear may disappear for good.

THE KERMODE BEAR

The Kermode (or "spirit") bear morph is found in the subspecies *Ursus americanus kermodei* of mid-coastal British Columbia. This is a white- or cream-coated bear with normal skin and eye pigmentation due to a recessive genetic trait, a single nucleotide replacement (Ritland et al. 2001).

Hornaday (1905) described how it came to be discovered: "In November, 1900, while making an examination of the skins of North American bears that were to be found in Victoria, British Columbia, the writer found a very strange specimen in the possession of Mr. J. Boskowitz, a dealer in raw furs. The skin was of a creamy-white color, and very small. Mr. Boskowitz reported that it had come to him from the Nass River country, and that he had previously received four or five similar skins from the same locality.

"Although this skin was of small size, and had been worn by an animal no larger than a grizzly cub one year old, its well-worn teeth indicated a fully adult animal. Believing that the specimen might really represent a new ursine form, it was purchased, and held for corroborative evidence. In view of the multiplicity of new species and sub-species of North American bears that have been brought out during the past ten years, it is not desirable to add to the grand total without the best of reasons for doing so.

"Four years have elapsed without the appearance of a zoological collector in the region drained by the Nass and Skeena rivers, and further evidence regarding the white bear of British Columbia was slow in coming. At last, however, the efforts of Mr. Francis Kermode, Curator of the Provincial Museum at Victoria, have been crowned with success, in the form of three skins in a good state of preservation. They represent two localities about 40 miles apart. The four specimens now in hand are supplemented by the statements of reliable persons regarding other white bear skins which have been handled or seen by them, and were known to have come from the same region.

"Following the route that a polar bear would naturally be obliged to travel from its most southern haunt in Bering Sea to the Nass River, the distance is about 2,300 miles. But the teeth of these specimens show unmistakably that they are not polar bears.

"There is not the slightest probability that albinism is rampant among any of the known species of bears of North America; and it is safe to assume that these specimens do not owe their color to a continuous series of freaks of nature. There is no escape from the conclusion that a hitherto unknown species of white bear, of very small size, inhabits the west-central portion of British Columbia, and that it is represented by the four specimens now in hand. In recognition of his successful efforts in securing three of these specimens, the new species is named in honor of Mr. Francis Kermode."

Marshall and Ritland (2002) noted that Kermode morphs are highest on the coastal islands, with Gribbell Island showing the highest percentage of white bears, along with the lowest genetic diversity among the bear population.

Klinka and Reimchen (2009) discovered that the Kermode bear's light coat allows it to stream-fish more effectively for salmon during daylight hours than typical black-coated bears, providing a small adaptive advantage during the salmon run season.

There was one recent sighting of a cream-colored "spirit" bear on Vancouver Island, by a couple hiking in East Sooke Park (Canwest News Service 2008). This was apparently the first time the morph had been seen that far south, assuming that it was not a true albino.

White non-albino bears have also cropped up elsewhere. In 2004, a black female was spotted with a white cub in Manitoba. Not long after, the female was killed by a car, so the cub was captured and placed in the Assiniboine Park Zoo in Winnipeg, where it resides today. This white bear, which they have named Maskwa, has a different mutation on the same gene that creates the white Kermode bear (NABC 2010). NABC also notes two other cases of white morph black bears: a subadult white male (nicknamed Halo in the media) was reported near Orr, Minnesota, in 1997, and in 2000, a black female was seen with two white cubs near Beausejour, Manitoba.

Brown Bear
Ursus arctos

The brown bear ranges across northern Eurasia and North America. Two subspecies are now recognized in North America: the grizzly (*Ursus arctos horribilis*) and the Kodiak bear (*Ursus arctos middendorffi*).

Kodiaks are the larger subspecies of the two, but are often matched (and exceeded) by the Kamchatka brown bear, the largest Eurasian subspecies. Brown (1996) noted that one wild-caught Kamchatka bear weighed 2500 pounds. The largest Kodiak bear was killed in 1952 by USFWS biologist Roy Lindsley, for the Los Angeles County Museum. It had a "green length" of 11 feet, 2 1/4 inches (Anon. 1953), a mounted height of 8 feet, 8 inches, and a weight of 1200 pounds (Bachay 1957).

The grizzly bear is known to attack livestock. Attempts to re-introduce the species in some western states has met serious opposition by ranchers, including both legal suits and illegal poaching campaigns. One USDA wildlife trapper noted that most grizzlies don't attack sheep, but once one does, it is almost impossible to get it to stop (Puckett 2008). He was then working on the largest sheep-killing spree he had ever seen: 71 sheep killed by two young grizzlies over a two-month period, most of which appeared to have been killed for "fun."

Attacks on humans are known. In the last decade, fatal attacks were reported from Alaska (Anon. 2003 and AP 2005), Alberta (AP 2005b, Callahan 2007, and CTVCalgary 2008), and the Yukon (Anon. 2006b).

Grizzly coloration can be various shades of white, blonde, brown, or black, often with pale-tipped guard hairs giving the "grizzled" appearance (also denoted by the name given to some grizzlies, *silvertip*). Historically, grizzlies with dark coats but pale facial hair were called bald-faced bears. I've seen a few mentions of albino grizzlies, but nothing concrete. One was described here:

ALASKA HAS A BEAR
Fairbanks, Alaska, *Daily News-Miner*, July 19, 1929

Wrangell, July 19.—Passengers aboard the steamer *Hazel B.*, arriving here from the Stikine river, reported having seen an albino grizzly bear at the junction of Claire creek with the Stikine. The boat was nearing Glacier near evening when the bear was discovered standing on a ledge almost directly above the boat, affording an opportunity for a good inspection before it made off into the woods.

(CC) Carl Chapman

Grizzly Bear

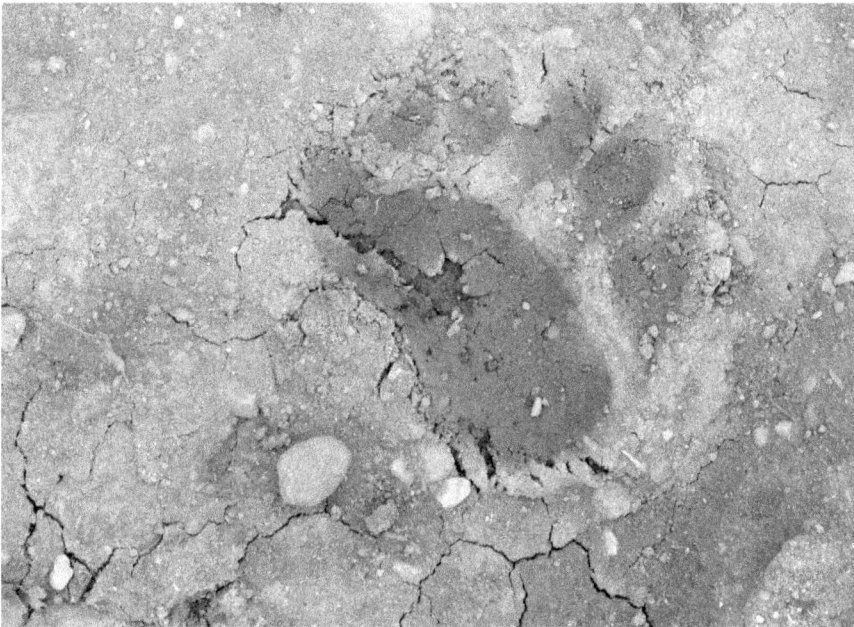

©Laura Lohrman Moore

Grizzly Bear Track

Captain S. C. Barrington described the bear as being large and creamy white in color with reddish ears and pink eyes.

Hybrids

In recent years, hybridization between brown bears and polar bears has become popular news fodder, as the grizzly appears to be expanding its range into polar bear habitat. Of course, the spark for such discussion was the confirmed grizzly-polar hybrid shot in 2006 in the Northwest Territories (Vanderklippe 2006; CBC News 2006). One Alaskan Fish & Game biologist, hearing about this hybrid, recalled seeing an odd polar bear skin with "lots of brown in it" at a well-known hunting guide's home (Rozell 2006). A bear shot in April 2010 by an Inuit hunter near Ulukhaktok, Northwest Territories, has also been confirmed to be a grizzly-polar bear hybrid. In fact, it is a second-generation hybrid, as genetic tests showed it had a grizzly father and a hybrid mother (CBC News 2010).

Hybrids between grizzlies and polar bears were known to occur in captivity. In one case, the Washington, D.C., National Zoo inadvertently bred their female Alaskan brown bear and a male polar bear, producing cubs in 1935 (AP 1935). Then, in 1956, a pair of this hybrid offspring bears produced twin cubs of their own (AP 1956).

Of particular interest to cryptozoology enthusiasts is MacFarlane's bear, which has been suggested to be a possible hybrid grizzly-polar bear. This specimen was shot in 1864 by two Inuit hunters at Rendezvous Lake, northeast of Fort Anderson Mackenzie, then acquired by naturalist Robert MacFarlane and sent to the Smithsonian (Goodwin 1946). MacFarlane didn't recognize any unusual features in the large yellow-furred bear, and it wasn't until it was examined by C. Hart Merriam, decades later, that the bear was considered noteworthy. Merriam placed it in its own genus (*Vetularctos*) based on its dentition. A few years ago, the television show *MonsterQuest* talked a paleontologist into examining the skull, and he gave his opinion that the skull was that of an ordinary grizzly bear. No morphological or genetic reanalysis of this bear has been published in the scientific literature, and it is considered by most bear biologists to be synonymous with the grizzly.

The Lava Bear, or Dwarf Grizzly

One of the most interesting folkloric varmints is the alleged dwarf grizzly bear that inhabited the lava fields of southeastern Oregon and possibly adjacent Idaho. While the brown bear is recognized to have a dwarf race in the Gobi desert, the miniature lava bear was said to be far smaller. There were captured animals and opportunities for biologists to look over the specimens, but there doesn't appear to be a published scientific record confirming or refuting this animal's existence as an actual race. There are too many unanswered questions. Could cubs have been used to create the illusion? What happened to the remains? Are there "lava bears" in museum collections that have been forgotten and overlooked?

The story of the lava bears, however far it went back locally, sprang to the national public's eye when writer and humorist Irvin S. Cobb took a trip that he would eventually memorialize in his book, *Some United States.*

Irvin S. Cobb Finds Sun Bear in Lava Rocks
Lebanon, Pennsylvania, *Daily News*, October 29, 1912

That the lava bear, provincially known as the "sand lapper," inhabiting the lava flow in the Fort Rock country, may prove to be the sun bear, believed for the last 100 years to be extinct, was the theory advanced by Irvin S. Cobb today on his return to Bend after a week's trip to the south, in the course of which the humorist has hunted bear and deer and fished in all the lakes along the way. Incidentally, Cobb lost twenty pounds of his former weight.

Cobb said today that he believed a diminutive species of grizzly found in the Seven Devils country in Idaho and the dwarf bear of the Fort Rock lava flows are identical and are the last survivors of the sun bear, a species formerly found over a large territory in the lava flows of California and Oregon. He is eager to get specimens of the tiny grizzly and hopes that the attention of the Smithsonian Institute may be sufficiently aroused to start a more scientific investigation of the subject than he has been able to make.

New Kind of Bear is Discovered in Oregon
Resembles Small Grizzly, United States Biological Expert Announces
Tyrone, Pennsylvania, *Herald*, August 2, 1923

The discovery of a. new species of bear was announced at a meeting of several hundred Department of Aviculture employees with Henry C. Wallace, secretary of agriculture, at Tacoma, Wash.

The new bruin has not been officially named, but he probably will be called the Lava bear, his known habitat being the lava beds

of southeastern Oregon. The discovery was reported by Director
Jewett of the United Stales biological survey. Mr. Jewett said that
about a year ago a settler killed the first of these strange bears and
brought the carcass to him. The animal had a shaggy coat very much
like a grizzly bear and despite that it weighed only 40 pounds it was
exceedingly ferocious.

Mr. Jewett determined that the animal was full grown and that
its characteristics were distinctly different from those of any variety
of bear known to him. He sent the carcass to C. Hart Merriam, former
chief of the biological survey and now director of biological investigation for the Rockefeller institute. Mr. Merriam is the recognized
world authority on bears.

Mr. Merriam confirmed the fact that the animal was different
from any other known species but not much was made of the matter
until two weeks ago when another animal of the same sort was killed
by investigators of the biological survey. The experts' theory is that
the Lava bear is a development from the grizzly in the district where
it was found. Food is very scarce there and just us the huge Kodiak
bear of Alaska is believed to be a grizzly developed to abnormal size
by feeding on the salmon, which abound in the rivers there it is
thought the Lava bear may be a grizzly stunted through many generations of scant feeding.

Its characteristics, however, are so different from any known
variety of bear that Doctor Merriam declares that it must certainly
be classed as a new species.

LAVA BEAR NO MYTH

Kingston, New York, *Daily Freeman*, May 31, 1924

Lava bears are no myth, declares the Portland *Oregonian*. A
couple of years ago Irvin Cobb came to Oregon to hunt the lava bear.
He didn't see one, and he was advised by many Oregonians who have
spent outdoors many years of their lives that there was no such
animal. However, Cobb met one chap at Fort Rock who informed
Cobb that if he met a lava bear he needn't shoot the creature. "Just
kick 'em to death," was the advice.

Well, there are lava bears in Oregon. They're not very numerous, but they do exist. One reason why few have been seen is that

they live in the lava beds where a man dare not penetrate more than a couple of hundred feet for fear of being lost and perishing.

Dan Godsil, game warden for Lake county, who is in Portland, says that he has soon two lava bears. One was trapped a year ago by an agent of the government biological survey. The lava bear is about the size of a house cat and is mostly head. It looks exactly like the regulation bear, except for its small size, and as the old-timer told Cobb, the lava bear doesn't have to be shot, for if the hunter can get near enough he can boot the bear to death. The bear caught last year was in the lava field northeast of Fort Rock.

Lava Bear, or Dwarf Grizzly, Is Captured

Port Arthur, Texas, *News*, June 4, 1924

Bend, Ore., June 4.—The Lava Bear has been caught.

Alfred Andres, for nine years a government trapper in the Silver Lake district, but recently relieved from duty, caught him and says he is merely a species of dwarf grizzly.

It is the first one known to have been caught alive. This bear, of a species long thought to have been extinct, is two years old, but he is only 30 inches long and 15 inches high.

The question of the Lava bear and whether or not it existed was brought into national prominence three years ago by Irwin S. Cobb, when Andrews took the famous writer on a trip in search of the strange animal. Now Andrews' nine years' quest has been rewarded. He found the trail and set a trap. One day after the trap was set he found the animal alive with both hind feet caught.

It is said the Smithsonian Institute has offered $2,000 for a live Lava bear, but Andrews declares he will not sell it. He is planning to tour the country with the animal, exhibiting him as a unique specimen.

Cage Life Liked by Dwarf Grizzly
Lava Bear, Again in Bend, Much Fatter Than
When First Taken Near Fort Rock

Bend, Oregon, *Bulletin*, October 17, 1924

Life in captivity is obviously agreeing with the lava bear captured last summer in the Fort Rock country, declare persons who are handling the little, but ferocious animal, again brought to Bend. The lava bear has increased in weight and is much fatter than when captured.

According to zoologists who have examined the dwarf bear this summer, the animal is about 17 years old, his handlers report. From Klamath Falls the lava bear is to be taken south to Los Angeles, then to the east, according to plans.

Members of the Bend Lava Bears, commercial club auxiliary, are having pictures taken of the odd Fort Rock grizzly, the only one of his kind in captivity. Managers of the bear plan to send some of these photographs to news picture agencies in various parts of the country.

So, what happened to this particular lava bear? I have not traced any news accounts of it, but did find an interesting record in the *29th Annual Report of the New York Zoological Society*, covering the year 1924. Among the zoo's "important acquisitions" for the year was "1 cinnamon 'lava' bear." The same bear?

EXPEDITION SEEKS PYGMY BEARS IN EASTERN OREGON
Fayette, Iowa, *County Leader*, August 30, 1928

Walla Walla, Wash.—To obtain living specimens of a 40-pound lava bear, known here as the sand lapper, and which lives in the ancient lava flows of eastern Oregon, a scientific expedition headed by University of Washington professors has left here for that region.

The diminutive bear is a species of grizzly ten times dwarfed and believed to be the remnant of the sun bear, thought to be extinct. At least three specimens have been killed during the last ten years. These were about 30 inches long and 18 inches high. Seven Devils country, where the bears have been seen, is a wild, barren and forbidding region, where lava and eruptions left grotesque formations. It is a safe refuge for the pygmy bears. Several thousand dollars is the price set for a pair brought out alive.

LAVA BEAR CAPTURED
El Paso, Texas, *Herald-Post*, December 5, 1933

Klamath Falls, Ore.—A lava bear, one of the most unusual and rare species of the bear genus, was caught in a coyote trap near here recently. The bruin weighed over 50 pounds.

Find Lava Bear

Monessen, Pennsylvania, *Daily Independent*, February 14, 1934

Lapine, Ore., Feb. 14.—A lava bear, smallest of the bear family in North America, was captured near here by Walter Gore and Roy Yeager. Full grown, it weighed only 30 pounds.

Rancher Attests Cobb's Veracity

La Crosse, Wisconsin, *Tribune and Leader-Press*, April 8, 1934

Klamath Falls—(AP)—The veracity of Irwin S. Cobb was attested Saturday when Ray Yager, rancher, arrived with an adult lava bear. The little animal, a miniature specimen of the grizzly, weighs 25 pounds. It was captured last January.

Several years ago Cobb visited the lava beds of central Oregon and then went east to write of the lava bears, without having seen a member of the rare species. For many years the tale was regarded as fictional.

Yager has a letter from Cobb congratulating him upon obtaining "living proof of the existence of a sub-specie of the common variety."

Trace is Sought of Two Extinct Lava Animals

Freeport, Illinois, *Journal-Standard*, February 11, 1936

Siskiyou, Calif., Feb.—(UP)—California university authorities have undertaken to reconstruct the existence of two species of animals that once inhabited, the great lava beds between the Sierra Nevada mountains to the north and the Cascades to the south.

These are the lava bear and the lava sheep. Both have become extinct. Scientists are convinced that at the time of the early settlement of California they actually existed, and they hope not only to establish this definitely but to find enough bones of the extinct animals to reconstruct skeletons and establish their relations with other species of bear and sheep in surrounding mountains.

All inhabitants of the country have been asked to send in any existing relics they may have of such animals, and eye-witness accounts of having actually seen them.

The lava bear especially is supposed to have been a small, brown bear, that spends its time in the caves and crevasses of the lava, that is exceedingly shy, and comes out from its cavern recesses only at night.

SAYS LAVA BEAR IS NOT EXTINCT
COTTAGE GROVE RESIDENT TELLS OF CAPTIVE
Bend, Oregon, *Bulletin*, April 6, 1937

Lava Bears are not extinct unless the one she saw in the McKenzie river country several years ago died from eating too many apples, writes Mrs. Ethel Gabrio of Cottage Grove in commenting on "The Lava Bear Trapper" editorial used in *The Bulletin* and reprinted by the *Oregon Journal* recently.

While Mrs. Gabrio was living at Rainbow on the McKenzie river in 1934 some men caught a female lava bear in the lava fields of the McKenzie summit, writes the Cottage Grove resident. The little bear was taken to Rainbow and placed in a flimsy cage, where it attracted considerable attention. Tourists coaxed the bear from the top of its cage with raw apples, which the diminutive bruin ate with great relish.

That night, the lava bear tore a hole in the top of its cage and disappeared. Lew Quimby, at that time postmaster and store keeper at Rainbow, offered a large reward for the little creature, but it was never found, Mrs. Gabrio reports.

The animal is believed to have gone back into the lava beds, and if it finally reached its native habitat, lava bears are not yet extinct, Mrs. Gabrio holds.

In a retrospective of Oregon desert stories, Alexander (1987) noted that for the 1933 lava bear exhibit, 8,000 people paid admission to a Portland park to see the creature. She quotes from a 1937 article, "Andrews was offered $5,000 in cash for the specimen but refused. His assistant sold it without his knowledge for $7,000. The tiny bear disappeared shortly after that and was reported to have found its way to a New York museum. So far as Mr. Andrews knows, that bear is now dead."

So, is there a lava bear in a New York museum collection? Did a dwarf race of grizzlies (or perhaps black bear) actually exist in the lava fields? Or was it a Barnum-style hoax? As I said, too many questions at present, but worth investigation.

Ghost Grizzlies and Other Remnants

While not cryptozoological, as they are a known species, the possibility of a straggling number of grizzly bears in the mountains of Colorado has intrigued a number of researchers, particularly as a female grizzly was killed when it attacked a Colorado bowhunter in 1979 (UPI 1979; Saile 1983). Interested readers should take a look at *Ghost Grizzlies*, by David Petersen, and *The Lost Grizzlies*, by Rick Bass. For more details on the extirpation of grizzlies in the southwest, see *California Grizzly*, by Storer and Tevis, and *The Grizzly in the Southwest*, by David E. Brown.

The grizzly was believed extirpated from Mexico in the 1960s, but biologists examined a bear skull shot in northern Sonora in 1976, and determined that it was a juvenile grizzly (Gallo-Reynoso 2008). On the opposite side of the continent, the former presence of the grizzly bear in northern Quebec and Labrador was controversial, with only circumstantial evidence reported (Elton 1954). Then a grizzly skull was dug up in 1976 (apparently a good year for grizzly skulls) during an archaeological excavation in northern Labrador (Spiess 1976; Spiess and Cox 1977), with further review of the total evidence suggesting that grizzlies survived in this barren region at least to the early twentieth century (Loring and Spiess 2007).

Polar Bear
Ursus maritimus

The polar bear was for many years thought to be a separate genus from the grizzly and black bears, but recent research shows that it is a member of *Ursus*, having a more recent common ancestry with the brown bear than was previously recognized. Polar bears live in the northern circumpolar region, and are found in Alaska's and Canada's northern coastal areas. They feed primarily on seals, but capture or scavenge other marine mammals opportunistically, and include other food in their diet when seals are unavailable.

Male polar bears get much larger than females. The largest polar bear so far discovered was shot in 1960 by Arthur Dubs on the arctic ice in Kotzebue Sound, Alaska. The hide was initially measured at 11 feet, 4 inches "from nose to tail" (AP 1961). Its mounted length was apparently slightly smaller, and it had an estimated live weight of 2,209 pounds (Ellis 2009). Ellis also stated that the present location of this mount is unknown. One news article I have seen (Park 1968) stated that the mount was displayed at Ripley's "Believe-it-or-not" Museum in New York. I do not

©Coverstock

Polar Bear

know if Ripley's still has this mount, though they do have a large central storage facility for artifacts and specimens not on display at their various facilities. Another large Alaskan polar bear mount can be seen at the Commercial Casino in Elko, Nevada, which is claimed to be 2200 pounds (and 10 feet, 4 inches). They nicknamed this bear White King, and it has been on display at the casino since 1958. Another large specimen was shot in 1959 by Mrs. Earl Anderson at Kotzebue, measuring 11 ½ feet and weighing 1600 pounds (Anonymous 1960).

Polar bear attacks are very rare. As Dr. Timothy Floyd (1999) noted: "Bear-inflicted human injury and death is rare. Brown bear attacks tend to be severe and to occur suddenly, without provocation. Black bear attacks usually result in minor injuries and tend to be predacious. Polar bear attacks are exceedingly rare, and the ferocity of polar bears has probably been overemphasized." Still, there are a few cases in the last couple decades worldwide, though only a few from North America, and most of those are little-known due to the remoteness of the locations. One involved a 64-year-old Nunavut woman, Hattie Amitnak, who tried to fend off a 250-pound bear that was attacking her fellow campers (McKibbon 1999).

At present, the only truly cryptozoological angle I've seen on polar bears involves a rarely reported "giant polar bear" that would dwarf a 2,000-pound bear. Such stories are attributed by the Inuit to the 10-legged bear in their mythology.

Gray Wolf
Canis lupus

The wolf is a wide-ranging and highly variable species, ancestral to our domestic dogs. Exactly how many subspecies there are is uncertain, with numerous revisions by splitters and lumpers over the years. Today, biologists debate whether some subspecies (particularly the eastern wolf and red wolf) should be given full species status, and how much hybridization with coyotes factors into the genetics of certain populations. For the purpose of this brief overview, all wolves will be considered part of the *Canis lupus* complex.

Quite a few historically described subspecies in North America are extinct. Those that would still be extant include:

Canis lupus arctos: Arctic wolf of northern Canada
Canis lupus baileyi: Mexican wolf
Canis lupus crassodon: Vancouver Island wolf
Canis lupus hudsonicus: Hudson Bay wolf

Canis lupus irremotus: Northern Rocky Mountain wolf
Canis lupus labradorius: Labrador wolf
Canis lupus ligoni: Alexander Archipelago wolf
Canis lupus lycaon: Eastern timber wolf
Canis lupus manningi: Baffin Island wolf
Canis lupus nubilus: Great plains or buffalo wolf
Canis lupus occidentalis: Mackenzie Valley wolf
Canis lupus pambasileus: Interior Alaska wolf
Canis lupus rufus: Red wolf
Canis lupus tundrarum: Alaska tundra wolf

Some biologists only recognize six of these as valid, lumping all previously described subspecies together in *arctos, baileyi, lycaon, nubilus, occidentalis,* and *rufus*. They argue that the entire historical range of wolves in North America can be covered by these.

Wolves do take livestock, though wolf kills make up a smaller percentage of livestock deaths than do natural causes. Attacks on humans are more commonly reported in Europe and Asia, though wolf aggression in North America does occur (McNay 2002). While there are historical accounts of human fatalities from wolf attacks in North America, these are unconfirmed and can often be attributed to newspaper sensationalism or propaganda being put forward by groups advocating government bounties. Still, in more recent years, there are a few cases where wolves were involved in human deaths.

In 2005, a 22-year-old engineering student, working at a mining camp in northern Saskatchewan, didn't return from a hike. Following his tracks for half a mile into the bush, mining company employees found his body covered in bites and partially eaten. Wolf tracks surrounded the body, and the men could see and hear glimpses of them in the brush. By the time the men returned from retrieving the RCMP and coroner, wolves had dragged the body a few yards further. Initially, a provincial wolf and bear biologist said that the physical evidence pointed to an initial black bear kill, but two years later the student's parents persuaded the government to reopen the coroner's inquest, bringing testimony from wolf biologist Mark McNay. The jury agreed with McNay that wolves were responsible, making this the first documented wolf fatality in North America (CBC News 2005, Anon. 2007, AP 2007).

More recently, a 32-year-old special needs teacher in Chignik, Alaska, was killed while jogging by a pair of wolves. Wildlife officers killed the two wolves they believed were responsible for the death (Halpin 2010; Anon. 2010).

©Christophe Namur

Gray Wolf

Reports of wolf-like canids raiding farms and prowling woods are fairly common in newspapers throughout the eastern states, long after wolves were supposed to have been hunted to extinction. It is likely that most involve feral domestic dogs. Of course, while the wolves were disappearing, humans expanded rural farming, and this opened up routes for dispersal of western coyotes into the eastern states, so they too are suspects in alleged wolf encounters. As known species, I won't reprint general wolf stories in the state sightings section, so just keep in mind that wolves were common villains in varmint lore, even after they had disappeared. There are a few interesting cases of confirmed out-of-place wolves, though. In one instance, a pack of six wolves were hunted down one by one in Carroll County, Ohio, in the early 1940s. They killed 100 sheep before being stopped. One of the animals was shipped to Stanley P. Young of the USFWS, who confirmed that it was in fact an eastern wolf. Their origin was never discovered (Stewart and Negus 1961). The New England states are close enough to Canadian populations that the occasional wanderer ends up there. In one 2008 case, Massachusetts biologists examined a livestock kill and misidentified it as a dog attack. They gave the go-ahead to allow the animal to be shot, and a subsequent necropsy of the animal confirmed that it was a true wolf (Anon. 2008).

British Columbia Coastal Wolves

Researchers investigating wolf genetics in British Columbia discovered that gray wolves along the coast were genetically different from inland wolves (Muñoz-Fuentes, et al. 2009). These coastal wolves exhibit distinctive morphology and behavior, and occupy a specific habitat (a temperate rain forest stretching from Alaska's Alexander Archipelago to Vancouver Island). Coastal wolves are smaller than inland wolves, and prey on the only ungulate to inhabit this region, the black-tailed deer. (This deer is, itself, the smallest ungulate species in North America.) When deer aren't available, the wolves include marine life in their diet, even fishing for salmon. Genetically, these wolves show greater differentiation from inland wolves than is found between some recognized subspecies (where there may be no genetic isolation). Rather than describe the coastal wolf as a new subspecies, the researchers suggested that it should be denoted an Evolutionary Significant Unit (ESU), as it may represent a species in the making.

Eastern Wolf

The eastern wolf (also known as the eastern timber wolf or Algonquin wolf) is today known from Algonquin Provincial Park, Ontario, and nearby regions. There

©Karel Broz

Gray Wolf

©Denis Pepin

Gray (Arctic) Wolf

is growing evidence that this wolf is a distinct species, possibly related to the red wolf (Wilson et al. 2003; Grewal et al. 2004). Unfortunately, eastern wolves are part of a confusing taxonomic debate that has significant political implications for conservation. Actually, it would be more accurate to state that it has been part of multiple debates (Schwartz and Vucetich 2009):

> A) Is the eastern wolf a subspecies (or perhaps just an ecotype) of the gray wolf, or is it its own species?

> B) If it is a distinct species, is the eastern wolf a result of hybridization between gray wolves and coyotes, or did it split from a common ancestry with the coyote?

> C) After the eastern wolf was almost exterminated from the Great Lakes region, are recolonized wolves in this area (particularly in the western Great Lakes region) actually eastern wolves, or do they have a different genetic composition? Can they even maintain a stable genetic population structure, given ongoing hybridization with western coyotes and gray wolves?

What makes these questions so difficult is that hybridization between coyotes or gray wolves is recurring throughout the eastern wolf's entire range. There have also been numerous cases of feral wolf-dog hybrids (deliberate or accidental releases) being found in the wild in the eastern wolf's range (Adkins Giese 2006; VT ANR 2007; Schwartz and Vucetich 2009), highlighting a possible contributing source of domestic dog genes into wild canid populations.

Hybridization certainly appears to be contributing to the appearance of distinctive races of eastern wolves. Researchers in Ontario (Wilson et al. 2009) confirmed three different wolves in that provice: a) the central 'Algonquin' type, which is the typical eastern wolf, b) the larger northern 'Ontario' type, also known as the 'Great Lakes wolf' (Rutledge 2010), representing hybridization between gray and eastern wolves, and c) the southern 'Tweed' wolf, which represents hybridation between eastern wolves and coyotes. The researchers note that the lack of gray wolf/coyote hybridization in western North America is almost certainly due to the absence of *C. l. lycaon*. Other researchers have suggested that hybridization (to some extent) is a good thing, as it may enhance the eastern wolf's ability to adapt to environmental changes (Kyle et al. 2006).

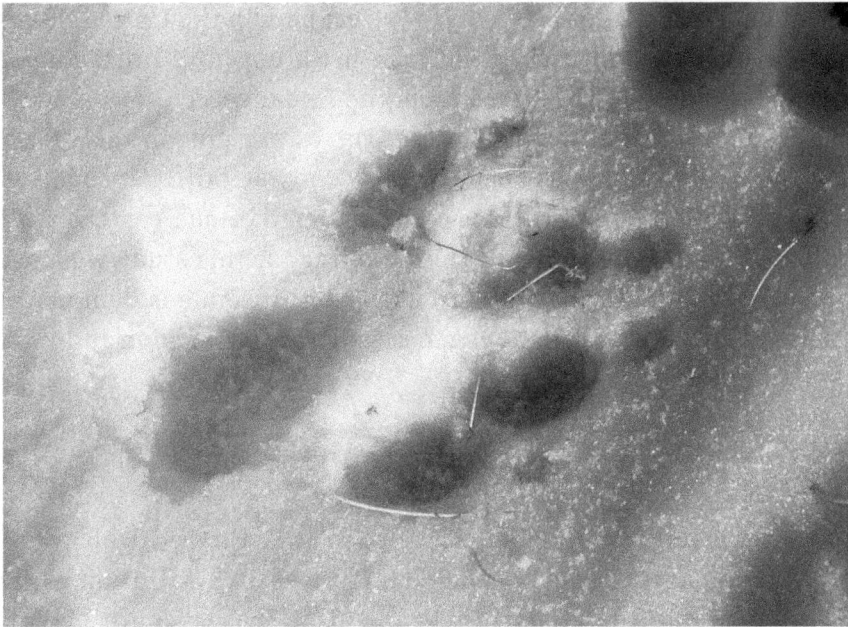

National Park Service

Gray Wolf Track

National Park Service

Gray Wolf Pack

The 'Great Lakes wolf' is of particular interest here in the United States, due to the successful rebound of wolves in recent years in the upper Midwest states, leading to this wolf population being delisted from the Endangered Species Act in 2007. Some biologists argued that most of these wolves are not, in fact, true eastern wolves, but rather hybrid canids (Leonard and Wayne 2008; Leonard and Wayne 2009), which could have implications for whether the true eastern wolf should still be protected. Others argued that the sampling for this study was insufficient (Mech 2009) and that historical 'Great Lakes wolf' specimens demonstrate early hybridization while retaining a distinct genetic identity (Wheeldon and White 2009; Koblemüller et al. 2009).

Red Wolf

By the 1960s, wildlife biologists recognized that the red wolf was in grave danger of extinction, due to overhunting, habitat loss, hybridization with coyotes, and heavy parasite infestations (Phillips et al. 2003).The last confirmed red wolves was removed from the wild (Louisiana and Texas) by USFWS in the 1970s. In 1980, it was declared extinct in the wild. Since then, a vigorous captive-breeding and reintroduction program has tried re-stocking the wolves into more sustainable environments. Success has been met in eastern North Carolina, though there are fewer than 100 red wolves currently in the wild there (USFWS 2010). More recently, an attempt to reintroduce the wolves into the Great Smoky Mountains National Park did not succeed. One of the biggest problems is that coyotes easily hybridize with red wolves, which could easily remove a true red wolf population in a matter of generations. One response to this has been to trap, sterilize, and re-lease coyotes back into an area where the red wolves are being reintroduced. The coyotes then maintain their territories as normal, creating a buffer zone for the red wolves to begin rebuilding their population with less chance of hybridization (Beck 2005).

For some time the proposal has been made that red wolves themselves are hybrids between coyotes and gray wolves, likely formed in the last 2500 years (most recently argued by Reich et al. 1999). As with the eastern wolf, there are political and conservation issues at stake, as it has been suggested that a hybrid may not be worth saving. Other biologists have argued that morphology, the fossil history, and genetics show that the red wolf and eastern wolf did not come from the gray wolf, but from a coyote-based lineage (Nowak 1992; Wilson et al. 2000). This may be one reason why hybridization with coyotes is more common with red wolves than Mexican wolves (Hailer and Leonard 2008).

Some researchers have suggested that eastern wolves and red wolves are essentially conspecific (Kyle et al. 2008), and that the current conservation agenda is based on the arbitrary premise that a certain geopolitical range should dictate conservation management decisions. They suggest that the eastern wolf should be utilized in conservation breeding, just as Texas mountain lion stock was used in Florida panther breeding efforts. They also note that coyote/red wolf hybridization shouldn't be viewed as entirely negative.

Several island parks have been utilized over the years to experiment with releases, and to allow carefully watched pairs to breed in the wild, providing viable offspring for mainland releases. Currently, there are a few red wolves in the St. Vincent National Wildlife Refuge in Florida.

There are, occasionally, reports of "red wolves" in other areas. In most cases, wildlife officers note that this animal was considered extirpated in the wild in 1980, and that there is little likelihood that red wolves from reintroduction projects would have wandered that far without being noticed. In 2007, photos were taken of one canid in northern Collier County, Florida, but a USFWS biologist pointed out some morphological discrepencies, suggesting that it was a wolf-dog hybrid (Staats 2007). In 2009, sightings and video of canids in Okaloosa and Walton Counties, Florida, spurred rumors of red wolves, but a Florida Fish & Wildlife spokesman suggested they were just coyotes (Harwood 2009).

Melanism in Wolves

Black wolves have been reported in North America for almost as long as there have been colonists. In fact, one early published description of the red wolf was of black specimens in Florida. William Bartram called them *lupus niger*, stating that they were "larger than a dog, and are perfectly black, except the females, which have a white spot on the breast" (Phillips et al. 2003). The black phase red wolf was not uncommon, with up to 25% of red wolves killed in the Arkansas Ozarks in the 1930s being melanistic (Gipson 1976). Some researchers have suggested that red wolf/eastern wolf hybridization with coyotes and gray wolves may have carried the black gene to those canids (Rutledge et al. 2009). Other researchers suspect that the mutation for melanism in wild canids originated in the domestic dog (Anderson 2009; Barsh 2009), and passed along through hybridization to the various wild relatives.

Melanistic canids represent one potential source of confusion for "black panther" witnesses. Just as yellow/tan canines are sometimes mistaken for mountain lions, black coyotes and wolves may be inadvertently described as feline. Many

Steve Maslowski, USFWS

Red Wolf

historical reports of black "beast" sightings are too vague to determine whether a feline or canine source is more likely.

The Great White Wolf

There are a few stories in the cryptozoological literature about large white wolves that some have theorized may be a distinct species, perhaps even a remnant population of a supposedly extinct canid. The main source for these reports comes from Ivan T. Sanderson (1974), who wrote a brief article on the subject in *Pursuit*. Two individuals related stories to Sanderson of such an animal. Tex Zeigler was filming in Alaska for an extended period, and brought back stories of a "great white wolf," which he thought might just be huge white solitary wolves that roamed the Alaskan wilderness. Frank Graves told Sanderson about a large white wolf-like creature that he saw in the Northwest Territories, in the Nahanni Valley. Sanderson stated that Graves described it as "an enormous white wolf, with very long, rather shaggy hair, but with a very wide head; and standing about three feet, six inches at the shoulder." (Graves' story is in the Northwest Territories section, as he related it to Sanderson.) Frank's guide told him that "these animals were much larger than any wolf; were 'loners'; avoided real wolves; had smaller ears and much wider heads; and rather short legs, with splayed feet. Their tails, he said, were very thick and more like those of otters, while they were scavengers rather than predacious animals. He also said that they were comparatively rare and lived up near the tundra, but that they sometimes came down in winter. Yet, he also affirmed that they were to be found all the time in this valley, and in some others to the west."

The name Waheela has been used in popular cryptozoological literature for these giant white wolves, but as this name originated in some early Michigan stories (Sanderson 1974), it does not appear to have ethnozoological relevance to the Alaskan/Canadian stories. Rather, it would be more appropriate to determine if these mystery canids are still reported from that region, and if there is in fact a correct ethnozoological descriptor.

Coyote
Canis latrans

The coyote can be found throughout all of North America, except the northern regions of Canada. The largest, usually northern, coyotes weigh more than 70 pounds and reach 5 feet in length. It does well in areas of human development.

One recent news item noted allegations of "giant" coyotes in East Texas. Trappers and farmers were running into 60- to 90-pound coyotes. Easy feeding on chicken farm discards and hybridization with dogs were suggested explanations, but no testing was being done (Leggett 2008).

Coyotes are common predators of sheep and small livestock, as well as pet dogs and cats. Attacks on humans are not as common, and are not usually serious. There are two documented fatalities by coyotes, however. A three-year-old California girl died from injuries sustained during a coyote attack (AP 1981), and more recently a 19-year-old woman was killed by a pair of coyotes while hiking in Cape Breton Highlands National Park, Nova Scotia (Vallis 2009).

As noted in the wolf section, there is a black coyote morph that crops up occasionally. Typically, it is described as black with a solid white chest. I have seen reports of this color morph from Personville, Texas (Mexia, Texas, *Daily News*, November 9, 1971), Louisiana (Ruston, Louisiana, *Daily Leader*, April 7, 1977), Georgia (WSB News 2008), and Arkansas (Gipson 1976). In the Louisiana case, outdoors columnist Glynn Harris spoke with Dr. John Goetz, a zoologist at Louisiana Tech, who stated, "You never see a black coyote in the rolling, open plains out west. Blacks seem to be confined to forested, brushy areas and it is apparently nature's way of protecting them. A black coat blends in with shaded woods while one that color in the open plains would stick out like a sore thumb." I suspect that the following article from Michigan could be explained by a melanistic coyote, as that species was just expanding into the southern portion of that state:

DOG, FOX, WOLF? STRANGE ANIMAL ON DISPLAY HERE
Ludington, Michigan, *Daily News*, July 3, 1930

(By Mrs. H. J. Gregory.)

Fountain, Mich., July 3.—Harry Myers, local merchant, had on display the skin of an animal no one knows what it is. It was killed two years ago by Herman Myers, brother of Harry.

This animal is neither dog, wolf nor fox, but has the markings of all three.

Has Head Like Dog.

The head is shaped like a police dog—pointed nose, large eyes. The body is long and has a bushy tail like a fox. The feet also are like the fox.

It is a glossy black and has beautiful fur, some places being mixed with white.

©Sascha Burkard

Coyote

©Robert Kelsey

Coyote

This animal killed 50 sheep for O. G. Puckett, near Sauble, always coming at night. A chase was planned and after eight hours with the help of fox hounds this animal was killed. No other has ever been seen near there, since or before. One was killed in Gladwin county which resembles this in description.

Is Taken to Ludington.

This beautiful skin was on display in Fremont until Sunday when Mr. Myers brought it to Fountain and has taken it to Ludington to place it in the Izaak Walton wild life show.

Mr. Myers was very desirous of having this mounted but for some reason the taxidermist was unable to do this.

A rug or neck piece will likely be made from the pelt.

Eastern Coyote

As the wolf was exterminated from most of eastern North America, western coyotes took advantage of this, moving to the east coast as habitat (through human development) opened up for them. Apparently, some coyotes were even introduced, either accidentally or deliberately, by hunters (Gompper 2002). Every once in a while, from the 1920s to the 1960s, newspapers printed articles about strange dog- or wolf-like beasts ("with bushy tails"), more often than not with actual carcasses in hand (and not uncommonly "puzzling the scientists"). Often, wildlife officials concluded these were coydogs.

While hybridization between coyotes and domestic dogs is known, this tends to be less common in the wild than typically believed. It usually occurs when coyotes wander far from a stable population, and perhaps find it difficult to find a suitable in-species mate. In any case, hybrids tend to be less capable hunters, and there are physiological and behavioral issues that make it harder for hybrids to successfully mate with full coyotes. While domestic dog genes have been found in some coyote populations (Adams et al. 2003; Schmutz 2007), there are physiological issues that make it difficult for hybrids to easily contribute to a coyote population's future generations (Gompper 2002).

What is more certain is that there are wolf genes moving into northeast coyote populations. Hybridization with eastern wolves occurs in a southern "hybrid zone" in Ontario, and wolf genes have migrated into the northern front (particularly Maine, New York, and New Brunswick) of the expanding coyote population (Wilson et al. 2004; Kays et al. 2009).

Whether or not there are remnant eastern cougar, it is amusing to examine the folkloric parallels between official and editorial proclamations during the pre-recognition period for eastern coyotes and those seen over the last several decades in regard to the possibility of mountain lions in the east. Of course, with the coyote, the burden of proof (i.e., a population explosion) eventually burgeoned to the point that wildlife officials had no choice but to accept it as a recognized species.

Domestic Dog
Canis lupus familiaris

It should be no surprise that domestic dogs, whether free-ranging or feral, are often mistaken for strange wild beasts. Reports of mountain lions, black panthers, maned lions, and more have been tracked down to mistaken identifications of dogs, even by people who have a fair knowledge of the natural world.

While free-ranging dogs are found in most states (those having some association with humans), not much is known about true ferals. One researcher, however, has discovered some interesting details about the "yaller dogs" that roam the pine woods and cypress swamps of the Savannah River region. These canines, now termed Carolina dogs, may be the oldest lineage of dogs in North America. They have a dingo-like appearance (short yellow hair, white feet, erect ears, and scythe-tail), are almost indistinguishable from certain Asian free-ranging dogs, and have several unusual behaviors not seen in typical domestic dogs. Genetic studies are still being pursued, and it is still uncertain where exactly they fall in the canine family tree (Mlot 1997; Handwerk 2003; Raisor 2004), but preliminary studies suggest that these are true primitive dogs. There is one report of a similar canine showing up in northern Florida (Simberloff et al. 1997), but they don't appear to be confirmed in any other regions.

This article describes the probable misidentification of a feral dog (perhaps even a Carolina dog) as an eastern cougar:

MYSTERY OF THE WILD ANIMAL AT LAST SOLVED

Rock Hill, South Carolina, *Evening Herald*, February 10, 1921

Fort Mill, Feb. 10.—Ed Scott, an intelligent Fort Mill Negro, be-lieves that he has the solution of the question as to the nature of the wild animal which has been seen to the northwest of Fort Mill in the

Providence section of Mecklenburg county, North Carolina, commonly referred to as the "Providence panther." For several weeks numbers of hunters from Charlotte, N.C., and the section to the south have tried to trail and capture or kill the strange animal whose presence and depredations are reported from day to day in different localities, the last authentic report coming from the farm of Will McKinnney, a few miles to the north of Fort Mill.

Ed is employed by a local furniture firm to drive their delivery truck which makes trips throughout this entire community for many miles around, and he is an enthusiastic hunter, having for years past covered all the hunting grounds of this section in the open season for any kind of game. He says that for a number of years past there has been a breed of dogs, probably of the shepherd breed crossed with bull dog stock, gone wild, in the thickets along the creeks near the old cemetery which lives a short distance to the northeast of Fort Mill. The animals are elusive and show a decided tendency to fight when their way of escape seems threatened. Litters of pups have been found by young boys of Fort Mill in small dugouts, and the little animals have shown fight and have been very hard to tame after being brought to the homes here.

Yesterday afternoon when returning to Fort Mill and near the location where the "panther" was last reported, he caught a glimpse of a swiftly moving animal, and caught up his gun for a shot at it, but the animal came clear of the brush before he saw that it was a dark brown dog, one of a wild breed of which he had known for several years. Except for his prior knowledge, he says that he might have taken the animal for almost any kind of a stranger. Ed was accompanied on his trip by a responsible white man who supports the statement of the deceitful appearance of the animal, and was followed by Bill Ross, a responsible and highly esteemed farmer of the community, who is inclined to the belief, after this experience, that the wild animal sought is simply one of these wild dogs.

The next article describes a large black animal suggested to be a "panther," but the presence of distinct claw marks on the tracks clearly identify it as a canine, not a feline. While dogs do not account for all "black panther" sightings, it is undeniable that they account for a significant number of them.

BLACK PANTHER IS RUMORED AT LARGE IN STATE

Newark, Ohio, *Advocate*, July 7, 1962

Springfield, Ohio (AP)—Folks in this area report an unlikely critter has been roaming the Champaign-Clark County area the past month.

First seen in Champaign County, it now has been reported seen just four miles north of here.

Everyone who has seen it says it's a black panther. And Deputy Sheriff William Cookus has tracks in a cornfield to prove some animal was there. The tracks measure over three inches long and two and three-quarter inches wide with distinct claw marks.

The animal has been described as all black, big as a large dog and with a long black tail.

Foxes

There are six recognized species of fox in North America:

Urocyon cinereoargenteus, Gray fox
Urocyon littoralis, Island fox
Vulpes [Alopex] lagopus, Arctic fox
Vulpes macrotis, Kit fox
Vulpes velox, Swift fox
Vulpes vulpes, Red fox

The red fox is the largest of these. While adults may typically run 10-15 pounds (similar to many gray foxes), the largest red foxes, in northern populations, may reach 35 pounds. I have yet to run across any cryptid foxes in North America, but foxes do play a part in spreading stories about strange animals. Pelage abnormalities, whether genetic or health-related, may make the fox unrecognizable to the general public, giving rise to rumors about "chupacabras" or other folkloric creatures.

Red foxes have several recognized color morphs. Melanistic foxes can be *black* (having all black guard hairs) or *silver* (having white-banded guard hairs). *Cross* foxes are partially melanistic, having yellow-reddish sides, a dark back, and a strip of dark hair crossing over from one front leg to the other. *Bastard* foxes are produced when a red morph breeds with a silver, producing a fox with increased dark

©Wesley Aston

Red Fox

©James M. Phelps, Jr.

Gray Fox

pigmentation (sometimes showing up as a gray-blue color), often black on the legs and belly (Cross 1941). Gray foxes also have a black phase (Long 2008), and the Arctic fox has a rare "blue" (smokey-gray) winter phase.

A *Samson* (or Sampson) fox is produced by a mutation that removes the animal's guard hairs, leaving behind the woolly pelt (Cross 1941). Similar Samson-like characters (not all of which have been evaluated for a specific mutation) have been reported in gray foxes (Fritzell and Haroldson 1982; Long 2008), coyote (Long 2008), marten (Manville 1961), and cottontail (Manville 1961b). Samson foxes have no value to the fur trade. I have seen one story about fox hunt groups releasing Samson foxes into an area in order to dissuade fur trappers, but I suspect that this is another variant of the false "game officials secretly release (you name the predator)" rumors.

Hypotrichosis, a medical condition resulting in hair loss that may appear similar to Samson animals, has been documented in Arctic foxes, particularly around warmer coastal regions (Hersteinsson 2007).

Mange (scabies) is found in foxes, coyotes, wolves, bears, and many other mammals. It is caused by small parasitic mites, resulting in thinning hair to complete hair loss, wrinkled and thickened skin, scabs and sores, and emaciation. It can be fatal (and has been noted as a probable cause of the decline in red foxes). Mange foxes are probably responsible for some "strange animal" reports. (There is at least one case where a mange wolf was mistaken as a hyena-like animal (Oakes 2006).) If there are multiple animals that are fully hairless, however, without any other signs of scabies, the hair loss may have a genetic origin, instead.

Hairless Canines

Over the last several years, there have been several sightings of hairless foxes, coyotes, or feral dogs around the country, with particular media attention in Elmendorf, Texas (Robinson 2004), Lufkin, Texas (Winthrop 2004; Johnson 2004), Glyndon, Maryland (WBAL News 2004), Raleigh, North Carolina (Mott 2006), Mount Pleasant, South Carolina (2006), and Steamboat Springs, Colorado (Weinstein 2009). This isn't a new phenomenon, as I've seen a few reports from decades ago about hairless foxes being shot by hunters.

In some cases, like the Lufkin, Texas, canine, the animal turned out to be an animal (this time a coyote) with mange (Kramer 2004). The Glyndon animal was obviously a mange-ridden fox. In others, where the animal otherwise appears to be fairly healthy, a genetic explanation like hypotrichosis may be more appropriate.

©Ronald Sherwood

Arctic Fox

©Max Allen Photography

Kit Fox

However, I've noticed that with some of these hairless foxes, the official expert often suggests that they are Samson foxes, pointing out the loss of guard hairs. I'm not certain, however, that the Samson mutation (as is usually noted in the fur trade) is the same as is seen in these hairless canines. Looking over the photo of a Samson gray fox on the ArcheryTalk.com forum for February 19, 2009, the pelt clearly has a wooly texture. This is very different from the sleek, almost hairless appearance we see in the North Carolina animal. Can the same mutation cause distinctly different phenotypes? Maybe. Is it possible that a different genetic condition is actually responsible? I suspect so, but I think this needs more research, preferably by someone with an academic interest (and the right tools for genetic analysis).

Cougar
Puma concolor

The cougar (puma, mountain lion, etc.) is a wide-ranging New World species which is of interest to biofortean researchers in several respects, particularly in regard to possible surviving populations in regions of supposed extirpation, bio-geographical anomalies, and the possibility of melanistic cougar.

Adult cougar can reach over 8 feet in length (females may be a third smaller), and be anywhere from 75-175 pounds. The largest recorded cougar was shot in 1917 by a government hunter near Hillsdale, Arizona, registering at 276-pounds (Busch 2004). Recent records include a 220-pound cougar shot in Colorado in 2001 (CO DOW 2002), 227-pound Montana cougar shot in 1970 (Busch 2004) and a 240-pound cat hit by a car in Arizona in 2007 (Whitehurst 2009).

I have seen one historical account of an even larger cat (*Salt Lake City, Utah, Tribune*, March 1, 1897): "Carl Welty, the fifteen-year-old son of Dr. Welty of the Shoshone agency, killed a monster mountain lion while hunting on Black mountain. The animal measured eight and a half feet from tip to tip, six feet three inches around the hips, four feet around the shoulders, three feet six inches high and weighed 300 pounds." While this is unconfirmed, it does offer reasonable proportions for such large cougar.

Natural prey for the cougar include deer, elk, mountain goats, mountain sheep, feral hogs, and smaller mammals. They will occasionally take pets and livestock. Chew (2009) noted one mountain lion in Washington that killed livestock without feeding on them, apparently just enjoying the attacks.

©Tony Campbell

Mountain Lion

Mountain lion attacks on humans receive global media attention, but there are very few recorded fatalities in recent years. In 2004, a mountain lion in the Whiting River Wilderness Park, California, attacked a bicyclist (who survived) after having killed a hiker (Watercutter 2004). In 2008, a New Mexico man was killed near his Pinos Altos home (AP 2008). In other cases, mountain lions have been able to habituate to urban settings without becoming aggressive toward humans (Nellans 2009).

Mountain lions range in coloration from reddish or tawny to slate-gray and dark brown. In certain western states, different colors may be seen in cougar from the same region. While melanism is not uncommon in several other wild cat species, it seems to be very rare in mountain lions, assuming it occurs at all. Most semi-credible reports come from South or Central America, and I am unaware of any absolutely confirmed black cougar here in North America. Rumors of them, certainly. But though that any natural history museum would be thrilled to acquire (and exhibit with all due fanfare) a true melanistic cougar, none have reported such a specimen.

There have been numerous theories about how a normal-phase cougar might be mistaken for a black one (usually involving environmental conditions like poor lighting or rain darkening the pelt), but this is an unlikely scenario in most cases. Mistaken identity of other species as "black panthers" is probably the most common cause, as there are any number of dark-haired quadrupeds that might given a panther-like impression to individuals unfamiliar with rarer species. What is most telling, though, is that the majority of "black panther" sightings suggest an animal in the 3-foot range, not including tail. That may be larger than most house cats, but it doesn't come close to the lengths of adult mountain lions. I won't say there isn't a real puzzle here, but I do doubt that mountain lions are involved in the black panther mystery as a whole.

While there may not be melanistic mountain lions roaming the hills, there was a confirmed white cougar. In 2005, a hunter shot a 7-foot-long white female mountain lion in north-central Idaho (AP 2005c). It was not an albino, as it had pigmented eyes and nose. (So, perhaps similar to the famed white African lions of Timbavati?) I have accounts on file of alleged white or albino mountain lions from Humboldt Co., California, in 1889; near San Antonio, Texas, in 1951; south of San Francisco, California, in 1958; near Belmont, California, in 1963.

In 2001, a white mountain lion was reported by an alternative newspaper as roaming the Red Rock, Nevada, area (Kiraly 2001). The detailed article was later (January 2, 2003) unveiled as "satire" by the editors. They thought it should have been obvious to readers (even after receiving calls from worried locals): after all, who had ever heard of an albino mountain lion?

National Park Service

Young Mountain Lion

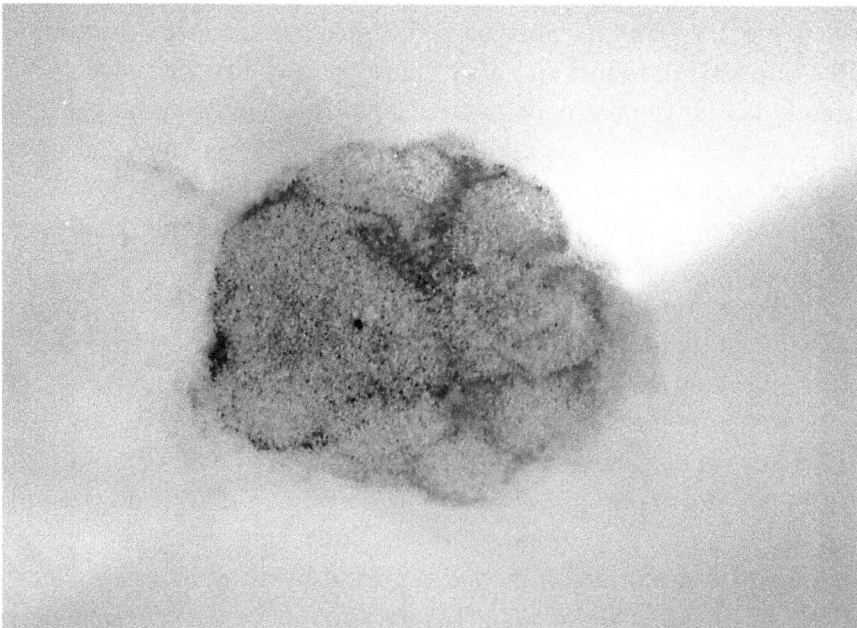

National Park Service

Mountain Lion Track

Eastern Cougar

The cougar can be identified with one very early strange animal account given by the French explorer La Salle during his journey through the Great Lakes region. His journals (Cox 1905) noted: "There are no wild beasts formidable to man. That which is called Michybichy never attacks man, although it devours the strongest beasts. Its head is like that of a lynx, though much larger; the body long and large, like a deer's, but much more slender; the legs also shorter, the paws like those of a wildcat, but much larger, with longer and stronger claws, which it uses to kill the beasts it would devour. It eats a little, then carries off the rest on its back and hides it under some leaves, where ordinarily no other beast of prey touches it. Its skin and tail resemble those of a lion, to which it is inferior only in size."

Cougar began to be killed off and driven out of eastern North America in the 1800s, and it didn't help that their primary prey, the white-tailed deer, was also under great pressure from unregulated hunting in this part of the country. While white-tails have successfully rebounded, reports of eastern cougar are still treated as unlikely by most wildlife officials. The only recognized cougar population in eastern states is the Florida panther of southern Florida. Of course, even the Florida panther has its share of controversy. There are plenty of sightings in Florida outside of the panther's officially recognized range, including one that was recently shot by a hunter in Georgia (Davis 2009; Pavey 2009). Some biologists debate whether the assumptions in place for Florida panther conservation management are even valid (Beier et al. 2006).

Discussion of the eastern cougar would require far more space than I can allot here. For those looking for an overview of such sightings and the debate over their legitimacy, I would recommend Bolgiano and Roberts' compilation *The Eastern Cougar* and Bob Butz's *Beast of Never, Cat of God*. This is a complicated debate, but it is good to see that there are some biologists willing to at least examine the evidence (Clark et al. 2002; Leberg et al. 2004). Unfortunately, there will always be someone who sees a large domestic cat and calls in a report of a mountain lion (e.g. Kenney 2008), but once you get past the obvious errors, there is far more of interest than the skeptics imply.

Of course, whether or not there are mountain lions living in the Appalachians, Adirondacks, or Ozarks, it is now well established that there is no eastern cougar subspecies. Culver (et al. 2000) showed that there is only one North American cougar subspecies (*P. c. couguar*), which developed as the species moved north from South and Central America. This study also noted that Florida panthers were highly inbred, but there is no valid claim for a distinct subspecific status for that population.

Maned Mountain Lions
I recently came across the following account of an alleged maned cougar:

CALIFORNIA MOUNTAIN LION WITH A MANE
Salt Lake City, Utah, *Deseret News*, March 15, 1909
Petaluma, Cal., March 16.—A California mountain lion measuring seven feet from the tip was shot on a ranch near here today by Robert Cook. For some time the farmers of this vicinity have been losing calves, sheep and other stock, and the depredations have been laid to coyotes. Cook was out with his pack of hounds looking for these animals when he trailed the lion to its den. One shot which lodged in the neck of the big beast killed it instantly. The lion wore an immense mane like its African relative.

Now, if in fact there was a genetic mutation in cougars for a mane, that could clear up a fair number of "maned lion" sightings reported in various parts of the country. (Certainly it could have implications for Ivan T. Sanderson's (1973) "ruffed cat" skins that he bought in a Colima, Mexico, market.) But, unless someone were to track down this specimen (if it still exists) and confirm that it was actually a maned mountain lion, I wouldn't use it as firm evidence. I have not run across any reports of ruffs or manes in cougar in the scientific literature.

Henry W. Shoemaker (1912) gave the story of Peter Pentz, a Pennsylvania mountain man who supposedly killed a maned mountain lion and its family in Clinton County in 1798. Unfortunately, we don't have Shoemaker's original sources, so it is difficult to determine the accuracy of this story. (Shoemaker had a penchant for turning folklore into "fakelore," as it is termed by professional folklorists. I tracked down "gorilla" sightings from Pennsylvania newspapers in the 1920s that Shoemaker spun into a story of romance and revenge in the mountains, and there is quite a difference between the source material and Shoemaker's story. So, while I am willing to assume there could be a nugget of truth to this maned cat account, I would not take all details as gospel.)

Onza
While my focus in this book is on the United States and Canada, there are several interesting mystery carnivores reported from Mexico. Anyone who has

©Linda More

Jaguar

Ron Singer, USFWS

Melanistic Jaguar

followed cryptozoology for a while should be familiar with the Onza, an alleged cougar-like feline with longer legs, a long lean body, a dark back stripe, and a ferocious reputation. A specimen shot in the western Sierra Madre in 1986 was tested and found to be genetically indistinguishable from a normal mountain lion, though its phenotypic differences might be a regional adaptation (Dratch et al. 1993-1996). Personally, I'd like to see further research into this, even if it is a mountain lion. With the technological advances over the last couple decades, I think there is more that could be determined. Is it just an unusual regional morph, or does it represent (as the British Columbia coastal wolves appear to do) pre-speciation?

Readers interested in the subject should find a copy of Robert Marshall's *The Onza*, or Neil Carmony's *Onza! The Hunt for a Legendary Cat*. Marshall's book is a favorite of mine, not least because it rekindled my interest in mystery animals when I ran across it in a library back in my high school days.

Jaguar
Panthera onca

The jaguar is the largest of North America's cats, reaching up to 350 pounds (Nowak 2005). Hunting and habitat loss have driven it out of the southwestern United States, and only 500 or so are still remaining in Mexico. The melanistic phase is a South American morph, but has been bred in captivity, so remains a potential source (though rare compared to the melanistic leopard) for escaped "black panthers." There is one camera trap photo of a melanistic jaguar allegedly from the El Fuerte river valley in northwest Mexico (Dinets and Polechla 2005), but jaguar experts question its authenticity (Johnson et al. 2009).

White jaguars have been reported from South America.

For biofortean researchers, the interesting questions about the jaguar in North America revolve around biogeography, so I'll focus here on individual states. Of course, the jaguar ranged throughout the southern United States up to the end of the Pleistocene, with fossils found as far north as Nebraska, Tennessee, Pennsylvania, Washington, and Oregon (Daggett and Henning 1974). Jaguar motifs have been recognized (sometimes speculated) in pre-European colonization artifacts and rock art in Alabama, Missouri, Ohio, New Mexico, Texas (Daggett and Henning 1974; Pavlik 2003).

Within historical times, the northern range has shrunk quite a bit, but there are interesting indications that a few jaguars may have been farther north than the American southwest during the very early colonial periods. (A few very early 1800s reports among the State Sightings could be jaguars, but it is often difficult to determine validity.) A good overview of possible and probable names for the jaguar among southwestern Native Americans is given by Pavlik (2003). For further information on jaguars along the US-Mexico border, see Brown and González's *Borderland Jaguars* (2001). The jaguar was added to the U.S. endangered species list in 1997.

California

Merriam (1919) noted several very early reports of jaguars in southern California. One significant account came from James Capen ("Grizzly") Adams, who reported seeing a pair and their young in the Tehachapi Mountains. Another report came from the *Personal Narrative* (1833) of a fur-trapper from Kentucky, James O. Pattie, who claimed to have killed an animal spotted like an "African leopard" on the islands of the Colorado River delta. Merriam also noted that the Kammei tribe in southern California had a name for the "Big-spotted Lion" that used to inhabit, though rare, the Cuyamaca Mountains.

Strong (1926) noted several stories from southern California's Cahuilla people. The most recent involved a jaguar killed around 1860, near Palm Springs, when it attacked a Cahuilla who was stalking deer while disguised with a deer's head.

Arizona

Jaguars in Arizona began declining in the 1800s until they had almost disappeared in the 1940s. Very rarely were they seen after that, and any jaguars noted were dismissed as transient wanderers. In 1996 two confirmed sightings spurred the interest of wildlife biologists. Trail cameras and track surveys from 2001 to 2006 then found evidence that a few adult jaguars were frequenting southern Arizona enough for some researchers to call them residents (McCain and Childs 2008). Other researchers argue that Arizona jaguars are wandering males that primarily occupy Mexican territory (Rabinowitz 2010), and that conservation efforts to designate protected jaguar habitat in the United States are counter-productive.

New Mexico

One early description: "In New-Mexico, there is found a beautiful animal, which, perhaps, may be the true Leopard. Its skin is a fine yellowish-white colour, spotted elegantly with brown and blackish spots. General Wilkinson carried one of the skins of this animal with him, from this country. They are said to possess enormous strength: relations are given of their carrying away the carcasses of horses and bullocks. If this be true, they must be larger and more powerful than the Panther, which seldom attempts to carry away any thing above the size of a hog, or large calf." (Dunbar 1805)

Today, there are sporadic reports in southwestern New Mexico, part of the territory kept by the few resident Arizona jaguars. Davis (2006) noted a confirmed (photographed) jaguar discovered in the Animas Mountains.

Texas

Bailey (1905) noted several cases of jaguars being killed in Texas up through the early 1900s. The last confirmed jaguars were killed in 1946 near Olmito, and in 1948 near Kingsville (Sinclair 2008).

Louisiana

Nowak (1973) reported a couple of early newspaper accounts (both in 1886) in which large felines were killed in Louisiana, which were referred to as "American tigers" (one noting it was another name for jaguar). Both were said to weigh over 200 pounds, but Nowak noted that the other details were vague, though one was specifically distinguished from the panther. Nowak also noted that jaguars had been killed in East Texas within 100 miles of the Louisiana border.

There is an earlier account of a Louisiana "tiger":

Camden, New Jersey, *Mail*, August 12, 1835

Strange visitors—A tiger, measuring eight feet from the snout to the end of the tail, was killed last week by a negro, within ten or twelve miles of this place. He dispatched him with a load of buck shot which he fired when the animal was about springing on him.— *Attakapas Gaz.*

Florida

Pleistocene jaguar fossils apparently are not uncommon, but I haven't seen any convincing accounts from historical times that weren't likely former captives.

Of course, if there's any place that a jaguar could survive for a while, it's southern Florida. One jaguar was shot near Felsmere, Indian River County, in 1968. It had been sighted in the vicinity for at least two years (Simberloff et al. 1997).

North Carolina

Nowak (1973) noted that there were a couple of very early references to felines in the Carolinas that were larger than panthers (mountain lions).

Brickell (1737), for example, wrote: "The *Tyger* is in shape somewhat like a *Lioness,* but has a short Neck. His Skin is most beautifully mottled with several kinds of spots resembling the *Panther,* only the former are not so round, nor have such different Colours. They are large, strong and swift Beasts, but are never to be met with in the Settlements, being more to the Westward, *viz.* on this and the other side of the Mountains, but are very scarce and seldom to be found in this Province, by what I could learn from the *Indians;* and in our Journy up towards the Mountains we saw but one. They have a great many young Ones at a time, and are very fierce and bold Creatures, and will spare neither Man nor Beast to satisfie their Hunger, as I have been inform'd by the *Indians* and some of the Planters who have seen and kill'd them."

Tennessee

Tennessee is well-known for its cave fossil jaguars, and one cave provided footprints of what is probably this feline, as well (Simpson 1941; McCrady et al. 1951). Simpson noted that the bones in Craigshead Caverns were unmineralized and "very fresh in appearance in spite of having soaked in limy waters." The footprint was found in soft, moist earth: "They were clear and sharp and looked as if they might have been made only a few minutes before their discovery, except that a small amount of loose, almost dry earth and darker dust had drifted into them." Simpson explored the possibility of historical eastern jaguars, but was unable to find any convincing evidence. He thought it might be possible for Texas jaguars to occasionally wander as far north as Tennessee.

McCrady et al. (1951) suggested that, because the two feline skeletons they reported were found far beyond the twilight area (over half a mile into the caves in both instances), the jaguars may have, "like cave bears, cave rats, and raccoons, frequented the inner recesses of caves."

Colorado

Seton (1920) uncovered one possible historical record: "Rufus B. Sage, while camped on Soublet's Creek at the base of the Rockies, head waters of the Platte, within 30 or 40 miles of Long's Peak and 2 days' march from Fort Lancaster, in December of 1843, says: 'One of our party encountered a strange looking animal in his excursions, which from his description, must have been of the Leopard family. This circumstance is the more remarkable, as Leopards are rarely found except in Southern latitudes. However, they are not infrequently met in some parts of the Cumanche country, and their skins furnish to the natives a favorite material for arrow-cases.' (*Rocky Mountain Life*, p. 347.) As Sage was quite familiar with panthers and bobcats this may have been jaguar or ocelot."

Northern States

Rafinesque (1832) wrote the following article, "On the large wandering Tigers or Jaguars of the United States." I was able to locate a few of the newspaper stories he mentions, and they are listed in the State Sightings section.

"The Jaguars are the spotted Tygers of America, found from Mexico to Paraguay. It was supposed that none were ever seen further north or with us; they are hardly mentioned in our Zoological books, and their casual visits disbelieved by many when they hear of them. But Humboldt has lately ascertained that the striped Tyger of India, often wanders to the north as far as Tartary and Siberia. I will prove that the spotted Jaguars do the same in America, and wander as far as Kentucky and Lake Erie in latitude 42. This always happens in summer, and is not at all extraordinary, since our summers are as warm as in the tropics, and these carnivorous animals are known to range very far in search of prey.

"Several instances of huge beasts having been seen in Louisiana, Arkansas, and Kentucky could be collected by enquiries among old hunters. When seen at a distance only, they are commonly mistaken for large Panthers, our unspotted Couguar. When seen too near, the boldest hunters are afraid of them. When shot, nobody knows them, not even the Indians; and the skins are sold high at once for side-saddles. Sometimes the account gets into some newspaper, but is usually disbelieved or soon forgotten.

"Harlan in his Fauna Americana only mentions that the Jaguar or *Felis onza* of the naturalists wander sometimes east of the Mississippi, which must be crossed by swimming. This animal comes as far north as Kentucky in lat. 38. While I was in Kentucky I heard of several having been seen and shot. Two of them, a male and female, did once make a stand near Russelville, and alarm many travellers, feeding on hogs, until a party of hunters went in pursuit of them, killed one, and drove away the other.

"Before that another had been shot on the 6th of June, 1820, by Mr. John Six, on Green River, 10 miles south-east of Hartford, in Ohio county. The skin was brought to Frankfort and an account given in the papers. This animal appeared to be a true Mexican Jaguar. The body was 5 feet long and the tail 2 feet. It weighed 150 pounds before skinning. The back and sides were yellow with black spots curiously arranged in several rows, a row on the back much larger and extending over half of the tail, which was rather slender, with very long hair at the end. Chin, belly, and feet white, ears small round black outside, white inside. Whiskers stiff 6 inches long, black with the end white.

"But another Jaguar still larger and of a different species has lately been seen as far as Lake Erie, and lat. 42. One was shot by the Seneca Indians, to whom it was totally unknown, another was killed in the Alleghany mountains of Pennsylvania, and an account given in the papers. These animals were totally distinct from the common Jaguar; they must have been wanderers from New Mexico or the Oregon mountains, and belong probably to a new species which I propose to call *Felis dorsalis*, owing to the black band on the back. There are several other species of Jaguars in South America, little known or not well distinguished.

"*Specific characters*, FELIS DORSALIS, Dorsal Jaguar. Of a grey colour, neck fallow, a black line or band all along the middle of the back, two rows of ringed spots on each side, black above, brown below. Total length 10 feet including the tail, body 6 ½, tail 3 ½. Very different from *Felis pardalis* by size four times larger, neck and back, &c."

©RobynRG

Canada Lynx

Canada Lynx
Lynx canadensis

The Canada lynx is a moderately-sized feline, adults measuring almost 3 feet in length and weighing 20-30 pounds (Long 2008), with some reaching more than 40 pounds. Found throughout much of Alaska and Canada, and northern regions from New England and the Great Lakes region to the Rocky Mountains and the Pacific northwest (per the USFWS). They may occasionally be sighted further south. There are older stories of lynx being caught in, for example, Pennsylvania (Shoemaker 1919), but there is a chance that some of these are mistaken bobcats, which can sometimes be as large as the typical lynx.

There are two currently recognized subspecies: *L. c. canadensis*, which is found throughout most of the range, and *L. c. subsolanus*, found in Newfoundland.

In many locations, the Canada lynx is a specialist predator on snowshoe hares, and there is a a long-studied boom-and-bust cycling of population numbers due to the predator-prey relationship. Lynx, however, will take down other prey when the need arises. They are fully able to kill white-tailed deer (Fuller 2004), mule deer (Poszig et al 2004), Dall sheep, and caribou (Stephenson et al. 1991).

There have been two reintroduction programs for Canada lynx. 83 radio-collared lynx were released between 1989 and 1992 in the New York Adirondacks. Of these, 32 were killed by cars or accidental shootings, while the rest just disappeared (Ryan 2008). They apparently wandered north, as no sign of lynx in the Adirondacks has since been found. Colorado Department of Wildlife began a lynx reintroduction program in the region north of Durango. This has been a more successful effort, and some lynx have been found to remain in Colorado, while others disperse elsewhere in the Rocky Mountains (as far south as New Mexico).

Canada lynx-bobcat hybrids have been discovered in the wild in Minnesota, Maine, and New Brunswick (Schwartz et al. 2004; Homyack et al 2008). First generation hybrids appear to be fertile. Physical characteristics included some from either parent (bobcat's spotting to varying degrees, lynx's black tail tip, etc.), as well as other traits that were intermediate between the two species.

Bobcat
Lynx rufus

The bobcat is found throughout much of the U.S., and in parts of southern Canada. Normally smaller than it's cousin the Canada lynx, the bobcat usually runs

©John Pitcher

Bobcat

Tom Smylie, USFWS

Ocelot

three to four feet in length, about two feet at the shoulder, and 20-30 pounds. The largest bobcat will, however, dwarf the largest lynx. A bobcat weighing 76 pounds was reported from Maine, and one weighing 69 pounds was recorded in Colorado (Jaques 1946; Rue 1968).

The bobcat's main prey are rabbits and rodents, but will take other small animals, including dogs, cats, and sheep. (They will also take carrion, so some "live-stock kills" may be misidentified scavenging.) Larger bobcats are known to take deer (Labisky and Boulay 1998) and pronghorn antelope (Beale and Smith 1973; Bright and Hervert 2005).

The bobcat typically runs varying shades of brown or gray with black markings, but bluish-gray and reddish specimens are known. Albinos have been confirmed throughout its range. Melanistic bobcats are also known, primarily in southern Florida (Ulmer 1941; Regan and Maehr 1990). Another black bobcat was observed in northern Florida (Hutchinson and Hutchinson 2000), and another trapped in New Brunswick (Tischendorf and McAlpine 1995).

Black bobcats are a common explanation for "black panther" sightings, but it doesn't seem to be a common morph outside of southern Florida. And, of course, most "black panther" sightings specifically mention a long tail, which isn't found in bobcats. Still, Bogue (2009) noted photographs of one very dark (but not necessarily black) bobcat that was seen in one California location with many alleged "black panthers" over the years.

As noted in the Canada lynx description, lynx and bobcat hybrids have been found in the wild. There are also records of possible naturally-occurring bobcat-domestic cat hybrids (Young 1958; Gashwiler et al. 1961), but these appear to be very rare, and there is no demonstration of fertility. There are several domestic cat breeds with superficial "bobcat" traits which were speculated to have hybrid ancestry (and marketed as such), but these have shown no genetic markers for bobcat parentage.

Ocelot
Leopardus pardalis

The ocelot is a small beautifully-marked cat that has a wide range in South and Central America, thinning a bit in Mexico, and now just reaching over into the southern border. It has a preference for dense scrub, which has been removed from most of Texas, so its former range has shrunk drastically in the United States.

Historically, ocelots were found in Arkansas, Louisiana, Arizona, and Texas. In fact, the ocelot was first described from southwest Arkansas, but soon disappeared (Bowers et al. 2001). It may have rarely wandered over the border from Mexico into California, but does not appear to have ever been considered a resident. There are Pleistocene fossils of ocelots in Florida (Kurtén and Anderson 1980), so the southeast was probably the extent of its northern range.

Ocelots in the U.S. today are only confirmed in the southern tip of Texas, and the population there is declining, with recent estimates of as few as 50 in the state (Tompkins 2010). A fairly large ocelot, measuring 4 feet, 6 inches, and weighing 45 pounds, was taken in 1950 in Donley County, Texas (Hock 1955).

There are sporadic reports in Arizona, with a few confirmed specimens and several reliable observations of ocelots over the last century (Grigione et al. 2007). It was noted that some of these may have been accidental or intentional releases, or they may have wandered in through subtropical vegetation corridors in certain river drainages.

Chapman (1894) made this interesting observation: "From many sources I have received information of the occurrence in Florida of a long-tailed, spotted Wild Cat, which may prove to be the Ocelot (*Felis pardalis*)." It doesn't appear that he ever followed up on this or published his data.

The 1965-6 television mystery series *Honey West* included a pet ocelot, which (no surprise) helped create a rising demand for ocelots in the pet trade. The species was added to the endangered species list in 1972, so it is far more difficult to acquire today, but they were quite popular at one time. I have seen a number of reports of ocelots accidentally or deliberately released (e.g., a 1968 leopard-like mystery cat in Michigan that was shot and identified as an ocelot), likely exacerbated by maturing ocelots that didn't stay as friendly as when they were kittens. (There were a number of incidents of pet ocelots attacking people, including babies and young children.) Obviously, identification of a wandering feline as an ocelot should be confirmed wherever possible with photos; I have one Knox County, Texas, account where the photograph of the shot "ocelot" clearly shows a bobcat with extensive spotting.

Margay
Leopardus wiedii

The margay is a smaller, arboreal kin to the ocelot. There is only one historical record of the margay in the United States: a specimen in the US National Museum,

Gary Halvorsen, USFWS

Jaguarundi

©GrabJ

Margay

collected in Eagle Pass, Texas, in 1852. This specimen showed some differences from Mexican margays, in that it had somewhat longer fur, and completely black spots, so it was given its own subspecific designation, *L. w. cooperi* (Goldman 1943).

A subfossil margay was found in Orange County, Texas (Kurtén and Anderson 1980). De Oliveira (1998) notes that some small feline fossils in Georgia and Florida may be referrable to the margay.

Jaguarundi
Puma yagouaroundi

The jaguarundi is a strange-looking feline, with otter-like lengthened torso and short legs. It has two color phases: reddish and grayish-black. (Some biologists suggest the gray and black phases are distinct, but I haven't seen genetic support for this published.) There is variance of shade in these color phases, as well as ontogenetic differences (younger light gray individuals may darken with age). Because the dark phase is more common, it was assumed to be the "wild type," but genetic research shows that it is a melanic mutation that became more common than its original coloration (Eizirik et al. 2003). The dark phase is of interest to cryptozoology, as it has been suggested that "black panthers" in the southern United States may be jaguarundi sightings (Moore, Jr. 2002; Wojcik n.d.). Moore suggested that because jaguarundi inhabit a wide variety of habitats throughout its range, it was a mistake for wildlife biologists to assume that they only stick to one habitat type in the United States. Moore had his own sighting near Port Arthur, Texas, and was able to identify it as a jaguarundi, having had hands-on experience with a number of wild cat species while working at a big cat refuge.

The jaguarundi has an extensive range in South and Central America, following both Mexican coasts north (de Oliveira 1998). It just crosses the border into Texas. Officially, only the two southernmost counties in Texas are thought to possibly have jaguarundi, and even then evidence is rare. Efforts to locate jaguarundi in these counties, after credible observations, have been publicized in the media (Brezosky 2004; Sinclair 2009), but I haven't seen reports of positive results.

Historically, it is undetermined if jaguarundi ever reached further into the United States than Texas. Fossils originally referred to jaguarundi in Florida are more likely margays (de Oliveira 1998). There are numerous modern sightings of jaguarundi in Arizona, but no proof of such (Hock 1955; Brown and González 1999; Grigione et al. 2007). There are also many sightings of jaguarundi in the southeastern United States, particularly Alabama and Florida (Britt 2004; Miller 2009).

Whether or not jaguarundi currently reside in Florida is up for debate. Simberloff (et al 1997) noted that stories of the animal's deliberate introduction there by a traveling author allege it took place as early as 1907 in DeSoto County and between 1934 and 1940 in Santa Rosa County. (See the Florida sightings section for the identity of this author.) Published reports came from Levy, Highlands, Lee, Hillsborough, and Osceola Counties in the 1940s. By the 1950s, sightings had been gathered throughout most of peninsular Florida. In the 1970s, sightings appeared to be in decline. However, recent research by Wojcik (n.d.) collected a number of Florida jaguarundi sightings as recently as 2004-2005.

Domestic Cat
Felis catus

The domestic cat is the main culprit when it comes to strange feline reports. Some people fail to recognize a large scruffy yellow feral or a big black free-roaming tomcat as domestic cats, leading to "black panther" or "mountain lion" reports. This has been going on a while:

> Kennebec, Maine, *Daily Kennebec Journal*, September 2, 1903
> A Bowdoinham man reports that he recently saw a large coon cat in the deep woods which had evidently given up a domestic life for one in the forests. He says that it was a very handsome animal but as wild as a hawk. Pretty soon there will be blood-curdling stories of a strange monster roaming the Bowdoinham forests.

Now, I realize that this is hard for some cryptozoology enthusiasts to accept, but we've seen far too many cases where a photograph or video has been taken of the suspected mystery feline, and it is obvious (to anyone with any experience with felines) that a large domestic cat is shown. Even when this is pointed out, the witness may not be able to accept it.

Part of the difficulty comes from the mental image of an animal that everyone subconsciously carries—and this differs from person to person. We don't immediately identify animals by looking at the whole, but often just by recognizing a few specific characters. So, if a character differs from the norm in some way, or what we *think* is normal, and does not jibe with our mental image, we may have trouble identifying it.

Another problem with mystery felines is that there often isn't an obvious visual clue for scale of the animal. In one case, a large black feline was videotaped walking through a field in White Lake Township, Oakland County, Michigan. Initially, the witness thought it could be 5 feet long and up to 70 pounds. Later, a wildlife officer used landmarks in the tape to determine that it was a large domestic cat, weighing about 20 pounds (Patton 2006; Wilkins 2006).

While there appears to be gathering evidence for very large feral cats in Australia (Naish 2007), it is less likely that this would be a comprehensive explanation for North American mystery cats. There is always the possibility, however, that a localized population might develop distinctive traits that causes confusion in ready identification (as has been seen in one UK population (Shuker 1990)).

Wolverine
Gulo gulo

The wolverine is the largest terrestrial mustelid, with large males running 35 to 40 pounds, and almost four feet in length. This is primarily a northern species, with most of the North American population living in Canada and Alaska. Wolverines are particularly associated with areas where there is deep snow cover during spring denning (Aubry et al. 2007), which is usually alpine habitat in the southern part of its range.

One reason so little is known about wolverine biology is that they are wide-ranging (some traveling hundreds of miles in a single month) and have a very low population density, so they are rarely encountered by humans. They are primarily scavengers, but will raid traplines and livestock, and are fully capable of taking down large prey when necessary. There are accounts of wolverines attacking floundering moose, and killing caribou and white-tailed deer (Grinnell 1920; Lapinski 2006).

This is another species where the primary biofortean interest is in its biogeography. This may involve extreme long-distance wandering or the possibility of relict populations in former habitat. I've run into several historical accounts of animals that may be wolverines in areas they wouldn't inhabit today. (See, for example, the New Mexico section.) There are specimens and photographs of wolverines in Colorado (Nead et al. 1985; Pankratz 2009), Wisconsin (Jackson 1954; Wydeven

©Zastavkin

Wolverine

National Park Service

Wolverine

and Wiedenhoeft n.d.; Nichols 2005), and Iowa (Haugen 1961), though evidence of breeding populations is lacking, and sighting reports throughout the upper midwestern states and Rocky Mountains.

California

Wolverines formerly inhabited the Sierra Nevada range as an isolated population (Schwartz et al. 2007), but was considered by wildlife biologists to be extirpated for almost 90 years before being a male was discovered in 2008 (Moriarty et al. 2009). It was confirmed to be related to the western Rocky Mountain population, so is considered a solitary long-distance migrant. (It is also possible that it was captured and released outside its natural range by someone, but there is no evidence of that.) In 2010, biologists noted behavior that suggested the wolverine was looking for a mate, but those are almost 800 miles away (Weise 2010).

Interestingly, there is anecdotal evidence for both a continual presence of wolverines in the Sierra Nevadas, and a historical range outside of those mountains in California (for example, see Jones 1955). In one paper, this body of observational data was used as evidence of poor scientific rigor when anecdotes were used as confirmative evidence (McKelvey et al. 2008). While I agree that anecdotes alone should not be used as confirmation of a species' presence, I do think they are more valuable than these authors suggested. There are other researchers who have been able to do much more with anecdotal data (Paxton 2009; Boshoff and Kerley 2010). Certainly, Groves (1988) was able to use observational data (through a questionnaire given to trappers and biologists) in a preliminary look at wolverine distribution in Idaho. Frey (2006) provided a very interesting framework for evaluating occurrence records (including observational data) for reliability, even using this to suggest that sightings of wolverines in the southern Rocky Mountains in New Mexico (where they are considered extirpated) should not be dismissed out of hand.

New York

An Adirondack naturalist and others claimed to have seen wolverines (Grondahl 2004), while state wildlife officials were highly skeptical, suggesting people were confusing fishers for wolverines. The article noted a few historical accounts (early 1800s) of wolverines being found in the region.

Michigan

Michigan, the "wolverine" state, has always been an uncertainty in the actual range of the species, and many wildlife officials argued that the animal was never native to the state (Robbins 2004). In 2003, USFWS officers stated they planned to begin an in-depth study of the species, including the question of whether it actually had lived in the Great Lakes Region (Anon. 2003b). They noted that there were notations of the animal in fur-trading records from 17 counties in the state, but no skeletal remains were known. A historian, Larry Massie, also made a case for the wolverine's early presence through historical records (Robbins 2004).

I have come across a couple of early Michigan account of wolverines, also:

Manitowoc, Wisconsin, *Pilot*, March 30, 1860

Strange Animal.—The Green Bay Advocate has an account of a strange animal being caught in a trap on the 15th of February, by a German living at Marquette, Lake Superior. It was about three and a half or four feet long, with black legs and head, with rather coarse hair, resembling that of a bear; reddish color on the sides, and dark brown on the belly, with whitish spots on the breast; head small, something like an ant eater's, nose pointed, tail short and bushy, legs short but very large, with claws like a bear. Nobody seems to know any about it but Mr. C. Henry; and he speaks of seeing the same description of animal on the coast of Labrador, and that there the French call it "Carcajou." He, also describes them as possessing great strength, and as being for their cunning much dreaded by trappers.

"MONSTER" SEEN NEAR ISHPEMING?

Ironwood, Michigan, *Times*, August 2, 1935

While residents along Lake Michigan are hearing stories of fabulous sea-serpents and strange marine animals, a "monster" of a different kind is occupying the attention of the upper peninsula.

It's a wolverine. At least, it appears like one say those who report having seen it.

Several residents north of the Straits have told of seeing the tracks of the mysterious animal or observing it in the woods during the past year. Every now and then it is reported in some different part of the country. The latest report comes from four Ishpeming men, Axel Anderson, John Bess and Ed and Swan Merila.

The men were fishing on the east branch of Fence river west of Republic where the strange animal made its most recent appearance. It boldly approached the entrance to their camp, it was said.

Students of natural history are becoming more interested in the reports from the upper peninsula and are wondering, if the animal is a wolverine, how it happened to be in Michigan since the last wild members of its species were supposed to have disappeared from the state many years ago.

Just as the USFWS began to get interested in the possibility of Michigan wolverines, one appeared in Huron County (AP 2004), almost 800 miles away from known wolverine habitat. It was photographed again in 2005, and in 2006 hair samples revealed that it was a female and probably related to wolverines in Alaska (Anderson and Langley 2005; Bohn 2006), suggesting it was not a natural immigrant. Later testing showed that initial results were mistaken, and that in fact it was related to wolverines in Ontario and Manitoba, so its origin remains a mystery (Shaw 2010). In 2010, its body was found by hikers, and necropsy showed that it died of natural causes (Bell 2010; Greenwood 2010).

North American River Otter
Lontra canadensis

Historically, the river otter was found throughout most of North America, but after more than a century of trapping, it was extirpated from a number of states. Today, after numerous reintroduction programs (Raesly 2001), river otters can be found in at least small numbers in every state.

New Mexico was the last to bring back the otter, with the importation of otters from Washington state into Rio Peublo de Taos in 2008 (Matlock 2008). One odd thing about this was that the article stated the last confirmed sighting of otters in New Mexico was in 1953. However, just a few years prior, DNA evidence of what may have been true southwestern otters was discovered in northwestern New Mexico (Polechla, Jr. 2005).

Generally regarded as playful and cute, there are actually quite a few stories about otter attacks in the newspapers, both on humans and on dogs (Lush 2001; Morrison 2006; UPI 2009b; Gomez 2010). Sometimes, the otter turned out to be rabid, but in other cases, there may have been territorial or other behavioral issues.

Most river otters reach 3-4 feet in length, including the tail. One larger example was caught in Maine:

GIANT OTTER TRAPPED AT PORTAGE BELIEVED BIGGEST EVER CAUGHT
Portland, Maine, *Press Herald*, December 4, 1949

Portage, Dec. 3. (AP)—The otter Walter Bolstridge trapped may have been the giant of the otter world.

Game Warden Wilfred L. Atkins said the animal's glossy pelt measured 66 inches long. The average otter is about 40 inches.

And Bolstridge said that before being skinned, the huge otter was about 76 inches long from the tip of its nose to the end of its tail.

State Fish and Game Commissioner George J. Stoble said an otter as big as Bolstridge's trophy may be a world record.

Bolstridge caught the otter recently on the Big Fish River, between Portage and St. Froid Lakes, in northern Maine.

Mink trapping was disappointing in the area in the month-long season that ended last Wednesday. Rain and snow hampered trappers.

North American Badger
Taxidea taxus

The badger is found throughout most of western and central North America and the Great Lakes region. These burrowing mustelids reach about three feet in length.

Biofortean interest in badgers particularly applies to "gravedigger" folklore from the Great Lakes region, where an unknown creature is said to burrow in cemeteries and dig into coffins. Other burrowing mammals may be responsible for some stories, but in several cases a badger is noted or described.

OTTER BE A RECORD—This 66-inch otter pelt shown with an average otter pelt held by Walter Boistridge, Portage, may be a world's record catch for otters according to state officials. Bolstridge trapped the animal recently on the Big Fish River.

The "Giant Otter" Trapped in Northern Maine in 1949

THE HIGH SCHOOL MUSEUM (excerpt)

Sandusky, Ohio, *Register*, February 14, 1919

By Edwin Lincoln Moseley

"In northwestern Ohio people tell about a strange animal that is rarely seen but which makes burrows larger than those made by a woodchuck. In several instances this mysterious creature has dug into graves. It has on its fore feet claws of enormous size and strength, and the ugly beast has such powerful jaws that no dog is a match for it. This description fits the badger, whose range formerly extended from Ohio to the Dakotas and beyond."

Reddish (erythristic) and albino badgers have been discovered (Roest 1961; Laacke et al. 2006). One case was described:

Warden Shoots Albino Badger

Bakersfield, California, *Californian*, November 5, 1941

For a moment last Sunday, C. L. Brown, Kernville game warden, thought he was seeing a white polar bear. He noticed the animal while driving through Kelso Valley with his daughter, Marilyn, who finally chased it into a hole among the rocks and discovered it to be an albino badger with pink eyes and milk white skin. The badger was dispatched and presented to J. R. Bechtel, Kernville sportsman, who is having it mounted. It is the first albino badger of record to be reported in this part of the state.

Unusual animal companionships are occasionally encountered in biofortean literature, though most involve captive animals. Here is one apparent case involving the normally solitary badger:

Curious Companionship of the Coyote and Badger.—I have occasionally heard 'cow boys' and others in Wyoming Territory speak of the existence of intimate social relations between the coyote (*Canis latrans*) and the badger (*Taxidea americana*). They report that the former not only travels in company with the latter, but

©Jan Gottwald

River Otter

©Richard C. Bennett

North American Badger

often feeds and protects him. Up to the summer of 1883, I regarded
these statements as the fruit of fertile imaginations, and as having
no basis in fact. Observation, however, has compelled me to believe
that intimate relations at least occasionally exist between these
animals. During the last season I had the good fortune to see the
coyote in company with the badger on three different occasions, and
once under peculiar circumstances. I was engaged in making geological observations about fifteen miles east of the Beaver river, in
Wyoming Territory, in a region of Miocene Tertiary bad lands. Sitting on a lofty butte examining some fossils, I saw, several hundred
feet below me, a coyote and badger walking together, and every few
minutes stopping and playing. The coyote would go in front of the
badger, lay its head on the latter's neck, lick it, jump into the air,
and give other expressions of unmistakable joy. Its antics with the
badger were very much like that of a young dog playing with another pup, or when meeting its master. The badger seemed equally
well pleased. This playing and fondling of each other was kept up
for over half an hour, and until they disappeared around the end of
a butte near by. Again, during the month of March, when camped
near the intersection of the Bridger road with the overland trail,
a still more curious example of this companionship occurred. At
early dawn I observed a coyote close to a pile of specimens, about
200 yards from my tent. The following night I placed some meat
on this pile and the coyote 'went for it,' about the same time in the
morning. I repeated this experiment, as my object was to observe
more closely its habits, and especially to ascertain how tame it might
be made by decent treatment. To my surprise, on the fourth morning the coyote was accompanied by a badger. The following morning
the coyote came alone, but on the morning after that the badger again
came along. After that neither coyote nor badger made their appearance, but why they abandoned my hospitable quarters is a mystery
to me. In these instances the coyote carried off the breakfast that I
had provided, and I could not tell whether or not it was shared with
the badger. It, however, became evident to me that these animals do
not associate by accident; they must have some affinity for each
other, or else they would not thus come together.

The coyote is naturally sociable. Often when I have encountered
him amid the wilds of nature he stood and gazed after me wistfully,

as if he meant to say, 'I wish I could have your company.' He soon learns to know that man is his enemy, and for his own preservation gives him a wide berth.

It is not at all improbable that future investigation may show this fellowship to be a case of symbiosis. It is hoped that this will call out others who have had more extended opportunities to observe the habits of these animals. One of the first from whom I learned of this intimacy between the coyote and badger was W. U. Hostile, of Lander, Wyoming Territory.—Samuel Aughey. (Aughey 1884)

Fisher
Martes pennanti

The fisher is a long, low-built mustelid, with males typically reaching three or four feet in length and 10-15 pounds. Their fur ranges from brown to black in coloration. Head, neck, and shoulders have tri-colored guard hairs, often giving a grizzled appearance. They are sometimes called fisher cats, and are commonly given as a possible explanation for "black panther" sightings by wildlife biologists in the northeast states.

There are cases where fishers are a distinct possibility for some of the bushy-tailed strange animal sightings recorded later in this book, particularly when the witness is unfamiliar with rarely-seen wildlife. In other situations, it is difficult to imagine that a reasonably well-informed individual, under good observational conditions, will mistake a fisher for a large feline. I encountered an adult fisher in Lancaster County, Pennsylvania, last year, watching it lope through a field into a small woods. Anyone who has kept a ferret will immediately recognize a musteline's distinctive gait.

Fishers were once widespread throughout northern forests in Canada and the United States (Powell 1981). Pressure from overtrapping and habitat destruction severely reduced populations in some regions, and extirpated them from others. Reintroduction campaigns and habitat restoration projects have been successful in bringing fishers back to some regions (Drew et al. 2003), though populations are still tenuous in some areas (Kontos and Bologna 2008). Schwartz (2007) reported that genetic studies found historical DNA present in certain post-reintroduction Rocky Mountain populations, showing that some fishers managed

©John Pitcher

Fisher

to survive in an area thought extirpated. (My own sighting of a Lancaster County fisher was of an animal descended from West Virginia releases that moved east, rather than the fishers reintroduced into northern Pennsylvania. Those, however, are moving east into New Jersey now (Burnett 2008).)

Fishers are well-known for their unique adaptation as porcupine-killers (Powell 1981). They take other prey, also, and have a reputation as cat-killers. Exactly how often they will take a domestic cat is up for debate, but it is likely that bobcats and coyotes kill far more free-ranging pet cats.

Albinos have been reported (Williams et al. 2006).

American Marten
Martes americana

The American marten is a smaller version of the fisher, found in the forests of Canada, Alaska, New England, and the Rocky Mountains. It has been reintroduced into areas where it was previously extirpated by overhunting, including parts of the Great Lakes region. It has a lighter coat coloration than the fisher, as well as an orangish throat and chest (Clark 1987). It preys on rabbits, hares, squirrels, mice, voles, and smaller animals.

While today most biologists consider the North American martens to be one species, there is some debate whether the Pacific Northwest/Great Plains population should be split back into a distinct species, *Martes caurina* (Carr and Hicks 1997; Small et al. 2003).

Albinos are known, one mount being displayed at the Museum of Yukon Natural History.

American Mink
Neovison vison

The mink is usually associated with water, and so is found throughout much of Canada and the U.S., except for the American southwest, which is too dry to support it (Larivière 1999). It feeds on crustaceans, fish, amphibians, small mammals, and waterfowl. Mink are prime targets for fur-trappers, but they are also farmed. The farmed mink is considered a domesticated animal, though wouldn't

National Park Service

American Marten

©Richard C. Bennett

Mink

usually be considered pet-suitable, like domesticated ferrets. Due to accidental escapes (and intentional releases by animal rights activists), domesticated mink have been released into the wild, and they have established at least one new population in Newfoundland. Hybridization with inbred domestic mink is causing serious problems for wild populations (Kidd et al. 2009).

Albino mink are known, and of course the fur farming community produces a wide range of color and pattern morphs. (One of the most distinctive is a "jaguar" morph, which is a white mink with black spotting.)

The sea mink is a now extinct relation to the American mink, twice as large, and probably a full species in its own right, *Neovison macrodon* (Sealfon 2007). It was overhunted into extinction in the late 1800s from its range along the rocky coasts of New England and the Canadian Maritimes.

Black-footed Ferret
Mustela nigripes

The black-footed ferret is an interesting case of a species that was highly controversial after its description in 1851 by Audubon and Bachman. Because no other specimens were found for the next 25 years, and the original specimen was lost in a fire, scientists at the time questioned whether it actually existed. In 1874, Dr. Elliot Coues published a request for more specimens in a popular hunting magazine, and received enough evidence to confirm the species' identity (BFRIT 2009).

The black-footed ferret once ranged throughout the Great Plains states, south into Texas, and north into Canada. It lived where there were large enough prairie dog towns to support a population. As agriculture moved in, and prairie dog towns were destroyed, ferret populations disappeared. Diseases were also a problem, as sylvatic plague attacked prairie dogs, and ferret succumbed to distemper.

Never common, the ferret rapidly declined in numbers. It was extirpated from Canada in 1937, and in 1974 the last known U.S. ferrets (in South Dakota) disappeared. Some thought the species was extinct (though Hillman and Clark (1980) noted there were unverified sightings), until in 1981 a Wyoming farm dog brought home a strange animal. This wild population was captured in the mid-1980s to start a conservation program. They have produced offspring that have since been reintroduced in several states. A small group was released into Saskatchewan in 2009. Unfortunately, several states in its former range do not have a large enough prairie dog population to support reintroduction efforts.

LuRay Parker, USFWS

Black-footed Ferrets

©Richard C. Bennett

Least Weasel

I have run across one early account that likely involves a black-footed ferret:

PRAIRIE DOGS HAVE A MYSTERIOUS FOE
Duluth, Minnesota, *News-Tribune*, January 6, 1907

Miller, S. D. Jan. 5—The prairie dog, the animal that has refused to move before the march of civilization, is now being hunted to his death by a mysterious, strange little animal not more than half the size of himself. The fur of the new conqueror of the dog is valuable, fetching $10 a pelt.

One man, whose section of land was being riddled with holes by the dogs, offered $4000 to have the pests killed off. No one would undertake it, as the pests were thought to be death proof until the little fur-bearing stranger came to live in their towns, when they began to move.

Other Weasels

There are three additional smaller mustelids native to North America: the ermine (*Mustela erminea*), the long-tailed weasel (*Mustela frenata*), and the least weasel (*Mustela nivalis*). Domestic ferrets (*Mustela putorius furo*) have occasionally been discovered in the wild, either escaped or released, but there are currently no introduced populations in North America. (There was an introduced population on San Juan Island, Washington, but it is apparently extirpated.)

The ermine is found in northern habitats of our continent, throughout Alaska and Canada and into the northern contiguous states. The ermine typically has a reddish-brown summer phase and a white winter phase. It is also called the short-tailed weasel, or the stoat. Hybrids between ermine and domestic ferrets have been produced in captivity (King 1983).

The long-tailed weasel is found throughout southern Canada and most of the United States. Northern populations molt seasonally to a white winter coat and a light-brown summer coat.

The least weasel is the smallest carnivore in the world. It can be found in Alaska and Canada, south into the midwest, the Great Lakes region, and the Appalachians. It also has a white winter pelage in northern populations. Sheffield and King (1994) noted that the least weasel's killing instinct is triggered by prey movement, not

©Geoffrey Kuchera

Striped Skunk

©Arlene Jean Gee

Raccoon

appetite, and it will continue to kill prey until exhausted. They also noted that a least weasel-domestic ferret hybrid had been produced in captivity.

Skunks

There are five species of skunks in the United States and Canada: the striped skunk (*Mephitis mephitis*), the hooded skunk (*Mephitis macroura*), the eastern spotted skunk (*Spilogale putorius*), the western spotted skunk (*Spilogale gracilis*), and the American hog-nosed skunk (*Conepatus leuconotus*). All are typically black and white. Albinos and various color morphs, such as brown and white skunks, are known (and often available captive-bred in the alternative pet trade).

One wildlife removal specialist trapped a hairless skunk in Oil City, Pennsylvania, in 2006. "On first glance, the creature enclosed in a small box trap resembled a ferret. Its bare skin contained numerous folds, and the telltale black and white markings (that would have enlivened its coiffure if it had one) were faintly visible on the skin, like a faded tattoo" (Boughner-Blair 2006). It was relocated.

Raccoon
Procyon lotor

The raccoon is found throughout most of the United States , southern Canada, and further north in the central Canadian provinces. Historically, raccoons were primarily a southeastern species, but as colonists developed the land and exterminated large carnivores like the wolf, raccoons quickly spread north (Zeveloff 2002). Raccoons are a major invasive species elsewhere in the world. Recent research has determined that several island raccoons in the Caribbean (originally considered distinct species and worthy of conservation concern) are probably early introductions from eastern North America (Helgen and Wilson 2003).

Albinos and a range of genetic color mutations are known (e.g., blonde, red, orange, black), several varieties of which are bred for the fur and pet trades. Hairless raccoons are also known from several states, including Texas (Michael 1968; Miller 2010), Oklahoma (UPI 2010), and Kentucky (KFWC 2007). As with hairless canines, some hairless raccoons have their origin in a disease like mange, but several of these raccoons appeared healthy in every way except pelage, suggesting a genetic mutation.

©Luis César Tejo

White-Nosed Coati

©Yahya Idiz

White-Nosed Coati

National Park Service

Ringtail

White-Nosed Coati
Nasua narica

This raccoon-relative is primarily found in Central America and northern South America, but its range extends into the woodlands of southeastern New Mexico (Geluso 2009) and southwestern Arizona (Lanning 1976; Kaufmann et al. 1976). Rarely, they have been reported from southeastern and northern Arizona. They often travel in large bands, though older males can become anti-social. Wallmo and Gallizioli (1954) noted two phases from Arizona, one pallid and one dark. This is a relatively recent invasion, with Taber (1940) noting that in Arizona, "During the past two decades, the coati has risen from the rank of the rare to that of a common part of the native fauna." Very rarely, coatis may wander north into Texas from Mexico (Taber 1940), though some researchers speculate these were released captives (Kaufmann et al. 1976).

The white-nosed coati is established in Florida, with records from Highlands, Glades, Dade, Palm Beach, Hillsborough, Okeechobee, and Hernando Counties (Simberloff et al. 1997).

In one case from Cortez, Colorado, investigator Nick Sucik (*pers. comm.*) noted that a strange small bipedal creature was almost certainly a misidentification of an out-of-place coatimundi. These are sometimes kept as pets, and it is not uncommon to run across sightings far outside of their native range.

Ringtail
Bassariscus astutus

The ringtail ranges north from California into southwestern Oregon, covers most of New Mexico, Arizona, and Texas, and parts of Utah, Nevada, Colorado, and Oklahoma. Ringtails venture into southwestern Wyoming and southern Kansas, and have moved into northern Louisiana and southern Arkansas (Long and House 1961; Poglayen-Neuwall and Toweill 1988; Tyler and Webb 1992).

Because ringtails have long been kept as pets, they occasionally are reported roaming areas outside their native range. (Ringtails and coatis are the first animals on my mental checklist of possible out-of-place species whenever I run across a strange critter sighting.) Goodpaster and Hoffmeister (1968), for example, reported four separate instances of ringtails discovered in Ohio.

Bangs (1898) noted a report from Florida: "Among other things I have heard of a Bassariscus being killed in the everglades, and a 'gray gopher' on one of the interior prairies. The 'gray gopher' I had a chance to run to earth, and it turned out to be a gray squirrel (*Sciurus carolinensis*), with a broken, stumpy tail, that was surprised and shot while attempting to cross a treeless prairie by the man who told me the story. It was mounted, and I traveled some miles to see it, because my informant had come to Florida from Minnesota and assured me that he knew a gray gopher (meaning *Spermophilus franklini*). The Bassariscus rests on the authority of W. F. McCormick of Cocoanut Grove, Florida, who is well informed, and who told me he had shot the animal himself."

I suspect the following may describe a ringtail:

WILD BEAST PUZZLES EXPERT
STRANGE ANIMAL FOUND IN NEW JERSEY HAS THE HEAD OF A RACCOON AND CAT'S TAIL
Waynesboro, Pennsylvania, *Press*, January 6, 1921

Cohansey, N. J.—A strange animal that has made even the oldest rivermen and natives pinch themselves in this dry era is puzzling trappers and hunters as to its species since its capture in the Cohansey River. The animal has the head of a raccoon, the tail of a cat and its body of a little more than a foot long is covered with gray and reddish brown fur. Its feet are black. It was captured in a scoop net at the edge of the river by Clarence Cheesman, who lives in a boat-house colony near the shore.

Opossum
Didelphis virginiana

Technically, of course, the opossum and the armadillo (following) are not carnivores, but they do show up as nuisance "varmints," and occasionally have been mistaken for something less innocuous.

The opossum is the only native marsupial in the U.S. and Canada. It is found throughout the eastern and central U.S., where it is moving westward and northward (McManus 1974). It can now be found in southern Ontario. This range expansion has been going on for well over a century. There was a time when its presence in south central New York was considered unusual (Loring 1899). It has

(CC) Mat Honan

Opossum

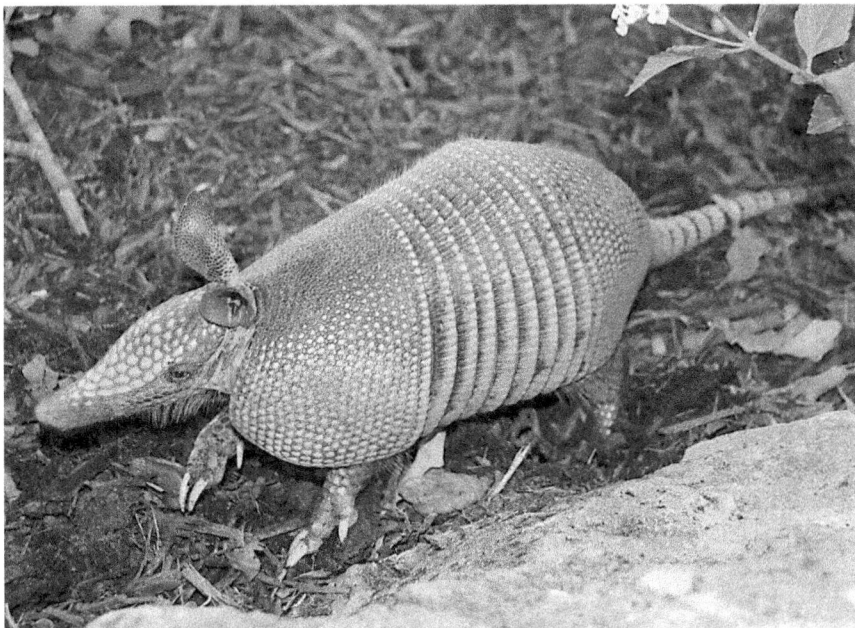

(CC) Phototram

Nine-banded Armadillo

been introduced in several locations along the West Coast, including southern British Columbia, and has been expanding its range there.

The opossum is typically gray. It has a short white underfur, tipped with black, and is covered with longer white guard hairs, creating the grayish effect. There are occasional black-phase opossums, which have black guard hairs. In some regions, these are known as midnight opossums (Branham 1995). Very rarely, there are brown ("cinnamon") opossums, where the black pigment in the hair is replaced with brown pigment (Hartman 1922; Cuyler 1934). Of course, the opossum may be lighter or darker brown, depending on whether the guard hairs are pigmented or white. An "orange" opossum was discovered in Ohio years ago, and is in a natural history museum collection in Dayton now; this may have been a variant of the brown mutation. There are known albino opossums, as well as albinotic opossums (where pigment is only found in the eyes and certain small patches of skin) (Campbell 1981).

Years ago I ran across mention of a talk given at a biology conference in Ohio that discussed biofluorescence under ultraviolet light associated with the opossum's skin and fur (Meisner 1983). Unfortunately, personal experimentation with a UV light and an angry 'possum in a car trunk did not result in a glowing marsupial, and I am unaware of published confirmation of this trait.

Nine-banded Armadillo
Dasypus novemcinctus

This armadillo is the only xenarthran in the U.S., although we formerly had giant ground sloths. They are insectivores, and reach up to about two feet in length. They will eat carrion, which probably contributes to armadillo road mortality, as they are hit by cars while feeding on roadkill. (Actually, many DOR armadillos are caused not by direct hits, but because the armadillo has a "jump reflex" when frightened. If a driver attempts to miss the animal with the tires by driving directly over it, the armadillo leaps up and smacks into the undercarriage.) They are considered nuisance animals because of their proclivity for digging, which causes crop and structural damage (Gammons et al. 2009), and because they act as a zoonotic disease reservoir.

Before the 1850s, nine-banded armadillos were not known in North America. As south Texas settlements formed, this lead to habitat change and removal of natural barriers, allowing the species to move north out of Mexico. By the early

1900s, it was found throughout southern Texas and was moving into New Mexico. By the mid-1920s, it entered Louisiana, and it was east of the Mississippi River by the 1940s. It was deliberately introduced from a personal zoo into southern Florida after World War I (Taulman and Robbins 1996), steadily expanding throughout the peninsula (McBee and Baker 1982). It wasn't long before it moved into Colorado (Meaney et al. 1987), Oklahoma, Kansas (Merriam 2002), Nebraska (Freeman and Genoways 1998), Missouri, Illinois (Van Deelen et al. 2002; Hofmann 2005), Tennessee, Georgia, Alabama, and South Carolina. Some of these locations, such as Illinois, are considered "pioneering zones" for armadillos, rather than holding established populations. Within the next few decades, it is expected that breeding populations will be discovered. (And, the "pioneering zones" will move accordingly.) Extralimital confirmations (whether natural wanderers or human transplantations) include Nebraska (Van Deelen et al. 2002) and South Dakota (Platt et al. 2009). I have newspaper articles of out-of-place armadillos from Pennsylvania, New York, Rhode Island, North Dakota, and other states, so one way or another, they do get around.

Long periods of cold weather and limited rainfall are both restraining factors to armadillo range expansion (Humphrey 1974), but it is likely that the species will continue to fill out those areas in suitable climates, possibly even into southern Ohio, Pennsylvania, and New Jersey. The rate of expansion has been estimated at up to 10 km/year (Huchon et al. 1999), and there could be as many as 50 million armadillos in the U.S. today (Merriam 2002). North American armadillos show a distinct "founder's effect" (less genetic variability), compared to South American populations (Huchon et al. 1999).

Nine-banded armadillos are a known zoonotic reservoir for both Hansen's disease (leprosy), and Chagas' disease (Paige et al. 2002). Armadillos have been recognized as a source of infection for Hansen's disease in Texas, Louisiana, and Georgia (Lane et al. 2006; Clark et al. 2008). Truman (2005) noted that the disease "is most common among armadillos in low-land habitats and may be rare or absent elsewhere." Loughry (et al. 2009) suggested that the disease is spreading to previously uninfected populations.

©Steffen Foerster Photography

Melanistic Leopard

©Steffen Foerster Photography

Leopard

Exotic Carnivores

Out-of-Place Exotics

Whether deliberate or accidental releases, or vagabond wanderers (*waifs*, in invasive biology terminology), there is no debate that non-native carnivores have found their way into the North American landscape ever since the early colonial period. In the majority of cases, the animal will be unable to survive on its own, or will be recaptured or killed. There is absolutely no truth, however, to the argument that states that 100% of all such animals will be unable to survive long-term in a new environment. (The varied vertebrate species introduced into southern Florida serve to render that argument invalid.) Factors including habitat, climate, prey availability, species adaptability, and the individual animal's background all play their part, and under the right circumstances, an exotic predator may cope on its own. Under certain conditions, a species might even become established.

This has been suggested a few times with regard to certain mystery carnivores. Wright (1972) suggested that leopard escapees could survive in the wild and be responsible for black panther sightings. Some researchers have suggested that escapees could have been established as early as the colonial period, with slave ships bringing in exotic animals for trading purposes. I've even seen rumors that black leopards were released up north back in the 1960s or 1970s by members of the Black Panther Party, feline mascots that outgrew their welcome and became too much trouble to keep.

There is a large alternative pet trade today, though it is rarely advertised to the general public. While a growing number of states are outlawing private menageries, it is still legal (usually requiring a permit) in many states for people to keep big cats, wolves, bears, hyenas, and a wide range of smaller carnivores. (Smaller species include kinkajous, tayra, genets, smaller canines, and quite a few small

©EcoPrint

African Lion

cat species.) Only when the animals are being bred or exhibited, or if they are a protected species, is a federal permit required.

Numbers in the trade are difficult to estimate. Tigers are one of the more popular big cats, and I have seen estimates that there could be 5,000 to 6,000 in private hands in North America. There may be close that that number in lions and leopards. Deliberate releases are probably lower than the animal rights groups would suggest, but there are a few accidental escapes every year from both private collections and public zoos.

What follows is a brief review of some of the more significant cases (i.e., with media exposure) of confirmed or alleged exotic carnivore escapes in North America. Several cases are also included in the state sightings section, where the origin of the animal may not have been immediately known.

A Few Confirmed and Alleged Escaped Exotics

1901 | Evansville, Indiana | Circus man Col. G. W. Hall was loading animals after a state fair when a leopard escaped. This caused a sensation in the newspapers, with several livestock kills chalked up to the animal. It was eventually cornered on a local farm and shot, but only after it had attacked two of the hunting party. (See EvansvilleHistory.net.)

1926 | Red Bank, New Jersey | A leopard escaped in early August from Oliver W. Holton's Middletown zoo, resulting in rumors of sightings from locations up to 100 miles apart. It was finally shot in mid-October after it was caught by an Island Heights farmer in an otter trap, who thought dogs were killing his ducks. (INS 1926; UP 1926)

1927 | Baltimore, Maryland | A hyena escaped from the John Hunt Circus, and was shot by policemen when it scared some milkmen. (Anon. 1927)

1936 | Bedford, Pennsylvania | A 175-pound leopard escaped from a roadside zoo, and was at large for three weeks before being shot by a hunting party. It was said to have "stalked two cows, six sheep and a flock of turkeys from farms in this district, carrying the kills to its lair on the mountain side." (Anon. 1936)

1950 | Oklahoma City, Oklahoma | In late February, a leopard escaped from the Lincoln Park zoo and managed to stay on the run for four days of massive

©Vova Pomortzeff

Tiger

hunts. It was eventually tracked down and captured with drugged bait. Unfortunately, the drugs were too much for the feline, and it died the next day. (AP 1950; UP 1950)

1950 | Mt. Savage, Maryland | A strange animal shot by a hunter was identified as a kinkajou by the local game warden. It was said to have escaped from a passing circus a week prior. (Anon. 1950.)

1951 | Mena, Arkansas | A hunter managed to track and kill a leopard that had escaped a week earlier from a circus train wreck near Mount Ida, Arkansas. The leopard was roaming the Ouachita mountains before being discovered. (AP 1951)

1955 | Oak Hills, Florida | Susie, an African leopard, escaped from the Kelly-Morris Circus grounds. She stayed in the area for at least 18 days, but I have not run across any mention of her recapture. (Anon. 1955)

1963 | Thousand Oaks, California | A black leopard escaped from its cage at Jungleland, and remained missing for three days before being discovered and shot. (Anon. 1963)

1974 | Mangum, Oklahoma | An illegal 4-year-old jaguar escaped from a private animal farm, and roamed southwestern Oklahoma for 5 days before being recaptured. (UP 1974)

1976 | Pittsboro, North Carolina | A jaguar escaped from the Carnivore Evolutionary Research Institute, and was loose for three days before recapture. (AP 1976)

1976 | Ringgold, Georgia | Seven African lions escaped from a private zoo in Cattoosa County. Eventually, six were shot, and one recaptured. (UP 1976; AP 1976b)

1995 | Hugo, Oklahoma | A tiger escaped from a wintering circus, and remained on the loose for 10 days before being recaptured. (Anon. 1995)

1999 | Jackson, New Jersey | After police followed up on sightings in local pineland, a young 431-pound tiger was discovered wandering Jackson Township. They

©Arnold John Labrentz

Serval

attempted to tranquilize it, but failing that, ended up shooting it to death. Its origin was never proven, but suspicion pointed to a private "tiger sanctuary," though they denied it was one of their tigers. The local Six Flags amusement park also denied it was one of their animals. In 2003, the private sanctuary was shut down by the state for unfit conditions, and 25 tigers were moved to a facility in Texas. (Yanchunas 1999; NYTNS 1999; Cosgrove-Mather 2003)

2002 | Quitman, Arkansas | Four adult African lions were killed in the woods near this central Arkansas town, unsettling the residents. A local exotic animal farm was under suspicion, but denied any knowledge. The operator suggested they were released by someone who had offered lions to him the previous week. (Pils 2002)

2004 | Fort Polk, Louisiana | A young "tiger cub" was reported in the thick brush near Fort Polk army base. Witnesses thought it was wearing a collar. It was estimated at 100 pounds. After a long search, the Army gave up, stating there was no physical evidence that the tiger had ever been there. (Moore 2004; Anon. 2004)

2004 | Shepherdsville, Kentucky | A 50-pound snow leopard escaped from a private facility, roaming for nine days before it was recaptured on a farm two miles away. (Hoffman 2004)

2005 | Ventura County, California | After a week of sightings, a Fish & Game officials tracked down and shot a tiger that was roaming the Simi Hills near the Ronald Reagan Presidential Library. Its origin was unknown. (Hernandez 2005; Wides 2005)

2005 | Atascosa County, Texas | In May, a rancher observed a full-grown tiger, "waist-high with orange fur and black stripes" along a creek on his property. Animals had been reported missing for six months. I have not seen any report that the animal was ever captured. (Chasnoff 2005)

2006 | Valley Center, California | Three separate sightings of a large feline were reported to officials, but a search turned up nothing. One witness described a tiger wearing a collar, another thought it was a striped mountain lion. Authorities decided it must have been a juvenile mountain lion, noting that local farm workers typically called any big cat "el tigre." (Anon. 2006c)

©Glenn Gaffney

Snow Leopard

2006 | Berlin, Alabama | Cullman County residents near Berlin reported seeing a "Bengal tiger," but officials were unable to find it after extensive searching. (Crutchfield 2006; Simms 2006)

2007 | Nye County, Nevada | Sheriff's Office received several sightings of a loose tiger, but deputies were unable to find evidence of the animal. It was suggested that a mountain lion had been misidentified. (Di Massa 2007)

2009 | Brazoria County, Texas | In December, Oyster Creek residents reported seeing a tiger roaming the area. Wildlife officers were unable to locate it. (Hewitt 2009)

The Indian or Small Asian Mongoose
Herpestes javanicus

While I don't intend to go into detail with each of the possible exotic species that could be encountered, it is worth noting this animal, as it is federally banned as a harmful invasive species in the United States. (In most legislation, the former species name is used, *Herpestes auropunctatus*.)

In the 1800s, mongooses were released in Jamaican sugar cane fields to control rats, and the results intrigued Hawaiian farmers. They brought the mongoose to several of the Hawaiian islands, where it promptly devastated the endemic birds. This caused enough concern that the Indian mongoose was banned from the United States in the early 1900s.

This didn't keep them out completely, however. Mongooses were sometimes kept as pets by sailors, who may not have realized that these were not allowed in the States. Stowaways were another potential source for accidental introductions.

Several mongooses were found on Dodge Island, where the Port of Miami is found, in 1977, and they may have arrived with a ship from the Caribbean. Wildlife officials quickly stepped in to eradicate them (Nellis et al. 1978). Even earlier, a *Herpestes* mongoose was found in a Kentucky haystack in 1920, and the remains of a mongoose were found in a golden eagle roost in California in 1937 (Van Gelder 1979).

While this particular species is banned, there are other mongooses, and some of these have been imported legally to zoos and private collections. (Meerkats, well-known for their film and television appearances, are a type of mongoose.) I suspect Federal wildlife officials would be very unhappy if someone tried to import another species within the genus *Herpestes*, however.

© Christophe Thélisson

Indian Mongoose

©Graham Taylor

Domestic Ferret

©Kristian

Caracal

©EcoPrint

Spotted Hyena

Varmint Folklore

In many instances, folkloric expression (and gleeful journalists) gave rise to names like "The Beast of Bladenboro," "The Thing of Sheep's Hill," or simply "The What's-It." In other instances, folkloric names were more colorful and descriptive. Here are a few of the names you may encounter in investigations (though not all are included in the sighting accounts section):

BLACK PANTHER: The Black Panther is the most commonly reported unknown carnivore here in North America. Sightings go back to the colonial period, though they aren't particularly common until the 1900s. There have been a few attempts to tie black felines into Native American folklore, but I have not seen persuasive evidence for this.

The word *panther* goes back to a time when naturalists differentiated between the panther and the leopard, both being spotted felines, but supposedly distinguished by variations in size, hue, spotting, girth, or skull characteristics. And, of course, the exact demarcation varied depending on the naturalist writing about them. Eventually, it was determined that a single (highly variable) species was responsible, and leopard won out as the common name of choice, with panther becoming a generic (often folkloric) term for certain larger felines. (And, of course, it lent itself to *Panthera* as the genus for extant big cats.) Taxonomically, there is no such thing as a "panther." It has historically been used as a generic for the mountain lion (today, especially for the Florida panther), but does not intrinsically refer to a specific feline or to a specific phenotype. (There are some eastern cougar investigators who promote the idea that *panther* was historically used to denote only a black feline. That is patently false.) Today, it is a recognized generic for tawny mountain lions and for melanistic leopards and jaguars.

The strangest thing about Black Panther sightings is that in many cases the described animal is smaller than one would expect with an adult mountain lion, leopard, or jaguar. (There are some sightings of much larger black cats, but by and large these are exceptions.) Witnesses may describe a length larger than a housecat, but that doesn't mean it is "cougar-sized."

There are more confirmed hoaxes and misidentifications in this category, as well. Physical evidence is very important in these investigations, as there are many known cases of witnesses mistaking domestic felines and canines as something more exotic. And we should not dismiss good evidence for misidentification just because we don't like the result: e.g., when one investigator argued that Black Panther tracks look like clawed canine footprints, not feline tracks.

There are several candidates proposed by wildlife officials and investigators for possible misidentifications, often varying due to geographic location: bobcats, fishers, and jaguarundi in particular. Domestic cats, dogs, and even black bears have been confirmed as the source for numerous sightings. In one interesting misidentification, "a road-killed carcass that had been run over many times, supposedly of a black jaguar, was collected in SW NM and sent to UNM Biology Department. The carcass was cleaned up, examined, and found to be of a mountain lion covered with asphalt or road oil" (James N. Stuart, NM Dept. Game & Fish, *pers. comm.*).

It is interesting to note, however, that the presence of mountain lions in a state doesn't mean there will necessarily be a strong tradition of black panther sightings. (Which might be expected if, for example, there was a stronger likelihood of viewing normal mountain lions under conditions that created the appearance of a black feline.) There are a number of western states with known mountain lion populations but few black panther reports, while there are eastern states with no officially recognized mountain lions (and if there are any unrecognized cats, there can't be many), but quite a few black panther sightings over the years.

While there have been a few attempts to link black panther sightings to maned cat sightings, there is very poor correlation. Given the number of possible classifications of strange feline sightings (stripes, spots, etc.), one could mix and match in any number of ways, but it doesn't actually answer any questions. It's a rather pointless exercise.

CATAMOUNT: While most researchers are aware that a species like the mountain lion is known by multiple common names to different people, it should also be noted that some of these common names may be given to multiple species. Catamount (derived from 'cat of the mountain'), for example, has been applied to both mountain lions and lynxes, and possibly even to bobcats on occasion. References to catamounts and wildcats in early reports should never be assumed to be a less likely candidate unless they are accompanied by detailed descriptions. Unfortunately, there are many early news accounts that make statements regarding a generically-named animal without adequate description to identify the species in question.

DEVIL MONKEY: There are a handful of sightings of black panther-like creatures that walked erect. These have sometimes been collected with certain kangaroo-like sightings, with the suggestion that they actually refer to an unknown gracile primate prone to leaping. Another trait that supposedly characterizes these animals is aggressive behavior (e.g. attacking cars).

DOG EATER: This is a common name for any nondescript beast that has a penchant for killing and eating dogs.

DOGMAN: Linda Godfrey (2003) has suggested that this half-dog, half-man creature from Michigan may be connected to Wisconsin's Beast of Bray Road and similar folkloric entities. I strongly suspect that there is quite a bit of confusion here regarding what may be several distinct categories of folkloric creatures, some strictly wolf-like, others ape-like, and a few sharing traits of both. It is also likely that there is quite a bit of misidentification with known species in this category as a whole.

DWAYYO: In November 1965, a pseudonymous caller told the police that he had a fight with a Dwayyo in his back yard. That triggered a series of newspaper articles and jokes about the vaguely described animal (May 1965a-h), as well as a few sighting reports of unusual canines. One person compared the newspaper sketch (likely an embellishment of the original report) to an Irish wolfhound on its hind legs. It's possible the original report was a hoax, with unrelated misidentifications following, but I also wouldn't rule out the possibility that the expanding range of the eastern coyote may have contributed. In an interview with the reporter, George May, Chorvinsky and Opsasnick (1988) discovered that the name Dwayyo was apparently the result of a malfunctioning teletype. While friends and family were convinced that May made up the whole thing, May insisted that the original police teletype regarding an unusual animal encouter was real.

GLAWAKUS: This is a region-specific portmanteau, supposedly created by a Connecticut journalist who was tired of writing about "wacky" panther sightings from Glastonbury and vicinity. Mangiacopra and Smith (1995-6) gave a good overview of the 1939 Glawakus flap, which in most accounts was a tawny cougar-like animal, though there were a few sightings describing a black feline.

GRAVEDIGGER: Gravedigger folklore involves a small animal that burrows underground, sometimes in cemeteries, where it indulges in ghoul-like behavior. In North America, it is usually tied to the appearance of badgers, though Bart Nunnelly (see his book, *Mysterious Kentucky*) has collected Gravedigger stories from the Bluegrass State that may indicate something altogether. I would want to see more research done on that before coming to any conclusions, however. In any case, Gravedigger is an immigrant descriptor, coming from European folklore, which was applied to convergent North American phenomena.

GREAT NAKED BEAR: This was sometimes called the Stiff-Legged Bear. One of the earliest accounts of this creature was written up by Heckewelder (1797), who heard about it from the Mohicans of the Hudson River Valley and the Delaware (Lenape). They said it had disappeared about two hundred years earlier, but was a terrible scourge to their ancestors. It was a very large creature, with only a little hair, "a man, or a common bear, only served for one meal to one of these animals," its teeth "could crack the strongest bones," and "it could not see very well, but in discovering its prey by scent, it exceeded all other animals." As its heart was "remarkably small," it was difficult to kill with an arrow, and "the surest way of destroying him was to break his back-bone."

In a later writing, Heckewelder (1805) told the story of one encounter:

"The Jagisho (or naked animal, or Bear, as some of the Indians call it) was an animal much superior in size, to the largest bear. It was remarkably long-bodied, broad down its shoulders, but thin, or narrow, at its hind legs, or just at the termination of the body. It had a large head, and frightful look. Its legs were short and thick. Its paws (the toes of which were furnished with long nails, or claws, nearly as long as an Indian's finger) spread very wide. Except the head, the neck, and the hinder part of its legs, in all which places the hair was very long, the Jagisho was almost naked of hair, on which account the Indians gave it the name of 'Naked.'

"Several of these animals had, before this time, been destroyed by the Indians, but this particular one had, from time to time, destroyed many of the Indians, particularly women and children, when they were out in the woods, getting nuts, digging roots, &c., or when they were working in the fields. Hunters, when fast pursued by this animal, had no means of escaping from it, except where a river or lake was at hand. By plunging into the waters, and swimming out, or down the stream, to a great distance, they effected their escape. When this was the case, and the beast was not able to pursue his intended prey any further, he would set up such a roaring noise, that every Indian who heard it trembled, with fear.

"This animal preyed upon every beast it could lay hold of. It would catch and kill the largest bear, and devour it. While the bears were plentiful, the Indians had not so much cause to dread the Jagisho: but when this was not the case, he would run about in the woods, searching for the track, or scent, of the hunters, and follow them up. The women became so much afraid of going out to work, that the men assembled to deliberate on a plan for killing him.

"This beast had its residence at, or near, a lake, from which the water flows in two different ways (or has two different outlets), one northerly, and the other southerly. The Indians being well informed of this circumstance, a resolute party of them, well armed, with bows, arrows, and spears, made towards the lake. They stationed themselves on a high, perpendicular rock, climbing up the same by means of Indian ladders, and then drawing these ladders up after them.

"After being well fixed, and having taken up with them a number of stones, the Indians began to imitate the voices and cries of the various beasts of the woods, and even those of children, in order to decoy the Jagisho thither. Having spent some days in this place, without success, a detached party took an excursion to some distance from the rock. Before they had reached the rock again, the beast had gotten the scent of them, and was in full pursuit of them. They, however, regained that position, before he arrived. When he came to the rock, he was in great anger, sprang against it with his mouth wide open, grinning, and seizing upon it, as though he would tear it to pieces. During all this time, numbers of arrows and stones were discharged at him, until, at length, he dropped down, and expired.

"His head was then cut off, and was carried, in triumph, by the Indians, to their villages, or settlements, on the North-River, and was there fixed upon a pole, that it might be seen. As the report of the death of the animal spread among the neighbouring tribes, numbers of them came to view the head, and to praise the victorious Indians, for their warlike deed."

While there were some academics who treated stories of the Great Naked (or Hairless) Bear as pure fiction, others suggested that these stories could refer to a real animal, even if the description has been muddled a bit over the centuries. One suggestion was that these were the last of the mammoth in North America (Siebert 1937). A more likely candidate was the grizzly bear (Godman 1826); after all, if the grizzly historically was found as far east as Newfoundland, perhaps an occasional wanderer ended up in New York?

Regarding the hairless nature of this animal, this may just be a folkloric attachment related to its "man-eating" nature. Siebert (1937) noted a Penobscot story from Maine describing how a bear turned into a hairless monster after it ate human flesh. There is a direct correlation here to cannibalistic Windigo and Stone Giant transformation folklore. There is also a possible folkloric connection to northeastern Asia, where there are stories of a "hairless polar bear" (Bogoras 1902).

Long-tailed Wildcat: This has been described as a bobcat-like feline (often noting a flattened face and tufted ears) with a longer ringed tail than would be found in a true bobcat. It is often suggested to be a domestic cat that has "gone wild," or a domestic cat-bobcat hybrid. I've discovered most reports to be in Pennsylvania and Ohio, but similar stories crop up elsewhere on occasion. There have probably been more physical specimens of this creature obtained than of any other mystery animal (as there are several accounts of it being shot or trapped, even exhibited), which makes it all the more frustrating that there don't appear to be any specimens on hand today for genetic analysis. There are quite a few small wild cat species around the world that can be confused with wild-living domestic cats (and even interbreed with them), and it would be a shame if there was an unrecognized small feline here that disappeared just because trappers thought they were picking up ferals.

Ozark Howler: The Ozark Howler was a name given to an internet hoax by a skeptic who intended to fool cryptozoology enthusiasts (Arment 2004). More recently, I have been told by researchers living in the Ozark region that there are actual stories of an "Ozark howler," but I suspect that there may be some confusion with the descriptor here (as the researchers in question were unfamiliar with the hoax). I have seen old reports that use the singular name Howler for unknown creatures that make strange cries in the night. It is possible that there could be an etymological convergence here, where the skeptic may have created a generic name that might have actually been used in prior folklore, but as of yet I have not seen confirmation that the term Ozark Howler was used historically as an actual name for a mystery animal. (And, frankly, I'd hate to see future researchers use the name Ozark Howler for future sightings if there isn't a confirmed history of the descriptor, unless it is specifically used by witnesses with no knowledge of the hoax sightings, which are probably still floating around the Internet.)

Ranger Bear: The American black bear is highly variable in appearance, and has often been differentiated by non-specific characters in folklore around the country. Different names may be used for similar variations in different parts of the country, or the same name may be given to somewhat distinctive variations in different regions. The *ranger*, or *racer*, *bear* is typically described as larger, longer-legged, and leaner-bodied, and its hair is lighter brown and coarse (Clapp 1868). It may have a white star or crescent on its breast. It is considered savage in nature. In one story from Maine, it had killed and was eating two normal black bears when a hunter came across it and shot it (Robinson 1907). It shares some characters with the *cow bear*, which is very large, often an older male, with a dark coat and a light throat patch. It can catch and kill cattle or moose (Traylor 2010). It may also be called the *cassenyie bear*, after the cassena shrub (Harper and Presley 1981). The *dog bear* may also be a variant of this, as it is aggressive, with long legs and a narrow pointed snout (Shoemaker 1917). A smaller variety is the *hog bear*, not weighing more than 400 pounds, with a snub nose and short legs. In some regions it is thought to be brown, in others it is black (Clapp 1868; Harper and Presley 1981; Shoemaker 1917).

SANTER: The Santer was a nondescript regional predator first reported in 1890 in North Carolina. The name derives from the verb, "saunter," and probably had its etymological roots in the southern black community. It is likely that several of our folkloric names originated there, but further research is needed. The Santer came to cryptozoological attention when Angelo Capparella III (1977) reprinted and discussed several news accounts. He spoke with a staff member of the Statesville *Landmark*, who suggested that the original Santer story was a hoax created by the editor, either because news was slow at the time, or to cause a scare in the local black community so as to keep men out of the saloons. Capparella was only able to find corresponding newspapers that reprinted or summarized the *Landmark's* original articles. I've located a few of the originals (though not all are available to me), and it looks like, if fact, several of the original stories were told to the editor by a local black man, William Newman. Some of it has the ring of yarn-spinning, but there are hints that perhaps the stories were based on some unusual tracks and sounds in the area (perhaps bear, or cougar).

SCOTLAND DEVIL: This name was originally applied to materialized spirits, but appears in a few early accounts (mid-1800s) from New England concerning unrecognized black feline-like animals (probably including fishers).

SHUNKA WARAK'IN: This is an Ioway descriptor of a hyena-like animal that snuck into camps to steal dogs, and "cried like a person" when killed. This story was told to Loren Coleman (Coleman and Clark 1999), along with the comment that it was known to the Ioway witness (from personal experience) and to his relatives in Montana and Idaho. (Note that the Idaho sightings section does contain the story of an animal with similar characteristics.)

This descriptor has been connected in cryptozoological literature to a taxidermy specimen of a strange animal first noted by naturalist Ross E. Hutchins (1977). Hutchins stated that his grandfather, who had a ranch in Montana's Upper Madison Valley, awoke to barking dogs one winter morning in the 1880s. He discovered and shot at a black wolf-like animal that was chasing the geese, but it got away. When it happened again, he managed to kill it. The animal was described as being high in the shoulders and downward sloping back. It was stuffed, and traded to a local museum. Photographs of the taxidermy mount show a long wolf-like face, long scraggly hair, skinny legs, and something of a slant to the body. It is difficult to say for certain, though, that it isn't a wolf that was badly mounted, or that had a health condition prior to its death. The museum exhibited it as a Ringdocus (a well-known strange animal folk moniker).

After the mount disappeared from view for many years, it was tracked down in 2007 and brought back to the Madison Valley for display at a local museum (Williams 2007). Williams gave a more detailed description of the animal, noting it "strongly resembles a wolf, but sports a hyena-like sloping back and an odd-shaped head with a narrow snout. Its coat is dark-brown, almost black, with lighter tan areas and a faint impression of stripes on its side. It measures 48 inches from the tip of its snout to its rump, not including the tail, and stands from 27 to 28 inches high at the shoulder. The mount is in amazingly good shape, showing no signs of wear and tear and retaining the color of the fur." As of yet, I have seen no indication that any analysis has been done on the mount to determine its exact identity.

SNALLYGASTER: This folkloric descriptor came into use in Maryland in the early 1900s, when newspaper writers tried to beef up circulation. It supposedly referred to a flying reptilian beast, but was occasionally brought up (along with the Dwayyo and Jaberwoks) whenever any mysterious creature was reported. In 1932, a Snallygaster was captured, and turned out to be a large owl. In 1944, an nondescript howling animal in Carroll County was termed a Snallygaster. The name is a variant of Snolligoster, a term of political derision (like Mugwump and Polliwog) in some southern states, dating back to the mid-1800s.

TEN-LEGGED BEAR: Stories of the Ten-Legged Bear (sometimes an Eight-Legged Bear) are known to the Point Barrow Inuit, and have been applied to sightings of a monster-sized polar bear (see Alaska sightings section). There are comparable myths both in northeastern Asia and in Greenland of giant polar bears with extra limbs (Murdoch 1886; Bogoras 1902).

There actually appears to be some overlap in monster bear myths in Alaska, with some stories referring to Kokogiak, the eight- or ten-legged bear, Nanorluk, a giant bear that chases down humans and swallows them whole, and Kinik, a bear too large to drag itself out of the sea.

TIGER: While there are quite a few stories of striped mystery cats in North America, "American tiger" was historically a generic name given to the jaguar. It originated, of course, from the Spanish *tigre*, given to patterned felines (while the patternless cougar was termed *leon*.)

WAHEELA: In cryptozoological literature, this is often referred to the Great White Wolf, but it appears to have a localized origin with Michigan werewolf-type folklore.

WAMPUS CAT: A generic feline descriptor with a strong folkloric component. Allison (2009) noted that in Indiana, it may especially refer to large gray or black felines with terrible screams that would occasionally stalk humans. Catawampus is also used sometimes. (In particular, I have seen it used as a folk descriptor for the ringtail.) There is a variant, the Gollywampus (or Gallywampus), which in Ozark folklore is sometimes described as a large aquatic mink-like creature (Shuker 1989).

VARMINTS BY STATE AND PROVINCE

ALABAMA

MAY BE THE WEST END PANTHER

Atlanta, Georgia, *Constitution*, February 25, 1894

A special from Preston, Ala., says some strange unknown animal is abroad in that section. The strange creature is about the size of a two-year-old calf and is very wide across the back. It is of a blackish hue. It makes a variety of noises. Sometimes it goes much like an owl, then it barks and whines like a dog. Occasionally, too, it caws like a crow. It is as wild as a deer, and although many have seen the strange beast, nobody has ever gotten a shot at it.

SEEKING STRANGE ANIMAL

Des Moines, Iowa, *Daily News*, September 10, 1916

Huntsville, Ala. Sept. 9.—(Spl.)—The reported presence of a strange wild beast, whose roaring has been heard at night, has greatly alarmed residents of various localities in the Meridianville and Hazel Green precincts, north of this city.

The animal has not been seen by any one so far as known and lacking definitive knowledge, there is much speculation.

The mountains are being searched by armed men, many of whom believe that the animal is a lion that has escaped from a circus.

PANTHERS REPORTED IN TALLADEGA FOREST

Dothan, Alabama, *Eagle*, July 14, 1949

Talladega (AP)—Several reports of black panthers roaming in Talladega National Forest have been received here recently.

Three young people who live in the Oak Grove community reported a "huge black animal" sitting on the walk in front of a residence July 2.

The animal was sighted about dusk, they said, and was emitting loud screams and snarls. It left soon afterwards.

A motorist said his automobile was attacked by a black panther on June 29. Witnesses said the automobile bore scratch marks which went through the paint to bare metal.

J. H. Morris, senior game warden and supervisor of the district, said, "There have been numerous reports of people who have seen black panthers and one has been known to attack a man in an automobile."

Morris said he was told a circus caravan had been wrecked in the area in the early 1930's and that several black panthers and mountain lions had escaped.

A mountain lion was killed in nearby St Clair county last year.

CAR IS ATTACKED BY BLACK PANTHER IN TALLADEGA CO.

Cullman, Alabama, *Democrat*, July 21, 1949

Montgomery—A coal-black panther, appearing to be six feet in length, recently attacked a moving automobile in Talladega County, Conservation Department officials have been told.

Fred L. Allen, of the Oak Grove Community, reports that he was rounding a curve on a country road about two miles north of U.S. Highway 241 one evening recently when his headlights caught a panther sitting on the bank by the roadside. Startled by the sudden glare, the animal lunged at the car, according to Allen, hitting it just below the window and leaving his claw marks all the way down the side of the vehicle.

Mrs. Leonard Smith, a resident of this neighborhood, says that she heard a panther screaming the same night

Panthers are reported in Alabama infrequently. They are most generally seen in Coosa, Talladega, and Shelby counties and, according to reports, are always as black as tar.

PANTHER SLAIN
Florence, Alabama, *Times Daily*, August 8, 1959

Fort Payne, Ala. (AP)—A female black panther, believed one of a pair that have preyed on livestock on Sand Mountain the past few days, was killed Friday.

Farmer Carl Justice, who lives near the Jackson County line, said he sighted the big cat on a ridge. As he fired, he said, another and larger panther leaped from theledge and escaped.

BOY, 12, HUNTING PANTHER, KILLS ALABAMA MAN
Pasadena, California, *Independent*, December 22, 1959

Warrior, Ala. (UPI)—A "black panther" hunt by an Alabama farm boy Monday ended in tragedy, the death of a man waiting on a roadside for an early morning ride to work.

Coroner J. O. Butler said the man, Joe Stovers, 51, was killed by a shot fired by a 12-year-old boy, Paul Higgenbotham, who was walking along the road with a .22 rifle before dawn to join young friends in a "black panther" hunt. He heard a noise on the roadside and fired.

There had been rumors circulating that a "black panther" was loose in the area, Butler said. He ruled the death accidental.

'BLACK PANTHER' ROAMING MOBILE
Pasadena, California, *Star-News*, August 15, 1961

Mobile, Ala.—(UP)—Police yesterday received their third report in as many weeks of the sighting of a "black panther" roaming the suburbs.

Mrs. P. A. Burnham told officers the animal chased her automobile in a remote area of the city Sunday night.

"It was loping along beside the car," she said. "I thought it was a dog at first and ignored it , but the funny way it was running caused me to look closer. I couldn't mistake it, it was the blackest thing."

Anniston, Alabama, *Star*, October 31, 1962

Chandler Mountain Monster?—Whatever the "thing" is, according to Dexter and R. T. Hyatt, it makes one of the strangest tracks of anything that ever crossed these parts. These tracks were found recently near Hyatt Gap on the eastern cud of Chandler Mountain. About two weeks ago, Henry Smith saw a black panther about three miles north of this spot. But panthers are old stuff on Chandler Mountain, and according to the Hyatt brothers these tracks in no way resemble the tracks of a panther.

REPORTS CONTINUE ABOUT BLACK PANTHERS
Tuscaloosa, Alabama, *News*, June 14, 1987

Several residents reported seeing a black panther in the Morris area in Jefferson County. Dogs and livestock were killed by the animal, and an adult pit bull wounded. Sightings also were reported from the vicinity of Nauvoo, in Walker and Winston Counties.

THERE'S SOMETHING OUT THERE
Mobile, Alabama, *Register*, December 1, 1997

After noting that some biologists suspected that jaguarundis were living in Mobile and Baldwin Counties, the newspaper received calls from dozens of individuals who claimed to have seen felines that were larger than housecats but smaller than cougar.

One man who lived near Dog River said he had watched them his whole life, seeing them near Fowl river and hearing their "screams" near Dog river. He said they were tannish, weighing about 35 pounds, with long tails and short legs.

Another man, an avid hunter, was out videotaping deer when he came across a large black cat in a clearing, with the article noting, "The black feline does appear to be muscular, with a long tail, but legs too long and head and ears too large to be those of a jaguarundi." The hunter went back to the area with a 20-pound stuffed bobcat and a 110-pound stuffed mountain lion for comparison, and the unknown cat appeared twice as large as the bobcat, in the 40-pound weight category. Jaguarundis are only known to reach about 20 pounds.

TALLADEGA PANTHERS REMAIN PHANTOMS
Birmingham, Alabama, *News*, July 12, 2003

"Black panther" sightings shook up residents of Talladega, but officials could find no trace of the feline. (Or felines, as multiple cats were reported by some.) A state biologist noted that they get about a dozen calls about panthers every year from across the state, but they've never panned out.

TALES OF BIG CATS IN STATE MORE MYTH THAN REALITY

Clanton, Alabama, *Advertiser*, February 11, 2008

A painter in northern Alabama wrote to a columnist about an interesting sighting he had after finishing a job: "It was dark, and as I pulled out of the driveway, about 50-60 yards in front of me, four deer ran across the road. My headlights were shining right on them. Then, about 30 yards behind the deer were two big black cats. They looked like they were about four feet long and their tails were about four feet long. I got a good look at them. Because I'm a painter, I know all about colors and hues. These cats looked like a cougar that had been spray-painted black."

ALASKA

"Fig. 414 . . . from Nuwuk is a newly made ivory figure, which is interesting from its resemblance to one of the fabulous animals which figure in the Greenland legends. It is 4 inches long and represents a long-necked bear with ten legs, an animal which the maker gave us to understand had once been seen at Point Barrow. The resemblance of this animal to the 'kiliopak' or 'killifvak' of the Greenland stories, which is described as 'an animal with six or even ten feet' is quite striking." (Murdoch 1892)

"*Wi'-lû-ghó-yûk* is the sea shrew-mouse—a small animal, exactly like the common shrew-mouse in size and appearance, but it possesses certain supernatural powers. It lives on the ice at sea, and the moment it observes a man it darts at him with incredible swiftness, piercing the toe of his boot and crawling over his body in a moment. If he remains perfectly quiet it disappears by the hole through which it entered without doing him any injury and, after this, he becomes a very successful hunter. If a man stir ever so little, however, while the animal is on him, it instantly burrows into his flesh, going straight to the heart and killing him. Hunters are very much afraid of this animal, and if they chance to come across a shrew-mouse on the ice at sea they stand motionless until the creature goes away. In one case, of which I chanced to hear at St. Michael, a hunter who was out on the sea ice in that vicinity during winter stood in one spot for hours, while a shrew-mouse remained near him, and the villagers all agreed that he had a narrow escape.

"*Az'-i-wû-gûm ki-mukh'-ti*, the walrus dog. This animal is believed to be found in company with large herds of walrus, and is very fierce toward men. It is a long, slender animal, covered with black scales which are tough but may be pierced by a good spear. It has a head, teeth somewhat like those of a dog, and four legs; its tail is long, rounded, and scaly, and a stroke from it will kill a man. The people of the islands in Bering strait told me that sometimes they see these walrus dogs, and that their walrus hunters are very much afraid of them; they also informed me that on one occasion a walrus dog attacked an umiak full of people and killed them all. . . .

"*I-mûkh'-pi-mi a-klan'-kun*, the sea weasel. The Norton sound people described this as a long, weasel-shaped animal found in the sea. They say it has black fur like the shrew-mouse with a white patch between its forelegs. This animal is also known among people living on the islands of Bering strait. There is no question that this myth has is origin in the sea otter, although the latter has been unknown in this region for a long period. Owing to its absence it has been invested with various supernatural traits, among which it is said to bring harm to lonely hunters when it finds them at sea. To this same animal may be ascribed also the *i-mum' tsi'-kak* or *i-mum' pikh-tukh'-chi*, a rare animal said to be like a land otter, but which lives in the sea and is taken by only the best hunters.

"*I-mum' ká-bvi-á-ga*, sea fox. This is described as being similar in appearance to the red fox, but it is said to live far out at sea and is very fierce, often attacking and killing hunters." (Nelson 1899)

WHAT IS THIS ODD ANIMAL?
LOS ANGELES MAN SEES LIVING SNOW PLOW IN NORTH
PURSUES STRANGE CREATURE FOR MILES ON FAST SHOES ALONG ALASKAN
LAKE BUT CANNOT OVERTAKE IT—WHO WILL PROPERLY CLASSIFY?
Los Angeles, California, *Times*, January 10, 1904

Alfred L. Dominy, No, 1269. East Adams street, came across a seemingly unknown creature in the Far North—a veritable living snow plow, partly like a bear and partly like a wolf, yet unlike either—and he could like to know what it is. Reference in last Sundays *Times* to recent animal discoveries in Alaska prompted Mr. Dominy to write of his experience in that direction. . . .

"In the spring of 1899 myself and partner were going up the Francis River," writes Dominy, "when we came across an animal which apparently had been drowned and left by high water on a gravel bar. It was about the size and shape of a small bear, would weigh about 300 pounds, snow white, hair about same length and thickness as a bear's, short stub tail, heavy neck, head, teeth and ears like a wolf, legs short, not over a foot in length, and feet and claws like a large dog or wolf. Mr. Weyhrich and I examined it thoroughly, but were unable to determine what it was. I am satisfied there are still living specimens of the same animal further to the north.

"The following winter while hunting near the lower end of McPherson Lake I came upon queer fresh track which crossed the trail where I had been not an hour previous. The snow was pretty deep, and owing to the animal's short legs it was dragging its body through the snow. Satisfying myself it was the same kind of animal I had found dead farther south, and being on a good pair of snow shoes, and this animal plowing through the snow, I thought it an easy task to overtake and capture the snow plow, as I called it.

"I found it could plow snow and then beat me. After following it the entire length if Lake McPherson, about seven miles, and two or three miles up a small stream that flows into the north end of McPherson, darkness overtook me and I had to camp for the long night. I saw tracks several times afterwards but never gave chase again.

"When I saw Indians the following spring near Deace Post, I described the animal to them, but could get no satisfaction out of them. They declared it was an Indian devil, said it had no heart and could not be killed by shooting: had the power of changing into the form of any other animal it chose. They told wonderful tales of its ferocity, and endurance. The Indians are very superstitious about it and seldom if ever visit this particular locality. It is carnivorous and not hybernating, as I saw signs of it during the entire winter, while the bear were taking their long sleep under the snow. What was it?"

Wolverine, Likely.

It is altogether likely that Mr. Dominy saw a wolverine, an animal now nearly extinct, and which is known in Alaska as "Indian devil," and by other names, but the creature referred to above may be something else and it would be interesting to have its identity established.

"Tarak told us Wednesday evening that the ten-footed bear lives mostly in the water like a seal. Looks like a polar bear all but the ten legs. When he walks on ice the five feet of each side track after each other so the bear makes a double track like a sled. Walking the bear often gets his legs tangled up; there are so many, he can't manage them all. Once a man was followed by a ten-footed bear. The man walked between two cakes of ice and the bear was caught in the crevice between them. If his feet had not become entangled he might have gotten off. As it was, the man speared him. When dying, the bear fell on his back, all his feet pawing the air. This is an old men's story. Tarak never saw such a bear or tracks." (Stefánsson 1914)

"From Wales to beyond Barrow the Eskimo have many tales of two monsters, a ten-legged white bear, *qoqogaq* or *qoqogiaq*, and a walrus dog. A Wales native living near Point Barrow, who claimed to have seen the bear, said that the distance between its ears is the full stretch of a man's two arms. It is so big and heavy that it can break through ice as thick as a man is tall. Sometimes it lies on its back and waves its ten legs in the air so that from a distance they appear like men in motion; hence hunters are warned to be careful if they see anything that looks like a man on the ice. In the spring and summer it lies in wait to drag the hunter's kayak under the water. Whether there is only one of these animals, or more than one, the Eskimo cannot agree. A party traveling eastward from Point Barrow in the autumn of 1913 heard one swimming beneath their sleds, and when they coughed loudly and moved away from the trail, the monster poked its head through the ice, which was too thin to endure its weight. Afraid lest it should follow their footsteps and devour them, they hurried home by a long and circuitous route.

"The monstrous dog that watches over some of the walrus herds is said to be larger than the largest bull walrus. When the animals are alone on the ice-floes they raise their heads every few minutes to guard against danger; but when their watchdog is near they sleep unconcernedly. It has a three-edged tail bristling with spikes with which it lashes the water if enemies are near, emitting at the same time a peculiar whistling noise. It never strikes a walrus unless one is refractory and will not obey its warning; for it feeds on seals and fish. In its ear dwells a tiny animal like a fox, which darts into the water when it sees a hunter in a kayak, springs at him from behind and kills him. A Point Barrow native who was hunting once saw this dog lash up the water and cause a herd of walrus to disappear in the foam. Another native who was going out on the ice to hunt seals heard it whistle, and, knowing it was dangerous, turned back home." (Jenness 1953)

ANOTHER COUGAR IS REPORTED

Fairbanks, Alaska, *Daily News-Miner* (excerpt), September 11, 1953

The mountain lion rumors brought an announcement this morning from Cecil Wells, of Wells Alaska Motors, that he saw a "huge, black cougar" in 1947 about 50 miles from here on the Livengood road. ...

"It jumped into the highway and stood there and looked at me," he said. "It was so big and black that I thought it was a bear at first. And then I saw there was too much daylight between the road and his stomach. It was a cougar, alright."

ALASKAN ADVENTURE (excerpt)

Port Angeles, Washington, *Evening News*, July 10, 1956

By Charles J. Keim

... This recalled the story an Anchorage guide with more than 25 years of experience told me when I encountered him near Nabesna in the Wrangell Mountains while on a recent grizzly hunt.

He said that a bush pilot friend of his spotted a strange animal walking along the coast when he was on a flying trip. The pilot brought his plane lower and got a good view of the animal which appeared to be a tiger. He believed that it was a Siberian tiger that came to Alaska via the ice floes.

Otto [William Geist, University of Alaska paleontologist] said that could be possible, but the tiger would have had to have come from the Mongolian side where they live.

He added that there once were tigers in Alaska before the Pleistocene period (the glacial epoch or ice age 20-30,000 years ago). How does he know? He has retrieved their preserved bones from frozen muck in various areas in the territory.

MONSTROUS POLAR BEAR ROAMS ALASKAN COAST

Yuma, Arizona, *Daily Sun*, March 11, 1958

Point Barrow, Alaska (UP)—A monstrous polar bear, described as "at least 30 feet" in length, is roaming the ice shelf along the coast of the Arctic Sea near here, three hunters said today.

Raymond Kalayauk, an Eskimo seal hunter, said he saw the giant animal twice while hunting along in the ice lead about five miles west of Point Barrow last month. But when he came home with the story, no one would believe him.

Then a week ago, Nathaniel Neakok and Raymond Ipalook returned from a seal hunting expedition and said they also saw the monster bear twice while they were camping about 12 miles northeast of Point Barrow.

Kalayauk said he first saw the head of the bear from a distance of 200 yards. He said he fell flat on his stomach and watched the huge creature for five minutes before it submerged in the frigid sea. Then it emerged again and he was positive of what he had seen.

"The head could easily reach five to six feet in length," Kalayauk said. "It was breathing heavily and the hide was dark." (Not all polar bears are white.)

Kalayauk said he had his .270 rifle ready in case the bear got too close for comfort.

Neakok and Ipalook said they both saw the bear emerging from the sea twice and, like Kalayauk, they watched it until it submerged.

The whole body had not been visible, they said, but judging from the size of the head, they estimated the oversized animal was "at least 30 feet."

When other hunters asked Neakok and Ipalook why they didn't shoot, they replied, "Are you crazy?"

In the spring of 1956, Paul Tazruk reported seeing a polar bear similar in size to the one the other three hunters said they had seen.

ESKIMO HUNTERS GET LOOK AT MONSTER BEAR
Idaho Falls, Idaho, *Post-Register*, May 15, 1958

Barrow, Alaska (AP)—Nathaniel Neakok, the mighty hunter of polar bears, has quit scoffing at reports about the great Kinik being seen in this northernmost region of North America.

A Kinik is the name Eskimos give to a bear they say is too big to come out of the water. Its size varies with the individual story. But all agree he is a monster of great size and strength and appetite.

Several weeks ago, Neakok laughed so loudly when told Raymond Kalayauk had reported seeing a 30-foot bear that his hearty guffaws echoed and re-echoed across the great, frozen polar wastes.

He's Believer

But Neakok isn't laughing anymore. He has seen a Kinik with his very own eyes.

This Kinik, Neakok says, was grayish white and only its head was visible as it swam through the water. It was so large he did not attempt to shoot it. Neakok said its head alone must have been five or more feet long—and almost as wide.

This was not the first time a monster was reported by respected men of the village. Floyd Ahvakana and Roxy Ekownna, elders in the Presbyterian Church and men of undoubted veracity, tell of seeing a tremendous sea monster in 1932 while hunting with a third Eskimo, now deceased.

All three thought it was a Kokogiak or 10-legged bear which occupies a prominent role in Eskimo legend.

Until now there have been many scoffers in the village, especially about Kokogiak. And that white men have been known to make reference to the "coming tourist season" or "another abominable snowman."

But since the respected Neakok added his testimony, the scoffers are strangely quiet. Even fearful. You don't even hear much about how the Arctic's strange mists distort distances or size, creating weird optical illusions. But you do here told and retold stories about Kokogiak, the 10-legged bear of Eskimo legend.

The stories could something like this:

"Once upon a time there was a lazy Eskimo area. He was the laziest man in the village. One evening after he had heard the village hunters tell of their experiences, the lazy one went out on the ice. He came to a large hole where some seal lungs were floating, showing that a large bear had eaten seal, leaving only the lungs.

"The man watched and waited by this hole and, sure enough, a monstrous bear came up. As he started out onto the ice, the man rammed his spear in first one eye and then the other, blinding the bear.

"However, the bear came right on out of the water and following this scent gave chase after the fleeing hunter. The man ran and ran, dodging among the humps of ice but he could not shake his pursuer. Finally, he saw ahead to towering walls of ice, with only a narrow corridor were between.

"Through this pass he ran but the bear, following close upon his heels, was too big and he stuck fast, unable to back out. The hunter ran around and succeeded in killing the giant, 10-legged bear."

Many Eskimos—possibly even some white men—still believe there are Kokogiak out there somewhere in the Arctic fastness.

And, says wide-eyed Nathaniel Neakok, a 30 foot polar bear too big to shoot.

CAT'S OUT OF THE BAG ON KENAI
FIVE PEOPLE TELL FISH AND GAME THEY'VE SEEN AN AFRICAN LION
Anchorage, Alaska, *Daily News*, July 27, 1995

Four people reported seeing a maned lion along the Sterling Highway on the north side of Kenai Lake. It was about noon, and they saw it from about 100 yards away.

Ten days later, a man reported seeing a female African lion ("way too big" to be a cougar) crossing the Snug Harbor Road on the south side of Kenai Lake. He said it was larger than a Great Dane, with "an S-shaped tail, powerful rear legs and a small, catlike head."

Animals considered extinct sometimes crop up in oral traditions, suggesting the extinction may not have been as far back as usually thought. While traveling in Alaska a few years ago, researcher John Warms (*pers. comm.*) talked with a Native Alaskan who mentioned that elders sometimes spoke of "hunters with knives for teeth," which the Alaskan took to mean sabertooth tigers.

ALBERTA

Hyena Reported to Be on Loose Near Edmonton
Lethbridge, Alberta, *Herald*, August 8, 1991
Hyena, Rats Stories Questioned
Lethbridge, Alberta, *Herald*, August 27, 1991

Residents near Gibbons, north of Edmonton, alleged that a hyena-like animal had been roaming the area for four or five years, noting that its cries were sometimes heard at dusk. One woman recorded the strange vocalization, "which resembles maniacal laughter." Officials thought it might be a mangey coyote. There apparently was also a sighting of a hyena-like animal near Stony Plain, west of Edmonton.

Edmonton Hunter Says He Shot 'Hyena'
Lethbridge, Alberta, *Herald*, October 1, 1991
Animal Shot Last Saturday Wasn't Hyena
Lethbridge, Alberta, *Herald*, October 3, 1991

Clarence Paley shot a mange-ridden wolf, and believed that it was the animal reported over the summer as a hyena. He said "that from a distance, the bald patches and missing fur on the wolf's back and sides gave it the appearance of having spots like a hyena." He shot the animal four days after a sighting 15 kilometers away.

One retired soldier who inspected the dead wolf said it looked like the animal he had seen. Two other witnesses stated that what they saw was very different from a wolf. Bill Kidd said, "I'm a prospector and I've seen lots of wolves. What I saw on my field was a hyena. It had spots all over and the front was a lot higher than the back."

ARIZONA

A Strange Animal
An Arizonan Hunter Kills One That He Can Not Classify
Fort Wayne, Indiana, *Gazette*, February 23, 1888

While in one of the principal stores of this tropical Western "city," writes a Winslow, Arizona, correspondent, a long, lean, lank specimen of humanity stopped in and asked the proprietor of the store if he was "on the trade." The storekeeper, who evidently had previous dealings with his customer, said he guessed he was if he could see any money in it. The frontiersman then without further remark unpacked a large and apparently heavy parcel he had brought in with him wrapped in the dilapidated remains of a Navajo blankets, and displayed to his amazed audience a beautiful striped hide, which, had it not been for its small size, would have been mistaken for that of a Bengal tiger. It was of a fine tawny color, with rich black stripes and marks and measured 7 feet and 3 inches from snout to the tip of its rather long tail. In addition to this novelty the customer untied the rawhide thongs that held another and smaller parcel, and from it he took a small china jar and a large bundle of rattlesnake skins. After having displayed the whole contents of the blanket, the storekeeper asked him what animal he had killed to get the tiger like hide from. The frontiersman before answering deliberately cut himself a large quid from the speaker's plug, which he had borrowed for the purpose, and then said that he had been on a prospecting trip into the mountains, and that, as he thought he knew of a good location for a mine, he had gone alone. On the third night out he had eaten supper and spread his blanket, preparing for a good sleep, when he thought he heard a noise like the purring of a large cat. Thinking that a mountain lion was pretty close to him, and knowing that the animals hide would bring him a few dollars, he drew his Winchester close to his side, and turning on his stomach, he watched for his visitor in the direction from which the noise had come. In a few moments he detected the form of the beast about twenty yards away in the timber, the flickering light of the fire making the outlines of the animal just visible, and awaiting a chance when it looked as though a shot could be sent into its body with advantage to himself, he fired and threw another cartridge into the chamber of his rifle before moving an inch. This most probably saved his life, for he had hardly got his weapon in order again before the supposed mountain lion, with a scream that the narrator said "would have turned the marrow in a dead man's bones," made a spring clear over the hunter's body, being probably deceived as to distance by the light of the fire, and alighted within a few feet of the now thoroughly scared prospector, who had sense enough left, however, to put a ball through the brute's head before it could recover itself for another spring. This shot was effective, going through the brain and instantly killing the beautiful creature. After a cursory look at the animal, and noticing that it was of a species he had never before seen, although he had spent thirteen years in those mountains and thought he knew all the animals infesting them, he turned over on his blanket and slept as well as he could until daylight. He then skinned the disturber of his sleep, and, knowing he could sell it at a good price to one of the dealers in Indian and other curiosities in the settlements, he started up and made tracks for Winslow. The rattlesnakes' hides he also wanted to dispose of to the storekeeper, who retailed them to visitors from the East, who had hat bands made of them to wear home and make their relatives believe they had met the snake and its lair, and after a fearful battle finally subdued it, taking its skin as a trophy. The china jar contained the oil of these disgusting reptiles, and the proprietor of the store told the writer that it met with a ready sale assisting people troubled with rheumatism, who used it as an ointment with which to rub the troubled member. The hide of the strange animal was purchased by the storekeeper, as were also the snake skins and oil; but although shown to some of the oldest Indians on the Navajo reservation, none could tell from what animal had been taken, neither did any of the volumes of natural history in the city gives a description of any animal even nearly like this one killed in the mountains of Arizona.

A RARE ANIMAL

Peach Springs, Arizona Territory, *Arizona Champion,* Nov. 4, 1888

While coming down Miller's canyon in the Huachucas a few evenings since, W. J. Burner saw, at a point about a mile above the Huachuca company's reservoir, an entirely strange animal to him. It had the head of a fox, neck like a deer, body of a dog, and a long bushy tail resembling a Newfoundland's. The color was a reddish-brown, and the animal about six feet in length from the point of nose to the tip of tail. Its height was from two and a half to three feet; a little dog which accompanied Mr. Burner made a dash at the animal, which lumbered off in an ungainly way. The brute was not more than 100 feet from Mr. Burner when discovered and he had an opportunity to inspect it closely. The animal did not show signs of fear, nor did it make any effort to attack him. Mr. Burner is a resident of the Huachucas and has lived a long time in Ramsey's Canyon, and never saw the like before, although a story is prevalent among the Mexican woodchoppers of the existence of such an animal in those mountains.

FINDS BLACK PANTHER

Casa Grande, Arizona, *Dispatch,* August 30, 1924

John Riggs, of Chloride, tells a story of the discovery by three Chloride boys of what is believed to be a black mountain lion. The young men, Chas. Shull. Jr., John Kay, Jr., and Kenneth Hill, were over near Pilgrim mine hunting rabbits, being armed with a .22 rifle, a few days ago, and scared a big black animal of the cat family, similar in shape and size to the California lion. They shot at the animal at a distance of 200 yards, but ineffectively, as the cat got away and entered a cave in the volcanic rocks. They followed the animal up, but were unable to induce it to come out of its hiding place, and darkness coming on they were compiled to leave the place. The animal is described as shiny black and is covered with fur, not hair. The question is, if the animal is a lion, where does it come from as it is an entirely new species of the tribe.

Has Extinct 'Hodag' Been Found?
Is Strange Creature Trapped in Hills Beast
Feared by Natives Three Decades Ago?
Casa Grande, Arizona, *Dispatch*, March 15, 1935

What it is? The reputedly vicious and extinct "hodag" (pronounced haw-dag) that natives of these parts occasionally glimpsed, greatly feared, and were never able to capture some three decades ago?

That is the question being provoked by a strange beast captured in the Casa Grande mountains four miles south of Casa Garude, by Dan Kinser, and now caged on the Pinson ranch. No one can tell, though the old timers claim it is the "Hodag."

Its snout is like the peccary's, its feet those of a bear, its ears those of a mouse, and its tail greater than that of a monkey—half again the length of its body and heavily furred. It is three times the size of a fox. Certain it is its like has not been seen in desert or zoo for many a year, if ever—at least not in Arizona.

One theory is that Kinser's strange beast has wandered up from the wilds of Central America, but among those who journey for miles to see it as word spreads of its capture, none have been able to speak with authority. Maybe tellers of the strange tales of the hodag of a past generation are at last to be vindicated.

In one respect, the strange creature differs greatly from descriptions of the famous "hodag", and that is that all of its legs are of the same length. The "hodag" is said to have had right legs longer than its left legs in order to facilitate it in traveling around the mountains which were its natural habitat.

Maybe It's Only an 'Edentata'
Picture in Webster's Dictionary May Give Clue
to Identity of Beast Believed to be Hodag
Casa Grande, Arizona, *Dispatch*, March 22, 1935

Perhaps it's not the Hodag after all, maybe it's only an Edentata.

That is the opinion of some valley amateur zoologists concerning the strange beast trapped in the Casa Grande mountains a couple of weeks ago. The important discovery of its probable identity resulting from a clue given by Lynn Morrill, valley rancher.

Mr. Morrill suggests that the animal may be a ground sloth, common in. South America where he spent some time in the oil industry.

The cry of the animal, described as similar to that of a baby in distress, first suggested the sloth to him, and recalled other similarities.

This suggestion led to an investigation with Webster's Dictionary as the research laboratory.

Under "ground sloth" was found mention of the group "Gravigrada" and under "Gravigrada" was found mentioned the word "Edentata" and there under "Edentata" was the picture of an animal which some of those who saw the captive beast declare is an identical picture of the animal; that is in all aspects except that the claws in the picture are longer than those of the "what-is-it?"

This, however, might be explained by the possibility of its claws wearing away on its long trek from South America.

The peccary-like snout, bear-like feet, mouse-like ears and long bushy tail are similar in both picture and animal.

The "what-is-it" was caught by Dan Kinser and is now caged at the Pinson ranch south of Casa Grande. At first, suggestions were made that it was the mythical Hodag, a muchly feared phantom beast which was frequently reported seen but never captured by natives of three decades ago.

Tucson, Arizona, *Daily Citizen*, June 22, 1946

Phoenix, June 22. (AP)—Irrigator W. L. Smith said today that he saw a husky black panther last night on the Isabell-Hartner ranch, near the Arizona canal, where he was irrigating. He said the animal came out of a near-by sugar beet field and followed him as he crawled to a fire he had started. He said he kept the flames going until daylight with strips of his shirt and coat. At dawn the panther slunk back into the field.

WHAT IS IT?

Yuma, Arizona, *Daily Sun*, March 11, 1949

After the fights Wednesday night, we were chewing the towel with some of the wrestlers, and Terry McGinnis came up with a very interesting story. Terry, you remember, is the big guy who nearly put

another series of joints in Vic Holbrooks hand in their semi-windup
bout.

Terry's interests are completely centered on wrestling, we found
out. He is an excellent pistol shot as well as a top wrestler. He does
quite a bit of shooting with members of the Los Angeles police department and keeps in practice by shooting rabbits on some of his
trips. Once in California he killed a bobcat on a trip and collected a
nice bounty.

Getting back to the story, Terry said that the other night on his
way to Phoenix, he started through a cut in Telegraph Pass and noticed two bright red objects resembling bicycle reflectors up above
the highway. The two objects moved, so he stopped the car and turned
his spotlight up the hill to catch an animal resembling a bob cat resting on a ledge. It was a large animal, weighing 75 or 100 pounds,
Terry estimated. It had a long tail, and the blamed thing was coal
black, he said. Apparently it made no effort to get out of the beam of
light, but Terry became so wrapped up in watching the animal that
he let his brake release a little and the car rolled back, taking the
light off the ledge. As soon as it was dark again, the animal bounded
down onto the highway and loped along until it came to a break in
the high rock walls looming on both sides of the road. Then it disappeared, leaving Terry wondering what the heck it was.

Got Any Ideas?

He spent the next day asking questions around town, trying to
determine if black panthers were common in this area. He found
out they aren't, so is completely in the dark as to the animal's identity. Just as luck would have it, Terry didn't have his pistols with
him, and didn't feel that he would come out so well in hand to hand
combat with the beast. Had he brought his pistols on the trip, wildlife authorities might have had something new to enter in their files.
It's a pretty sure thing that a marksman who can bring down a rabbit by spotlight can knock over a big animal sitting still, and Terry
was not just a little bit angry at himself for not being armed.

Anybody have any ideas as to what he saw in the pass? We would
all like to know.

'Black Panther' Reported Invading Woman's Yard

Tucson, Arizona, *Daily Citizen*, May 6, 1971

A Tucson woman said she saw a "long, big and black" cat that looked like a black panther. She watched it go through a field at about 5:30 pm, from 300 feet away. An Arizona Game and Fish wildlife manager said there had been three sightings of a "black mountain lion" in the Catalina Mountains the past two years.

Big Cat: Leopard or Jaguar Eating Dogs Here?

Tucson, Arizona, *Daily Citizen*, September 17, 1976

State wildlife officials stated that what may be an exotic big cat has killed and partially eaten at least 40 dogs in northwest Tucson over the last 14 months. The predator kills the dogs quickly with a bite on the neck, then eats the entrails. Locals who have seen the animal say it is big and black. According to a wildlife specialist at the Arizona-Sonora Desert Museum, there had been reports of a mysterious black feline in the area for a few years.

Bill Quimby: It's a Mysterious Cat

Tucson, Arizona, *Daily Citizen*. September 29, 1976

The outdoor editor noted some of the difficulties in trying to determine what, exactly, was killing dogs in northwestern Tucson. Some kills were coyote, others bobcat, but there were other kills and sightings that didn't match known native species. One off-duty policeman with a degree in wildlife management watched a large black feline stalking his Brittany spaniels before he chased it off. Officials had difficulty hunting for the cat because they usually weren't told about the events until days later.

Bill Quimby: The Cover's There

Tucson, Arizona, *Daily Citizen*. October 6, 1976

The outdoor editor walked with a state game specialist along a three mile desert wash from the Catalina foothills to near the Rillito River, in the center of the region where the mystery feline had been killing dogs. Only a few vague 5-inch "catlike" prints were found that might indicate a large feline in the area. Quimby did note that the trek "convinced me that it's entirely possible for an animal such as that area's mysterious black cat to live almost undetected by humans."

Mystery Beast Kills Another Dog

Tucson, Arizona, *Daily Citizen*, October 14, 1976

After conducting an autopsy on a poodle recently killed by the mystery predator, a University of Arizona animal pathologist ruled out coyote or feral dog. Dr. James Shively called it an "efficient killer," as "it broke no bones, but it was powerful enough to kill the dog instantly." State game specialist Bob Hernbrode used skulls of different predators to try and match the teethmarks. He also noted it couldn't have been a dog or coyote. A jaguar skull matched the teeth marks, but Hernbrode "was reluctant to say the killer animal was a wild jaguar that had strayed up from Mexico." The article noted that witnesses described the animal as "a large black or steel-gray cat with spots."

No 'Mystery Beast' Says Game Agency

Tucson, Arizona, *Daily Citizen*, October 26, 1976

Killer Cat May Be Wild Coyote

Yuma, Arizona, *Daily Sun*, December 1, 1976

Private, state, and federal trackers searched for the alleged "black panther" in Tucson. Unable to find it, wildlife officers suggested that the attacks were from coyotes and dogs. (One tracker had noted two large black dogs in the area.)

Tucsonan Sees Panther

Casa Grande, Arizona, *Dispatch*, October 17, 1981

A Tucson woman claimed to have seen a black panther "as big as a German Shepherd dog" stroll through her yard in the Catalina foothills. She told her husband, who admitted he had seen it a few weeks earlier.

More Lions Spotted in 1998, Including Rare Black Cat

Santa Fe, New Mexico, *New Mexican*, February 9, 1999

Rare Black Lion Lurks in Hills

Lake Havasu City, Arizona, *Today's News-Herald*, February 9, 1999

There were at least four sightings of a "black mountain lion" in and around Prescott, Arizona, reported to Arizona Game & Fish in 1998, according to the state's big game specialist, Art Fuller. One group of women watched a black feline near the tennis courts in Prescott, while a couple of hunters reported seeing one on Mingus Mountain.

PROLONGED DROUGHT PUSHES WILD CREATURES INTO TOWNS
Casa Grande, Arizona, *Dispatch*, April 3, 2002

Article notes that a "black mountain lion" was seen by Ray Wilson walking along the dry gully behind his Prescott, Arizona, home.

Newton (2005) notes, "In April 2003, a 'black lion' with a cub was seen near Mayer, in Yavapai County."

LOCAL WOMAN IS AMONG FEW TO SEE RARE BLACK MOUNTAIN LION
Prescott, Arizona, *Daily Courier*, February 10, 2004

Brenda Patton reported seeing a sleek, black long-tailed feline "about the size of her 100-pound dog." She was pulling out of her driveway one winter morning and saw it just a few feet away, stopping to watch it as it loped off. Arizona Game and Fish noted that someone had reported "a black lion with a cub in the Mayer area last April."

ARKANSAS

A Christmas Story of Panthers and Grief

Fayetteville, Arkansas, *Northwest Arkansas Times*, December 16, 1972

The columnist gave details of a black panther sighting by several individuals. They watched it as it played in a field along an old cut bank for about ten minutes before it disappeared into the brush. They described it as a large cat the size of a German shepherd, but lower to the ground, with a long slender tail. Police found tracks, but were not sure if they were feline. "Mrs. Hegre and her husband checked the area and found, where the animal had stretched against the bank, the marks of large sharp claws in the frozen clay." (The language is uncertain, but appears to mean there were feline-like scratch marks, as opposed to claw marks appearing on the tracks like a canine print would show.)

Newton (2005) states that in 1977, "a maned lion and a black panther prowled the woods around Dierks, Arkansas."

Robert Prevo (2001) theorized that "black panthers" in Arkansas were jaguarundi, though noting that average reported sizes were sometimes larger than the species is known to grow. One sighting report he collected involved a raccoon hunter in the 1970s who treed a large cat in the Black River Bottoms. They shot the cat, but left it to continue the hunt. The witness stated, "It was a big black cat, with a long round body and an even longer tail. It hissed and growled at the dogs, scaring some of them off, but the rest wouldn't back down. It was a little bit bigger than a bobcat, but it had shorter legs. ... It was smaller than a cougar, but bigger than a bobcat. It had this long tail; I swear that tail was five or six foot long. It had a small head, and short legs, and smooth fur all over its long body."

Cryptozoology artist Bill Rebsamen reported his cousin's sighting of a black panther in the Arkansas River bottoms along the Arkansas and Oklahoma border, not far from Fort Smith (Arment 2000b). His cousin noted that a hunter friend had also seen it in the same vicinity. He pointed out that there are plenty of deer, rabbits, and feral hogs in the river bottoms.

FRANKLIN COUNTY'S WILD CRYPTO ZOO

Charleston, Arkansas, *Express*, January 10, 2006

By Tal H. Branco (Reprinted with permission.)

During the past four years The RFP Research Project has received numerous reports from residents of Western Arkansas concerning wild animals they have seen that did not resemble any known native or exotic species.

One such report was received from a life-long resident of Franklin County in early August of this year. On August 17 the writer drove to Ozark to meet the witness. It was a coincident that Lucille Elders' Barnes News column in the Spectator that day mentioned another sighting of an enigmatic animal by a Cravens resident. Ms. Elders herself had seen one of the animals about a year ago, and wrote about the encounter in her column dated September 15, 2004.

The witness the writer had arranged to meet in Ozark is Harley Edgin who lives northeast of Ozark on the north side of Highway 352. The property consists of about 200 acres which is used for cattle ranching. In the fall of 1991 Mr. Edgin and a friend were driving across the property checking on the cattle when they observed two animals standing in the field watching the vehicle. Mr. Edgin stopped the truck, and looked closely at the animals. He and his passenger had initially thought they were seeing two coyotes, but they soon realized they could not identify the animals. The animals sat down on their haunches and nonchalantly watched the vehicle for a few minutes. During that time Mr. Edgin picked up his binoculars and looked intently at both animals for a short time. He then passed the binoculars to his passenger. The animals then ran to a barbed wire fence along the north side of the pasture and stopped beside it. At that time the animals were about 100 yards from the vehicle. The animals stood for a few minutes as the witnesses alternately continued to watch them through the binoculars. After a few minutes the animals turned and ran under the fence and into the woods.

Mr. Edgin stated the animals were unlike any he had ever seen. He said that one animal was about 5 inches shorter than the other, the larger being about 3 feet tall at the top of its shoulder. Both were about the color of a bobcat, with heads that seemed oversized for the bodies. The head and nose were noticeably more like a canine's than a feline's. According to Mr. Edgin the most noticeable feature on the head was the animals' ears. He stated the ears were unusually large for the size of the head. The ears were reportedly mule-like, about 6 inches long, with tufts of hair growing from their tips. He noted that the neck was also thick and muscular. He stated the animals had tails that were catlike, but longer than a bobcat's and shorter than a cougar's. He particularly noted the tails were rounded on the ends, and not pointed like a dog's.

He said the animals possessed massive chests and very muscular front legs. Their back legs were noticeably shorter, causing the back to be sloped like that of a hyena. When the animals ran, he stated they "pushed off" using their front legs, landed on all four feet, and continued their travel in that springing fashion.

Mr. Edgin stated that on three other occasions from 1995 to 1997 he saw one of these same mysterious animals. He said that two of those sighting occurred on a tract of land south of the road from Watalua to the junction of Highway 219. On each occasion Mr Edgin was driving across the open field on that property and had a different passenger each time. Each time the occupants of the vehicle had an unobstructed view of the animal, and it appeared to match in all respects the animals previously seen by Mr. Edgin. During one of those encounters, the passenger saw the animal first and initially thought it was a deer, then decided it was a coyote, and finally decided he could not identify the animal. Mr. Edgin then attempted to run the animal down in the open field with his truck. He stated his truck came within 20 feet of the animal, but he was unable to hit it. After the animal ran away, the passenger was still in awe of what he had seen and quietly told Mr. Edgin, "There ain't nothing right about that thing."

During Mr. Edgins' third encounter with one of the unknown animals on the same property, it was standing in the edge of the woods beside the pasture. He and a passenger watched it for about three or four minutes before it walked into the woods.

Mr. Edgin left and came back with a another friend who brought along a gun. They reportedly found the animal's tracks in loose soil and leaves inside a cedar glade. The three men noted that the animals' front paws left tracks that were generally round like a cat's, and were about three inches in diameter. The back feet left similar but smaller tracks. Mr. Edgin stated the tracks were unusual in the fact that the front paw tracks were about 12 to 14 inches apart in the side-to-side direction, while the rear paw tracks were only about half that distance apart in that direction.

Mr. Edgin's last sighting of the animal occurred in 1996 or 1997 when one of the animals crossed Highway 219 in front of his truck just north of I-40.

Mr. Edgin had previously arranged for the writer to meet with other area residents who have also seen animals they could not identify. The writer spent two days interviewing those witnesses.

The close-up encounter that Geraldine Wyers' described was ominous and frightening. Her detailed observations portrayed an aggressive animal that has been similarly described in reports from Montgomery County Arkansas and from LeFlore and McCurtain Counties in Oklahoma.

Ms. Wyers stated that in mid-summer of 1994 or 1995 she drove alone to a small branch below a stock pond dam on the parcel of land on which Mr. Edgin had previously seen two of the animals. She went to the property to destroy a beaver dam on the branch, and to attempt to kill the beavers that had built it.

She said she parked her pickup truck in the edge of a field within about twenty yards of the beaver dam. When she exited the truck, she left the driver's side door open. She stated she destroyed the beaver dam and went back to the truck to get a .22 caliber, semi-automatic rifle which she loaded with ammunition. As she walked upstream away from the truck to find a place to conceal herself and wait for the beavers to return to the dam site, she heard a strange bird-like sound coming from across the branch. As she slowly walked toward the sound, she began to hear loud moaning sounds. She thought the sounds were coming from some kind of animal in great pain. As she slowly and carefully walked toward the sounds they diminished in volume. When she reached a point about 20 yards from the truck, she saw an animal crouched and watching her from the other side of the branch. She was less than 10 yards from the animal at that time. The bizarre and intimidating appearance of the animal reportedly caused her to freeze in fear. She said the animal's description generally matched those given by Mr. Edgin. She stated the animal's head was in fact massive, but it had canine upper teeth that extended two or three inches outside its mouth, and she saw blood on those teeth. She stated she could clearly see the animal was intently watching her and it was poised as if ready to spring in her direction. She had the distinct impression the animal was in no way afraid of her or the rifle. She was carrying the rifle at her hip in both hands and pointing forward. She momentarily considered drawing the gun to her shoulder and emptying the 19 shot magazine of long rifle bullets into the animal, but she quickly decided the animal was too large and too close for her to ensure she could get off enough shots to kill it before it reached her in a full charge.

She began to carefully, and very slowly, walk backwards toward the truck without taking her eyes off the animal. She was moving her feet a few inches at a time, and it seemed to her she would never reach the safety of the truck. When she finally did reach the truck, she stepped beside the opened door and slowly placed her right foot on the floor board. Only then did she briefly take her eyes off the animal to reach for the door handle. She immediately grabbed it, sat down on the seat, drew her left leg inside the cab and slammed the door shut. The second the door closed she heard and felt the animal's impact against the outside of the door. She frantically started the engine and raced through the property to the main entrance gate. She opened the gate and hurriedly drove to Mr. Edgin's home, not remembering whether or not she had closed the gate to prevent the cattle from leaving the pasture. She then found the driver's side door was bent from the animal's impact.

The ferocious appearance and lack of fear of humans which Ms. Wyers observed were also noticed by an Ozark businessman who had a close encounter with one of the animals in Johnson County. The man stated he was in his pickup truck and stopped at a highway intersection north of I-40 when one of the animals walked closely past the driver's side of the truck, around the front of the truck and back down the other side. The animal reportedly stopped just past the end of the truck where it was brightly illuminated by the truck's taillights and brake lights where the driver was able to see it in his passenger side mirror. According to the driver the animal remained at the rear of the truck for nearly a minute, but it would turn its body to shield its eyes from the vehicles that were passing on the highway in front of the truck. This particular witness stated he clearly saw both upper and lower canine teeth outside the mouth of the animal. He said the animal's body and head were massive, and the jaws appeared to be wide and powerful. He stated the top of the animal's head was just below the bottom edge of the window openings as it walked past the truck. His description of the animal generally matched those given by Ms. Wyers and Mr. Edgin, although there was one notable exception. This witness estimated the animal's weight to be close to five hundred pounds. The animals seen by the other witnesses were estimated to weigh about 250 pounds. This witness was, by his own admission, very fearful of the animal, even though he was inside his pickup with the windows up during the encounter.

On a spring day about 1991, an animal similar to the ones described by the witnesses was also seen near the Belt Cemetery north of Ozark by two local women. Ruby Tolton and a relative had gone to the cemetery to do clean-up work when they saw a deer run from the woods into an open field behind the cemetery. When the deer reached the center of the field they saw an animal burst from the woods in hot pursuit. The women were very puzzled because they could not identify the obvious predator. They could plainly see that it was not a dog or coyote, although it appeared to be some type of canine, rather than a bobcat or mountain lion.

A similar animal was seen by an employee of a local utility company when it crossed Highway 219 about three miles north of I-40. The witness stated the animal ran across the road toward the creek on the west side. The man stopped his vehicle and got out of the truck to try to get a better look at the animal. He saw it standing and watching him from about 100 yards away, but the man could not identify the animal. As he watched, the animal sped across the creek, and with apparent ease and obvious agility, scaled the steep bluff on the other side and disappeared into the woods above it.

One of the witnesses vividly recalls driving to a local farm with her small children in the 1980s to pick purple hull peas, and being told by the farmer (now deceased) to park close to the pea patch and to watch her children closely. When she asked why, he told her he had recently seen a large, aggressive animal he could not identify near the pea patch. He said the animal was intimidating and showed no fear of him.

In the Cravens area northwest of Ozark the local residents have also reported seeing an enigmatic quadrupedal animals, although the descriptions of the animals seen in that area are basically the same, the descriptions are significantly different in some respects from those previously recorded.

The most pronounced difference between the descriptions from the two areas is in the length and size of the neck, and the angle of the neck in relation to the body when the animal is standing alert.

Ms. Lucille Elders, her son and another relative who have seen the strange animals in that area at close range report that the animals' necks are disproportional long for the size of the animals' bodies. Ms. Elder's son, who saw one of the animals at night on two occasions, particularly noted that the animal's long neck was held in a very upright, and somewhat awkward looking position when the animal stood watching him. He stated the neck was very thick and the head was "fat." He estimated the top of the animal's head was about 3 feet from the ground when it was watching him.

One resident of the Cravens area reported that the animal he saw had long ears. The animal seen by Ms. Elder was reported to have "small like ears."

Anecdotal reports from this area indicate the strange animals have been recently seen by other residents. One rancher reportedly saw such and animal resting on a hay bale in his pasture. According to some residents, these animals have been seen in the area for generations.

Based on the variations in the descriptions of the animals seen in Franklin County, it seems there are either two separate types of enigmatic animals in the area, or there is one species that manifests very unusual changes in its appearance while growing to maturity.

The animals will probably never be accurately identified—or properly classified if they are in fact an unknown species—until their DNA profile has been examined by professionals in that field. While

DNA can sometimes be obtained from a wild animal without it being killed, the odds are in this case that one of the animals will be shot by a farmer or rancher, or will be run over and killed while crossing I-40. If such an event occurs, and the animal still can't be identified upon close inspection, the preserved carcass would be of great interest to science and to the world in general.

BEASTS ATTACK FAULKNER COUNTY MAN
Conway, Arkansas, *Log Cabin Democrat*, January 6, 2008

Tim Ledbetter was walking several of his small dogs in the woods behind his home (about two miles east of Conway), when some of the dogs were attacked by four strange canines: "Four large wolflike creatures Ledbetter described as too heavily built to be coyotes and with hair too long and 'woolly' to be wolves were 'all over' one of the miniature dachshunds." Ledbetter used a branch to club off the animals, and grabbed his dog to carry it home. Three of the animals continued to follow him, backing him up against a fence, knocking him down, and attacking him. One of his basset hounds jumped in and knocked the animals off, allowing him to get up and make it home. He thinks the animals may have been some sort of wolf-dog hybrid, but hasn't been able to find pictures of similar canines. A state wildlife official suggested they were just feral dogs.

BRITISH COLUMBIA

MUSEUM WANTS CLOSER LOOK
North Island, British Columbia, Gazette, June 19, 1968

John McCormick reported seeing a black cougar the previous fall, and the provincial museum expressed interested in it if any hunters managed to take it. McCormick said, "I was driving to Port McNeill one evening, and the animal crossed the road about 50 yards ahead of me. ... At first I thought it was a bear and didn't pay too much attention, but when I realized what it was I looked more closely. I don't think I've ever seen a more beautiful animal." McCormick noted that someone else had seen a similar animal near Woss Lake five years prior.

CALIFORNIA

"I have several times heard of some large animal of the cat kind said to differ from the cougar. One was reported to have been seen in California by some mining acquaintances I made there. It was described as stouter than the cougar, deep chested, with a dark tawny mane! Lately a very intelligent man, Mr. Samuel Woodward, of Shoalwater bay, W. T., informed me that he had seen in that neighborhood an animal standing higher upon its legs than a cougar, with erect ears and a short tail. The Indians of the Willamette have a story of some terrible animal inhabiting the woods bordering the Columbia on the south, which is not a cougar. It may be that there are imaginary differences, but the subject deserves investigation. Perhaps these animals are straggling specimens of the northern lynx." (Suckley & Gibbs 1860)

Coleman (2007) noted an early memoir describing a distinctive feline that was hunted for killing livestock in Lake County in 1868. After treeing the animal with hounds, Archie McMath shot it. It weighed over 300 pounds and measured eleven feet tip to tip. McMath stated, "It was built entirely different from a panther, it was very heavy in front and light behind with black stripes along its shoulders and back and down the fore parts. The rest was of a yellow cast color. The black was on his back and shoulders and it resembled a mane. The hair was longer there."

San Francisco, California, *Bulletin*, February 12, 1870
A Mysterious Animal.—The Elko *Independent* of the 9th inst. says: Mr. Durgan informs us that the remains of W. W. Lair, who was supposed to have perished by cold, have been found and examined. Mr. Lair's death is now attributed to a strange wild animal

which has made its appearance in flocks in Bruno district. No one knows what species of animal they are, as none have been killed, and their appearance and habits puzzle the oldest mountaineers. Mr. Lair's clothes were found torn into strips and ribbons by the teeth and claws of some wild animal, and many of the bones were broken short off. A few days ago one of these animals chased a prospector in Bruno, and made a sharp fight with three or four men before he could be driven off. It seems to care but little for pistol shots, and pistols were all the men were armed with. They eat mule flesh, but will not touch horses or cattle, and seem to have a keener appetite for man than animal flesh. For want of a better title the whites in that district have given them the name of man-eaters. Some think these mysterious animals are a species of wolves, and others a species of wild cat. Whatever they may be, they carry terror to every lonely traveler who finds them in his path.

San Francisco, California, *Bulletin*, December 28, 1872

A Nondescript.—The following account of a curious animal captured in Calaveras county is copied from the San Andreas *Citizen* of the 21st instant: "the Leonard boys last Friday night, as the moon was peeping over the distant mountains, treed an animal, with the assistance of their dogs, which was at first thought by the boys to be a coon. They discharged ten loads of buckshot from a shotgun into the body of the creature before he came to the ground, and then he wouldn't give up the ghost, and lived until the boys reached home, a quarter of a mile distant, they continually banging his head against stumps and rocks as they went along, and then he finally succumbed. Robert brought the head, tail and one of the fore feet of the defunct quadruped into our sanctum yesterday, and they were examined by hunters, professional men, printers and others, but a conclusion as to what species of an animal it was could not be obtained at the time we went to press. The head resembled that of a fox, the tail that of a monkey, and it was web-footed. Indeed it was a wonderful animal."

Mendocino, California, *Dispatch and Democrat*, March 6, 1896

There has been some curiosity over a strange animal that has appeared in Rider's gulch, on Wages creek. It has a gray appearance and closely resembles the hyena. Those who have traveled say the same animal is seen in Utah and think this one has been driven out, or ran away, from its numerous relations and found its way to this locality. It is considered cowardly and treacherous and there may be a reward offered for its capture.

A Rare Animal Attracts Attention
Jackson, California, *Amador Ledger*, October 7, 1910

A rare animal has been on exhibition during the week at Mat Zwinge's place, that has attracted considerable attention and no one seems to be able to classify it. The animal was shot by Melvin Zwinge on the Walter cattle range on Blue creek and is apparently about two thirds grown, as another similar one was with it at the time, very likely its mother and considerably large. The animal has been variously named as a Pine Martin, Otter, Weasel, Civet Cat, etc., but though it has some of the characteristics of all of these, it does not exactly answer to any of them. It answers the Natural History description of the Ichneumon of South Africa more nearly than anything else.

The animal had been thrown away after being shot by Zwinge, as worthless, but Jas Nuland happened to be on a visit to the camp notices its peculiarities, had it skinned and brought the skin down with him.

Mat Zwinge, who is an expert in that line has mounted the animal and many have since examined it.

One feature, the stout tapering tail, that marks most animals of this class is wanting, and the tail in this case is long and slim, ending with a plume like the lion. Neither is it a furbearing animal, the hair being extremely coarse, like a dog's. It has short rounded ears closely fitted to its head, and the long pointed nose and long neck of the weasel. Its measurements as it now stands mounted are as follows:

Length from tip to tip, 47 inches; length of tail, 22 inches; body, 25 inches; from shoulder to tip of nose, 9 inches; from shoulder to ear, 6 inches; height, 11 inches; girth, 12 ½ inches; end of nose to tip of ear, 6 inches. It is a dull grayish color above shading to coal black beneath, the legs being black and much longer than the otter family. It has cat claws, five toes, and is a tree climber.— *Prospect.*

KILLS FAMOUS BLACK WILDCAT

OAKLAND MAN WITH PARTY THAT BAGS LONG SOUGHT GAME

Oakland, California, *Tribune*, November 15, 1910

Recently, while S. C. Scott of the Farmers and Merchants' Savings Bank was spending a few days hunting in the Bar Mountain district of Calaveras county, accompanied by W. S. Ryland, a Stockton man, and Ralph B. Day, a mining engineer of Cleveland, Ohio, met and captured the oft-reported black wild cat. The beast put up a fierce fight and the combined forces of the three men, fully armed, were required to subdue the animal. This was the only known black wild cat in existence, and the passing of this monster, weighing some three hundred pounds, this particular specie without doubt ceases to exist.

The above named gentlemen have had the skin dressed, mounted, and have shipped same East to be presented to the Smithsonian Institute at Washington, D. C.

The struggle took place some fifteen miles from camp and just before dark. As the hunters started on the return trip two enormous rattlesnakes were discovered asleep on the trail.

WILDCAT IN A CITY PARK

Benton Harbor, Michigan, *News-Palladium*, January 21, 1911

A giant wildcat that had been slaughtering quail and cottontail rabbits in the Golden Gate park for the last three years was trapped recently by George J. Barron. The park gamekeepers and others often caught sight of the destructive beast, but it was so wily that they never could got an opportunity of drawing a bead on it. The great cat was thought to come from the Santa Cruz mountains.

The other day the lad rigged up a plain box trap and baited it with a piece of fresh beef, setting the contrivance in the shrubbery near the chain of lakes. He returned several hours later and found the big cat snarling in the trap.

Prof. William G. Blunt, curator of the natural history department of the museum, was sent for, and after a close examination of the beast at a safe distance pronounced it a fine specimen of the hybrid wildcat of California.

As hybrids of all kinds in the class of mammals are barred from the museum, the cat was killed and buried at the foot of a Monterey cypress. The animal weighed 40 pounds, and was one of the largest specimens that the professor had ever seen.—San Francisco *Chronicle*.

MONSTER STRIPED WILDCAT NABBED IN COYOTE TRAP
Oakland, California, *Tribune*, December 4, 1925

Tonopah, Dec. 4.—Secured in a strongly barred parking case in a store here is a monster wildcat of irregular type which was captured in a trap set on the outskirts of this city.

The animal is larger than either a lynx or a wildcat of ordinary type and has stripes like a tiger.

A few nights ago a coyote of large size with half of its body devoured by some unknown animal was found in the trap. The trap was reset and this morning the strange animal on exhibition here was found secured by a foreleg.

PANTHER, BENGAL TIGER, MOUNTAIN LION REPORTED
Van Nuys, California, *News*, March 15, 1934

Terror reigned supreme in the vicinity of 4500 Laurel Canyon boulevard Wednesday morning when a resident reported seeing what he thought was a panther in a tree. He found his dog at the foot of a tree and called the dog off when he saw the animal, then reported to the Valley police division.

Soon another report came to the police division from another citizen who said he saw a Bengal tiger in the wash between Riverside drive and Moorpark street in the vicinity of Laurel Canyon boulevard. The last report from still another citizen was that the animal was a mountain lion.

Police officers headed by Captain E. E. Hack of the Valley division were unable to locate the animal, although the territory was thoroughly gone over.

BLACK PANTHER HUNTED AT MARIN COUNTY ARMY POST
Oakland, California, *Tribune*, August 27, 1952

The Army, already plagued by flying saucers, had a new mystery by the tail today. An animal described as a "black panther" is loose at Fort Barry on the Marin County side of Golden Gate Bridge.

The beast escaped into tall grass on the Army post early today after it had been shot and wounded by a military policeman who spotted it by flashlight.

The Army was taking no chances on the animal, verified or not. It notified 40 families in the Fort Barry housing area to keep their children out of danger, and not to be alarmed if they heard shots.

Some eight persons were reported to have seen the animal. It was described as black and larger than a police dog. Tracks this morning were found around the housing project.

Military policemen, armed with shotguns searching for the animal, said the tracks had the appearance of those of a mountain lion. They are "four or more inches" in diameter.

The animal first was discovered by Tony Costello, 16, son of Chief Warrant Officer Clyde D. Costello, who lives in an Army housing project on the base. Tony heard it trying to reach bait in a coon trap he had set. He summoned his mother and other housing project residents.

"It sure looked black—it looked just like a slouching black panther," Mrs. Costello said.

"The military police thought it was some sort of joke until they saw it themselves."

Young Costello not only insisted he had seen two of the animals earlier, but that a mother and five panther pups came on a garbage raiding expedition several days ago.

When he told his parents about it, they thought he was joking, he said.

But last night, with the animal present and accounted for, Mrs. Costello notified the Army.

She summoned neighbors, including Lieut. Harper Mack, and his wife, all of whom saw the animal. Meanwhile, six military policemen arrived.

Sgt. 1/c Robert G. Long and Corp. Roger F. McGuire from the Presidio of San Francisco joined the search. They spotted it near the fringe of the post's housing area.

As Sergeant Long fired the animal fell but got up again and thrashed about in the deep grass. Soldiers, reluctant to follow a wounded animal into the deep grass at night, abandoned the search until morning.

As the lieutenant in charge of military police last, night said: "They might as well have reported Bengal tigers—they would have been just as unbelievable, but we are on the job to stamp out any menace to our security."

'BLACK PANTHER' STILL ROAMS MARIN
Oakland, California, *Tribune*, August 28, 1952

That "black panther" still roamed the Marin County hills today, defeating the best efforts of the U.S. Army to capture it.

The animal was seen Tuesday night by Mrs. Clyde Costello, of Camp Barry, and her son, Tony, 16, who alerted other residents of the Marin Army post and military police.

Everyone who saw it insists it was big and black.

Carey Baldwin, director of the San Francisco zoo, opined it might be a black mountain lion, a freak which would be worth as much as $6000.

Children were kept indoors as 14 armed soldiers went on a safari through the tall grass but no trace of the black animal was spotted.

BLACK PANTHER HUNTED NEAR PORTERVILLE
Oakland, California, *Tribune*, September 20, 1954

Porterville, Sept 20—(UP)—Sheriff's deputies were investigating today reports from several persons that a "black panther" was roaming the foothills east of here.

Deputy Sheriff Pay Hayes said tracks found in the area indicate there is a "pretty good sized animal around here." He said there have been several reports of geese being killed by an unidentified animal.

The persons who saw the "panther" said it looked similar to a mountain lion only it was black and very large.

'BLACK PANTHER' SEEN NEAR VISALIA
Eureka, California, *Humboldt Standard*, September 21, 1954

Visallia (UP)—The "black panther" that was reported roaming around Tulare County foothills was believed today by authorities to be a "sun-tanned" mountain lion.

Sheriff's deputies said several persons in the Lindsay district told them they saw the animal, described as "big and black." During the past few days, an unidentified animal had killed a number of geese in the area.

A mountain lion tracker from Springville has been called in by the sheriff's office to dispose of the animal. Local hunters said this was the first lion seen in the area for three or four years.

'Black Panther' Hunt Leads Only to Predatory Dog

Fresno, California, *Bee Republican*, September 21, 1954

Porterville, Tulare Co.—Following a report that a black panther was seen by motorists shortly after Sunday midnight at Plainview west of Strathmore, sheriff's deputies and game wardens made a preliminary investigation, but failed to find anything but the tracks of a large animal.

Game Warden Ross Welch of Porterville said the tracks were those of a large dog which had been killing geese in the area.

The reported panther also chewed up several dogs and carried off a litter of puppies.

A dragnet for the beast was set by the searchers. Luman Gage of Springville and his lion hunting dogs were alerted for a tracking mission, but without result.

'Panther' Hunt May Be Held Near Orinda

Oakland, California, *Tribune*, October 22, 1954

Orinda, Oct. 22.—(AP)—A state trapper will be asked to lead a safari into the hills between Orinda and Moraga to catch a black panther two men say they saw.

Bill Lance of Moraga says he took a rifle shot at such an animal yesterday but missed. Since then he and George Lavey of Lafayette, who was with Lance, have decided the cat may be a valuable fugitive from a carnival or zoo and should be trapped alive if possible. They plan to contact the trapper today.

Lavey said the two men spotted the panther, described as "coal black and five or six feet long" off Moraga Road between Orinda and

Moraga. They drove to Lance's house to obtain the rifle and returned. When he fired at the animal it fled over a hill.

Afterwards the men searched and found a paw print apparently made after last Monday's rain. The print of the pad measured three inches across.

DAILY KNAVE
Oakland, California, *Tribune*, April 14, 1955

If you're out El Cerrito way, watch out for the gosh-what-is-it.

The strange critter made its first known appearance in the quiet community the other 2 a.m. as L. G. Andrian was sleeping at his home al 7930 Terrace Drive.

A dog spotted the gosh-what-is-it and, sensibly cornering it, proceeded to bark with really impressive violence.

Andrian, awakened, peered out the window. He saw The Thing and bounded to the telephone. In seconds, he was telling El Cerrito Radio Dispatcher Jim Barker all about the creature.

"Gee whiz!" gasped Barker in a strangled tone. "I'll send someone!"

When Andrian got back to the window, the pooch, baffled, was slinking away, leaving the unintimidated what-is-it quietly standing in a corner. "I don't believe it!" Andrian assured himself, groping for a flashlight.

He played the light on the creature. It was just as he thought, only more so.

When Patrolman Lawrence Hamm arrived, the animal had disappeared down the street.

Andrian described the critter as having a flat light-gray face; four legs, on two of which he stood and walked; a brown body with rounded shoulders; and lacking a tail. Its height: about 16 inches.

The El Cerrito man is sure it wasn't a skunk or raccoon, and a monkey seems ruled out by the lack of a tail.

El Cerrito police have just one question to ask concerning the what-is-it.

What is it?

FIGMENT OR FACT

Oakland, California, *Tribune*, July 4, 1956

It could be that the big black '"panther" seen by Orinda's Scott Tyler and Paul Locklin, both 10, was the real thing.

"I will go along with Scott and Paul," writes Mrs. B. P. "That old panther is no figment of the imagination of some little boy. It was around the hills in Oak Knoll last year and everyone was hunting for it on horseback and otherwise. Believe me, it is BIG."

Now that a Sleepy Hollow Lane neighbor has also spotted a large dark object moving about on the hillside, the consensus is that the boys knew whereof they talked.

What's more, on Oct. 22, 1954, Orinda's Bill Lance and Lafayette's George Lavey reported a '"coal black panther, five or six feet long" off Moraga Road between Orinda and Moraga. When they shot at the animal with a rifle, it bounded away, leaving behind only the print of a pad measuring three inches across.

In any event, Mrs. Harl Tyler reports, the incident has had a salutary effect. "Neither of the boys," she says, "now has the smallest desire to sleep in the great outdoors."

'BLACK PANTHER' SEEN ON COAST

San Mateo, California, *Times*, September 25, 1956

Redwood City, Sept. 25—A group of Sharp Park residents spotted a mountain lion, or a black panther, or a pussy cat—something, anyway—near the end of Moana drive in Sharp Park yesterday afternoon.

Sheriffs deputies who investigated said they got about as many descriptions of the beast as there were people in the group. There was agreement on only one point—the animal was black.

An excited resident called the sheriffs office at 2 p. m. and reported that a black panther had been spotted. Deputies John Kemp and Gerald Moorman raced to the scene.

There they came upon a group of armed men who said they had flushed the critter out of the brush and had fired two poorly-aimed shots at it.

Descriptions of the animal, the deputies reported, ranged from a large house cat to a mountain lion which stood more than two feet high, had a long tail and weighed more than 60 pounds.

One of the men allowed as how the beast had a tuft of hair at the end of his tail. There was a great difference of opinion about that. Another said the lion was accompanied by two cubs.

The deputies searched around the area for awhile and, after turning up only a couple of field mice, called off the hunt.

'IT RUSTLED'

BLACK PANTHER ALARMS DUARTE

Pasadena, California, *Star-News*, June 14, 1957

By Bill Mayer

The San Gabriel Valley's roving, 4-foot-long black feline, seen in Monrovia Monday and in Arcadia Wednesday, turned up in Duarte yesterday and terrorized residents with angry roars.

At 1:30 a.m. Mr. and Mrs. Warren W. Stevens of 160 Brightside Ave., Duarte, called the Sheriff's Temple City substation to report that an animal had been near.

They were awakened, they said, by a "tremendous roar" in their backyard.

It "Rustled"

Peering out an open rear window, they couldn't see any thing, but they heard "something rustling."

Four deputies were despatched to the Brightside Ave. address, and while they were talking to the Stevenses also heard a loud roar—but this one was from across the street.

"It was not a dog, cat or any domestic animal," Deputy Oscar Desilver declared. "It was at least a bobcat. Maybe something bigger."

The animal remained as elusive in Duarte as it has been in other places where it has been spotted. The deputies searched by foot and in their cars for an hour, but there were neither roars nor any other sign of the big cat.

Vanishing Act

It has been seen at intervals and very briefly for eight days, usually in the foothills springing from behind some brush and vanishing completely.

Nobody is quite sure what species of animal it is. Mr. and Mrs. Ken Stanton of Mongrovia, who saw it first, described it as about 4-feet long and black. Monrovia police and Arcadia officers concurred in the description.

It has been suggested that the animal may be a black panther, although black panthers are not native to this hemisphere.

Arcadia Officer Lawrence Whittington, who got close enough to fire a shot at it Wednesday back of 1926 Highland Oaks Drive, thought it might be an oversized bobcat.

No Tracks
Black Panther 'Seen' Again
Pasadena, California, *Star-News*, July 1, 1957

Arcadia. The vanishing black panther seen and then not seen in the foothills here for the past 3 weeks, was spied again yesterday.

Milton Green, 524 Gloria Rd., told police he saw the animal about 2 a.m. yesterday in his front yard. It ran around to the back and vanished, he said.

Investigators were unable to find the "panther" or its tracks.

Black What?
Pasadena, California, *Independent*, July 31, 1957
By Russ Leadabrand

A dark-complexioned animal, sometimes referred to as a black panther, has been seen repeatedly in the foothill areas out around Monrovia, Duarte and Azusa.

Experts of various kinds have been brought in to track the animal.

The results have been negative.

We had a talk with some mountain folks the other day and brought up the subject of black panthers. All we got was a laugh.

"Well, what is it then that these people are seeing?" I asked.

One Fish and Game Dept. warden told me that he had gone out and studied the track found after a black panther appearance.

"There is no doubt about it, they were cat tracks. But since all cats make the same kind of tracks—all the way from a house cat to a mountain lion—the only thing we had to go on was the size. I've tracked a lot of cats in my time. This track probably belonged to a large house cat. Might have weighed 15 pounds. But no bigger than that."

"House cats do go wild," he went on. "In fact it happens all the time. And when they do go wild they are as wild as anything born that way. They hunt mostly at night along the edge of the foothill communities, but because they are, or have been, domesticated, they can find their way about the cities with no trouble."

"Could the tracks have been made by a bob cat?" I asked.

"An adult bob cat would be bigger than that and not as dark unless it had been caught in a forest fire and burned," I was told.

"A mountain lion track, even a small one, would have been almost as large as a man's hand," the warden told me.

He was convinced that the black panther was little more than a feral house cat, a big fellow, but no panther.

"Panthers don't belong in this country. If one had escaped from a zoo or carnival we'd have heard about it. Furthermore a panther would be so lost in our kind of woods and mountains here it would go crazy and would end up being shot in the middle of some city. It couldn't cope with our range here."

"Well, do we have mountain lions in our hills here?" I went on, trying to rub a little bit of ferocity in a waning story.

"Lots of 'em," Warden Wes Mongey told me. "I've seen tracks on several of the truck trails above Monrovia and Glendora. Bob cats and bear, too," he told me. "There are a lot more wild animals in these mountains than most people realize."

But a black panther?

Sorry, not this time.

'PANTHER' BELIEVED FAR BACK IN HILLS
Pasadena, California, *Star-News*, June 25, 1957

Monrovia. This city's black panther, which has been seen recently by residents of the foothill area, probably is far back in the hills today.

This was the opinion of Police Chief Grant Petersen, who resumed search for the cat Saturday with the aid of trained lion-hunting dogs.

Petersen said the dogs caught the trail of what they thought was a bobcat.

He said he plans to discontinue the search for the cat until additional reports are received.

Pasadena's 'Panther' in Altadena?

Pasadena, California, *Independent*, April 19, 1958

An elusive "black panther" seen in the upper Hastings Ranch area Wednesday may have visited Altadena yesterday.

Sheriff deputies at Altadena station said a 4-inch-wide paw print made by a large animal was found near a house under construction at the north end of Kinneloa Canyon road.

Investigators agreed it was probably made by a "cat."

They said the paw print showed claw marks very plainly.

The animal was not sighted, as it was Wednesday in the Hastings Ranch area when two carpenters and a truck driver reported seeing a large panther-like animal.

Otis Parham, 24 W. Harriet St., Altadena, said he saw the cat "sitting up like a big old house cat." He said the cat lay down in a bamboo patch.

He watched it for about five minutes, then slipped off and phoned police.

A hunt with high-powered rifles failed to turn up the elusive panther.

Is Dairy Valley 'Lion' Rustler in Disguise?

Long Beach, California, *Independent Press-Telegram*, August 17, 1958

Dairy Valley—In the cold light of day, reports of a half-grown lion "with a huge mane," seen ambling across the intersection of South St. and Studebaker Rd. here, Friday night, didn't get the lion's share of attention from sheriff's deputies.

After taking photographs of a pawprint of some unknown animal near the site of the reported lion-sighting, deputies sent the film off to a commercial lab. The prints will be ready on Monday.

A veterinarian is to study the track photographs to see if he can identify them when they return from the photo shop.

Deputies from the Norwalk Sheriff's substation did look for a lion in the area Saturday morning after Jerome Franke, 21, and Larry Sayers, 15, both of Whittier, insisted they saw a lion there about midnight Friday.

No lions have been reported missing in any area nearby.

However a small calf has vanished from a dairy near where the "lion" was seen.

Deputies have been alerted to look for the missing calf and/or lion. But a western story-reading deputy suggests another line of investigation.

"Look for a rustler disguised as a lion."

'PANTHER' JUST A BLACK CAT
Pasadena, California, *Star-News*, March 12, 1959

Glendora's "black panther" has turned out to be a large alley cat.

The feline was killed this morning by a State Game Warden Wes Mongey, who said it had apparently reverted to a wild state.

Mongey was called by Glendora police after complaints from residents of the South Hills area that a "black panther" was roaming the hills behind the Glendora dairy.

The "panther" was a black tom cat, approximately 1 ½ feet long.

Alarmed residents on Linfield street had been watching the cat through binoculars for five days and were certain that it was the "black panther" which had been reported in the Monrovia hills about a year ago.

BLACK PANTHER SEEN NEAR OXNARD AFB
Oxnard, California, *Press-Courier*, September 23, 1959

Camarillo—Ventura County's elusive black panther has apparently shifted his nightly forays from his Box Canyon home to territory north of the Oxnard Air Force Base.

The black cat last night was reported in flight after being sighted by Ralph Ward, 78 Calle El Valeador, Capehart Housing District.

This Panther Just a Tabby
Oxnard, California, *Press-Courier*, September 28, 1959

The black panther reported near Camarillo every night for the past three nights showed up on Hueneme road at 9:45 p.m. Thursday.

At least that's what a passing motorist told Highway Patrolman Harrold Moberly. Officer Moberly called the sheriff's office. Deputy John Atkinson was sent out and soon radioed that he had found a black and white pussy cat, but no panther.

Black Panther Reported in Highland Park
Pasadena, California, *Independent*, July 20, 1959

Another black panther was reported on the loose near here yesterday.

But the panther—if that's what it was—managed to keep clear of a team of policemen sent out to track it down.

Mrs. Ann Kitson and her daughter, Beverly, 18, said they spotted the beast perched in a tree across the street from their home in the Highland Park district. A large spread of brush and scrub oak on the slopes of Mt. Washington faces the Kitson home.

Police searched the hillside, found nothing.

Earlier this week, a woman and a girl who live in Box Canyon, northeast of Los Angeles in Ventura County, notified authorities they had spotted a black panther. A search failed to turn up any trace of such an animal.

Black Panther in Box Canyon Eludes Hunters
Oxnard, California, *Press-Courier*, July 20, 1959

Box Canyon—A black panther eluded weekend hunters who stalked it with high-powered rifles in the Santa Susana area.

The big cat was reported seen several times last week during daylight and at night.

Box Canyon fire captain Carl Friddle believes the reports are based on fact. Residents reporting the animal, he said, are reliable witnesses.

It is believed that the animal was routed from his mountain lair
July 19 by a 40-acre brush fire.

POLICE HUNT FOR LION IN HILLS ABOVE CANOGA PARK
Van Nuys, California, *News*, August 13, 1959

A lion hunt was under way last week in the hills above Canoga
Park near the west end of the San Fernando Valley.

Mrs. Blanche Muller and her daughter Katherine, 15, said they
sighted a mountain lion Thursday morning from their home at 23175
Cohasset St., Canoga Park.

Watch Through Glasses

The cat, which they described as about two feet high and four
feet long, with a long tail was on a grass-covered hillside area across
the street about 300 yards away, they said.

Mrs. Muller said they watched it through binoculars and called
her husband, Earl, who was mowing hay.

Muller called police and Officers Robert Rackerly, Fred Cochran
and Daniel Danko went to the scene with a high-powered rifle.

The Mullers said the black mountain lion went over the brow of
the hill and disappeared about 30 minutes after they first saw him.

The police drove into the hills but were unable to find the cat.

Expressed Fear

Cochran left the car with the rifle to continue the hunt on foot.

The Mullers, who raise calves, expressed fear the animal might
be after their livestock.

Muller said he could have easily shot the lion from his farm, but
did not do so because of a city ordinance forbidding discharging firearms.

Ralph Rowe, manager of the Mission Riding Stables, 23030
Sherman Way, told police one of his horses had teeth marks on its
neck about three weeks ago.

Rowe said he has been searching for the animal for the past few
nights, but has not sighted it.

Police said people in the area told them that some sheep and
calves in the vicinity have been killed during the past few months.
The area is permeated with cattle and sheep ranches.

Sees Little Danger

About a month ago a lion was reported in the Box Canyon area, just north of the Canoga Park sighting. At that time Ventura County sheriff's deputies investigated, hut were unable to find the animal.

Police said lions are sighted in the hills from time to but there is very little as they usually stay far away from civilization.

DEPUTIES REPORT SEEING "PANTHER" BUT NO DEER

Oxnard, California, *Press-Courier*, August 8, 1960

Ventura—Two, off-duty deputies from the sheriff's office were among the deer hunters roaming the hills of the county over the weekend, who found more game than they were looking for.

Lt. John Chamberlin and Lt Merle Hollis were scouring the hills between Highway 101 and Lake Sherwood, just off Triunfo road, at daybreak yesterday morning, when they spotted a large black panther. They got three shots at it, but their aim was poor.

Lt. Chamberlin said it was "really big." However he couldn't come up with an excuse for the cat getting away. In fact, he couldn't even give an excuse for failing to kill a deer.

BLACK PANTHER SIGHTED BY MOTORISTS ALONG HIGHWAY 99

Lodi, California, *News-Sentinel*, October 17, 1961

An animal described by two Modesto couples as a "300 pound black panther with green eyes the size of quarters" was sighted along Highway 99, a mile south of Mariposa Rd., shortly after 1 a.m. Sunday.

It has not been seen since.

However, sheriff's deputies who talked to the startled motorists found several paw marks, measuring three inches in diameter near a large hole under the freeway fence in the immediate area. There were no other signs of the presence of the jungle cat.

Mr. and Mrs. Ed Mitchell, and Mr. and Mrs. Cal Veale told officers they were returning to Modesto from Sacramento when they spotted the panther along side the highway. (The Modesians were described by authorities as apparently quite reliable.)

"We had a good look at it as we passed by," Veale insisted. "It turned its head toward us and we could see large green eyes about the size of a quarter."

Capt. Glenn Whitesell, area director of the State Department of Fish and Game said San Joaquin County is not the habitat of the black panther. But he agreed that an escapee from a zoo or circus train could possibly be in this area.

James Fahey, superintendent of the Micke Grove zoo and park said he has not had a panther at the zoo, and the nearest place where they are kept is in Sacramento— "and we have had no word of one breaking loose."

Fahey said the average weight of a grown black panther is between 180 and 200 pounds. Such animals in this part of the country are "extremely rare," Fahey said, and if one is wandering about he is undoubtedly an expensive possession.

He also might be quite hungry.

PHANTOM PANTHER HAS GLENDALE ALL AFLUTTER
Pasadena, California, *Independent*, May 2, 1963

A phantom "black panther" that may or may not be lurking in the hills around the Verdugo Woodlands area had the city of Glendale aflutter last night.

The community was split into factions of panther believers and non-believers. It was strictly take your pick who was more convincing.

The non-believers faction was led by Glendale police who said if there is anything loose among the hillside subdivisions and apartment houses, it's a large dog.

The believers included 12 people who insisted any resemblance between the animal they saw and a dog was strictly coincidental.

It all started Tuesday night when police had several calls of a "panther" loose behind the 1000 and 1100 blocks of North Verdugo Road, near Glendale College. One woman reported it had growled at her window.

Six officers, one aimed with a high-powered rifle, set out on safari. Officer William Stewart came closest, within about 50 feet, and reported it looked like a large dog.

"We're convinced it's nothing but a dog, probably about as big as a good-sized collie," said police Lt. Dan Curtis.

Len Edwards, photographer for a twice-weekly Glendale newspaper, wasn't convinced. He insisted he saw an animal that was some kind of wild cat. But it darted away before he could snap its picture.

Three-toed footprints, about the size of an outstretched hand, were reported found in the hills yesterday.

Experts from the Glendale animal shelter measured the prints and concluded anything that makes marks that big is 20 feet long and 8 feet high.

SHERMAN OAKS PANTHER HUNT CALLED OFF
Oxnard, California, *Press-Courier*, November 18, 1963

Sherman Oaks (AP)—Police say a search for two black panthers reported seen in the Sherman Oaks area was called off without any trace of the animals being found.

The panthers were reported early Sunday afternoon by Mrs. Margarie Muntean, 39, who said she had spotted them through binoculars. She said she first thought they were bear cubs.

Officer Waid Woodruff of the Van Nuys police, who was first to arrive at the scene, said he saw one of the panthers but didn't go after it because he was armed only with a revolver.

Panther For Sure

"It was a panther." Woodruff said. "It was about 75 yards way and I watched it for about five minutes. There was no doubt about it."

Scores of policemen worked their way through the heavy brush but no trace of the cats were found.

Sgt. Ken Bouey said "We've called off the hunt for now. We didn't find any trace of panthers. There were none reported missing, and one of our men saw a dog up there that could have been mistaken for a panther."

SEARCH FOR BLACK PANTHERS
Pontiac, Illinois, *Daily Leader*, November 22, 1963

Sherman Oaks, Calif.—(UPI)—Police got another telephone call Thursday from a frightened housewife that brought a current "black panther" scare in this area to a climax.

The woman said she saw two of the beasts slink under a partially finished house. Officers arrived on the run, shotguns in hand. They found two house cats under the house.

There was a widespread hunt in the San Fernando Valley last weekend because of several reports of loose black panthers.

BLACK PANTHER REPORTED SEEN NEAR CAMARILLO
Oxnard, California, *Press-Courier*, January 1, 1964

The Ventura County black panther was reported by a woman who watched it through binoculars on the hillside near her Camarillo home. A deputy was unable to find tracks of the animal.

PANTHER IS WINGED BY DEPUTY
Long Beach, California, *Press-Telegram*, June 2, 1964

Ventura County deputies responded to a woman's sighting of a black panther, and they saw the animal moving toward them on the hills when they arrived. One deputy shot the animal with a rifle, but they were unable to locate the body and assumed it got away. A Northrup security guard reported seeing a similar animal the next day.

PHANTOM CAT TRAVELING FAR AFIELD
Oxnard, California, *Press-Courier*, November 2, 1964

Sammy Bradshaw of Meiners Oaks reported seeing a big black cat in the Ventura River bottoms. A woman claimed to see the "black panther" in the hills of north-central Ventura.

SHORT TAKES, BY DAVE WHITE
Oxnard, California, *Press-Courier*, January 18, 1965

A black panther was seen twice in the Conejo Valley, first by two girls on a picnic in the hills, and then by a woman in the same area. A state hunter was searching for the animal with his dogs.

Loren Coleman (1979) noted sightings of a large black feline in December 1967 and January 1968 in Ventura County. Several sightings in the 1970s were also noted, including a rancher near Mt. Diablo watching one walk near his home (leaving 5 inch tracks and scaring his German shepherd) in 1972, and a naturalist at Las

Trampas Regional Park observing one in company with a tawny feline as they chased deer.

Wild Lion on Prowl?
Oakland, California, *Tribune*, March 3, 1973

A curator at the Alexander Lindsay Junior Museum in Walnut Creek gathered information on five "well substantiated" sightings in the prior year of a large black feline. The reports came from "the Walnut Creek hills, Danville, on the slopes of Mt. Diablo and in Las Trampas Regional Park." One witness said his Irish setter treed the "coal black" animal, which was as big as his dog and had a "rope like tail," before it managed to escape. Two Danville women saw it drinking from a swimming pool before it loped off into the hills.

'Black Panther' is Mountain Lion
Fremont-Newark, California, *Argus*, December 21, 1973

California Fish and Game warden Dennis Rankin stated that a "black panther" reported prowling in Niles Canyon was actually a 90-pound female "rare black mountain lion," and was therefore a protected species.

Letter to Editor: That Black 'Panther'
Oakland, California, *Tribune*, December 25, 1973

Gary Bogue, curator at the Alexander Lindsay Junior Museum, urged locals not to panic at reports of the "black panther," which he believed was a melanistic mountain lion. He stated that it had never attempted to harm anyone or any livestock in three years of sightings. It had only been seen hunting rabbits and deer. He noted that there was at least one sighting of the animal with a cub.

Black Panther Reports Doubted
Hayward, California, *Daily Review*, December 27, 1973

A California Fish and Game regional manager, Jack Fraser, said he did not believe there was a truly black feline roaming the Milpitas and Fremont areas. Instead, he thought it was just a normal mountain lion.

PANTHER HUNT FINDS ONLY BIG 'DOG TRACKS'
Modesto, California, *Bee*, January 22, 1975

After another panther sighting (a motorist spotted a 5-foot long black cat with a long tail run into the brush near the river by the county airport), game wardens and a wildlife biologist searched for the animal along the river banks. The only tracks found they determined to be canine.

WEST COVINA HIT BY BLACK PANTHER SCARE
Long Beach, California, *Independent Press-Telegram*, October 11, 1975

A black panther was reported from the hills near Gaster Wilderness Park, prompting an official search by ground and helicopter.

MODESTO MAN SEES MYSTERY PANTHER AGAIN
Modesto, California, *Bee*, January 15, 1976

A black panther believed to live in the Tuolumne River bottoms was seen by a Modesto man who awoke early one morning to his squealing rabbits and squawking chickens. He turned on the outside lights to see the animal jump over two fences (one 3-foot high, one 4-foot high), then run off towards the river.

HILLS ARE ALIVE WITH SOUND OF RUMORS
San Francisco, California, *Chronicle*, November 7, 2004

There were two separate sightings of a black panther by hikers in the Las Trampas Regional Wilderness. A wildlife biologist with the East Bay Regional Park District said they get one or two sightings a year. He noted locations where sightings have occurred: "In the Bay Area, black panthers have been reported by hikers at Sunol Regional Wilderness, Chabot Regional Park, Carquinez Strait Regional Shoreline and the Marin Headlands, as well as at Las Trampas."

IS THERE SOMETHING LURKING IN OUR CANYONS?
Torrance, California, *Daily Breeze*, August 28, 2006

This article discusses a variety of large feline accounts in the area, with mention of one "black panther" sighting. In 2004, Lisa Baginski was walking her dogs on a Palo Verdes Estates trail when she encountered a large "shiny" black feline with a long "bristly" tail. She was about 30 feet away when it jumped from the brush, ignoring them and stretching itself. She stated: "It just bounded out from up above us into the sun. The tail was absolutely huge and it was very long. If it had stoodup (on its hind legs), it would have been over 6 feet tall." She estimated it at between 150 and 200 pounds.

The Mysterious Black Panther Makes a Rare Appearance, Scout Reports
San Francisco, California, *Chronicle*, April 13, 2008

An avid hiker and Bay Area wildlife expert, John Balawejder, was hiking with a friend at Point Reyes National Seashore when they saw a large black mountain lion. "We came up a short rise through a grassy swale, and then, looking up, saw a large, jet-black mountain lion calmly sitting, eyes half asleep looking out at us from about 30 yards away. ... My friend and I stood there, stunned. It then started to slink away from us in a large semi-circle, attempting to hide in the grass. ... This lion was not darkish, not a brownish-tawny like some I've seen since, but jet black."

Black Cougar, or Cat, Prowls Near Forestville
Santa Rosa, California, *Press-Democrat*, May 2, 2008
Black Cougar Caught in Photo?
Santa Rosa, California, *Press-Democrat*, May 3, 2008

Don Callen captured images of a large black feline with a game trail camera on his 52-acre wooded property near Forestville. Callen made estimates on the ground that suggest the cat "would have come up to his mid-thigh." Two experts viewed the photos, calling the enticing, but not detailed enough to make a solid identification. An image of the feline published in the newspaper is clearly not a bobcat, as it has a long bushy tail, or a cougar, as it has the wrong facial profile and build, and is fairly bushy all over. There is nothing to indicate that it isn't a large domestic house cat.

'Black Panther Sightings' in Bay Area Parks
Santa Rosa, California, *Press-Democrat*, November 13, 2008

Writer Tom Stienstra summarizes the theories and official explanations for alleged black panther reports in California, noting some eyewitness accounts, and listing a few parks that are worth visiting for potential sightings. One witness, Larz Sherer, stated: "I was curious about EBMUD's protected watershed off Redwood

Road in Castro Valley, so I obtained a permit and checked it out. ... I decided to navigate into the gully, walked maybe 30 or 40 feet to the east and suddenly found myself locked eyes with this big black cat. It was roughly 50 feet from me, through several barriers of logs and overgrowth. The first thought is that it looked like a panther, but the weird thing is that sort of animal should be in Africa, not the East Bay. It was so out of place."

BEAR, BLACK LION - IMPOSSIBLE? NOT NECESSARILY
San Francisco, California, *SFGate.com*, August 9, 2009

Writer Tom Stienstra detailed two black panther sightings from the week prior. "Lynn Reed, a ranch owner, avid hunter and wildlife expert, said he and his wife watched what appeared to be a black mountain lion for more than 10 minutes in the foothills near Dublin in Alameda County." Reed said it was very black, with a head the size of a cantaloupe. "It was 3 feet long, the tail 2 ½ feet, maybe 60, 75 pounds."

The following day, another was spotted by a surveyor, Art Whitten, in the San Ramon hills, from about 100 feet away. He had seen mountain lions before, and said this was about the same size.

COLORADO

Dog With a Human Face
Strange Monstrosity Seen by Many Persons in Colorado Hills
—Attempt Capture in Vain.

Rake, Iowa, *Register*, June 15, 1905

Buena Vista, Col.—A strange animal roaming the bills in the vicinity of Wildhorse, a station on the Colorado Midland railway, two miles west of this city, has been seen a number of times by various people and has been described differently by each one.

The most startling of all, however, was the experience of a prominent ranchwoman last evening. She was driving slowly along the road across the Arkansas river, from where the animal has its lair, when her horse suddenly shied, almost throwing her from the rig. She was horrified to see, a few feet ahead of her and in the middle of the road, the monstrosity.

It was about the size and build of a full-grown greyhound and of a drab color, its glistening sides being covered with black spots as large as silver dollars. It had a long, smooth tail and the woman declares it had an almost human face, and a bristling red mustache ornamented the proper place upon its physiognomy. The eyes were close together and deep set and its ears stood erect and were very pointed. After a moment it uttered a piteous cry and slunk away through the brush, turning at the top of the hill for a last look. It stood erect on its hind feet, punctured the rarefied atmosphere with sounds that reverberated among the crags and compelled a pace on the part of the usually staid horse that was a revelation to the driver. A number of hunting parties have tried in vain to kill this animal and efforts are being made to capture it alive.

Michael Newton (2005) noted that a Utah hunter, Bruce Hartman, supposedly shot a black cougar on December 8, 1912, south of Gunnison. Current whereabouts of the remains are unknown. From personal research, I have seen a record of a Bruce Hartman who lived on a ranch in Gunnison County, Colorado, in the early 1900s, so perhaps this is a lead worth investigation by someone in that region.

DOG OR LION? BEAST LOOSE IN EL PASO COUNTY
Denver, Colorado, *Post*, July 14, 2008

A strange animal leaving 6-inch paw prints was seen by multiple witnesses, but there was serious debate over whether the animal was a maned lion, or just a big dog.

CONNECTICUT

Hartford, Connecticut, *Weekly Times*, February 24, 1855
"Scotland Devils."—We remember many years ago having heard of an animal, termed a "Scotland Devil," being seen in various parts of this State and Massachusetts, but had supposed this "Devil" and his race extinct ere this, till we heard the following:—A gentleman (we do not now recollect the name,) was riding on horseback on the road leading from Fisherville to Thompson Depot, a few evenings since, when suddenly he heard a most unearthly yelling a few rods in advance of him. Putting his horse at the top of his speed, he determined to discover the author of the music, but could not get nearer than five or six rods. The gentleman accordingly stopped his horse and dismounted, when, strange to say, the object of his curiosity stopped, but continued his yellings with unabated ardor, at the same time whisking his tail violently. He then endeavored to approach the object on foot: it stood its ground for a moment, and then galloped off to the West, making the woods echo with its screeches. The animal is described as of a jet black color, with long, straight hair, about 12 or 14 inches high, and proportioned very much like a cat.—*Killingly Telegraph.*

'VERITABLE SCOTLAND DEVIL' KEPT 'EM HOPPING IN '69
Hartford, Connecticut, *Courant*, August 14, 1959
Recent reports of roaming killer "panthers"—which some argue are wild dogs—and other strange animals in Connecticut is not new. In fact, they're "old hat." Here's a report from the *Tolland County Journal* of Aug. 21, 1869:

Union—The people of this town who reside in the vicinity of Lake Mashapaug have of late been considerably frightened by the appearance among them of a strange animal, believed to be the veritable "Scotland Devil."

Whatever it is, it seem endowed with a peculiar power to frighten nervous people.

One man, a short time ago, when about a half mile from home, jumped a wall to get across lots, home: and as he got over the wall he saw the object of fear on the other side, near him, and without waiting for an introduction, or even to see what kind of an object the intruder was, he started for home, leaving some very long tracks, his hat, and a pair of straight coat tails behind.

When he reached home, he did not stop to unlatch the door, but burst the latch and went in head first. Of course he is not able to give a very minute description of the monster.

'Severely Scared'

Since then, the animal has been seen by a boy sixteen or seventeen years old, who met him in the road while driving a load of hay, and though he admits being severely scared, he is able to give a better account of the animal than any of the others. He describes the animal as being about the size of a large Newfoundland dog, black, with head shaped much like a cat's, very large feet and long claws. He says the animal came toward him until he shouted at it, when it turned, jumped the fence, and took to the woods.

Sound in Night

Quite a few have heard it in the night, which time it appears to take to travel about in to procure food.

Some sheep and lambs have been killed lately, and some are ready to attribute the work to this fellow: and in Stafford, and also in Brimfield, calves have been reported to have been killed by it.

There is an old saying, "Give the devil his due," but in this case it is not certain but that he gets more credit than belongs to him. That there is some animal about which does not belong to the locality and is hardly up to the times in civilization, is pretty evident.

GLAWAKUS SAFARI BALKED BY WARDEN
BULLETS, NOT A MONSTER, KILLED GLASTONBURY DOG,
HE RULES, HALTING 'JUNGLE' TREK
HUNTERS TO GO IT ALONE
STORIES OF ROVING BEAST MAKE EXPERT THINK IT A
MOUNTAIN LION BROUGHT IN BY FLOOD
New York *Times*, January 21, 1939

From a Staff Correspondent

Glastonbury, Conn. Jan. 20.—after much brave talk and many menacing flourishes, Glastonbury officials made a last-minute back-down on their proposed safari today into the nearby jungle country to seek the marauding "glawakus," that elusive creature whose nightly prowlings have terrorized the countryside for weeks.

Local nimrods, annoyed over the creaking and groaning of the official safari machinery, will take to the field alone tomorrow in a final effort to exterminate the "animal" which has become the town's chief topic of conversation. Some of them, in fact, started on the hunt today.

An impartial survey of the glawakus situation indicates that there is a good reason to believe that the glawakus, whatever it may be, is really roaming around hereabouts.

There were further reports today from purported eyewitnesses that some large, unidentifiable animal tracks have been discovered in the Buckingham section. That district is convinced that a death-dealing creature foreign to these parts is on the loose; the rest of the town seems about evenly divided between amused skepticism and annoyed resentment that the matter has not been settled.

Body of Dog is Exhumed

The failure of local officials to start their much-heralded hunt today resulted from a finding made by the county game warden, Charles Allshouse, that bullets and not the fangs of some monster had killed Christopher Bigdotta's hound dog Tex last Wednesday.

At an early hour this morning one of Mr. Allshouse's deputies went to the Bigdotta farm on Matson road, exhumed the dog's body and examined the "fang" marks on its head. The deputy pronounced them to be bullet holes.

"I won't start an official hunt on that evidence," said Mr. Allshouse. "I've got to have definite evidence."

Chief of Police George Hall, who has received several complaints concerning the glawakus, lifted his eyebrows when told of the game warden's findings. The police investigation had ended with a finding to the effect that some strange creature had nearly bitten off the dog's head. After a conference with Officer Cornish, who comprises the rest of the Glastonbury police force, Chief Hall announced: "It's up to the game warden. I won't take anything on myself."

The Hurricane is Blamed

Meanwhile, the testimony of witnesses and an analysis of available evidence have led the more sober minded citizens, and particularly the hunting fraternity, to believe that the glawakus really is a mountain lion, escaped from some menagerie and blown into Glastonbury with the September hurricane.

Glastonbury spreads along the easterly side of the Connecticut River, just south of East Hartford, and during the hurricane and the flood which followed, nearly everything except mountain lions blew into the town from foreign parts. It was reported at that time that a wild animal farm in New Hampshire had lost a mountain lion and a panther in the storm.

Some of the townspeople hold other views, notably Pat Hall, soda dispenser at Franklin's Drug Store, next door to the police station. He accepts the game warden's bullet findings in the case of the dog and also accounts for the tracks of some cloven-hoofed creature found near the scene of the attack. It is his belief that the glawakus is a kind of Pan-like creature, with goat's legs, which carries a revolver or a machine gun.

Others think that it may be a wild boar.

Expert Goes on the Hunt

Probably the best qualified expert in Glastonbury, who is also one of the best in the country, is Jim Laneri, the man who piloted Mr. and Mrs. Martin Johnson's plane in Borneo and the East Indies several years ago. He has hunted everything from rabbits to elephants, and yesterday he searched through the woods for the glawakus.

"I think it is some kind of a cat, probably in escaped mountain lion, or something that may have wandered down from Canada," he said today. "I have hunted these woods since I was able to hold the gun, it is hard to account for the stories told by people I know any other way. If Wells Strickland says he saw an animal that looks like a mountain lion, then I believe him."

Mr. Strickland, a leading citizen of Buckingham, saw the creature in the woods while he was looking over his timberland and described it as a brown cat, about three feet long, with a long tail.

George Cavanna, a Glastonbury butcher, said his mother saw the animal a few weeks ago, and her description of it tallied with Mr. Strickland's.

Mrs. Charles Bell, wife of a lumber company executive, who lives in the Hopewell section, said that her prize greyhound dog came home about three weeks ago clawed from head to foot and had refused to venture near the woods since.

A wood chopper, little given to storytelling, said today that he was setting a trap in the woods when he saw a huge animal crouching nearby. His dog rushed for it, beat a retreat, and animal leaped nearly fifteen feet and bounded away.

The etymology of the word "glawakus," it was explained today, derives from the first three letters of "Glastonbury," and the first syllable of "wacky," with the Latin "us" added to distinguish it from the plural "glawaki" in the event that more than one glawakus should be discovered.

GLAWAKUS 'SAFARI' TRICKS A BLUEBIRD
IT IS LURED FROM A HOLLOW TREE AS COW BELLS
OF EXPEDITION RING OUT THROUGH WILDS
MONSTER'S 'TEARS' FOUND
OR THAT'S WHAT BLUE SUBSTANCE IN SNOW IS SUPPOSED TO BE—
BEAST ELUDES GREAT HUNT
New York Times, January 22, 1939

From a Staff Correspondent

Minnechaug Mountain, near Glastonbury, Conn., Jan. 21 (By Courier)—After an arduous trek over the difficult Minnechaug Mountain terrain, a Glastonbury expeditionary party on safari after the mysterious glawakus paused on the mountainside this afternoon to study field exhibits and develop photographs in a dark cave.

The glawakus continues to elude the party, but great interest is being shown in one of the exhibits—a frozen blue substance found in snow and believed to be glawakus tears. This tends to support the theory of Chief of Police George (Cape) Hall that the mystery creature which has been frightening the countryside may not be so much an animal as "some kind of a Finkelstein monster, half man and half beast."

Its tears are thought to be tears of laughter.

The expedition after the glawakus was early disrupted at one point in the joy of discovering the first bluebird of the season in a hollow tree a little east of Roaring Brook Gorge. This resulted from a most curious circumstance.

An advance party in the gorge had paused under a hollow tree to study snow crystals and theorize some more about the glawakus. The police chief was maintaining liaison with the main party on another hill through means of an intricate system of cow-bell tinkles. He and the cow-bell operator for the other party were tinkling away when suddenly a frantic peeping was heard in the tree overhead, and the bluebird popped out of a hole.

It is the belief of Town Prosecutor John Royston, assistant leader of the expedition and chief of the scientific staff, that the little bird had bedded down in the hollow tree for the Winter and that the insistent tinkling of cow bells startled it into believing that Summer pasture-time was at hand. The chief quickly brought out his candid camera and pulled the trigger. These are the photographs being developed in the cave.

The expedition got an early start from the Police Department building on Main Street, Glastonbury. The departure was touching in the expressions of sentiment exchanged between the hunters and the stay-at-homes, some of whom appeared to have strong doubts concerning the outcome of the safari. Small boy volunteers were sternly rejected by Chief Hall. His final word as the party moved off was:

"Tell the family I'll be home for supper."

Chief Wears Gold Braid on Hunt

The equipment of the expedition was most diverse. Chief Hall led the advance party garbed in a gold-braided dress uniform. He spurned arms, but carried his police whistle in his hand during the most difficult passages. Town Prosecutor Royston was clad in a blue ski ensemble. The prosecutor was prepared to enact a defense role by means of a chromium-plated notary public's seal punch, which he had slung at his belt.

Accompanying the advance party as gun bearer was Dominick Manfredi, who constitutes the entire police reserves of Glastonbury Center. He does not have a uniform yet, but he wore his badge on his sweater and carried a mail-order rifle which Chief Hall confiscated recently from a woodchopper, the latter having bought it for the purpose of settling some argument with his wife. The other hunters, wearing a variegated collection of colorful shirts and caps, brought shot guns and large stoves.

Very little provender was carried as the party intended to live off the country eating checker berries, nuts and edible barks. During the perilous ascent of the mountain side frequent pauses had to be made, while all members quaffed strong draughts of "spirits of ammonia" to revive their flagging morale.

It was a matter of regret to everybody that Charles Allshouse, county game warden, abandoned the expedition for want of confidence in the existence of the Glawakus. It was privately whispered among some of the members that his withdrawal really resulted from some aspersions which had been cast on his dog, Lead.

Lead was obtained by Warden Allshouse from a breeder residing high in the Ozark Mountains, and the animal's prowess as a crack tracking dog had been widely acclaimed. A mixture of Irish setter, Airedale and Ozark coon dog, he was reputed to climb trees and to be able to follow the coldest scent.

Early this week Mr. Allshouse took the dog out on a small safari of his own after the Glawakus. The dog soon found a scent and betrayed extreme agitation. Bounding and baying over hill and dale, with the safari party stumbling hot behind him, he ended up whining in front of a stone wall, with his nose glued against a small hole about an inch and one-half in diameter. It was after that that the game warden decided that the Glawakus was no concern of his department.

Meanwhile, as the present safari party rests in the cave, regaling each other before the cheering fire with holly stories of adventure on the trail, the mystery of the glawakus grows deeper and deeper.

Is it a mountain lion? It is a bob cat? An ape-like monster? Is it at all? These are still hotly debated questions in Glastonbury.

The newest report comes from a resident of Hartford, who telephoned a Hartford paper that his two brothers had sighted a black panther today in the Buckingham district of Glastonbury. He said that his brothers were hunting when their dog suddenly flushed from the woods only a few feet away a sleek black animal about three feet long, with a tail about two feet long. One of the men started to shoot but his gun jammed and the animal bounded away. The story is receiving wide credence among many people of the district tonight.

WILD ANIMAL REPORT SETS OFF GRANBY AREA SEARCH
Hartford, Connecticut, *Courant*, December 27, 1957

Granby, Dec. 26 (Special)—Several hunters searched rain-drenched woods in West Granby this morning for a wild animal reported in the area. They returned without finding a trace.

Mrs. Hugh A. McIntyre of Barkhamsted Road reported seeing a sleek, black, panther-like animal on the north side of Route 20 at about 10 a.m. The animal had the characteristic slink associated with cats. It proceeded across the field at a slow walk "as though out for a stroll." Tall grass obstructed her view, she said. She estimated the animal to be about four feet long with a long tail.

Chief Constable Harrison Hotchkiss was one of those who search of the animal. He drew on his experience of a half century ago when he hunted pumas in Idaho.

From Mrs. McIntyre's description, Hotchkiss believes she saw a puma or mountain lion. These are the same as the catamount which roamed New England in Colonial days, Hotchkiss said.

About 30 years ago a catamount was killed in Barkhamsted, Hotchkiss said. He noted that game wardens had been told of the report and efforts are being made to positively identify the animal.

Lyle Thorpe, director of the State Board of Fisheries and Game, commented, however, that panthers are extremely rare. "If there's one in the Granby area," he said, "it got there only by escaping from a wild animal show, and I don't know of any wild animal shows which have been traveling around here lately."

May Be a Bobcat

Ralph L. Emerson of the Emerson Wild Animal Farm in Newington said that Buddy, a 200-pound panther, and Bill, a 650-

pound lion, were safely in their cages. When asked if a puma could have wandered into the area, he gave a flat "No." He said that perhaps the animal could be a bobcat. The largest one he had ever seen weighed about 40 pounds.

Pet Turned Killer?
Marauding 'Cat' is Panther
Hartford, Connecticut, *Courant*, October 31, 1962

By Robert E. Sheridan

Southbury—Lyle M. Thorpe, director of the State Board of Fisheries and Game, said Tuesday the marauding animal reported in the Woodbury-Southbury-Oxford area in recent weeks is probably a panther.

Thorpe said he is "convinced" the panther (a black leopard) was brought into the state as "someone's conversation piece" and has escaped.

"Its owner, of course, is not anxious to identify himself," Thorpe said.

Not Bobcat

Board workers have talked with all persons who have reported sighting the animal. The investigation here is under the direction of Theodore B. Bampton of Bethlehem, district supervisor.

All "credible reports" indicate the animal is a panther and "not a bobcat, Canada lynx, lion, cheetah or catamount." Thorpe said.

Persons who have seen the animal describe it as black, about three feet long and with a tail about the same length which turns up at the end.

"The cat has small ears, walks with its body close to the ground and is extremely fast," Thorpe said. That description has "no resemblance" to bobcats or other animals native to Connecticut, he said.

Fare: Game

The panther is common in southern Asia. Large males may exceed 100 pounds. They feed on antelope, deer, sheep, goats and birds and are known to have a special fondness for dog flesh.

Older leopards and panthers may become man-eaters when they are no longer able to catch animals.

In recent weeks, marauding catlike animals have been blamed for killing cattle and pheasants in the Southbury-Oxford area, attacking two Great Danes in Harwinton and leaping from a tree to attack a horse ridden by a Southbury girl.

Sunday night an animal attacked Edward R. Sanford of Thomaston, scratching his back and chest and tearing his clothes.

Monday morning a donkey was killed at a farm owned by Jerome Shaw of Southbury. State Police said the animal was hamstrung—crippled when the hamstring muscle in its leg was torn by the attacker.

Thorpe urged caution in attributing all the reported incidents to the panther or any other single animal.

"When something like this happens," he said, "everyone reports seeing the animal. Most of these reports turn out to be erroneous," he said.

Granby Cat?

Thorpe said the situation was similar to the "Granby Cat" incidents in the summer of 1959. That animal was also described as a panther or catamount, but was never found.

"At that time," Thorpe said, "people saw the Granby Cat everywhere."

Thorpe said he doubts the Great Danes in Harwinton were attacked by the panther. The horse presumably attacked in Southbury "bore no scratches or marks" he said.

Board officials and state police continued their search for the animal Tuesday. Police were equipped with riot guns and were aided in the search by hunting dogs.

Big Black Cat Spotted by Farmer in Hartland

Hartford, Connecticut, *Courant*, November 2, 1962

By Joseph A. O'Brien

East Hartland—A farmer of the Washington Hill section told State Police Thursday morning he had seen a large, black catlike animal in a pasture near his home. State Police searched the area but found nothing.

Theodore Le Geyt, 60, said he saw "some kind of animal other than a dog" from a kitchen window about 7:30 am. In recent weeks, a large black cat has attacked residents and livestock in southwestern Connecticut and has been tentatively identified as a panther.

The East Hartland area where the animal was reported is about 35 miles overland from the Woodbury-Southbury-Oxford area where a similar animal has been reported. The last report from that area was Monday.

Heifer Missing

Le Geyt told State Police the animal came from the direction of a pasture where nine heifers are kept on the farm of his nephew, Dwight Le Geyt of Le Geyt Road. The heifers were counted but none were missing or injured. Dwight Le Geyt said about six weeks ago a black angus heifer disappeared from his pasture and has not been found.

The area is not far from Granby, where sightings of a panther or catamount were reported in 1959 but never verified, and cattle killed.

Theodore Le Geyt said he saw the animal, "came up across a brow of a hill," 500 feet from the farm house. He said it moved swiftly "with a trotting stride, but a different kind of a trot than a dog."

He described it as "jet black" about three feet long and "all of two feet high."

La Geyt said the animal moved toward him at first and he couldn't see the tail. Then it turned and "covered a lot of ground" as it disappeared over the hill on a far side of the pasture. He said the animal had a tail that seemed more than two feet long.

Trooper Victor Zordan of Canaan Barracks searched the pasture area and nearby woods but found no tracks or trace of the animal. The farm is about 10 miles from Pleasant Valley where a resident reported seeing a catlike animal weighing close to 100 pounds in his garage one night in May of 1960.

CAT'S MEOW FAILS TO HALT COLEBROOK WEED STALKER
Hartford, Connecticut, *Courant*, November 12, 1962
By Joseph O'Brien

Colebrook—A Winsted couple told State Police late Sunday afternoon they saw a large black cat in a field off Pinney Road here. They watched it for more than five minutes as it crept through grass and halted at times to watch them.

"I've never seen anything like it before," said Gus Carnevalini of Wakefield Boulevard, Winsted. "It must have been a panther. It was definitely a cat."

Carnevalini and his wife were out for a drive Sunday when they saw the large cat-like animal in a field behind a snow fence about 4:30 p.m.

Behind Fence

They stopped their car and watched the animal that was lying behind the fence about 150 feet from the road. The field is about a mile from Rt. 44. Residents of the area said it is often frequented by deer at night.

In recent weeks, a large black cat reportedly attacked residents and livestock in southwestern Connecticut and has been described as a panther.

On Nov. 1, a farmer of East Hartland reported seeing a large, black cat in a pasture near his home. State police checked that area but found nothing.

Identity Questioned

Theodore Bamptom of Bethlehem, district supervisor of the State Board of Fisheries and Game, said Sunday it is "unlikely" the animal seen here is the same one reported in the Southbury-Woodbury area in October.

Mr. And Mrs. Carnevalini said the black cat they saw Sunday was about the length of a German Shepherd dog but "low to the ground and long."

Animal Startled

Carnevalini said the animal was lying in field grass near a vacant house and as he slammed his car door, it moved. "It dragged along in the grass," he said. "Its tail looked about a foot long."

Mrs. Carnevalini said when her husband made a noise like a cat growling, the animal stopped to look at them. As they watched, it moved across the field and disappeared at a stone wall.

No Sleep

The couple said at first they did not plan to report the sighting because of possible disbelief, but Mrs. Carnevalini called State

Police when they arrived home because feared the animal might harm children of the area. "I won't sleep tonight," she said.

State Police said they would check the area today.

FACT OR FANCY? BLACK, SLINKY 'PANTHER' BEGINS FALL PEEK A BOO
Hartford, Connecticut, *Courant*, October 23, 1963

By Joseph O'Brien

Norfolk—It's been seen again—the slinkey, black panther-like animal—or a reasonable replica.

A Norfolk resident said Tuesday he saw a low, black, catlike animal with a tail nearly three feet long on Rt. 44 Monday night.

A year ago, a large black cat reportedly attacked residents and livestock in Southwestern Connecticut and was described as a panther.

Irving Christinat of West Norfolk and his wife were driving east on Rt. 44 west of Norfolk Monday night when an animal "big as a good sized dog, solid black with a long tail, ran across the road. I never saw anything move so fast in my life," said Christinat.

His story prompted Joseph Veneziano of Hubarek's Market here to say he had also seen a similar animal in the North Norfolk area several months ago while delivering groceries.

But Veneziano said he "didn't want to say anything" at the time. He said the animal was "short to the ground" and black with a long tail.

On Nov. 11, 1962, a similar animal was reported seen in a field in nearby Colebrook about five miles from the Monday night's sighting. At that time, Mr. and Mrs. Gus Carnevalini of Wakefield Boulevard, Winsted, said they had seen a large black cat in a field and watched it for five minutes as it crept through grass and halted at times to watch them. They said the animal was about the length of a German shepard dog but "low to the ground and long."

Christinat said Tuesday the animal he saw ran in front of their car about 15 feet ahead on Rt. 44 and headed for the Backberry River.

"It was just a streak," he said. "Big and fast with a heck of a long tail."

DELAWARE

The Fence Rail Dog has been noted in cryptozoological literature (Newton 2005, for example), but appears to just be a canine spook story from the Frederica region of Kent County. (Portsmouth, NH, *Herald*, July 19, 1952)

GUARD COMES FACE-TO-FACE WITH TIGER

Huntingdon, Pennsylvania, *Daily News*, September 6, 1960

Wilmington, Del., Sept. 6—A guard at the Hercules Country Club reported he came face-to-face with a large animal early today on the club grounds and state police said that from his description it appeared to be a tiger.

State Police sent six men to the scene to begin a search and asked local radio stations to broadcast a warning to parents to keep their children indoors.

Donald Steen, a guard at the club owned by the Hercules Powder Co., said he met the animal face-to-face as he came around the corner of a building at 2:29 a.m. He shone his flashlight in the animal's eyes and backed away.

Steen said the animal turned away for a moment but then started toward him again. Steen ran into a nearby building and watched through a window as the animal walked away.

Two state troopers from the Penny Hill Barracks, investigating an accident Monday night, later reported they saw a large animal walk across the highway about a mile from the country club. They believed it was a dog.

Delaware state police asked Pennsylvania state police to check zoos and traveling shows to see if any tiger-like animal is missing.

Early today an unidentified man who lives near the club telephoned the New Castle state police barracks and reported he heard a wild animal "holler."

FLORIDA

Lion Reported Roaming Pinellas Wilds
—Hunter Organize—
Circus Cub Escaped Years Ago, Is Story

St. Petersburg, Florida, *Evening Independent*, May 27, 1935

Clearwater, May 27.—Big-game hunting, similar to that indulged in by sportsmen in Africa, promises to become one of the leading outdoor sports in middle Pinellas this summer, judging from conversations overheard in the sheriff's office at the court house.

Sheriff E. G. Cunningham said that reports had come to him of an animal resembling a lion that had been haunting the woods in the vicinity of the Crosley grove, a couple of miles east of Dunedin, and he had an idea that he might join a party intent on the capture of the beast the first chance he got.

Those familiar with the fauna of the sub-peninsular express the opinion that the animal seen a short distance from Clearwater is a panther, one of the family that once roamed the hammocks of Pinellas. Others insist, however, that the tracks of the animal are different from a panther's, being about 27 inches apart and unlike those left by the big cats of this last frontier.

The animal, whatever it may turn out to be, has been seen on several occasions near a dense bayhead not far from the county warehouse used for storage of machinery and road material in the commissioner's district No. 4. It was sighted two or three weeks ago by a party of young men hunting alligators in a pond. When gunfire began, the animal was heard to snarl as he leaped over the scrub on the way to his home in the bayhead.

Appearance of the strange animal has started a lot of stories on their rounds. Louis Kearney, who has hunted everything from trouble on a phone line to a panther, remembered about hearing that a lion cub escaped from a circus exhibiting at Tarpon Springs years ago and was never recaptured. This might account for the remarkable tales told about the Dunedin beast—but Sheriff Cunningham will probably help to solve this mystery at the first opportunity.

WOOD, FIELD AND STREAM:
20-FOOT RATTLESNAKE AND A BLACK PANTHER
BLOWN IN BY WIND IN EVERGLADES
New York Times, March 23, 1958

By John W. Randolph

Flamingo, Fla., March 22—Strong winds blew out the fishing here today but blew in a twenty-foot rattlesnake and that old American favorite of song and story, a black panther.

Nobody could fish except a couple of Yankees who insisted on trying the Bay of Florida, and they didn't catch anything except a few piddling redfish. Three gentlemen who shall be nameless here sat for a couple of hours in the blast on a nearby canal and took a couple of spotted weakfish and three or four jack crevalle.

Otherwise there were just too many waves for anybody except people who crave waves. The result was that the milkshake bar at Flamingo Lodge got a big play. The wind therefore is responsible for the unique rattlesnake and the black panther.

I have investigated both creatures carefully and have affidavits from all sides to prove that they exist. At least I could get affidavits, probably, if it were necessary.

The twenty-foot rattlesnake has been seen by at least twenty persons hereabouts and all of them have reputations as implacable non-drinking prohibitionists, they said.

Ernest Christiansen, the chief naturalist of the Everglades National Park, in which this fishing establishment operates, has been checking the story for months. He says rattlers grow big down here, but are not supposed to be that big. Yet he has too much testimony on this one to disbelieve.

Mrs. Bob Knight, the wife of one of the owners of the place, saw the snake on a road running along the Coot Bay Canal and says it was twenty feet long and as thick as a railroad tie. She said it reached clear across the road.

This set her husband on fire at the time because as a boy outside of Miami he had run a snake farm. He went out to find this snake and kill and skin him, but the closer he got to where she had seen it the less he believed the story. By the time he got there he didn't believe it at all and didn't bother to go into the thick mangrove brush after the serpent. He wasn't afraid to go into the brush: just didn't think the snake was there.

Capt. John Scudder of Homestead, a charter skipper here, saw the snake on the same road and said it was twenty feet long and as thick as a log. He was in a weapons carrier but didn't dare run over such a big snake with it. Picked up a big chunk of rock from a nearby excavation and dropped it on the snake. Said it didn't bother the snake. Doesn't want to see it again.

Christiansen says rattlesnakes more than ten feet tong have been seen in this part of the southern Everglades many times. No rattlers have been seen on the land, filled in by the Park Service, on which the Flamingo establishment stands.

But the black panther has been seen on that land. Biologists, naturalists and other austere scientists say there are no black panthers in America. But that is negative stuff and no red-blooded American boy will believe it.

One who has seen this black panther is Mrs. Stella Creer of Liverpool, England, a waitress in the Flamingo's fine dining room. It stood in the headlights of her car, not more than eight feet away, then strolled off casually. Mrs. Creer looks you straight in the eye to say this, and everybody knows that Britons never shall be slaves.

Christiansen testifying again: This part of the Everglades probably has more panthers, or mountain lions, or cougars, than any part of the United States except Arizona. There are more deer in the mangroves here than anybody realizes, and panthers love deer. They also eat rabbits, and there are plenty of rabbits. They are very fond of devouring bobcats, and bobcats abound. They esteem raccoons as Mississippians esteem corn bread, and there are raccoons here until the imagination reels.

But there are not supposed to be any black panthers—just as there are not supposed to be any around Bear Mountain, N. Y., or middle

New Hampshire, to name just two places where they have been seen by men of faith recently. Christiansen is beginning to wobble; it is plain that he thinks there may be a black panther around here.

If the wind blows a couple of more days here there won't be any question about either the black panther or the twenty-foot snake. It is a rare soul who can challenge universal convictions, and this is a country of strong opinion.

BLACK PANTHER PROWLING NEAR MIAMI?

Fort Pierce, Florida, *News-Tribune*, August 11, 1958

Miami (AP)—A black panther reported on the prowl just north of Miami.

Mrs. Amos Wells, 26, who lives in a jungly area, says she saw it her yard Saturday night. It was "black and shiny, big as a Collie," she reported.

Mrs. Wells said she was washing clothes behind her house when she heard a sound "a cross between a cat's mew and a baby's cry."

"I turned," she said, "and there this thing stood. Then it screamed and I screamed and ran in the house."

A boarder, J. P. (Tex) Ramsey, came out with a shotgun, heard rustling in the bushes and leveled a blast into the darkness.

"Then on Sunday morning," he said, "I found where some big animal had pushed through the sawgrass."

A neighbor, Mrs. Charles B. Corby, reported she has been hearing a panther screaming at night for some time. "I've lived here since 1918," she said, "I've seen black panthers and heard them and I know this is a panther. They scream just like a woman in pain."

The State Game and Fresh Water Fish Commission and the Humane Society promised an investigation.

Tom Shirley of the game commission said, "if it's a panther, it's dangerous. When a panther gets that close to people he's hungry.

"Of course it may not be a panther. People's eyes sometimes play tricks."

Researcher Brad LaGrange (2000) noted a black panther sighting from near Ft. Meyers sometime between 1959 to 1961. Brad's aunt and uncle were driving along a private overgrown road in the scrub forests when a large black panther crossed the road in front of them.

Brad LaGrange (2001b) spoke with a witness who claimed she woke one night thirty years prior, as a teenager, to look out her window to see several armadillos in the 4-5 foot range digging around in the shrubbery. They looked just like the smaller armadillos, with long narrow ears, round armored backs, and tails. She was sure they were not wild hogs. This was in a small town in the Tampa Bay area of Florida, and her block was "surrounded by woods and old orange groves." The animals came back several times over the next few weeks, and her mother and sister also saw them, but they didn't realize there was anything unusual about them.

BLACK PANTHER SIGHTING REPORTED
Sarasota, Florida, *Herald-Tribune*, May 2, 1978
NO BLACK PANTHER SIGHTED
Sarasota, Florida, *Journal*, May 2, 1978
Deputies and Animal Control officers searched for a reported black panther near Canary Street, close to the ballfield on Route 72. Deputies said panther reports weren't uncommon in that area. No animal was seen by the officers.

MYSTERY CAT SEEN ON ROAD MAY BE A RARE JAGUARUNDI
St. Petersburg, Florida, *Times*, December 8, 1984
Pasco County Sheriff's Department suggested that a jaguarundi was responsible for a report that came in from a motorist who struck a "cat that stood about two feet tall, had a very long tail and was 'almost black.'" The accident was on County Road 54 in Land O'Lakes, and the cat didn't appear seriously hurt as it disappeared into the woods. The Florida State Museum in Gainesville hoped to compare fur found on the car's bumper (collected by a deputy) to samples in their collection. The article notes that naturalist Wilfred T. Neill disclosed in a February 1980 *Pasco Times* column that A. Hyatt Verrill released jaguarundis in Florida near his Chiefland home. (This clears up one mystery for biofortean researchers who had seen Neill's previous claim that a famous writer released jaguarundis, but without releasing the writer's name. Verrill was one suspect discussed by Lisa Wojcik (n.d.).)

Many Theories Offered for 'Goat-Grabber'
Gainesville, Florida, *Sun*, July 16, 1989

An unknown animal carried off over 30 baby goats over a period of time from an Alachua County farm. Finally, Ellen Hodge caught a glimpse of a "panther-sized cat with a long tail" walking toward her nervous goats. She thought it might be a Florida panther, though it would be far north of its recognized range, while others suggested it might be a jaguarundi. A local biologist noted that jaguarundis prefer smaller prey than goats. Another farmer nearby had lost goats, but thought it more likely that bobcats, coyotes, or dogs were responsible.

Panther, Dog or Jaguarundi? Animal Sightings Spark Rumors
Ocala, Florida, *Star-Banner*, November 15, 1993

An Interlachen resident reported a "big cat" with a long tail and a height of about three feet. He blamed it for the death of a neighbor's cat. Officials said it wasn't likely to be a cougar or panther, though it might be a jaguarundi. A deputy found tracks with claw marks, suggesting it was a dog.

Boy Runs from Wild Animal That Could Be a Jaguarundi
Daytona Beach, Florida, *News-Journal*, June 3, 1998

A Deltona boy said he was chased by a large dark feline he thought was a panther. The animal control officer was skeptical, but after speaking with the boy thought it a credible story, though not likely to be a Florida panther. He went out to look and got a glimpse of a long feline with a long tail and dark gray coat, about 60 pounds in weight. He suspected it was a jaguarundi.

GEORGIA

A Strange Animal

Atlanta, Georgia, *Constitution*, October 17, 1882

From the Sumter, Ga., *Republican*.

Policeman J. W. Cobb, while fishing in the Muckalee one day last week in the rear of Ed Littleton's place, was startled by the appearance of a large animal on the opposite side of the creek which came out of the jungle of the swamp. He says that it was about three feet high, six or seven feet long, with a short tail and spotted all over like a hyena. The animal looked at him, stretched itself on its legs, opened its mouth which was armed with formidable tusks about as long as a man's fingers. The animal went behind a tree and peeped at Cobb for a few seconds and then disappeared in the canebrake, and Cobb made tracks for the city at race horse speed, although he declares that he was not scared a particle, was as cool as he was when he captured General Walker. He says that he does not know what kind of an animal it was, at first he thought it a tiger, then a hyena, or a catamount, and it might have been a leopard.

The Chase of a Night

In Which the City Marshal Acted as Leader

Atlanta, Georgia, *Constitution*, February 15, 1889

LaGrange, Ga., February 14.—[Special]—A strange creature made his appearance in the vicinity of LaGrange several weeks ago. The negroes grew excited over its supposed "dog eating" propensities, as many of their canines disappeared. The excitement finally spread among the white citizens, since an attack was made last week on Mr. O. A. Dunson's hogs. Several hunting parties have been

After the Supposed "Hyena,"

and on Thursday night last it was started, by our city marshal and a few others, on the square near a beef market. A pack of the best fox dogs in the county ran the animal for miles, and finally lost the trail. Since then the wildest rumors have been afloat, and the negro population have made it convenient to stay in doors at night.

Another Chase.

It was reported last night that the "hyena" had again attacked Mr. Dunson's hogs, when our city marshal and others quickly called together the pack of bounds and started for the scene, one mile from town. Only a few minutes sufficed to strike a hot trail, and then excitement began. The whole town was aroused from its slumbers by the yelping of dogs and firing of guns and pistols. Our efficient marshal led the chase on horseback, while others followed in hacks and on foot. Many of our old citizens brought out their muskets and horse pistols and run their breath out to keep up with the dogs. The chase grew more and more exciting, and at last the hounds came to a stand still in a skirt of woods back of the old Ben Hill place. The dogs had

"Treed" the Varmint,

and on dashed our gallant city marshal on horseback, with cocked revolver in hand. He was the first man on the spot, and just as he was about to annihilate the varmint that had caused so much commotion and excitement in our city, he discovered a fox skin hanging to a limb.

A few mischievous boys had given the joke away before the marshal got back to town and now it is not safe to mention "dog-eater," "loena" or "hyena," as the varmint had been variously called. Much relief is felt that the varmint has at last been discovered, and the boys will have to hunt up a new trick to have any more fun with the city marshal soon.

A TIGER LOOSE IN GEORGIA

Hartford, Connecticut, *Courant*, March 20, 1890

Savannah, Ga., March 11.—Tomorrow will be a great day in Bulloch County. The people of the little town of Easton and vicinity will go on a tiger hunt.

The animal that is at large is a fine specimen. For years he has viewed mankind in captivity, and peered forth upon the world from behind the iron bars of a museum cage. Recently the little show of which he was the crowning attraction, passed by the borders of Bulloch County. In an evil moment, through carelessness or malice aforethought, the tiger was allowed to gain his freedom. From the canvas to the wild delights of untrammeled liberty in the Georgia pine forests was but a few bounds. The tiger is old. He is perhaps nearly toothless, but these two facts have not allayed the public fears. All the women and children in a circuit of ten miles are afraid to stir out of the house. On Tuesday, a young man riding across an unfrequented field met the tiger face to face. He turned, spurred his horse fiercely, and went flying toward home and safety with the tiger in pursuit. The animal he rode was of good stock, and, finding that he was being distanced, the tiger let out a roar of anger, and, turning, retreated to the forest. So far, no human lives have been lost, but the tiger has committed considerable depredations among livestock.

THIS IS A "WOOG"

A NEW AND TERRIBLE ANIMAL IN JACKSON COUNTY

Atlanta, Georgia, *Constitution*, February 18, 1895

Jefferson, Ga., February 17.—(Special.)—Mr. J. G. Mauldin and Mr. S. P. Miller say that there is a strange animal near Jefferson whose home is on the headwaters of the mile branch. He is a long, keen animal with a long bushy tail, large flat feet, a small, keen neck, little ears and takes nocturnal perambulations through forest and meadow, seeking prey and devouring chickens, sheep, pigs and things of that kind. He has frightened the people badly over about Mr. Mauldin's and they are almost afraid to venture out at night for fear they will be attacked by this strange varmint. He has frightened the dogs so badly in that community that Messrs. Mauldin and Miller say they can't get them to even hunt a partridge or rabbit. This animal makes a circuit every night and sometimes they say he even roams in the suburbs of our city. The varmint is a strange thing and Billy LeMaster has named it a "woog."

It has even reached the ears of the people of Monroe that there is a "woog' about here and next week Charley Felker is coming up with a large pack of hounds and one night next week the whole country will turn out to hunt him. Mr. Miller says if they succeed in capturing him that his hide will be stuffed and sent to The Jackson Herald office for exhibition.

A Strange Animal
That is Creating Terror Near the Town of Comer
Atlanta, Georgia, *Constitution*, July 6, 1896

Comer, Ga., July 5.—(Special.)—A report reaches here from Five Forks that a very frightful and ferocious beast was seen by some children in a field near that place last Tuesday evening, and that their dog has not been seen since a combat with it. Late in the evening it was seen again and heard in a low undertone howl near Harry Foot's house. Will Gholston, a colored man living at Five Forks, claims that there was something very unusual in a swamp near Harry Foot's house as he passed there about dark Tuesday night. Its track is very much like that of a baby foot, except there is an imprint of sharp toes.

In a path near that place has been found the blood, horns and a few bones of a goat. As there are no goats near that place it is supposed that this animal brought its food from a pasture near Mr. Jim Faulkner's, where it was seen the morning before, and where several pigs have been destroyed.

A crowd of men were on the hunt with guns, lights and dogs Tuesday night late, but the dogs would not do more than track to the edge of a big swamp wherein it is supposed this animal at the time was stopping.

A cow which has lately been running in this swamp came home not many days ago with the flesh on her shoulders, one of her ears and flanks badly torn, having lost a great deal of blood.

BULLOCH ALL AGOG OVER HER MYSTERY

STRANGE ANIMAL CHOOSE HEADS OFF OF DOGS
AND FRIGHTENS PEOPLE—ELUDES CAPTURE

Macon, Georgia, *Telegraph*, April 14, 1919

Statesboro, April 13.—Bulloch County's "wanderer," in the nature of some unknown wild animal, has evidently moved his place of habitation from the vicinity of Brooklet to the outskirts of Statesboro. For weeks this mysterious animal visited the little town of Brooklet, literally chewed off the heads of dogs and frightened citizens, but for several months it had not been heard of, and it was thought some of the shots fired at it had found their mark and the animal made its way into the swamp and died. But for the past few days an animal has been alarming the citizens about three miles south of Statesboro, and has come up as near as the McDougald farm. While no dogs have been killed, the animal has terrified some of the farmers, especially the negroes. One negro tenant while plowing observed his mule becoming frightened and upon looking around saw the animal with its front feet on the fence. "Mister," said the negro, "that mule thought he could beat this n— running, but he found out he couldn't."

Upon reporting the incident several went to the place and large tracks were found around where the negro had been plowing. Today a party of hunters went out to see if they could find the animal. The dogs did find the track of something, but what it was is not known, as it did not come out of the swamp.

ANOTHER MYSTERY BEAST REPORTED; DEVOURS DOGS

Atlanta, Georgia, *Constitution*, June 14, 1923

Chipley, Ga., June 13.—(Special.)—For the past week a beast of mystery has kept the negro population of the little mountain town of Chipley terrorized.

It was seen by a number of reliable negroes, as well as several white citizens. So far, it has confined its nocturnal attacks to dogs. More than a dozen of these family pets have been slain and devoured by the animal.

The tracks of the beast are somewhat similar to those of a dog but are very much larger. Those that have been favored with a view of the monster declare it is much larger than a dog, and travels with an entirely different gait.

Night after night during the past week it has appeared near first one house and then another, attacking and eating up the dogs on the premises. On Monday night it was shot at by J. T. Cands, a farmer living several miles west of Chipley. He was not certain that either of the shots took effect.

Last night the town marshal and some other prominent citizens sat up and watched in vain for his appearance, so it is possible that Mr. Sands fatally wounded it. The darkies all hope so, any way, for they are tired "staying in o' nights."

Killer Roams in Darkness

Long Beach, California, *Independent Press*, April 27, 1958

Atlanta (UP)—A mysterious beast that cries like a woman screaming, mauls dogs and livestock and gets whole towns in an uproar is on the prowl again in the South.

It has unbelievable speed, leaps high fences and picks lonely, gloomy spots for its ferocious kills. Its howl in the darkness, like that of the Hound of Baskerville, sounds like the voice of evil.

The mystery beast is heard more often than seen. But no one doubted that three dogs found killed near Cedartown, Ga., were victims of a huge vicious beast. Then a woman's bulldog was dragged from her yard and killed.

It seemed clear that the mysterious beast that has struck in South Georigia, Bladenboro, N. C., and near Yemassee, S. C., in the past five years had come to Polk County.

David Jones of the New Harmony Community said his dogs bayed the beast in a ravine and his light revealed— "You wouldn't believe me if I told you"— something like a panther.

Soon Nathaniel Carter heard a sound "like a baby crying or a woman scorning" and saw what he first thought was a big monkey, and then decided it was a black panther.

Last week an animal resembling a "big black dog" attacked a motorist near Rome, Ga.

The beast leaped against the side of her car and left muddy pawprints, she said.

Some observers believe the beast, wherever it turns up, is a hybrid of bobcat, panther or whatever—and imagination.

The "Edison Ghost," the big shaggy beast that whizzed by a farmer five years ago in southwest Georgia, was something different. This beast, which ran at incredible speed and leaped the highest fences, kept posses busy for days. Finally they caught a long haired goat.

MYSTERIOUS BEAST LOOSE IN GEORGIA
Charleston, West Virginia, *Daily Mail*, May 2, 1958

Atlanta (AP)—The mysterious beast that has frightened residents of Atlanta's northwest section is still a mystery today.

Hundreds of policemen and civilians, some armed with shotguns, searched a wooded section last night after residents reported seeing a critter that was big, black and mean.

Several hunters maintained that only a panther could have made the four-toed tracks that were found. Police said they measured 4 ½ inches in diameter.

But patrolman Walt Jones, who used his bear-hunting dogs in the search, said in his opinion there was no panther in the area.

He called off the dogs at 12:30 a. m., soon after they treed a black house cat.

While the hunt was in progress residents of the section kept children indoors and some maintained an armed vigil. Several shotgun blasts were fired at shadows and objects moving across yards.

One member of the hunting party, Eugene Pyle of suburban Hapeville said he flushed the beast from a creek bed and got one shot at it. However, he said, so many persons were in the area he was afraid to chance a second shot.

Several persons who saw the critter said it walked with a noticeable limp, indicating that two policemen who fired at it Wednesday may have scored a hit.

The mystery beast has been reported seen in various places in northwest Atlanta in recent weeks, leaving in its wake dead chickens, dogs, rabbits and pets.

Shuker (1989) noted a witness who lived on a Savannah-area farm in the 1960s. Her family knew there was a large feline living in the woods nearby, as they had seen it several times. After a few attempts to lure it in, they eventually managed to shoot it, but apparently didn't keep it. It was described as the size of an Alsatian, and bluish-black in coloration, but medium-sized dark spots could be seen on the coat.

Newton (2005) noted that a hunter, James Rutledge, claimed to have killed a black panther in the Stockbridge area in 1975, but refused to say where he had buried it.

Shuker (1989) noted a witness, Gerald Cameron, who said he had recently seen a black panther several times, near Dean Forest Road, Savannah, and near Arnesville. Cameron stated it was 4 to 4 ½ feet long, 2 ½ to 3 feet tall at the shoulder, and may have weighed 100-140 pounds. He described them as inhabiting the remote woodlands, and very shy.

Godfrey (2006) recorded the 2005 case of an erect creature with wolf-like features from a Georgia swamp "about seventy miles south of Savannah, about twenty miles inland from the coast."

Woman Reports Large Black Cat Near Grovetown
Augusta, Georgia, *Chronicle*, October 1, 2008

A woman reported seeing a 100-pound black cat near the pond at the John Deere plant near Grovetown, Columbia County.

Two Say They Saw Panther in Oconee
Athens, Georgia, *Banner-Herald*, February 20, 2009

A bus driver and a brush-cutter claimed to see a black panther in the area between Bell's Lake and an elementary school. Some small animals, including a calf, had been reported missing in the area.

LION SIGHTING REPORTED

Gainesville, Georgia, *AccessNorthGA.com*, June 9, 2009

A motorist reported seeing a lion standing on the shoulder of Dahlonega Highway in Hall County near the Lumpkin County line. Deputies were unable to find a trace of the animal, but the driver stood by his statement.

HAWAII

While not technically part of North America, it would be a shame to leave out Maui's 2003 "black panther" scare. Obviously, there are no native wild felines in Hawaii, but there had been stories of a big cat circulating for several years on the island of Maui. In 2003, several individuals in the lower Olinda section of Maui claimed to have seen a large black feline. Large paw prints, clawmarks on trees, and dead deer in the area suggested there was in fact, something roaming the forest, but state wildlife officials were unable to trap it. A 60-lb. Labrador retriever survived an attack with wounds that were "consistent" with an attack by a big cat.

Exactly what the animal was became the focus of speculation and rumor. Officials had to point out that, despite rumors, there had been no circus train wreck in the area. Some suggested that a melanistic jaguar or leopard was the culprit, brought to the state illegally as a cub, which was released or had escaped. One official said it was more likely to be a smaller cat, like a jaguarundi.

After several months of sightings, the state brought in an Arizona Game and Fish Department big-cat expert, Bill Van Pelt, who spent some time on the search. While he was unable to catch the animal during his visit, he left convinced that there was a big cat of some sort on the prowl. Some residents were still unconvinced, saying that roaming pit bulls were responsible for the deer kills and that local black Labradors might be mistaken for a panther. While fur was found, it was not a high quality sample, and the results of DNA testing kept getting put off by the lab for "higher priorities" (Hurley 2003a-n, 2004; Kubota 2003a-c, 2004; Manson 2003).

It appears that the creature, whatever it was, disappeared after 2003. The Department of Forestry and Wildlife's mystery cat website (http://state.hi.us/dlnr/dofaw/wild/) was never updated after September 2003. The last sighting posted there (September 25) gave the following description:

"People driving on a road in Olinda just a quarter-mile downslope of the previous day's sighting, saw the animal crossing the road at about 5:30 a.m. They turned around and went back where again saw the animal.

"That same evening, at about 8:50 p.m., in the very same location, the cat crossed the road going the opposite direction. Two women passing in a car saw it from about 7 feet away. The described seeing an animal about 7 feet long with a long tail, black coat and yellow green eyes, a flat face and ears not prominent. A DLNR wildlife official responded to the call within 40 minutes and spent the next hour and a half searching the area with infrared equpment. Although no cat was spotted, nor any wild deer, animals in a pasture we seen grouped closely together."

Sightings were reported in the newspaper through October 2003, with Van Pelt noting that in an area as rugged as Olinda, small as it was, a large cat would have plenty of food, water, and places to hide for a long period. A second expert, visiting in October, agreed that a big cat was responsible for the sightings. When sightings dropped, however, state officials decided to call off the hunt in November until better evidence was turned in. One Utah State University professor, a part-time resident in Hawaii, gave a presentation at a wildlife conference comparing the sightings to the "Loch Ness Monster."

A sighting of a big black cat was reported in January 2004, in the Wailea area of South Maui, 15 miles away from Olinda, but a state biologist thought that a misidentified dog was likely responsible.

IDAHO

Ogden City, Utah, *Standard,* January 22, 1909 (excerpt)

"Mr. Briedenstein is making a good record for himself," said Mr. Kniepp, "and the stockmen of that section appreciate it. A few weeks ago he killed a strange animal which has not yet been classified by either himself or the oldest inhabitant thereabouts. It was an animal considerably larger than a gray wolf, its pelt being more than five feet in length. The sides and legs of the beast resembled a coyote, but the feet were those of a black Newfoundland dog. There was also a heavy strip of thick, curly black hair down the back and the same was also scattered over the body to some extent. The tail, however, was neither dog's nor coyote's, being only a few inches in length and covered with short hair. The animal was skinned with a view of having it mounted, and the pelt and skeleton has been forwarded to the biological survey for identification and mounting. It attracted much comment from everyone who examined it."

FREAK WOLF IS A RARE SPECIMEN
OFFICIALS OF BIOLOGICAL SURVEY EXPRESS OPINION.
STRANGE CREATURE OF ENORMOUS SIZE AND CURIOUS
APPEARANCE—NOT A HYBRID.

Ogden City, Utah, *Standard*, April 8, 1909

The district forest officers have received confirmation of their belief that Forest Supervisor E. Grandjean's freak wolf which he recently shot within the Boise national forest and subsequently forwarded to the biological survey at Washington, D. C., is indeed a rare specimen of quadruped.

The strange creature was of enormous size, for a member of the wolf family, with its back and other portions of its body covered with a heavy growth of black hair, resembling somewhat the coat of a Newfoundland dog, except that it was heavier and coarser.

The most curious feature in connection with the animal's appearance was the fact that it was "bob-tailed." Old hunters and trappers in that part of the country, who examined the beast, stated that they had never heard of nor seen anything like it in all their experience. Even the Indians in that region were unfamiliar with the species.

It will be recalled in connection with the early reports published upon the return of the Lewis and Clark expedition that Captain Clark mentioned several times the discovery of rude Indian drawings upon rocks and ancient skins of well known animals of the Rocky Mountain region: also a few that were not known to the people of that time. Among these was a giant dog wolf of ferocious appearance and enormous size. A curling shaggy mane was represented as extending down the back of the animal and in one instance it was pictured bearing a young deer in its jaws, illustrative of its size and strength.

It is possible that the specimen mentioned above is one of the rare descendants of an almost extinct species of wolf which once infested the Rockies, terrorizing the Indian inhabitants and remaining long in their traditions and picture records.

The officials of the biological survey at Washington, D. C., express their opinion that the animal who's skin and skull was sent by the supervisor of the Boise forest, was not a hybrid and state that the only specimen that resembles it at all closely is one which came from the Priest river forest in northern Idaho. They are anxious to secure further skulls and pelts and offer a good price for same.

ILLINOIS

It Must Be A Tiger
A Wild Beast Which is Terrorizing McDonough, Illinois
Decatur, Illinois, *Morning Review*, June 9, 1891

Lewiston, Ills. June 8—The piteous screams of a horse in the stable brought Frank Chatterton, a farmer of Bernadotte, to the scene Sunday afternoon. As he entered the lot a huge beast sprang from the stable door and, after bounding into an adjacent field, crouched low, uttering deep growls, while its long tail waved slowly to and fro. Chatterton was horrified, and, fearing either to advance or retreat, gazed helpless at the big beast, which in a few moments slunk away into a patch of timber.

Feeding on a steer's carcass

The horse lay on the stable floor weltering in a pool of blood which flowed from a dozen wounds. The animal's sufferings were soon ended by a ball from a rifle. Chatterton declares that the brute he saw is not a panther and that he believed it to be a tiger. The excitement is augmented by the report that John Hulvey, residing some miles from here, came across a large animal in his field Thursday, which was feeding on the carcass of a steer. The animal was disposed to show fight, and Hulvey retreated in haste. The animal's screams have been heard and its huge tracks have been found in the Spoon river bottoms.

Terrorized the County

for the last three months McDonough County has been terrorized by this creature. A party of hunters surrounded the brute in the crooked Creek bottoms a few weeks ago, but the dogs would not attack it. The men caught a glimpse of the animal, and were so frightened that they gave up the chase. About three years ago a menagerie, while crossing the crooked Creek bottoms, was caught in a storm. A cage containing a tiger was overturned, and the tiger escaped. This is believed to be the animal which is now terror rising this section.

TERRORIZED BY STRANGE BEAST

LARGE WILD ANIMAL FRIGHTENS FARMERS WEST OF MILLSTADT

Belleville, Illinois, *News Democrat*, August 7, 1914

A strange wild animal is at large on the farm of Jacob Dannehold on the Columbia road. Some say it looks like a lion and others have not been able to get a close enough look at same to describe it. Mr. Dannehold says one night last week he heard a commotion in his yard and got up to see what was going on. Some large animal disappeared in the darkness when he appeared and the next morning he found that one of his hogs had been attacked. Mr. Zoeller, a neighbor, relates that he got up one night when he heard a noise and saw a strange animal on the bank of his pond. At first he thought it was a calf, but its actions induced him to get his gun. He fired two shots at it and the animal ran, emitting a howl. Mr. Zoeller would not venture out again. John W. Rodemich also reports that while unhitching his three mules from the plow one evening near the timber the animals scared so that he had considerable trouble to manage them. Upon looking about he saw a strange animal jump into the bushes.

The Dannehold timber is a good place for wild animals to hide during the day, as there are numerous sinkholes into which they can crawl for a considerable distance and would be hard to find. A number of farmers and citizens are contemplating a hunt for the animal.—Millstadt *Enterprise*.

LEOPARD OR WILD CAT OR PANTHER OR LYNX, WHICH?

Alton, Illinois, *Evening Telegraph*, February 13, 1917

A party of young people while skating on the pond on the Ike Parsell farm near the Grange Hall on last Monday evening were badly frightened by the appearance of an unidentified wild animal. The animal came to the creek near the pond to get a drink and then came towards the young people screaming as it advanced. The animal was given possession of the premises as rapidly as the skaters could vacate. As the party of skaters left the pond they were followed for some distance by the animal.

Joe Cunningham, who, on account of having his skates strapped to his feet was the last to leave the pond, and gives a very accurate description of the animal as being as large as a shepherd dog, and of a gray color with black spots, with long whiskers alongside of its mouth.

The skaters went to the home of James Walsch near by who took his gun and dog and with the boys in the party went in search of the animal. He was soon chased up by the dog which chased it to the Phil's Creek timber. Here the dog was evidently whipped by the animal, as it came back howling and would not again take up the trail.

Chester Darr, Fred Springman and others have seen the animal during the past few weeks. The animal somewhat resembles the one which has seen east of town quite often a year ago.

Lion Hunt in Champaign County
Decatur, Illinois, *Daily Review*, June 6, 1917

Champaign, June 6—Farmers living in the eastern part of Champaign county engaged in a. lion hunt this morning. The animal was discovered about 8 o'clock and the country side was aroused. It is not known where the animal came from. The lion trainer's carnival company showing here was called to lead a search for the lion late this afternoon when it was learned definitely that a lion was at large.

Tuscola Men Fail to Kill That Lion
"Hunt" Near Camargo Monday Night Futile
Decatur, Illinois, *Daily Review*, June 20, 1917

Tuscola. June 20.—Two automobile loads of men from Tuscola drove to the vicinity of Camargo Monday night to join in a lion hunt. For several days there have been reports that the lion has been seen in the timber along the Embarrass river near Camargo, and it is supposed that it is the same animal which was reported to have been about Champaign several days ago.

On Sunday night while driving near the old fair ground west of Camargo. Dewey Parker and Glenn Hunt of this city saw the animal cross the C. I. and W. railroad track. As neither of the young men were armed, they did not stop to investigate, but almost blew the muffler off their car by opening her up in an effort to get away.

On Monday evening these two young men headed the party of men who, heavily armed, drove back to that locality. It was thought by members of the party on Monday night that the animal was caught sight of and several shots were fired "on suspicion," but the lion was not bagged.

Among those who were in the party were Lee Ingram, Dewey Parker, Roy Moulden, Frank Weathers, Glenn Hunt, Lloyd Dickinson, Nick Flesor, Smith Williams, Fritz Jacobson and others.

Lion Attacks Monticello Man
Thomas Gullett, Butler at Allerton Home, Grapples With Beast, Which Escapes Posses—Hunt Animal
Decatur, Illinois, *Daily Review*, July 14, 1917

Monticello, July 14—Residents in the neighborhood of the Robert Allerton farm five miles west of this city are terror stricken today is the result of an attack by a lion upon Thomas Gullett, butler at the Allerton mansion, on Friday afternoon.

Lion Seen for Five Days
The lion has been seen near the Allerton farm for the last five days. On Friday afternoon a party of Monticello hunters went to the Allerton place to hunt the beast.

Pounces on Butler
While the hunters were in the dense woods which surrounds the place the Butler Gullett went to the rear of the garden to pick flowers when without warning the lion sprang from a clump of shrubbery directly upon him.

....

Gullett grasped the animal under the chin and began choking it and having to defend himself as best he could. They both rolled over on the ground several times when suddenly the lion sprang away and disappeared again in the shrubbery. Mr. Gullett was able to get to the house and a doctor was summoned and the neighborhood. ... The hunters came in and formed again but were unable to locate the animal.

Badly Wounded From the Attack

Dr. ... who dressed the unfortunate man's injuries found that he was scratched about the breast and that one of his limbs was terribly lacerated. Mr. Gullett had two convulsions from the result of the wounds and fright shortly after the accident.

Tried to Bait Animal

John Phalen, the manager of the place came to the city and procured a large quantity of fresh meat which was placed in the dense timber along the paths and men were stationed there till last night. The animal did not appear. ...

Coon dogs from White Heath were put on the trail late in the evening but were unable to track the animal.

Killed Sheep

A large flock of sheep that are on the place have been the source of the animal's food supply so far as can be learned.

Seen at ...

The party of hunters did not return until after 1 o'clock this morning. On the way home from the hunt ... Lee B. ... said the lion crossed the road in front of their car and stopped just at the side. They stopped the car and got out to get a shot at it when it jumped the fence and disappeared in a cornfield.

Resume Hunt

The hunt was resumed today. It was surprising to see the larger caliber of big guns that were preferred when the matter of firearms came up and the hunters are fairly well equipped in that respect.

Same Old Lion

It is presumed that this is the lion that for the past two months has been reported east of Champaign in the river bottoms near Mahomet.

The place the animal has selected in this ... must be one of the best hiding places in central Illinois as the timber and undergrowth around the mansion ... that a man can hardly penetrate it and can see in it only a short distance.

....

Here is a chance for some intrepid hunters to come to the rescue. The field is open ... as the womenfolk of that neighborhood are badly alarmed. Mr. Gullett said that he was stooping down when the animal sprang on him.

Allerton offers $250 Reward to Slayer of Lion

Monticello, July 14—Robert Allerton has offered a reward of $250 to anyone who will kill the lion on his farm. ... Parties who will help hunt the lion are asked to come to Monticello Sunday morning by 8 o'clock—that is if they hanker after lion shooting and can shoot.

An effort will be made to cover the Allerton farm in a few days. ... A large part of the farm is dense forest and is invested with rattlesnakes and other animals.

LION DODGES 300 ARMED MEN
PICKETS ON DUTY AT ALLERTON ESTATE, NEAR MONTICELLO
Decatur, Illinois, *Daily Review*, July 16, 1917

Monticello, July 18.—If somebody doesn't kill that lion pretty soon, Monticello is going to send a hurry-up call for Teddy R.

Though pickets were on duty all day Monday on the Robert Allerton estate, no trace of the lion was found. The pickets will remain on duty day and night at the edges of the 300 acres of dense timber. Needless to say, if Mr. Lion shows himself, he will be promptly peppered. There's $250 in it for the man who brings him down.

One From Chicago?

It was reported Monday that Robert Allerton was arranging to have a male lion from Lincoln park, Chicago, shipped to Monticello in care of a keeper. It was thought that the Chicago lion might attract the other and thus effect a capture.

The lion hunt staged here Sunday will be long remembered. Hunters from Decatur, Lincoln, Mt. Pulaski, Champaign, Danville, Springfield and other cities within a radius of 100 miles, arrived in Monticello. They were armed with everything from an air rifle to an army musket of the vintage of '61.

Fifty Feet Apart

Of this number, 300 men who were armed with good guns, were selected. They were lined up fifty feel apart and started through the timber. All went one direction so there would be no danger of shooting each other. Not a shot was fired all day. The lion was not seen.

Paul Bear created some excitement by shooting off a gun accidentally. He was showing his wife what he would do to the lion and he thought the weapon was empty. It exploded and part of the shot went through the hood of a passing automobile.

Seen Again

The lion appeared within one ... of the spot at which she attacked Thomas Gollop, and was plainly seen in a field adjoining the residence by the housekeeper and two maids who were on the roof.

Operate as Army

Captain Herbert Walsh and Lieutenant J. M. Donahue, veterans of the Spanish war, and commanding local National Guard companies, were placed in command, and a strict discipline maintained. Orders were given to keep dressed to center of line and keep muzzles of guns depressed. A skirmish line of hunters, with a breadth of over half a mile advanced on the thicket. A line of communication by armed motor patrols and telephones was attempted, but once the skirmishers were swallowed in the thicket all attempts at communication failed.

Once engulfed in this mass of underbrush, the men were unable to see more than ten feet ahead and by continual shouting was the line kept intact. For nearly eight hours the posse stumbled over twisted vines and half-hidden logs. Despite this and the fact that numbers of barbed wire fences crossed, no accidents occurred nor was one gun accidentally discharged.

Find Fresh Trail

Once hopes were high as a wallow recently tenanted by the lioness was, discovered. When the posse first assembled at the picturesque residence of Mr. Allerton there were civil war veterans, with their old rifles; mere boys with smooth bore shotguns bigger than themselves; others with rifles with bayonets fixed, and some carried old squirrel guns of bygone days.

Say It's Lioness

All persons who have seen the animal declare it is a full grown African lioness, which is supposed to have escaped from a carnival company in Champaign, Ill., six weeks ago, for which a reward of $500 had been offered at that time.

Thomas Gollop, the butler, is not as severely injured as reported, having been knocked down by the beast that made no attempt to injure him further. The locality in which the hunt was made is along the Sangamon river, and rattlesnakes were encountered frequently.

One Roving Lion Only a Collie Dog

Chicago, July 16.—The "lion" reported at Ravinia, a suburb, Saturday night, by frightened automobilists, was a largo collie dog whose coat had been trimmed to resemble a lion. In the dusk the deception wan easy, but the illustration was dispelled in the cold light of day which disclosed a pink ribbon tied on his tail.

ROOSEVELT ASKED TO HUNT LION
MESSAGE SENT TO OYSTER BAY—ROBERT ALLERTON, MONTICELLO
MILLIONAIRE, SAYS HE IS POSITIVE LION IS AT LARGE
Decatur, Illinois, *Daily Review*, July 17, 1917

"To Theodore Roosevelt,
    "Oyster Bay, N. Y.
"You are invited to come to Decatur immediately and hunt the big lioness which is roaming through the 300-acre timber on the

Robert Allerton estate near Monticello. If you come, a big hunting party will be organised under your direction."

This message was sent out through the Decatur office of the Western Union Telegraph company today to Theodore Roosevelt. Up until 3 o'clock Tuesday afternoon no reply was received. It was regarded as unlikely that Col. Roosevelt would be able to come.

Footprint Found

Finding of a huge footprint in an oiled road adjoining the Allerton estate near Monticello increased the belief of that vicinity Tuesday of the fact that a big lioness is roaming through the underbrush and timber.

The print of the paw was taken up on a shovel and conveyed to the Donahue drugstore at Monticello where it was placed on exhibition in the window. The claws appear to be about four and a half inches long, and the men who have seen it declare there can be no doubt but that it is the footprint of a king or queen of beasts.

Keeps Off Hunters

Robert Allerton, owner of the ... acre estate on which the lioness is supposed to be, issued an order Tuesday that all hunters should keep off the place. Mr. Allerton wisely believes that promiscuous hunting would likely result in a serious accident of some sort.

Practically all the men employed on the estate, fifteen or more, are on the lookout for the lion and will continue a still hunt until they kill the animal or find out that "there haint no sech thing."

Mr. Allerton himself is keeping a steady watch on the roof of his mansion. The roof overlooks most of the estate and with the aid of field glasses Mr. Allerton is able to see nearly everything that goes on. To those who have scoffed at the idea of there being a lion in that vicinity is pointed out the fact that Mr. Allerton has offered a reward of $250 for its death or capture.

Allerton Convinced

"I am firmly convinced that the animal seen near my house is a lioness." sad Robert Allerton, the Monticello millionaire, over long distance telephone to The Review Tuesday afternoon.

"My butler and two or three other people have seen the lioness. They couldn't all be mistaken about it. I believe that the lioness is no longer on my property, but is probably roaming along the Sangamon river. I would not be surprised to hear that the animal is in the vicinity of Decatur. My land is along the Sangamon and there is an undergrowth all along the river until Decatur is reached."

Here He Is, In Decatur

When Lynn Tustison, 258 East Marion street, and Earl Cavanaugh, 1102 South Franklin street, were out in a boat Tuesday morning they decided that the weather and other climatic conditions made a swim advisable. They started to land their boat about a half mile north of the county bridge, when they saw a slinking form creeping through the underbrush.

Without more ado they hastily climbed in the boat and rowed for home as soon as they could with the result that the lion missed meeting them. They are positive, however, that it was the lioness which has been causing the residents of Monticello to lose sleep recently.

HEAR LION GROWL AT ALLEN'S BEND
YOUNG MEN AND GIRLS OUT FOR CANOE RIDE GET GOOD SCARE
Decatur, Illinois, *Daily Review*, July 23, 1917

Two or three young men and as many girls who were out for a canoe ride on the Sangamon river Sunday evening, got the scare of their lives.

They were paddling around Allen's bend when they heard a rumbling growl, which they were certain came from the famous Monticello lioness.

They heard the growl four or five times from the dark bushes along the bank, and they whisked their canoes around and hiked back up the river. As they paddled they looked furtively around, as though they expected to see the lioness swimming after them.

One of the young men was still pale when he got home. He says he will do no more paddling on the river till they kill that lioness.

COULDN'T LOCATE FAMOUS LIONESS

CATTLE UNMOLESTED NEAR WHERE BEAST WAS SUPPOSED TO
HAVE BEEN SEEN—BIG DOG ATTACKS ONE MAN'S AUTO

Decatur, Illinois, *Daily Review*, July 30, 1917

Decatur may have to send for Buffalo Jones, comrade of Teddy Roosevelt, to kill that lioness after all.

Though an automobile party reported that an animal which they took to be a lioness attacked them about four miles west of the city, Sunday night, investigation Monday by two or three armed parties of men failed to disclose the whereabouts of the beast. Some say the "lioness" may have been a cow which strayed on the road.

Cattle Nearby

In fact, there were no tracks in the vicinity which would indicate that a lion had been there. Twenty or thirty cattle were browsing peacefully in a nearby pasture. The cows or calves had not been molested. That lioness, if it is a lioness, has done a heap of fasting, for so far as can be learned, no animal has been killed by it.

One well known Decatur autoist says that a big animal sprang at his car on another road, Sunday night. He at first thought it must be a lion. But then he heard it bark.

Police Didn't Shoot

But some of the Decatur police say they saw the lioness, Sunday night. It was pretty dark at the time but they swear it was a lioness. They didn't shoot at the animal because they were afraid their revolver pellets might not puncture its hide.

Friends of Officer Fred Meese nearly "ragged" him to death because he said he saw big tracks and heard something growl near the Twin bridges a week ago. Now he has Captain Whitten, Officer Frank Lynch, Officer John Howley, Officer Charles Morgan, Deputy Sheriff Clarence Tandy, Pr. W. J. Davis, and several others for company, for they went Meese one better Sunday night and actually saw the lion, they assert.

Stopped Game of "Rummy"

It was about midnight when Earl Hill and Chester Osborn of 1183 West Main street went to police headquarters and headed off an exciting game of "rummy" by telling the police that they had actually been attacked by the lioness about four miles west of Decatur.

According to their story, and they were sincere in it, the two families, including Mr. and Mrs. Hill and their two children and Mr. and Mrs. Osborn, started out about 10:30 to drive to Niantic. The two men were in the front seat and the women and children in the rear. When about four miles west of Decatur and just east of the Glasgow farm the lion, which had been crouching beside the road, tried to leap into the automobile.

All the occupants of the machine saw the beast just before it made the leap, and they were badly scared. The machine was running about twenty miles an hour. "Go," shouted Hill to Osborn, who was driving the car, and he opened the throttle wide. This is believed to have been all that prevented the animal from getting into the car. One of the men said they got away so fast they couldn't tell what kind of an animal it was. As the machine shot forward the brute struck the side of the car and fell to the ground.

Notified Police

Mr. Osborn kept on going for several miles, and then it was decided to return to Decatur and notify the police so that a posse could be organised to kill the beast. They came back the same route, but they traveled as fast as the machine would ramble as they passed the place where they saw the lion.

Reaching police headquarters they told their story. There were no scratches on the side of the auto as reported but there was a smudge of dust.

Captain Whitten and the other officers and Mr. Davis left in two automobiles at 1:30 o'clock Monday morning and went out to try and round up the animal.

Captain Whitten and Officer

Lynch first went with Hill and Osborn to the place where the lion had last been seen. Captain Whitten says he heard something moving in the brush and pretty soon saw a pair of blazing eyes staring at the auto lights. The officers had only their revolvers with them, so they decided not to attack the lioness and they got into the car and rode back to town after heavy rifles. One of the party saw the lioness climb up the embankment at the side of the road as the machine drove off. When they returned with five or six others they were unable to find any further trace of the lioness. Clarence Tandy and John Howley went through the field where it was supposed to be and the others scattered along the edges, but the beast had disappeared.

Thinks Lioness May Be a Cow

"There may have been a ghost, but I don't believe there was a lioness," said Charles S. Deardorff, watchman at the Decatur post office, who was out on the West Main street road, Monday morning at the spot where a lioness was reported to have attacked an automobile.

"I couldn't see a track in the road anywhere," he continued. "There were some cows strawing nearby and one of them may have collided with the automobile. I haven't heard of the supposed lioness eating anything."

How to Capture Him

Editor, *Review*:

I believe if someone would drive an empty lion's cage with the door open, along the country road, especially if accompanied by a calliope, that the lion would capture himself. No doubt he mistook the automobile for a circus wagon. Pity the state of the poor abandoned lion, lonesome, footsore, and weary. Give him a ride.

Four Paw.

Firemen After the Lion

Bert Smith and Richard Carson of the No. 1 Fire company, went out to look for the lion Monday afternoon. In explanation of the fact that no stock, calves, sheep, or dogs have been missed, the other members of the fire company explained to the hunters that this lion lives on snakes, turtles and fish. The Pulmotor is in readiness to go out to the hunters should they find the beast and have heart failure.

MORE REPORTS ABOUT THE LION

Decatur, Illinois, *Daily Review*, August 1, 1917

"I am satisfied that there isn't any lion on my farm along the Sangamon river and hasn't been any there."

This statement was made by Thomas Jacobs, owner of a 300-acre tract of timber about six miles southwest of Decatur at 1 o'clock Wednesday afternoon. Mr. Jacobs and ten other farmers from the vicinity had just staged a hunt through the entire timber.

Calf Found Dead

Mr. Jacobs was asked about the truth of a report that a calf had been killed by a lion.

"One of my calves had been missing a week," said Mr. Jacobs. "We found it last evening dead and partly eaten. I believe that dogs could just as well have eaten it as a lion. We looked everywhere for traces of the lion, but without any result. The only evidence I have in support of the lion story is a statement by one of the farm hands that he saw it. He may have been excited at the time.

"But I'm satisfied there isn't any lion on my property and never has been."

There was a report in Decatur Wednesday afternoon that the lion had been shot, but it was evidently without foundation.

Farm Hand's Story

James Rutherford, farm hand on the Jacobs farm, saw that lion Tuesday afternoon and he says that it was a lion, too. He is positive that he saw the beast stalking quietly across an old gravel pit on the Tom Jacobs farm, about six miles southwest of Decatur, about a mile south of the river, on a line almost directly south of Wykles station.

Mr. Rutherford was on his way to join the men in the hay field on the Jacobs farm in the afternoon and as he went past the gravel pit he saw the animal walk across it not over 50 yards away. He is positive that it is a lion. He says he has seen lions in circuses and that the animal he saw was not only like a lion but that it was a lion.

Tuesday evening just at dusk Morris Brown, who lives about a mile and a half south of the Jacob's place, heard a noise in one of the fields near his place and went out to investigate, taking a gun with him. He saw some animal slinking away in the corn and shot at it.

Later a group of men were sitting near a hedge talking the situation over when they heard a roar which seemed to come from just over the hedge behind them. All plans to catch the beast vanished from their heads as they "beat it" up the road.

George Hill

George Hill, 1283 West Macon street, went out to the Jacobs place Tuesday evening after hearing about the lion and found Tom Jacobs out milking and, like the pioneer of old, Tom had his gun leaning against the fence handy to his reach in case of an attack

Didn't Disturb Goose

"What I cannot understand about that lion," said Attorney Lee Boland Wednesday, "Is why he did not kill a goose which was sitting on a nest within six feet of where he was supposed to have been Sunday night.

"I was out with Robert Watt Monday morning on the West Main street road, where the attack on the automobile was said to have occurred. Beside the road is a ditch in which the lion was said to have been lying before it leaped out at the auto.

"Just a short distance away was a goose sitting on a batch of eggs. The goose has been there a week or more, and was still there Monday morning. If there was a hungry lion about, he'd have made small work of that goose. We looked around for lion tracks and found a lot of cow and hog tracks."

Reported at Niantic

The lion was reported to have been seen at Niantic about a week ago. The matter was investigated but no lion was found.

Tiger Reported Seen By Farmers
Ruralist Near Hamel Organizes Posse
to Hunt Mysterious Animal

Edwardsville, Illinois, *Intelligencer*, December 22, 1933

A "tiger hunt" is on in Hamel township this afternoon by a posse of farmers who have been observing a mysterious animal in the community during the past few days. The animal is described as having yellow and black stripes, as long as a large police dog but not so high.

Plans for the hunt were made today after Alfred Koch, a young farmer observed the animal at 12:30 o'clock this morning. He was on his way home when it was encountered along the highway about one and a half miles southeast of Hamel. Mr. Koch was within 20 feet of the animal and by the headlights was able to get a good look at the varmint.

As he approached, the animal posed on its haunches, appearing ready to spring if attacked. Mr. Koch hurried home, obtained a shotgun and returned to the scene. When he arrived the animal could not be found. Previously Mr. Koch and his father, Henry Koch took shots at the animal but the range was too great and it escaped.

Edward Hulett, Henry Rode, Ferd Suhre and August Wilkening are others in the neighborhood who are reported to have observed the animal during the past week.

On one occasion the animal was observed feeding upon a dead cow in Silver Creek Bottoms. Farmers say the cow had died in the creek bottoms. They were certain the animal had not killed the cow.

From several sources today it was learned that farmers in a wide area have been moving about their places with more or less caution after dark. For the time being they have given up walking along highways after dark.

HUNTERS UNABLE TO FIND TRACES OF WILD ANIMALS
Edwardsville, Illinois, *Intelligencer*, December 23, 1933

A group of Hamel township hunters spent yesterday afternoon in Silver Creek Bottoms in search of strange animal that has been observed by farmers at different place during the past few days. Alfred Koch, member of the party, said that no traces of the varmint were found.

Efforts are being made today to locate other hunters with dogs able to pick up "cold trails" of animal and trace them to their lairs. It is said there are only a few dogs here that are able to follow old trails of various animals.

On the trip yesterday the hunters covered about two miles along the creek where the animal might be in hiding during the day.

STRANGE BEAST REAPPEARS IN WHITESIDE COUNTY
Freeport, Illinois, *Journal-Standard*, June 30, 1938

Sterling, Ill., June 30.—Sterling's wild beast made a sudden appearance the other afternoon near the home of Lyman Simpson, three miles southwest of Rock Falls. The animal, seen by a number of people, put a dog to rout and was pursued by several men with guns, but made its escape through a field of oats.

The animal was first sighted by several small children playing in the road close to the Simpson farm. They ran home, badly frightened.

While the men were seeking weapons, William Monnier of Morrison, who was a guest at the Simpson farm, and Clarence Simpson, both boys of 14, leaped on their wheels and raced down the road.

The dog sighted the animal first, turned tail and beat a hasty retreat. Both boys saw it, turned and hastily rode back to the Simpson home.

They described it as larger than an ordinary mastiff dog with white shoulders and yellow spots on a tan background.

The night before the animal killed a calf belonging to a neighbor of Mr. Simpson. It was seen at that time, but by the time pursuit was arranged it had gotten away.

Posse To Hunt Mystery Beast

Amarillo, Texas, *Daily News*, September 20, 1946

Oquawka, Ill., Sept. 19 (AP)—Mysterious wild animals, which Sheriff George Voorhees said, have terrorized this entire Mississippi River community, will be hunted Sunday by a band of armed men and dogs.

Reports of the depredations of the beasts—at least two have been seen together on nocturnal prowlings—have persisted and grown for about six weeks until officials became convinced of: their existence.

The hunt was ordered by Sheriff Voorhees and State Conservation Officer Roy Hoskins.

"I have no idea what they are or how many," Hoskins said.

Those who have seen, the animals gave this general description: A dark, cat-like creature which slinks along on rather short legs, about seven feet long from nose to tail and 25-30 inches tall.

No one has been attacked, but farmers have lost large numbers of chickens, and turkeys.

BEASTS STILL HAUNT VILLAGE

Oklahoma City, Oklahoma, *Daily Oklahoman*, September 23, 1946

Oquawka, Ill. Sept. 22 (AP)—A band of 150 armed men beat the bushes over a wooded area three miles square Sunday in a futile effort to locate the mysterious cat like beasts which have terrorized this village of 800 in recent days.

The huntsmen, accompanied by nine dozen assorted bird dogs and coon hounds, found no trace of the animals.

Three airplanes flew low over the densely timbered tract in which the beasts were believed to have their lair. This effort also failed to bring the animals into the open.

At least two of the animals—described as about 7 feet long and 2 feet tall—have been seen together, Sheriff George Vorhees said. Farmers have reported the loss of considerable poultry. No humans have been attacked.

District Conservation Inspector Guy Taylor, North Henderson, Illinois, inspected several spoors located by huntsmen.

"They're dog tracks," he said.

Taylor added, however, he was convinced from interviews with persons who reported seeing the beasts that there were "animals out of the ordinary" at large.

Local talk has identified the beasts variously as pumas, cougars, and giant lynx and some say they may have been set loose by an itinerant circus because it couldn't get meat to feed them.

MONSTER HUNTERS CONFUSED

Oelwein, Iowa, *Daily Register*, September 23, 1946

Oquawka, Ill.—(UP)—The Oquawka monster hunters were confused today. They couldn't even find out if they were hunting one monster, two, three, or a whole pack.

Sheriff George Voorhees, who led a posse of about 200 worried citizens in a fruitless hunt for the beast Sunday, said he had started out to capture one animal.

"Now we have several, maybe four," he said.

Some of the women who claimed to have seen one of the beasts "face to face" said it was 30 inches high and about seven feet long.

Voorhees said that when he first heard about the beast, or beasts, he thought it, or they, might be a wildcat or a panther.

He said he had no further plans for trying to capture the monsters. Sunday's search was aided by two airplanes which scouted the area.

"I don't know whose idea that was," he said. "I don't think they are going to be up today. No one has told me so, anyway."

Farmers have reported that the beasts have killed chickens, turkeys and dogs, but they did not come out during the hunt.

Mrs. Sarah Keath was the latest to have reported meeting one of the monsters. She said she was up on her roof a few nights ago and she bumped into the beast.

It appeared to be as much surprised as she was, she said.

The animal fled without making a sound.

Mystery Beast Roams in Illinois

El Paso, Texas, *Herald-Post*, September 26, 1946

Momence, Ill., Sept. 26—The "Beast of Momence," still at large, was believed today to be an escaped circus animal.

The view was advanced by Chicago animal experts even though Momence townspeople barred their doors and windows last night, and 100 others kept a fruitless, armed vigil for the "ferocious beast more than six feet long."

Only one person so far has seen the beast, variously believed to be a giant lynx or cougar or puma. He is Paul Therein, 27, son of the Momence postmaster, who claims he was attacked and clawed on the hand during a midnight encounter Monday.

Therein said he inflicted a slashing breast wound with a butcher knife before the beast fled.

'Monster' Seen While Fishing, Boy Declares

Racine, Wisconsin, *Journal Times*, September 29, 1946

Centralia, Ill.—(AP)—Add to the lurid reports of roaming, preying monsters from Iuka and Oquawka, Ill., and Lebanon, Ind., story by James Langenfeld, 14, that he encountered at a distance—a beast "seven or eight feet long" while fishing at nearby Crooked Creek.

The Centralia High School youth told Police Chief O. T. Bounds, who passed the report along to Sheriff Homer Lewellen, since the reported beast was purportedly at large in the sheriffs bailiwick.

Bounds reported the youth said the beast was stalking a crane, that it saw it and began making sounds that were neither mews, squeals, growls or barks. It was gray and furry. The chief said the boy added:

"I thought I ought to go home." And he did, but quick.

MYSTERIOUS 'BEAST' TERRORIZING FARMS IN AMBOY DISTRICT
FOOTPRINTS SEEN; WILD SCREECHES CAUSING ALARM NEAR CITY
Dixon, Illinois, *Evening Telegraph*, December 11, 1946

Foot prints of a huge animal which are said to resemble those of a black panther, have been discovered in several places south and east of Amboy, according to reports received at the office of Sheriff Gilbert Finch, where a request has been made for high powered rifles and ammunition with which to dispatch the much-feared animal. Terrorizing hair-raising cries of the supposedly wild animal reported to have been heard after dark, are said to have been the cause for extreme caution being observed by farmers residing in that section of Lee county and reports today indicated that few venture from the security of their homes after dark.

The first reports of the presence of the beast were heard the latter part of last week when stories began circulating about a large animal which was night prowling and was said to have attacked several dogs and killed one horse. The rumors have gained wider circulation.

Print in Schoolyard
Mrs. Clara Walker, teacher of the Binghampton rural school, was reported to have discovered a foot print of the animal in the school yard, and it was further stated today that she was hauling the smaller children in her school to their homes at the close of school each afternoon to safeguard them from attack. Others reported having traced the huge foot prints in the Ives pasture field south and east of Amboy, near where the mysterious beast is purported to have its den and nightly emerges to wander over surrounding farms.

Sheriff Finch today stated that he had not been called upon to organize a posse to take up a search for the animal, but told of having received requests for powerful firearms. ...

Open Searches for "Huge Black Beast" North of Sublette

Reports Describe Wild Beast as Black, Like Large Jet Panther

Dixon, Illinois, *Evening Telegraph*, December 12, 1946

A huge black beast which roams the countryside north of Sublette and southeast of Amboy has frightened farmers to the point that some are afraid to go to their barns at early morning hours to do the customary chores. Several have seen the animal and are able to give a good description of it and the entire countryside is united in the hope that it may very soon be destroyed or captured.

Sgt. George Ives of the state highway police force and Stoddard Danekas, deputy state conservation inspector, have entered the investigation into the reported presence of the beast, which has been described as a black panther. Sgt. Ives yesterday questioned several farmers living in the locality where the animal had been reported as having been seen and later stated that he was satisfied that some unusually large animal was on the loose in the section.

Leonard, Leffelman, who lives about six miles south and east of Amboy, stated that the beast ran directly into the side of his car on the night of Nov. 29 as he was returning home from Amboy with his wife.

Describes Beast

"The animal, jet black in color, about five feet in length with a long black tail, was running across the road and ran into the front of my car. I got a good look at it and it looked more like a black panther or a member of the cat family than anything I can describe," Leffelman told Sgt. Ives.

Asked if he stopped his car to investigate, Leffelman replied that he was too frightened at the size of the animal, adding that the force with which it crashed into the front of the car, jarred the machine. He described the beast as standing about three feet high.

That the beast has been inhabiting a section of woods and swampy land along Green river and east of Binghampton, was deducted from other reports gathered in the investigation yesterday afternoon. Raymond Turner was reported to have seen the large black animal walking through a swampy section as he was working in a field on his farm during mid-summer. At the time the beast showed no indication of fear, but looked at the farmer and continued on its way.

Many Stories Given

Mrs. Clara Walker, teacher at the Binghampton school, stated that she had heard several stories concerning the mysterious animal and had taken the precaution to take some of the smaller children in her school to their homes in her automobile, fearing for their safety on country roads. Parents of the children are also taking the precaution of taking children to and from the school by automobile, since the reports of the animal have gained wide circulation.

Raymond Eccles told Sgt. Ives of having seen the animal on the night of Dec. 3rd while he and other coon hunters were using dogs along Green river ditch east of Binghampton. According to Eccles the dogs started the mysterious animal on a run and chased it to the bridge which spans Green river at Binghampton. The animal crossed the bridge. ...

"BLACK PANTHER" GOES SOUTH TO GLASGOW, ILL.
Dixon, Illinois, *Evening Telegraph*, July 26, 1947

Glasgow, Ill., July 26—Appearance of a "black animal", possibly a panther, south of here, precipitated plans for a round-up hunt Sunday by 500 members of the Sportsmen's club of Glasgow.

Although the mysterious black animal is the objective, the club said, "foxes will be fair game"

Plans for the hunt specified that shotguns will be used, but no rifles.

Two former servicemen said they had seen the black animal about the size of a large dog, leaping from tree to tree and emitting cries which sounded like human screams.

ILLINOIS POSSE HUNTS 'MONSTER'
Charleston, West Virginia, *Daily Mail*, July 28, 1947

Glasgow, Ill. (UP).—Members of a 50-man hunting party said today that the "monster" which has been reported near here during the past 10 days may be a black panther.

The posse went "monster hunting" yesterday but came back empty handed. The men reported they had found tracks about three inches in diameter. The tracks a cat's but were larger.

They were found in mud along a creek.

Sheriff to Hunt Strange Animal Killing Livestock
Portsmouth, New Hampshire, *Herald*, December 6, 1950

Princeville, Ill., Dec. 6 (AP)—A Peoria county farmer asked the sheriff's office for help today in tracking down a huge mysterious animal that has killed more than 50 head of his livestock.

Fred Perdelwitz said the marauder has an "awful roar"' that sounds something "like a lion's," and that it has left tracks "as large as your hand."

He told the sheriff that since the animal first raided his farm several weeks ago, he has lost 42 pigs, four calves, several lambs and a dozen chickens. Some of the livestock, Perdelwitz said, simply disappeared but that he had found badly-slashed carcasses of some of the lambs.

Perdelwitz said the killer has made almost nightly raids on his farm and that he saw it one night in a field. He said the animal "circled me several times, setting up an awful roar," and finally fled.

Plane Search Organized for Jungle Beast
Eureka, California, *Humboldt Standard*, January 5, 1952

Kansas, Ill., (AP) — Several Edgar county farmers will go on an airborne safari today to track down two animals variously described as wildcats, jaguars and lions.

The flying farmers, who own their own planes, will be assisted by farmers on the ground who also will stalk the animals who have been slaying sheep, screaming at night and leaving large pawprints in the snow.

Farmer Charles F. Schneider, 50, said eight of his sheep were killed by a beast he described as weighing 200 pounds. It looked like a jaguar, Schneider said. He said the animal was accompanied by its mate. It looked like a wildcat.

Mrs. Lela Allen, who lives nearby, said animal's screams have awakened her at night. She's also reported seeing large tracks in the snow.

Allyn Adams, 22, said he came face to face with one of the beasts.

"He ran. So did I!" Adams said.

"His eyes were as big as dollars. He looked like a lion."

REPORT BLACK PANTHER NEAR WEST FRANKFORT
Harrisburg, Illinois, *Daily Register*, December 09, 1952

West Frankfort, Ill. (UP)—Hunters have reported a "black panther" in the West Frankfort area for the second time within two weeks.

Policeman Ralph Anderson said today two hunters reported seeing the animal about two miles east of here Sunday.

Anderson said Bruno Broy, West Frankfort, and Norman Siegal, St. Louis, reported they saw the panther running across a field in 20-foot leaps when they alighted from their car to hunt rabbits. They said the animal disappeared up a draw and they tried to head him off but couldn't find him.

The hunters described the animal as about two feet tall and at least four feet long with a long black tail.

Anderson said he has had no reports of farm animals being molested, but Broy and Siegal reported they had come upon the partly devoured carcasses of two raccoons near where they saw the animal.

About two weeks ago, two West Frankfort high school boys reported seeing an animal answering the description of the panther about four miles north of where Broy and Siegal said they saw the animal.

Several months ago about a dozen persons reported seeing a "black panther" near Centralia. Game wardens searched unsuccessfully for several days.

MYSTERIOUS BEAST KILLED BUT IDENTITY REMAINS A MYSTERY
Dixon, Illinois, *Evening Telegraph*, May 27, 1953

Springfield, Ill. (AP)—This is the unfinished story of one of those mysterious Illinois beasties that yowl in the night and frighten dogs, livestock or people.

It was shot fatally through the head. Usually the creatures are phantoms that manage to elude the hunter.

The mystery, however, remains much alive. What was the critter?

For at least three years, it roamed the Macoupin Creek area, south of Rockbridge in Jersey county. Residents said it clawed their dogs. It picked up one load of No. 8 buckshot, a post mortem showed, as well as the fatal slug from the rifle of Louis Seehausen, who lives southwest of Rockbridge.

A district game warden, Carl L. Keehner of Jerseyville, is as much in the dark as anyone about the four-footed creature's identity. He insists it's not a bobcat. He has seen bobcats in Southern Illinois in recent years.

Keehner gives this sketchy data on the animal, hanging by its front paws when he inspected it: 4 feet 3 inches from front to rear claws, bushy tail 18 inches long, dark brown thick pelt.

One other identifying mark was found. It was the buckshot which farmer George Grizzle said came from his gun in 1950.

The creature's final excursion coincided with a coon hunting walk by Seehausen. His dogs treed the animal as night fell. The animal descended and was re-treed several times.

Finally, it left the security of the woods for a pasture where the dogs brought it to bay. Seehausen flashed his light on it, and saw it erect on its haunches, striking with its claws at the dogs. Then Seehausen brought it down with a shot through the head.

DECATUR DIARY: CATS, COUGARS, PANTHERS
SEEN BY DECATUR AREA RESIDENTS
Decatur, Illinois, *Herald and Review*, June 22, 1975

The columnist discussed the "black panther" flap of 1955 in the area "east of Lake Decatur between Illinois 105 and Rea's Bridge." Game Warden Paul G. Myers thought he shot the animal in the flank in October of that year, but it got away.

BLACK PANTHER REPORTED
Edwardsville, Illinois, *Intelligencer*, February 7, 1956

La Salle, Ill. (UP)—Farmers armed themselves Tuesday to protect their cattle from a black panther reported to be roaming near the banks of the Illinois River.

'BLACK PANTHER' A 45-LB. TOMCAT
Cedar Rapids, Iowa, *Gazette*, March 18, 1956

Spring Valley, Ill. (UP)—A "black panther" reported prowling the fields and menacing livestock near here has been identified as a 45-pound jet-black tomcat.

Joyce Boyer, 14, sighted the cat when she rode a horse past a gravel pit where the "panther" was first seen.

Joyce called police, who said the black cat, about three feet long and two feet high, was accompanied by two other cats.

A witness, Bill Chambers, reported to researcher Loren Coleman that on June 2, 1963, while driving near his farm in Champaign County, he spotted a black feline in a field, about 300 yards away. He stopped and watched it for ten or fifteen minutes, trying to get a shot with his rifle, but the cat kept moving around without giving his a chance to shoot. Chambers checked the area the next day, and by measuring the clover height, estimated the cat was 14-15 inches tall at the shoulder and about four-and-a-half feet in length. Tracks in the wet ground were two and five-eights inches, and showed no claws. (Clark 1993)

WRITER RECALLS TALE
Carbondale, Illinois, *Southern Illinoisan*, July 6, 1969

Recent sightings of a black or dark brown panther in Pinckneyville spurred the article's author, Art Reid, to note rumors years earlier of a large black cat with a tail "at least five foot long." Hunters searched for it but only came up with tracks that turned out to be canine.

SEE BLACK PANTHER IN HUNT AREA
Newton, Illinois, *Press*, February 19, 1970

Mrs. Donald Miller reported a black panther from the area "about three miles west of Hunt." She said her family had heard the animal previously, but only saw it that Tuesday.

LETTER TO THE EDITOR: WHAT MANNER OF BEAST?
Carbondale, Illinois, *Southern Illinoisan*, June 11, 1970

Several witnesses saw a large black feline, "at least 5 feet long with a tail as long as their arm," at different times in the previous month. One was a hunter who came across it feeding on a deer.

OFF THE BEAT: THERE'S SOMETHING OUT THERE
Decatur, Illinois, *Review*, September 23, 1976

The columnist, Bob Sampson, noted several sightings of a panther southwest of Decatur. One man, James Scott, saw it one morning, describing it as "black, coal black. I'd say it was about five foot long, maybe a little longer with the tail. And it sure was pretty." Sheep had been killed in the area, but the farmer thought it was a pack of dogs that got them.

PLAINFIELD, ILLINOIS: BIG BLACK CAT
Reported in "SITUations," *Pursuit*, Spring 1978
From: Joliet, Illinois, *Metro East Herald News*, March 10, 1978

On March 6, 1978, William Hughes saw a "pure black" feline as large as a "police dog" stalking something near an industrial warehouse where his wife was a security guard. On March 8, his wife saw the animal and called police and wildlife officers. The cat's prints in the snow were "larger than a size-five woman's shoe."

I received the following via email and published it in an issue of *NABR* (Arment 2000b):

"I am a Geologist with a Master's Degree and am skeptical of many 'crypto' creatures, but I will relate one that I believe has some merit.

"I got my MS at Southern Illinois University at Carbondale, which is just north of the massive Shawnee Nat'l Forest. The locals talk of a 'woods cat' or sometimes a 'timber cat,' much bigger than a housecat but with a long tail, so definitely not a bobcat—we have a few of those too, most people know them by sight. These 'woods cats' are anywhere from brown to tiger striped, and about the size of cocker spaniel—say 25-30 lbs. I ran into a taxidermist in Olney, Illinois, a rural So. Illinois town, who claims he mounted one for a guy many years ago—he was about 80 years old in 1985, when this conversation took place, plus or minus a year or two. He claimed that he was aware of them as a boy, and his father told him that the Indians had brought them up from Mexico. I later took him a book hat had a picture of a jaguarundi in it and he said 'that's it!' Lots of good people in Southern Illinois claim to have seen these. As a geologist who works in the oil fields, I get in some remote backcountry quite a bit, and I make a point of asking farmers or rural residents about these cats—*many* have personally seen them.

"They may just be feral cats, but I though I would write you to let you know in case you didn't have any knowledge of these critters."

UNKNOWN ANIMAL KILLING DOGS
Pacific Stars and Stripes, December 23, 1986

An unknown feline ("possibly a black leopard") was reported killing dogs in East Carondelet. USDA wildlife trackers helped police search for the animal.

LARGE, BLACK, 'CAT-LIKE' CREATURE STALKS IN SOUTH CENTRAL BURR RIDGE
LaGrange Park, Illinois, *Suburban Life*, February 25, 1989

Residents of Burr Ridge reported a large black feline prowling the area. A cast of a 4 x 5 ½ inch print was shown to zoo curators, who were unable to say if it was canine or feline. Police said the depth of the footprint in the snow suggested the animal was between 125 and 150 pounds.

BIG CAT: ELUSIVE BLACK PANTHER SEEN AGAIN
Decatur, Illinois, *Herald-Review*, September 24, 1989

Chuck Hudgins and his son were driving south of Millstadt when they heard shots and saw a large black cat cross the road in front of them. A St. Clair County animal control officer searched the area with sheriff's deputies, but only caught a glimpse of the animal's eyes reflecting back from a flashlight, before it disappeared.

TIME TO CALL INSPECTOR CLOUSEAU
EISENHOWER 'PANTHER' IS BLACK INSTEAD OF PINK, BUT IT LEAVES AUTHORITIES RED-FACED AFTER A SEARCH
Chicago, Illinois, *Tribune*, August 9, 1994

Officers searched near the Eisenhower Expressway for a reported black panther without success. It was seen by a State Trooper and several commuters. Radio reports triggered clogged traffic and one pileup, as drivers slowed down to watch the hunt. Authorities suspected it might be a dog, as one witness said it was black with white spots. The State Trooper who reported it described it as a large black animal, but insisted it was not a dog.

SIGHTING OF BLACK PANTHER IN ITASCA REMAINS A MYSTERY
Chicago, Illinois, *Daily Herald*, August 12, 1994

Two construction workers reported seeing a large black cat in a corn field near a school. Other reports followed, including one from a state trooper. Animal control was unable to find it.

Mountain Lions in our Midst?

Decatur, Illinois, *Herald & Review*, October 20, 1996

This article mostly reviews cougar sightings in Illinois, but notes a few instances of black feline sightings:

Residents east of Lake Decatur reported a low-slung black panther in 1955. Game Warden Paul G. Myers believed he may have shot it along the Sangamon River.

In 1970, a Decatur woman and her daughters saw a large black panther and a half-grown cub. They blamed it for killing 40 of their chickens.

Southern Macon County residents reported a black panther in 1976.

Clarksdale Woman Logs 'Panther' Sighting

Decatur, Illinois, *Herald & Review*, May 12, 1999

A Clarksdale woman, Bev Ray, reported seeing a large black panther walk across her yard one morning. She said, "It was jet black, and it was just calmly walking along, not running or anything. And it was big. I saw it from less than a few hundred feet away, and it was really, really big. It walked through my yard and disappeared into some timber nearby." She and her husband kept a variety of pets, including 14 Rottweilers, and said the cat was larger than her biggest dog. The feline's tracks measured 4 inches across.

Newton (2005) noted that in 2001, Carl and Debbie Johnson reported that a "black cougar" was living on their Sheldon, Iroquois County, property.

Curious Over Cats

Alton, Illinois, *Telegraph*, March 15, 2004

In an article dedicated to "normal" cougar sightings in Illinois, the journalist mentions that there were a couple of "black panther" sightings in Hardin the previous month.

Some Strange Sightings

Freeport, Illinois, *Journal-Standard*, June 10, 2005

Two farmers in Afolkey, in separate incidents, said they saw a large two or three foot tall tawny feline. One described it as "tall, long, slender, tawny in color and having abroad, triangular face with dark fur around its face—fur that looked like a mane, he said, and a 'J-hooked' tail, possibly bushy on its end." The other farmer didn't see anything like a mane.

BLACK PANTHERS? ELUSIVE CREATURES SUBJECT OF MUCH SPECULATION IN FULTON COUNTY
Canton, Illinois, *Daily Ledger*, March 25, 2006

A sighting of a black panther by a couple just outside Bryant leads to discussion with investigator Russell Bedwell about similar reports in Fulton County through the years. A few hotspots mentioned include the Rabbit Run Road area and the Marbletown-Anderson Lake region.

ELUSIVE PANTHER SPOTTED IN SOUTHWEST ELGIN
Chicago, Illinois, *Courier News*, March 24, 2009

A large black feline (known in the area as the Plank Road Panther) was seen by two co-workers behind the Otter Creek Shopping Center as it walked along a line of trees and bushes lining a road between farmland and the Elgin Country Club. The witnesses were able to estimate its length (including body and tail) from its position and landmarks as it walked, later measured by the newspaper reporter as seven feet in length.

COUGAR SIGHTING NEAR NAPERVILLE?
Chicago, Illinois, *Daily Herald*, January 9, 2010

In a letter to the editor, a woman claimed to have had a full view of a large black feline with a "huge black tail" walking in the snow in the woods off the side of the road. She said the cat was "at least three feet in length." This was just past the intersection of Naperville and Warrenville roads.

INDIANA

Mitchell, South Dakota, *Daily Republican*, January 22, 1890
A strange animal, supposed to be a spotted leopard that has escaped from a show, is committing many depredations in the vicinity of Scottdale, Ind., and making night hideous with its screams.

A Wild Beast At Large
School Children Narrowly Escape an Attack
Hamilton, Ohio, *Daily Democrat*, December 21, 1894
Uniondale, Ind., Dec. 21.—The children of George Ditzler, a farmer south of this place, and their schoolmates, while returning from school, had a narrow escape from attack by an animal, which, judging by the description given by the oldest child, is a mate for the one killed by the Green brothers while coon hunting, some weeks ago, and which is supposed to be a cross between a leopard and a tiger. The body of the dead animal was sold to the high school at Bluffton.

The remaining animal has been seen several times, but in spite of a general hunt by the people of this community, who surrounded the thickets and woods, it has escaped capture. The Ditzler children were terribly frightened, but they managed to reach home unharmed. Last night several men beat up the country where the animal was seen by the children, and while the trail was very plain in a cornfield they were unable to trace it to its lair. Another general hunt is now being arranged. There is great uneasiness in the neighborhood.

A FIERCE WILD CAT
KILLED TWENTY DOGS AFTER BEING CHASED HALF A MILE
Newark, Ohio, *Daily Advocate*, December 29, 1899

Dresser, Ind., December 29.—People in this vicinity are much disturbed by the presence of a wildcat, which has killed a score of lambs, calves and hogs. Night before last a party of hunters, with 30 dogs, gave chase to the animal, locating it three miles west of here, and chasing the animal for a half a mile from tree to tree. The cat finally descended and killed and crippled 20 dogs, after which it again sought safety in flight and distanced pursuit. The cat is supposed to have escaped from a show wrecked on a train near Foster last summer.

STRANGE WILD BEAST WITH EYES LIKE BALLS OF FIRE CAUSES TERROR
San Jose, California, *Evening News*, June 13, 1908

Akron, Ind., June 12.—A strange wild beast, with a cry that resembles the shriek of a frightened woman and with eyes that shine in the darkness as if they were balls of fire, is causing terror among the farmers who live near the woods and swamps northwest of this place.

The beast was first heard by the farmers ten days ago. Its cry was described as weird and piercing. One man said he first thought it was the shriek of a woman but as the cry came from a swampy woods he concluded he was mistaken. Next day he investigated and found tracks large and broad enough for a bear's. Several men procured dogs and guns and trailed the beast along roads and across fields, but finally lost its tracks in a swamp.

Saw Two Fiery Eyes
Another search was made last Saturday, but without avail. On Monday night scores of people heard the cry in the woods a quarter of a mile west of town, and on Tuesday morning fresh tracks were found.

No one seems to have heard the beast on Tuesday night, but on Wednesday about dusk, Miss Clara Orr, who lives one mile north, heard a commotion in the pigpen. When she went out to learn the cause of the disturbance she was startled by seeing a strange animal

with eyes glowing like coals of fire among the hogs. She screamed
and ran to a neighbor's, Jacob Saygers. He and his hired man, armed
with shotguns, at once went to the pen. The strange animal was still
there. Both men fired at the same time.

A Screech and Snarl

There was a screech and a snarl, and a dark mass leaped the side
of the pen. The men fired again, but the beast went on towards the
woods apparently not crippled. The pellets in the cartridges were
small and probably only stung the animal.

Tracks were found leading from the pen to the woods, but although a party of hunters searched diligently until dark the strange
animal could not be found.

Inquiry has been sent to several circuses, in the belief that the
beast has escaped from a menagerie. One man, who got a glimpse of
the animal, said it was as large as a leopard, another believed it was
a huge Canada lynx, but his belief was discredited by examination of
the tracks. Some people are of the opinion that the "shrieker" is the
same animal that was seen in the "Hell's Neck" part of southern
Indiana several months ago and that it has worked northward in the
woods on the banks of the Wabash and other streams. It is about
250 miles from the "Pocket" to this place.

The whole neighborhood is excited over the strange animal, and
timid people will not go near the swamps or woods at night.

THE GREAT "WHAT-IS-IT" ROAMS NEAR ALBION
Fort Wayne, Indiana, *Journal-Gazette*, January 19, 1913

Albion, Ind., Jan. 18.—A monster animal that has been guessed
to be everything from a lynx to a lion has been sighted several times
during the past week north of the city, and, though all but captured
several times, retains its liberty to the terror of the countryside.

Winford Bowen, a farmer, the first to see the creature, came upon
it in the woods on his farm and tracked it to its lair. He returned
later and placed a trap in front of the lair and was watching results
when the strange beast, carrying in its teeth a ground hog that had
mistaken the warm weather for that of early February and had come
out to see its shadow, stepped into the snare. After a struggle the

brute tore itself loose, leaving in the trap one of its toes, which is as large as a man's thumb and is tipped with a claw an inch in length.

With R. D. Currell and "Red" Fitzpatrick, Mr. Bowen said a number of traps in the vicinity of the den, but the mysterious animal evaded them on its next appearance. Poor marksmanship on the part of the trio of watchers permitted its escape. Barking like a dog and running with incredible speed, it vanished, but through its tracks in the snow, it was tracked to a cave on the Kilpatrick farm. It is still there, so far as the farmers know, for countless traps that have been set all about the place have failed to bring results.

MAKES ITS HOME IN DESECRATED GRAVES
MYSTERIOUS BEAST RENDS COFFINS AND SCATTERS
HUMAN BONES—ROARS SCARE WOMEN.
Raleigh, West Virginia, *Register*, March 21, 1913

Shelbyville, Ind.—The most uncanny sensation that has stirred Shelby county in recent years has taken a firm grip on residents near the Patterson cemetery, east of Lewis creek. Graves in the cemetery are being despoiled by some animal or agency, and none of those who has taken an interest in the situation has been able to solve the mystery.

Burrowing into the graves, the beast rends coffins, scatters bits of broken wood, spinal columns, arms and leg bones on the surface and roars like a lion when intruders approach too near the scene of its operations. At first the belief prevailed that woodchucks were doing the work, but this theory has been abandoned, as none of the animals have been seen in the cemetery.

The situation was first discovered by women who went to the cemetery to put some graves of relatives in better condition. They discovered a great hole in one of the graves. They started to investigate, but were greeted with a roar from the grave which sent them from the cemetery with quickened steps and blanched faces. They have not since returned.

Men of the neighborhood then went to the cemetery armed with picks, shovels and guns. They found bones and pieces of coffins scattered over the graveyard. They dug into several despoiled graves and found they had been used as the home of some animal but they have been unable to determine its species.

The work was pursued with vigor until the men struck a grave where it was feared the despoiler was at work. There the men lost their nerve and some declared they were as badly frightened as the women. The work of trying to unravel the mystery is still on and the despoiler is as busy as over.

Tipton County Farm Boy Is Attacked by Strange Animal; Saved by Dog

Kokomo, Indiana, *Tribune*, August 27, 1930

Tipton, Ind., August 26.—His faithful dog was all that prevented Frank Tomlinson, 11-year-old son of Harmon Tomlinson, residing southeast of Tipton on the Rolla. L. Hobbs farm, from being torn to pieces by a strange wild animal which attacked the lad when he was driving the cows from pasture to the house for milking Monday evening.

The boy is suffering from a broken collar bone and numerous bites and scratches as the result of the attack and his clothes were nearly torn from him in the encounter.

The strange animal was described by the boy as having the appearance of a darkish yellow dog, similar to an Airedale, with hair standing up all over it and ferocious in appearance. It is believed to be the strange animal seen by several persons in Madison township last season and seen by G. C. Boyd, Tipton barber, who was driving after night and saw the animal slink across the road.

Beast Makes Attack

Young Tomlinson, whose mother is dead, resides with his father and his grandparents, Mr. and Mrs. Harmon Tomlinson. He had been sent to the pasture for the cows and had taken a strong stick and his dog along. While passing through the woods, he met the strange beast which rushed at him, knocked him down and tore at him with teeth and claws. It was then the dog attacked the beast, grabbing it by the leg and keeping up the attack until the animal turned and fled. It jumped over a five foot woven wire fence on top of which were three strands of barbed wire. The beast cleared the fence easily and disappeared.

This strange animal, which has been seen by several persons, is said to resemble a wolf and yet is unlike a wolf. Some believe there is more than one such animal abroad and wolf hunts have been started to capture the animal animals. It is thought by some that the beast may be a dog reverted to a wild state.

While extremely painful, it is not believed the boy's injuries will prove serious.

STRANGE ANIMAL ELUDES CAPTURE; TRACKS ARE LARGE
Valparaiso, Indiana, *Vidette-Messenger*, June 22, 1932

Logansport, June 22—(UP)—The Crystal Springs Farm is the purported habitat of the latest of a series of strange animals reported over Indiana. These animals, by the accounts from different sections, have ranged over most of the realm of natural history, past and present.

Only tracks of the beast, and its depredations, are tangible. It has neither been seen nor heard, although almost every day evidences of its slaughter have been found.

The tracks are large—too large for those of any but an extraordinary bobcat—and apparently those of some member of the feline family.

For two weeks hunters and trappers have tried in vain to capture the animal. It eludes all snares and baits, but continues in the proximity of dwellings. On one occasion it leaped from a tree to the roof of a farmhouse, awakening occupants—another indication of its feline characteristics.

MONSTER HAS HIM BAFFLED; BUGS NO HELP
Valparaiso, Indiana, *Vidette-Messenger*, August 30, 1946

Lebanon, Ind.—(UP)—Harry McClain said today he couldn't shed any light on the monsters reportedly roaming Indiana's countryside until the summer lightning bugs shed their tail lights.

McClain dropped the rifle he's been using in his two week search for the alleged monsters and took up a spray gun loaded with DDT last night to hunt lightning bugs.

He said he couldn't fire at themonsters until he sees the whites of their eyes—and the bugs are interfering.

"They keep lighting up and I can't see the eyes of the animal I'm looking for," McClain said.

Going to Try DDT

"So I'm going to get 'em out of the way with DDT. Then I'll catch the monster—or whatever it is."

McClain said he'd seen what he believed were the bright eyes of a wildcat or panther on a wooded farmland near here. But he's never been able to get a shot at the beast which allegedly "cries like a baby and kills livestock."

All of his traps have failed, too, despite savory bait including beefsteak and a live chicken.

He's Stumped

"This creature is about the slipperiest thing I've ever met," he said.

In fact, he added, it may have slipped all the way to Franklin, Ind., where residents believe they have seen paw-prints six inches across and three feet apart.

A highway worker said he discovered the prints which were unlike "anything I've ever seen before."

Several persons ventured that the prints were those of a bear.

This theory was dismissed, however, after it was discovered that the beast had wallowed in the mud.

Bears, authorities said, are too clean to stoop to such undignified pastimes.

HUNTER 'SLAYS' MONSTER; CORPUS DELECTI MISSING

Valparaiso, Indiana, *Vidette-Messenger*, September 5, 1946

Lebanon, Ind.—(UP)—The Lebanon "monster" is dead, Harry McClain has announced, but the search goes on for his carcass.

After a 15-mile chase through the wilds of Boone county, McClain said his assistant, Roy Graham of Indianapolis, plugged the monster with a 30-30 rifle.

"It was definitely a black panther," McClain said proudly. "We chased him out on the tip end of a big tree and he fell in a creek after Roy shot him."

The mud was so sticky and the water so deep, McClain added, that it was impossible to recover the body of the panther.

"He's probably floated into the next county by now." McClain said.

It's All Done Secretly

"The residents of Lebanon need worry no longer," said the ex-big game hunter. "If anything else shows up to scare people, it'll just be imagination."

"We didn't tell anybody what we were doing last night," McClain said, "and there wasn't anybody around to bother us."

McClain estimated the panther weighed about 80 pounds and had eyes that glowed "like fire." He said he was "coal black all over."

"He looked ferocious in that tree," he said.

Residents of Lebanon have complained for weeks that a monster that "cries like a baby and kills livestock" was at large. Mayor Louis Sterling commissioned McClain to hunt down the monster.

Memory Vague

McClain said the monster died "near the Calvin Lennox farm," but he didn't remember the name of the creek.

McClain and several thousand other persons have tramped over the Lennox farm and Joe Tyre farm recently looking for the creature which McClain termed "the slipperiest thing I ever hunted."

Chicken-baited traps also failed to lure the panther.

Harry said the Lebanon resident gave him $200 Monday because "he knew I was going to catch it."

"I'm going to rest a while now," said McClain.

"I hear there are a lot of other snakes and monsters loose in Indiana." be added. If there's an emergency, I might start out again."

"Frankly," he said. "I'm a little skeptical about some of these wild rumors."

Richmond, Indiana, *Palladium-Item*, August 1, 1948
(Reprinted in August 1, 2006 issue)

Richmond police officer Louis Danels, his wife and small daughter have seen Wayne County's mystery varmint.

Danels saw the strange animal near Doddridge Chapel a week before farmers around Fountain City complained to Sheriff Carl Sperling about a "strange thing scaring the wits out of our cattle, hogs and sheep."

"It was awful," Danels said, "I just didn't mention it to anyone at the time, but now that other folks are seeing things and talking, I will talk, too."

Danels said he was taking his family for a ride on Doddridge Chapel Road ... "when suddenly the strangest, most vicious looking thing walked toward my car down the center of the road. It had long front legs, a large head with pointed ears, and small glittering eyes." Danels said the monster had narrow hips and short back legs that made its spine sloop down in a peculiar manner. "We got within 10 feet of it and it ran off the road into the weeds. It was something out of a horror movie. We all remarked that it was the most evil looking creature we ever had seen."

Danels said the animal looked a bit like a hyena. "Those farmers aren't just dreaming. There really is some strange critter roaming around Wayne County. I saw it and so did my wife and our daughter."

Wayne and Union Counties Agog Over Reports of Mountain Lion
Kokomo, Indiana, *Tribune*, August 6, 1948

Richmond. Ind., Aug. 6 (AP)—The mountain lion hunt is going again on the rolling plains of Wayne County.

Sheriff's Deputies Jack Witherby and Ryland Jones picked up the scent by telephone last night. But they didn't find either the lion or the fishermen who started running when they saw.

The latest report came from two families who were fishing quietly near the falls on Elkhorn Creek, just off Highway 21, 5 miles south of Richmond. They said they saw a big tawny animal rise near a brushpile.

That broke up the fishing. The anglers, young and old, took off without looking back. They stopped at the Ivan Toney farmhouse long enough to tell Toney and call the sheriff's office. They were too excited to give their names.

"It looked like a lion," they reported before they fled.

Find Large Tracks

Deputies Witherby and Jones said the only tracks they found were some large tracks in the sand near Elkhorn falls.

The lion stories are scarcely new to the deputies. Farmers in these parts have long been keeping their shotguns handy for any chance encounter with the beast—whatever it might be. It's been variously described as a mountain lion, hyena and mixture of dog and coyote.

Excitement reigned up around Fountain City, six miles north of here a year ago, after farmers began reporting strange cat-like calls in the dead of night that frightened stock and horses as well as humans.

And then one night a group of boys headed home in their car after attending a late show in Richmond, saw a huge black shape leap from the center of the road and disappear into dense underbrush.

Sheriff Carl Sperling probed the strange visitations, then dropped his investigation, when no trace of the animal was found.

Two weeks ago Dorten Moore, who lives five miles from where the animal first was reported, complained of the strange deaths of seven hogs in a few weeks.

Moore reported the snouts of the pigs had been crushed. He told Sheriff Sperling that in each case the killer had eaten only the hearts and livers.

On the strength of Moore's report, the sheriff and three deputies spent an uncomfortable night on the sloping tin roof of a farm outbuilding on the Moore farm. "All we saw was the moon," says Sheriff Sperling.

Hears "Death Call"

But on the next night, Moore heard the strange "death call" again. He armed himself and rushed to his hog lot. There, still warm, was another dead hog, its snout crushed and its heart and liver gone.

Moore reports that at the time he was accompanied by a neighbor's dog. As they stood there, he says, a noise caused him to turn. He saw an indistinct shape pass. The dog gave chase, but turned back quickly and ran to the house. The dog has since declined to enter the field.

Louis Danels, Richmond policeman, relates how he saw a strange animal 15 miles southwest of here last summer. Danels, with his wife and daughter, were riding on a side road one Sunday afternoon. "The animal was in the road in front of me," he says.

"It had wide shoulders and narrow hips. Its front legs were longer than its rear ones, making it back slope down to the rear," he added.

Danels said the animal approached to within 10 feet of his car, then jumped a farm fence and disappeared into a woods. "I got out of there, and fast," the officer says.

A recent appearance of the animal, or another one strange to Indiana, was down by Roseburgh, in Union county. Cliff Fath and Charles Cornelius, conservation officers, saw it. They described it as a mountain lion. A posse was formed and the hunt was on.

Sighted during the hunt, the animal escaped under a hail of bullets.

MORE REPORT SEEING LION IN WAYNE CO.
Logansport, Indiana, *Press*, August 7, 1948

Richmond. Ind., Aug. 6 (AP)—The year-old hunt for the mountain lion of Wayne county took on a new twist today. In addition to their hunt for the lion, Sheriff's deputies were hunting for the two families that last reported seeing it.

The report came in last night by telephone, but the caller was so excited, he forgot to leave his name. At any rate, here's the story as he gave it to Sheriff's Deputies Jack Witherby and Ryland Jones.

The two families were fishing quietly near the falls on Elkhorn Creek, just off highway 21, five miles south of Richmond. Suddenly a big, tawny animal rose from behind a brush pile near them.

The anglers all agreed that it "looked like a lion," but they didn't stop to investigate. They took off for civilization, stopping long enough at a nearby farmhouse to put in their anonymous call to the Sheriff's office.

Deputies Witherby and Jones found some large tracks in the sand near the Elkhorn falls, but nothing else.

HEAVILY-ARMED POSSE STALKS STRANGLE 'CATS' NEAR RICHMOND
Kokomo, Indiana, *Tribune*, August 11, 1948

Richmond. Ind., Aug. 11—(AP)—A posse of heavily-armed farmers and townspeople hunted today for a large animal or animals believed responsible for numerous killings of livestock south of here recently.

Residents of the area who claim to have seen the marauders describe two animals, apparently roving the countryside together. One is described as tawny in color, resembling a lion, and the other as smaller and black.

The posse concentrated its search yesterday in the area around Abington, just north of the Wayne-Union county line. Farmers living near there had reported several stock-killings in the past few days.

W. R. Emslie of the Richmond Palladium-Item, who accompanied the hunters, said tracks of large animals other than dogs or livestock were found in the area. The animals themselves were not sighted.

Yesterday morning a 14-year-old farm girl living three miles south of Richmond told how she saw the two animals as she took the cows to pasture. The description given by the girl, Barbara Perkins, tallied with those of other persons who say they have caught glimpses of the beasts.

Farmers in the areas where the panther and/or lion have been seen said their horses and cattle seemed afraid to leave the barnyards, and even their dogs stayed around the house. Emslie said farmers in the Abington area were even carrying rifles when they went to the mailbox.

Man Shoots at Black Object; Believed Varmint
Richmond, Indiana, *Palladium-Item*, August 12, 1948

Two charges were fired from a shotgun at a large black animal in the yard of the Robert Martin home near Middleboro at 11 p.m., Wednesday, as Wayne County's "varmint" hunt continued.

Martin told Deputy Sheriffs Jack Witherby and Ernest Russell, and W. R. Emslie, Palladium-Item reporter-photographer, that he had fired two charges through a screen of an upstairs window at the animal. He said he had fired from a distance of about 50 feet.

"The dog made such a noise that my wife and I were awakened," Martin told the officers. "I looked out the window into the moonlighted yard and saw the black animal. I fired twice. At the second charge, the animal reared up and then fled through a gate leading away from the yard," he added.

The two officers immediately began a search for the animal, now believed wounded. The Martin home is located one mile north of Middleboro.

Earlier Hunt

Several hours earlier Wednesday night, a hastily organized posse failed to locate anything near the sewage disposal plant south of Richmond after Russell Eggers, 1223 South Twenty-second street, reported he had seen a strange animal there. Eggers was accompanied by his son, Russell, jr.

Only victim of the hunt near the sewage plant was a 4 foot black snake which fell from a tree onto the shirt of Deputy Sheriff Russell, who promptly killed the reptile.

Mrs. Robert Brinker, residing at the liberty Pike, is taking precautions against running afoul of the animal or animals reported seen in her neighborhood.

For Mrs. Brinker keeps her trusty rifle handy and she knows how to use it. In the yard of the home is a target practice range and Mrs. Brinker keeps her sight eye in trim by almost daily practice on the range. She has quite a walk from the house down a winding road to the mailbox on the pike and she takes her rifle with her, "just in case."

"Whether all these stories are true or not, I don't know. But I am not going to take any chances," Mrs. Brinker said.

The young mother used to put her nine-month-old daughter out in the yard for a while during the day, "but I don't do that any more."

So watch your step, varmints, or Mrs. Brinker might use you for target practice.

FARMERS SEARCH FOR MOUNTAIN LION

Cambridge City, Indiana, *National Road Traveler*, August 12, 1948

A 3-hour mountain lion hunt was staged in Union county last week. Tracks were found, but the animal was not seen.

Clifford Fath, Fayette county game warden, and several Union county farmers hunted from 9 p.m. to midnight, seeking the animal seen a few days ago by Mr. Fath on the road between Roseburg and Quakertown. The animal also has been reported seen in Wayne county.

The only thing they found were tracks in the sand at Silver Creek that measured six inches long and such tracks are not those of any of the wild life living in this section of the country.

Mr. Fath says he is receiving stories every day from farmers claiming loss of livestock, and a gnawed leg of a cow was found near the tree where the animal was thought to have been hiding in the hunt a few days ago.

One farmer said a 1,000-pound bull was killed, and Office Fath saw a dead hog that had been killed with claw marks on its head and its liver ripped out.

MARINES JOIN HUNT FOR MOUNTAIN LIONS
Kokomo, Indiana, *Tribune*, August 12, 1948

Bedford, Ind., Aug. 12.—(AP)—The Marines are hunting Indiana's mountain lions today.

The lion scare, up to now concentrated mostly in the Richmond area, spread yesterday to the Crane naval ammunition depot. And the armed Marine patrols at the depot have been authorized to shoot the animals—if they can find them.

Four men told of seeing a pair of mountain lions within a two-mile area of the huge ammunition depot west of here Tuesday. One of them, Andrew Street, said he had hunted mountain lions while in the army and knew one when he saw it.

Street said he spotted the animals lying in a field and approached to within 20 feet of them. One of them jumped up and ran, but the other just stayed there and looked at him. Street left.

First Sergt. M. N. Sutherland of the Crane marine detachment said the alarm was spread about 3 p.m. Tuesday and that patrols were notified to be ready to shoot the lions.

WILDLIFE EXPERT JOINS HUNT FOR STATE'S 'BIG CAT'
Kokomo, Indiana, *Tribune*, August 13, 1948

Richmond. Ind., Aug. 13—(AP)—H. B. Cotting-ham, a state conservation department employe, joined today in eastern Indiana's mountain lion hunt. After making plaster casts of tracks found in the region where the animal had been reported he seemed impressed.

Cottingham, an expert on wildlife, said the tracks led him to believe that "some kind of big cat" had been roaming the countryside.

For several weeks Wayne county authorities have been receiving reports of a "varmint" prowling south of here, killing livestock and terrorizing farmers.

A late report indicated the animal may have migrated to a territory northeast of Richmond.

Robert Martin, a farmer who lives north of Middleboro, about six miles northeast of here, notified the sheriff's office he has sighted a strange animal in his yard and had fired on it from his bedroom window.

"I know I hit him but I'm scared to go out in the yard to see what I hit," Martin reported. He said the animal "reared up" and disappeared in the darkness.

Deputies were sent to the Martin farm but they found no trace of any wounded animal.

Black Panther Scare Hits Capital; Police Skeptical

Valparaiso, Indiana, *Vidette-Messenger*, September 24, 1949

Indianapolis, Sept. 24—(UP)—A black panther was reported loose in southwestern Indianapolis early today and four heavily-armed police cars were assigned to search for it.

Authorities were skeptical of the report, however, despite, "hundreds" of telephone calls from residents who said they heard "mournful howls" in their back yards.

The first report came from Mrs. Edna Lawson, 25, who told officers she saw a "jet-black panther."

She said the animal "stood there looking at men and howled" when she stopped her car at an Illinois Central railroad crossing.

The 'phone calls began coming in after a radio station broadcast Mrs. Lawson's story.

Police said "that cat would have had to be in 15 different parts of the city at the same time for all of them to have seen it."

Four patrol cars toured the area until dawn.

"I won't believe there is a panther until I see it," Police Lieut. Jack Small said. "If there is one it probably holed up somewhere."

Reports of panthers, tigers and leopards are not unusual in Indiana and Illinois. Several safaris, have been organized recently to hunt down real or imagined big game but none of the animals was ever sighted.

BLACK PANTHER HUNTED NEAR INDIANAPOLIS
AFTER REPORTS OF 'JET-BLACK' ANIMAL SEEN

Davenport, Iowa, *Democrat and Leader*, September 25, 1949

Indianapolis.—(UP)—Sheriff James Cunningham said Saturday night he felt certain a black panther, reportedly prowling in a wooded and hilly section in southwest Indianapolis, would come out of hiding Saturday night, if it is there at all.

Mrs. Edna Lawson, 25, reported seeing a "jet black" panther Friday night. A search was organized but no trace of the beast was found. However, it was "sighted" at a half-dozen different places at the same time, police said.

Cunningham said a visiting big game hunter, who volunteered his services, looked over the terrain where Mrs. Lawson said she saw the panther and reported the country was well suited for a wild animal's lair.

"There has been a 'cat' in there," said John Nordyke, who described himself as a New Mexico hunter who has shot wild animals all over the world. Nordyke would not elaborate on what kind of a 'cat' he thought it was.

But he said the need for food and water, both available nearby, would be enough to lure the animal out of hiding as soon as night falls.

There are 27 hogs, a number of cattle and plenty of water close to the spot where Mrs. Lawson reported seeing the panther. The livestock belonged to a farmer who did not object to having it used as bait.

"We'll be there when it happens." said Cunningham. He said his deputies, on the job since Friday night trying to track down the prowling beast, were instructed to "move in" when the animal comes out.

Cunningham said the story Mrs. Lawson told Friday night could be possible. "But we are discounting the other stories told about seeing the panther," he said.

The sheriff said, if it had not been for Mrs. Lawson's story, he too might have abandoned the search as city police did earlier Saturday.

Mrs. Lawson said she saw the panther near an Illinois Central railroad crossing. A canal once ran thru the area and the country now is "wild."

Mrs. Lawson stuck to her story Saturday despite the fact that another woman said the beast probably was her dog.

Mrs. Ruth Chasteen told police that her black Newfoundland dog was running loose in the area where Mrs. Lawson reported seeing the panther.

"I questioned my wife's story at first," said Robert O. Lawson, 32, "but Edna isn't the kind to spread such stories without some basis."

Mrs. Lawson said she saw the animal's yellow eyes blinking and said it had whiskers "like a cat. It was the size of the largest police dog I've ever seen."

One thing still wasn't settled, Cunningham said. He said an investigation showed that no circus caravan moved thru here Friday night, and that no circus has reported it lost a panther near here.

Mrs. Lawson said she passed a circus truck shortly before sighting the animal.

BLACK PANTHER APPARENTLY IS GENTLE NEWFOUNDLAND
Edwardsville, Illinois, *Intelligencer*, September 27, 1949

Indianapolis, (UP)—The phantom black panther that terrorized southwestern Indianapolis this weekend apparently was a big gentle mutt named "Gypsy Blossom," police said yesterday.

Police and sheriff's deputies organized a full-scale search Saturday after a housewife, Mrs. Edna Lawson, reported seeing a panther that apparently escaped from a circus truck.

Early yesterday, a neighbor of Mrs. Lawson, Clifford Chasteen, reported that maybe his big Newfoundland was the object of the hunt. Gypsy Blossom, he said, was nearly as tall as a man when she stood on her hind legs. It would be easy to mistake her for a panther.

Deputy Sheriff Larry Slater agreed that he probably was right and called off the search.

FOLKLORE
Anderson, Indiana, *Herald*, March 9, 1952

by William Wade

"Making fun of a popular belief or story without investigation is just as ridiculous as reaching a decision about anything else in an arbitrary manner," Uncle Ezra said.

"When I was a boy in a sparsely settled section of Indiana people began to tell of having seen what looked like a black panther leap across the road in front of them at night. Some even said they had heard it scream.

"Two boys, Lem and Slim, thought it was ridiculous and funny. So they picked up a stray dog coated its hair with tar and turned it loose near where the panther was supposed to have been seen.

"A few nights later they were walking along a road in the vicinity. On a low stone wall to the right they saw what looked like a black animal.

"Lem nudged Slim and laughed.

"'That's our dog.' he said. 'Let's go over and see how it's doing.'

"As they came closer to the animal one of them yelled suddenly in terror:

"'It IS a panther.'

"They turned and ran. One of them swore that as he looked over his shoulder the animal bounded along the wall after them.

"A few nights later a farmer was passing the spot with a 38 caliber rifle after investigating the killing of two of his sheep.

"His dogs scared out something. He thought it was possibly a lynx or small wildcat. He saw two eyes gleam out at him from a tree and fired, getting it between the eyes.

"Down flopped a black panther.

"Nobody was ever able to explain how it got in the neighborhood. The farmer turned pale when he told about it afterward. He said it was only by chance he was carrying a 38 instead of a 22 caliber rifle."

SON OF DEPUTY SHERIFF SIGHTS STRANGE ANIMAL

Anderson, Indiana, *Daily Bulletin*, October 1, 1955

Interest in the community around New Columbus, 5 miles south of here on Ind. 109, in hunting a strange animal that has appeared in the area during the past week has increased as the result of a new report received last night that Allen Simmons, 11, son of Deputy Sheriff Robert Simmons, sighted the unusual beast yesterday afternoon.

Young Simmons was in a field on his father's farm, which lies along Fall Creek between Brown and Main St. Rd., about 1/4 mile west of New Columbus. Simmons told his father that the animal darted out of a nearby woods, stared at the boy a few seconds, then turned and ran through a field in which weeds are about two feet high, apparently leaping as it ran.

The Simmons boy is familiar with woodland animals and says the strange beast was not a woodchuck, coon or fox. It was about 2 ½ feet long, and had a long tail, snub nose and was about the size of a large dog. It was a dark gray in color, but was neither a dog nor a large cat, according to the Simmons boy's version. After his son saw the strange animal, deputy Simmons made an extensive search for the beast, but found no trace of it.

First reports of the animal were made to Sheriff Joe Brogdon when it was seen by Fred Beaman, a boy, near his home at the west edge of New Columbus. The animal has been cited by two other persons in the locality. Farmers are keeping a watch for the strange "visitor." Thus far no reports have been received from farmers about any livestock being bitten or attacked by any type of animal.

'Phantom' of Fall Creek Makes New Appearance Friday
Anderson, Indiana, *Herald*, October 2, 1955

A new appearance by the "phantom panther" of Fall Creek Valley was reported yesterday by the 11-year-old son of Deputy Sheriff Robert Simmons on the Simmons farm near the junction of Main St. Rd. and Co. Road 600 South, 1 ½ miles east of New Columbus.

The beast has been the object of an intensive search by farmers of the neighborhood and the sheriff's office since it was first reported about 10 days ago.

The Simmons boy, who reports he is familiar with the appearance of common woodland animals of this area, said the beast he saw Friday was not a fox, woodchuck, raccoon or any other common animal. He described it as about 2 ½ feet long, with a long tail and snub nose, dark gray in color and about the size of a large dog. He said it darted from a wooded area, stopped, stared at him for a minute and then fled through a field of weeds about two feet high, bounding high into the air as it ran for cover. The Simmons boy said the animal had the appearance of neither a cat or a dog.

The "long, gray animal" has been reported by other residents of this community, the sheriff's office reported that strange at tracks had also been located in the area.

WILDCAT KILLED IN INDIANA
Anderson, Indiana, *Herald*, January 12, 1956

Madison, Ind. (UP)—A mysterious animal killed near here was believed to be a "wildcat or bobcat," according to Jefferson County Sheriff Harold Raisor.

The strange "varmint" had been roving around the Jefferson County area since last fall.

It was killed when struck by a car driven by Pleasant Point teacher, Carroll Thurman.

The animal had a dark ridge up its back of light brown fur. It was 42 inches long, weighed 20 pounds, had sharp fangs and a long tail.

"BLACK PANTHER" SHOT BY LAFAYETTE HUNTERS
Logansport, Indiana, *Pharos-Tribune*, November 20, 1956

Lafayette, Ind. (UP)—An elusive "black panther" which has frightened Tippecanoe County residents for weeks may have a handful of shotgun pellets in its body today.

Donald Dubes and Warren Gates of near Lafayette said they were hunting pheasants Monday on a farm near Americus when a big cat peered at them from a clump of tall grass.

Dubes fired a charge of No. 6 shot at the animal. The men said it fell to the earth for a moment, then bounded away toward the Wabash River.

"There's no doubt about it," Gates said. "That was a black panther. It was about four feet long and about 10 inches thick at its broadest part. Black as my new car."

The men said they got 'coon dogs and tried to trail the animal but the dogs declined.

Ask State Aid

Logansport, Indiana, *Pharos-Tribune*, August 5, 1957

Petersburg (UP) - Pike County farmers indicated today they will ask help from the Indiana Conservation Department to track down a large animal believed to be a panther which was blamed for killing pigs in this area. Farmer Marvin Goodrid said he saw a black panther cross a road in front of his car last week.

Huge Cat Killed by Car a Mystery Genus to Viewers
It is Believed to Have Been Miami Bend Fowl Stealer

Logansport, Indiana, *Press*, August 24, 1956

Pennsylvania railroad employes, who last fall had the county stirred up with reports of a large catlike creature seen lurking, along the tracks, last night came up with another such creature. This one shouldn't scare anyone though. For one thing, it's much smaller. And besides that, it's dead.

Just what the animal was is debatable. One man said it was just an ordinary housecat gone wild, while another claimed it was a bobcat.

Still another person, evidently very wise, said that since it had been found in the vicinity of the river, it might be a channel cat.

The best guess seemed to be that it was a cross between a housecat and a bobcat. Although evidently still, young, it weighed about 11 pounds.

Paws Very Wide

It also had very wide paws and a long thick tail.

The cat is no longer among the living simply because it couldn't match the 40 mile an hour speed of R. D. Leslie's car.

Leslie, who lives at the east end of. Miami Bend, spotted the animal running down his lane. He speeded up, took careful aim, and that was the end of the cat.

Leslie said that he had lost 16 guineas and one hen to the cat in recent weeks, and that his neighbors had lost another half dozen guineas to it.

This could explain the cat's weight, but there is still the mystery of the wide paws and bushy tail.

BLACK PANTHER HUNTED BY POSSE
Hammond, Indiana, *Times*, January 26, 1958

Boonville, Ind. (INS)—What is believed to be a black panther was sought by a posse headed by Warrick County Sheriff Robert E. Shelton today.

Ed Hernz, of near Paradise, said he thinks the animal killed 35 chickens on his farm and that the remaining 15 chickens now are nervous at night and are roosting high in trees.

Mrs. Walter Brink, a neighbor, said she saw the animal which was about five feet long and 18 to 20 inches tall and State Conservation Officer Paul Sanders found huge paw prints near her home.

Sheriff Shelton said he heard reports that an animal trainer lost a black panther in Warrick County three years ago while transporting a load of animals.

POSSE HUNTS 'PANTHER' IN SO. INDIANA
Logansport, Indiana, *Pharos-Tribune*, January 27, 1958

Boonville (UP)—An all-day search for an elusive "cat-like beast" which has been roaming in rural areas near here ended Saturday without a trace of the animal.

About 25 Warrick County residents, armed with shotguns and rifles and led by Sheriff Robert Shelton, searched a rugged stripmine area in a "five or six mile" radius without results. Warrick is in southern Indiana.

Paul Sanders, a state conservation officer who accompanied the posse, said it failed to find tracks indicating the beast was in the area. He said the hunt would be "discontinued until another report comes in and then we will get right on it."

"All available help will be mustered when we need it," Sanders said. He said Richard Poehlein of Boonville volunteered to let officers use a pack of hounds especially trained to "track down such an animal."

The latest report of the animal was received Friday when William Willis, who owns a farm three miles west of Paradise, said he saw it in a field next to his home.

The mysterious animal has been reported seen on at least five occasions and is blamed for the depletion of livestock and chickens on farms in the Boonville area. It has been described as "cat-like, about five or six feet long, 20 inches high, black, with a long, curly tail." Many residents believe it to be a black panther.

John Hedges reported to authorities he lost 26 lambs to the beast last week. Ernie Winchester said four of his lambs were missing. Both farmers live within a few miles of the home of Mrs. Walter Brink of Paradise.

The hunt began when she reported spotting the animal from the window of her home shortly after 35 chickens disappeared mysteriously.

RESUME HUNT FOR 'PANTHER'
Logansport, Indiana, *Pharos-Tribune*, January 29, 1958

Boonville (UP)—Warrick County authorities Tuesday resumed their search for a "panther-like beast" which has been terrorizing farm communities in this area for the past few weeks.

Sheriff Robert Shelton and state conservationist Paul Sanders led the search Tuesday. They said a pack of dogs struck a trail through the woods and found "fresh cat tracks" in mud leading up a gulley.

The animal, described as a "cat-like beast" which fits the description of a "black panther," has been blamed for the disappearance of chickens and livestock in this area. It has been sighted at least a half dozen times.

NIGHTTIME SAFARI IS BEING PLANNED IN PERRY COUNTY
Anderson, Indiana, *Daily Bulletin*, May 31, 1961

Lilydale, Ind. (UPI)—Another safari into the hills around this Ferry County community is shaping up, but the next one will be at night.

About 70 volunteers, armed with high-powered hunting rifles, combed some 1,000 acres near here Tuesday in a search for a "black jungle cat" which several residents have reported seeing.

The hunters didn't find the cat but they found tracks which Conservation Officer Loren Howe said indicated the presence of a "large cat" in the area.

Howe said he would stage another hunt in the near future. He said the second hunt would be held at night, the most likely time for the animal to be on the prowl for food.

One of the persons who spotted the beast earlier was the Rev. David Brown who spent 18 years in Africa as a missionary. The minister, who saw the cat three times and fired at it without hitting it, said he believed it was a black leopard or panther.

Brown described the cat as being four or five feet long and about 30 inches high.

'STAKEOUT' FAILS TO TRAP 'LION'

Logansport, Indiana, *Pharos-Tribune*, June 29, 1962

Huntington, Ind. (UPI)—Farmer Ed Moorman and animal fancier Everet Wedmeyer spent Thursday night in a barn adjacent to a lot containing 55 prime hogs waiting for the vicious marauder that has killed at least 12 pigs, but it didn't show up.

Armed with 12-gauge shotguns loaded with shells containing .00 shot, the two men were waiting for just what, no one knew.

Wedmeyer, who has hunted big game and just recently donated a pet bear to the Indiana Conservation Department for an exhibit at the state fair, thinks the animal, which eats just the hearts and livers and drinks the blood of the hogs, is an African lioness.

Other theories were that it was a mountain lion of the type prevalent in Western states.

Sheriff Harry K. Walter limited the number of men on stake-out to two in order to keep them from getting in each other's way.

The two men expected the animal to appear Thursday night since it has not eaten since early Monday morning, at which time it killed two 100-pourid pigs.

So far the Moorman farm, R.R. 7, Huntington, has been the only place the beast put in an appearance.

Wedmeyer said if the animal did not show up, it was likely that Moorman wounded it with a chance shot Monday and killed it, or that it was driven away.

Moorman said he was afraid of the beast, not for himself, but for his three children aged 6, 5 and 4 years old.

Give Up Hunt for Beast in Huntington Area

Logansport, Indiana, *Pharos-Tribune*, July 3, 1962

Huntington, Ind. (UPI)—The search for a wild animal believed responsible for the killing of 11 hogs has ended in failure, authorities said today, but the elusive "cat" was still at large.

Weary hunters gave up the search and vigil Monday night, about a week after deputies, farmers and adventurers set out to bag what some experts believed was a lion.

Farmer Edward L. Moorman reported the loss of the hogs near here. He also reported the beast charged and scratched him when he surprised it sleeping in a woods and he said he got off one shot from a rifle.

But that apparently was the last anybody had seen of the creature.

"If anything turns up, we'll get right back on the trail," said Sheriff Harry K. Walter. He added he believes the beast is "either dead or has taken off for places unknown."

El Paso, Texas, *Herald-Post*, September 28, 1962

An armed search for a female black panther continues in Markleville, Ind. The big cat is believed to be looking for her cub which was found in a barn. Mrs. James Hines, who has the cub, alerted authorities when she heard the mother panther screaming at night.

'Autopsy' Reveals Panther Is Kitten

Anderson, Indiana, *Herald*, October 7, 1962

The "panther" is dead.

A supposed baby panther found in a barn near Wilkinson last week, has died and an autopsy has revealed it is just a common kitten.

Purdue University wildlife professor, Charles Mumford, said there "isn't much doubt about it," after he performed an autopsy on the dead animal.

The cat was found at the James Hines farm after some residents reported a black panther had been prowling the area. The kitten was believed to be an offspring of the panther.

Sprinkle (1994) reported on a few instances of black panther sightings. His own sighting was in 1977 near Walton, Indiana, when a large black feline crossed the road in front of his car. Around 1985, a woman claimed to have seen one near the town of Flora.

MONSTER CAT STALKS THE MIDWEST
Chicago, Illinois, *Tribune*, October 31, 1985
INDIANA HAS PANTHER BY THE TALE
Chicago, Illinois, *Tribune*, September 6, 1985

Since August, police received several reports of a large black animal in the vicinity of Michigan City, Indiana. In one case, a "utility company employee said he saw a 'very big' animal with a long tail leap nimbly over a tall wire fence while carrying a dead domestic cat in its mouth." A search later found part of the dead cat, but no trace of the mystery animal.

In most cases the animal was described as a large black feline, but in one instance, the witness called it "a jackal-like creature with large Mickey Mouse ears." There was speculation that this was the same animal reported not long before in Manchester, Michigan. Other officials suggested people were just seeing "large dogs or shadows."

Brad LaGrange (2000b) reported on a 1978 black panther case from Perry County, near Oriole, where a farmer and his Army buddy friend took a shot at a 5-6 foot long sleek black feline that had been prowling the fields. The cat was hit late one evening, but the M1 carbine used was not immediately effective, allowing the panther to escape into the woods. By the time they came back to try and track it, they noticed circling buzzards and decided that the panther had been killed. They did not try to recover the body.

Brad LaGrange received an email (Arment and LaGrange 2000) from a witness who claimed to have seen a mongoose or fisher-like animal "playing" with a group of squirrels. The witness had been deer hunting in Crawford County when he heard the animals in the trees above. It was dark in coloration, and larger than the squirrels, but didn't try to attack them.

Big Cats Can Cause Ruckus

Elkhart, Indiana, *Truth*, July 5, 2005

The article discusses several local cases where big cats were seen or inferred. One involved a black feline: "One afternoon about 2 p.m., Melody spotted a black panther-like cat slowly meandering down the middle of the county road by her house. It was in no hurry so she had plenty of time to observe it from her window. She said it was taller than her black lab dog and had a long tail."

Big Panther-Like Cat Blamed For Killing Pigs in Owen County, Ind.

South Bend, Indiana, *Tribune*, February 10, 2006

A woman in Owen County, Indiana, heard her German shepherd running around their pig pen and went out to investigate. She saw a large black feline, larger than her dog, jump out holding a small pig in its mouth. It dropped the pig and ran off. She and her husband found six more fatally mauled pigs in the pen.

Area 'Black Panther' On the Prowl Again

Bloomington, Indiana, *Herald Times*, June 29, 2007

An Ivy Tech student reported seeing a large black feline behind his Wapehani Hills apartment.

Woman Reports Finding Leopard in Floyd County

Jeffersonville, Indiana, *News-Tribune.net*, August 21, 2008

Barbara Johnson, a Floyd County resident living near the Horseshoe Casino, had seen a "leopard" several times around her home for the past year. (No description of the cat is given, so no mention of whether it was spotted or black.) On one occasion, she saw it with young. After seeing the cat take down a deer, she tried to find someone (animal control or state wildlife officers) to trap and remove it, to no avail.

DNR: Loose Panther Unlikely

Fort Wayne, Indiana, *News-Sentinel*, March 19, 2009

Several New Haven residents reported seeing a black panther, but wildlife officials said it was unlikely, as they would have seen evidence if there had been one roaming the area.

IOWA

A Strange Monster
Iowa People Frightened by a Beast
That Has Its Home In a Swamp
Oxford Junction, Iowa, *Mirror*, August 14, 1890

Independence township, Polk County, is wild with excitement over the appearance of a strange animal in the Kress neighborhood. The beast makes night hideous with its roaring. As near as can be located it has taken up quarters in a creek which passes through the farm of George Powles. The animal remains secreted in the under-brush of the stream during the day and at night wanders forth in search of food. Its voice is described by those who have heard it as a cross between the roar of a lion and the screech of an enraged panther. Several persons who, while walking or driving along the road near, have heard the roar, and becoming almost panic-stricken, can not be induced to pass over the route again. One young man who, while accompanying a lady friend home recently, heard the appalling sound but a few yards distant, and being unarmed, rather than pass over the same route again, he made a detour of nearly ten miles to reach his home. It has at last become a matter of so much vital importance that the farmers have organized for a hunt, and invited the Independence militia and the gun club to join in the festivities.

Escaped Leopard Roaming in Iowa
Hull, Iowa, Sioux *County Index*, November 25, 1921

Glenwood, Ia., Nov. 22 (Special).—Has the black leopard which escaped from a Kansas City zoo several weeks ago in its meanderings

274

reached southern Iowa? Several people the last week while hunting along the Missouri river bottom near here report having seen an animal which they all describe as of the feline species and as big as a shepherd dog. Two who saw the strange animal declared it is a panther. Tracks of the animal have been followed and a half dozen intrepid Glenwood men armed with shotguns and rifles took up the trail and followed it for many miles, but lost it in the timber.

STRANGE ANIMAL IS CAPTURED
NEAR FOX FARM AT CLEAR LAKE

Mason City, Iowa, *Globe-Gazette*, November 4, 1929

Clear Lake, Nov. 4.—An animal, appearing to be a cross between a coyote and a police dog, but showing greater ferocity than either was captured near the Windsot silver fox farm Sunday by Ed Teake and is now in captivity there. For some time it had been noticed prowling about the farm where it had been stalking and eating pheasants. After being confined to a pen at the fox farm, it attempted to escape by gnawing the bars and woodwork. It greatly resented the approach of persons to the cage, growling and snapping when anyone drew near.

SEES "PANTHER" IN IOP AREA

Burlington, Iowa, *Hawk-Eye Gazette*, November 9, 1946

A large cat-like animal was reported seen in the IOP area Friday night by Clifford E. Plumly, 1315 S. Leebrick street. He said he believed it was a black panther.

Plumly asserted he had been visiting a nephew, Robert Madden, who lives in the eastern part of the area, and was driving toward the Middletown gate when the beast suddenly appeared in the road.

"I had to slam on the brakes to keep from running into it," said Plumly. "I stopped the car about 20 feet from the animal and it looked right into the headlights. I'm certain it was a panther, Its black fur glistened in the light."

The motorist described the animal as about 5 feet long and with short legs and a bushy tail.

Plumly said the incident occurred about 10 p.m.

LION AND PANTHER HUNT SLATED NEW YEAR'S DAY

Burlington, Iowa, *Hawk-Eye Gazette*, December 28, 1946

Columbus Junction—A lion and a giant black panther will be the quarry of eastern Iowa hunters Wednesday, New Year's day, as a mass hunt is conducted for these animals which have been reported seen in this vicinity recently.

Walter Meredith, Columbus Junction gasoline station operator, is organizer of the hunt, and hunters will gather at 9:30 a. m. at the Columbus Junction fairgrounds. Plans call for most of the timberland and fields between this town and Wapello to be covered by the huntsmen, a territory of about 15 miles. It is also planned to swing through 3 or 4 miles of land south of Columbus Junction.

The beasts, thought by some persons to have been released by an itinerant showman last summer because of the meat shortage, have been reported seen by several residents of this area, including a Presbyterian minister, and a coon-hunter who said he was attacked.

Clayton Hartsock, Wapello, asserted his 150-pound coon hound dog was clawed by "a full grown male lion" during a recent night coon hunt.

Hartsock said he, too, was attacked by the lion and he crashed his gun stock against the head of the animal. As he hit the beast, Hartsock declared, he fell to his knees and the lion jumped over him and headed off into the darkness.

The dog, badly clawed, is recovering.

Rev. R. J. Bloomquist, Pastor of the Wapello Presbyterian church, reported several months ago that he saw a black panther while driving along a nearby highway.

"I saw a strange animal stalking towards me on the shoulder of the road," Bloomquist said.

"It's body was low to the ground, and noting that it was about 5 feet long, there was no doubt in my mind that it was a panther."

J. B. Hawkins, Wapello, is in charge of vigilante hunting groups in his vicinity. Future reports of seeing or hearing the beasts are to go directly to Hawkins, who will summon hunters living in the area of the report.

It is expected by Columbus Junction residents that about 100 hunters will be on hand for the venture New Year's day.

Seeks Panther, Shoots Wolf

Burlington, Iowa, *Hawk-Eye Gazette*, January 7, 1947

Following tracks reportedly made by a giant coal-black panther in the IOP area early Monday morning, George Killinger, special game warden at the ordnance plant, shot and killed a 40-pound prairie wolf late Monday afternoon.

Accompanied by Earl Welch, IOP agronomist, Killinger followed what he said was a distinct trail in the snow for 3 miles before getting within shooting distance.

Darrell Wolfe, IOP bus driver, and Frank Wimler, a laborer, reported to army officers at the plant Monday that they saw a huge beast lope across the road in front of their bus, that it looked like a black panther. Killinger and Welch received permission to track the animal, which had been reported seen in the plant area several times in the past year.

Killinger said Monday night that the wolf must have been the creature seen by the IOP employes, that Welch and he stuck to the trail made by the animal seen by the 2 who made the report.

Muscatine Posse Hunts a Panther

Waterloo, Iowa, *Daily Courier*, November 12, 1948

Muscatine, Ia.—(UP)— A posse of about 20 men led by Muscatine County Sheriff Fred Nesper combed the Mad Creek area north of here Thursday night for a "panther" after a hunter reported seeing the animal.

G. L. Thompson said he saw the animal while hunting Thursday afternoon about two and one-half miles north of here.

He said he first noticed tracks approximately six inches wide and followed them.

The animal was just ready to leap into a group of pigs when Thompson saw it and fired with his rifle but missed, he said. Thompson said animal appeared to be about seven feet long and "coal black."

Authorities believe the animal may be the same one reported in this area about a year ago.

At that time posses searched for it but no one saw the animal.

Police believe the animal, if a panther, may have escaped from a circus.

TALL TALES TOLD ABOUT "MYSTERY BEAST" WHICH
MANY REPORTED SEEING AT DIFFERENT LOCATIONS

Muscatine, Iowa, *Journal and News-Tribune*, December 30, 1948

You could almost fill a stadium with the people who saw "the mystery beast" in 1948.

It was one of his best years.

He pranced in the beams of automobile headlights.

He planted in the soft earth tracks as big as dinner plates.

His great tail switching from side to side, he stalked little pigs.

He mauled dogs.

With a single blow of his great paw, according to rumors, he felled sheep and cattle and ripped away the warm flesh. Some of the county's more imaginative citizens said the beast slew thousands of dollars worth of valuable livestock in eastern Iowa. This never got into the papers. Farmers blamed dogs. The laws of Iowa do not permit payment of domestic animal claims on losses caused by lions, it was pointed out.

These were some of the tales (tall ones), that circulated freely during the year in the country stores and in the town and city streets of Muscatine and Louisa counties.

Some of the accounts of the appearance of the lion, black panther, or what have you, came from persons known to be competent witnesses and not likely to confuse a big dog with a panther.

Yet the stories could never quite be confirmed by persons who investigated. Here are some more "facts" covering the activities of the animal:

Uttered Hideous Cries

The great slavering brute made the night hideous with his cries.

Boys stayed away from basketball practice. Occupants of lonely farm houses shivered.

Mothers quieted their crying children with the warning: "Be good or the lion will get you."

All of Muscatine county at times became a veritable lion's den and everyone in it modern Daniels.

Wire services chronicled the nocturnal prowlings for the information of all the world. Editors as far away as Chicago called local officers at 3 a. m.

Unlike the leopard the lion demonstrated he could with the greatest of ease "change his spots."

Sometimes in the glow of headlights he would be yellow like an African lion. The next observer would see him slink as black as coal across the highway.

His wanderings followed no pattern. His raids were as hard to predict as those of a task force.

One night he would be in southern Louisa county; the next, in Muscatine county.

Over Wide Area

His appearance late in the year became still more frequent and wider spaced. He began to cover hundreds of miles between performances.

He set the record Nov. 17 when with a magnificent effort he loped in a single night from Freeport, Ill., to Napa, Calif.

But his favorite stamping ground continued to be in Muscatine and Louisa counties. He made his lair along Mad creek north of Muscatine. Here, oftimes, he would yawn, stretch out in the warm sun, and purr away the long afternoons.

He was stalking a litter of little pigs in this very area Nov. 11 when a Muscatine man glimpsed him. The man fired but missed watched the beast vanish in the shadows.

The author of "The Hound of Baskersvilles" could have woven the adventures of the lion into fine mystery story.

The late Sir Arthur Conan Doyle would have delighted to point out Mr. Sherlock Holmes and Dr. Watson bending over odd marks in the sand.

Holmes would be saying: "Watson, what do you make of these marks among the lion tracks?"

"Holmes, they appear to me to be marks left by a man walking on tip toe."

The fog that frequently hangs like a pall over Mad creek would swirl around the great detective as he replied:

"That man was not walking on tip toe. That man was running, Watson, running for his life."

But stories as wild as any ever dreamed up by Doyle have been prevalent in Louisa county ever since the lion first appeared.

When an aged Inmate wandered off from a county home, one man expressed the opinion that "the lion got him."

It is to Louisa county, also that the world is indebted for the story of "The Coming of the Lion." This is the way the story goes:

From A Circus

The lion rode into Louisa county in a cage of a small traveling circus in 1946. Business had been bad for some time. Then came the meat shortage. The proprietor no longer could afford to feed his pet. He did not want to kill it.

One night on a lonely road he opened the cage door and shooed the animal out.

Those who believe this story, believe also that the circus man will want his lion back as soon as meat prices come down again.

Meat prices are falling now.

Some dark night in 1949, these folks predict, an aged and battered circus truck will rattle to a halt in the shadowy woods that border Mad creek.

A flash light will glimmer briefly as a thick steak plumps down on the floor of an open cage.

A voice will plead: "Here kitty, kitty, kitty. Here kitty, kitty, kitty. Here kitty, kitty, kitty."

A cage door will clang shut.

Next day posters will appear on Second street announcing the presence on the river front of "the greatest show on earth."

The prowling of the lion will be over.

"MYSTERY" BEAST IS SIGHTED NEAR JOHN HOLLIDAY FARM
Muscatine, Iowa, *Journal and News-Tribune*, August 9, 1948

That "mystery beast" which has been reported sighted at various points in Louisa and Muscatine counties during the past year and a half is abroad again.

It was sighted Sunday night in Port Louisa township by Leonard Peterson of Muscatine in Port Louisa township, near the John Holliday home, about 10 miles from Muscatine, according to a report by Earl Harrison of Route No. 6, Peterson's brother-in-law.

Harrison, who lives a short distance from where the mystery beast was sighted, related that his brother-in-law was returning to Muscatine Sunday night, and caught the animal in the headlights of his car. Peterson returned to the Harrison home and reported sighting the animal, which was reported to be catlike form, about 2 ½ feet high and five feet long. It had smooth dark covered hair.

The animal was gone when Peterson returned over the highway. However tracks about 4 inches across by five inches long were found today for a distance or several hundred feet along the highway where the animal was sighted.

Harrison reported that so far as he knew, there had been no reports of residents in the area losing chickens, but said that a large shepard dog belonging to Mrs. Chet Martin, living nearby, has returned home several days ago badly mauled, as if it had been in a fight with some animal. It is thought possible that the dog may have encountered the mystery animal.

Panther "Roars" at Auto Driver
Waterloo, Iowa, *Daily Courier*, October 26, 1950

Des Moines—(UP)—Paul Hildreth, 19, said Thursday a black panther put its paws on his car and "roared" Wednesday.

Hildreth said he heard the animal scream in southern Polk county and then spent 45 minutes in his auto while the animal prowled about.

He said the big cat once came up to the car, put his paws on it and roared.

Mrs. Anna Mae Cooley, Hildreth's sister, said the animal had been reported seen by several other persons last summer and she had seen it once.

She told the sheriff's office it was "black as coal" and was about three feet long with a long tail.

The sheriff's office said it had no previous reports of the animal.

Panther Hunt Organized in Polk County
Mason City, Iowa, *Globe-Gazette*, October 26, 1950

Des Moines, (AP)—Polk county sheriff's deputies plan to go panther hunting Thursday—right in Polk county.

They will investigate reports that a black panther has been seen in the vicinity of S. E. 14th street and the southern border of Polk county.

"It's been more or less a joke all summer," said Mrs. Anna May Cooley in reporting the animal, "but today my brother came face to face with it."

The brother, Paul Hildreth, 19, said he had been working in some timberland on his father's farm when he heard a "shrill scream." He said he saw the panther on a nearby hill.

"It walked right toward me," Hildreth said. "I got in the car and shut the doors. He put his paws on the car and roared."

Mrs. Cooley said others have reported seeing the animal during the past summer. Some cattle have been killed, bearing evidence of claw marks, she said.

She described the animal as "coal black, 3 feet long and with a tail just as long."

A 2nd report came Thursday morning from Robert Halterman, 33, of Des Moines. He told the sheriff's office he saw a black panther in a tree while duck hunting at Goodwin lake southeast of Des Moines. He said a black tail "about the length of my arm" was hanging from the branches.

HUNT BLACK PANTHER REPORTED AT DES MOINES
Mt. Pleasant, Iowa, *News*, October 27, 1950

Des Moines, (INS)—Polk county Sheriff Howard Reppert turned big game hunter today after a black panther was reported running loose near the southern edge of the county.

First official report of the beast was made last night by Mrs. Charley Hildreth who lives in the vicinity. Mrs. Hildreth said her son was chased into the family car by the animal. The youth, Paul Hildreth, 19, said the animal was coal black and about three feet long in body.

Mrs. Hildreth said "five good-sized" calves have been killed in the vicinity in the past two or three months and have been "partly eaten."

Mrs. Anna May Cooley, who also reported the presence of the beast, said it has been "hanging around all summer." She reported authorities in neighboring Warren county are also planning a hunt for the animal, which young Hildreth said kept him trapped in the car for nearly 45 minutes yesterday.

'Black Beast' Mauls Dog Near Des Moines
Winona, Minnesota, *Republican-Herald*, November 21, 1950

Des Moines—(AP)—Law enforcement officers, dogs and an airplane went on a "safari" today to hunt down an elusive black beast that is giving northwest Des Moines the jitters.

Several people have reported seeing the animal in the past several weeks, and from their descriptions officials said it could be a black panther. Tracks measuring about four inches across have been found.

Kenneth Sondeleiter, who operates a zoo, warned people to beware. He said panthers sometimes attack people "just for the sport of it."

Deputy Sheriff Max van Rees warned residents of the area to use caution when they are out of doors.

Several mothers asked yesterday whether they should let their children go to school. Deputies told them to "use your own judgment."

The latest encounter with the animal was reported yesterday by T. L. (Mike) Lester. He said his dog—115-pound Tennessee boar-hunting hound—fought with the beast while Lester was coon hunting.

Lester said he approached within 30 feet before scaring the animal away. He described it as "a black, shiny animal with a long tail." He added that it was "larger than the dog."

The dog is four feet long and 28 inches high. Though badly clawed, the dog will recover, Lester said.

Expert Archer Joins Hunt For Iowa's Black "Panther"
Strange Creature Roaming Wooded Areas
in Vicinity of Des Moines
Lowell, Massachusetts, *Sun*, November 21, 1950

Des Moines, Ia., Nov. 21 (UP)—An expert archer joined the hunt today for a black "panther," roaming wooded areas near Iowa's largest city.

Charles Austin, 27, civilian radio operator for the state highway patrol, said he hoped to stop the "panther" in his tracks and end the forays that have baffled and alarmed residents of the Des Moines area for days.

Dog Loses Out

He unlimbered his bow and arrow after coon hunter T. L. (Mike) Lester reported that his prize hound had come out second best in a battle with the "big, black shiny" beast.

Lester said he was coon hunting yesterday when "all .... broke loose" in a weed patch into which his Tennessee hound "Old Jack" had nosed. Lester ran up to the thicket, and found his dog locked in a fierce struggle with the strange animal.

Lester, who said he didn't carry a rifle when hunting coons, shouted and the beast fled into a brushpile. The dog limped up to his master with a torn and bloody coat.

Sheriff's deputies, already searching for the "panther" which various residents had reported sighting recently, hastily formed a posse to search for the animal that attacked the dog. They found nothing.

Gives Warning

Deputy Sheriff Max Van Rees and Kenneth Sunderleiter, operator of a private zoo here, said the coon hunter's description tallied with that of a black panther or leopard. Sunderleiter said he believed it might have escaped form a traveling circus or carnival, although no such escapes have been reported.

Sunderleiter warned residents of the area to beware. He said panthers or leopards attack "just for the sport of it."

Dozens of persons telephoned authorities to ask if it was all right to go outside their homes and send their children to school.

Austin, who often has killed deer with his bow, said he would take "two or three other fellows" with him and operate separately from the sheriff's detail.

"On this type of hunt a bow and arrow would be an advantage because they create less shock than a bullet," he said. "The fire of a gun excites an animal and makes him mad, while an arrow doesn't.

"Besides," he added, "there would be more glory in killing a panther with an arrow."

Posse Bags 1 Fox, 1 Hunter—No Panther

Mason City, Iowa, *Globe-Gazette*, November 21, 1950

By William Lovell

Des Moines, (U.P.)—A horde of hunters, crashing through woods and fields near Des Moines killed a fox and wounded a fellow hunter Tuesday but failed to find Polk county's prowling "panther."

Orville Jones, 46, was hit in the mouth by stray shotgun pellets when a group of hunters opened up on what they thought was the mystery beast reported seen several times in an area about one miles northwest of Iowa's largest city.

When the fox was sighted, the nearly 30 rifle-bearing members of the posse opened up with a barrage described by one reporter as "like the Battle of the Bulge."

The fusillade of shots brought down the fleeing animal.

Jones lost a couple of teeth and was taken to a Des Moines hospital where his condition was reported not serious.

The hunters, armed with everything from pistols and shotguns to rifles with telescopic-sights, closed in on the frightened fox when an airplane circling, overhead began waggling its wings and diving frantically. A deputy in the plane reported by radio he saw "something."

"There he goes," someone shouted.

"That's him all right."

Guns began popping and 30 or 40 shots rang out as the fox raced along the edge of a cornfield.

Otherwise, the 125 or so hunters and a score of dogs failed to flush anything but a rabbit.

The big hunt was ordered by the sheriff's office after Coon Hunter T. L. "Mike" Lester reported that his prize hound had come out 2nd best in a battle with a "big, black shiny" beast.

Charles Austin, 27, civilian radio operator for the state highway patrol, said he hoped to stop the "panther" in his tracks and end the forays that have baffled and alarmed residents of the De Moines area for days.

He unlimbered his bow and arrow after Angry Coon Hunter T. L. "Mike" Lester reported that his prize hound had 'come out 2nd best in a battle-with a "big, black shiny" beast.

FARM RESIDENTS REPORT THEY SAW IOWA'S 'BLACK PANTHER'
Sheboygan, Wisconsin, *Press-Telegram*, November 22, 1950

Des Moines, Ia.—(UP)—A "panther" hunt in the heart of the corn belt was spurred today by two farm residents' reports that they saw a "long, shiny black animal" darting through their fields.

Searchers led by Deputy Sheriff Max Van Rees took up their safari again despite the fact that a 125-man hunt yesterday bagged only a little red fox and one of Orville Jones' front teeth,

But Van Rees said he would try to restrict the posse to a few hunters, directed by a low-flying plane. He called off yesterday's hunt after Jones, one of the task force of nimrods, was wounded by shotgun pellets in a wild round of firing at the hapless fox.

Late yesterday, however, deputies rushed to the farm home of Mrs. Ed Lockner on the outskirts of this city of 180,000 population when she reported "a long, shiny black animal just chased one of my cows." Mrs. Lockner said the beast scurried off across a field when a farm hand ran up to investigate.

"It was much too big for a dog, and certainly was not a cow," she said. "It resembled pictures of panthers I've seen."

Deputies found huge cat-like tracks but did not attempt to follow them.

Joe Woods telephoned authorities to report he saw "a shiny black animal with a long tall" slinking around his stock farm near Urbandale, a Des Moines suburb.

He said the animal was only 30 feet distant when he spotted it. It ran.

Previously, the mysterious animal had frightened several persons in wooded areas near the city, and a hunter's prize coon hound was attacked and mauled.

Hunters from miles around, armed with everything from pistols and shotguns to expensive big game rifles, joined the posse yesterday and thought they had hit the jackpot when a spotter plane frantically wagged its wings and dived.

The long "skirmish line," which the posse had formed in the woods for safety purposes, disappeared in a moment. Men raced into the field the plane was circling, someone shouted "There he goes!", and shots crackled, like a string of firecrackers.

Jones fell, slightly wounded. The fox fell, dead.

More "Panther" Tracks Found Near Des Moines
Cedar Rapids, Iowa, *Gazette*, November 27, 1950

Des Moines (AP)—The "long black animal" that had been making frequent appearances left some tracks on a farm northwest of Des Moines, the sheriff's office reported Sunday.

Mrs. Laura Kimmel told the sheriff's office that some rabbit hunters found a number of large, fresh tracks.

This is in the same general area where organized hunts have been conducted unsuccessfully to trap the "cat-like" creature which has become known as the "black panther.

Marion, Ohio, *Star*, December 31, 1951

Beastly labor relations developed during a union feud with the Oliver Corp., Charles City, Ia. Instead of showing up for work, some 1,400 members of the leftwing United Electrical Workers suddenly heard about a black panther—and went out to hunt him. They couldn't find him for days. All the while, the union leader insisted that beating the bushes for the black beast was a "public service." The chase ended when the union contract was signed.

The panther is expected back at the contract's expiration...

"Panther Hunt" May Close Plant
Cedar Rapids, Iowa, *Gazette*, January 20, 1951

Charles City (UP)—Iowa's wandering black panther, which has been seen just about everywhere else, turned up in a labor dispute Friday.

Charles Hobbie, international representative of the United Electrical, Radio and Machine Workers of America, said the entire Oliver Corporation plant here may be shut down Monday while members of the independent union go "panther hunting."

The foundry of the farm equipment plant was shut down Thursday and Friday in a dispute over piecework rates. Hobbie said the entire local voted Friday "to organize a mass posse to hunt the big black panther near Bassett, beginning Monday morning."

Latest report of the panther, which has been the object of organized hunts near Des Moines and elsewhere, has come from nearby Bassett.

Hobbie added, however, that "if the company wants to help find the panther it might be found the first day."

TRAPPED CAT MAY BE 'BLACK PANTHER' OF MILLS COUNTY
Council Bluffs, Iowa, *Nonpareil*, February 3, 1954

Has the "black panther" of Mills County finally met his nemesis?

Only time can tell—but a cat of the size and color of the animal has come to the end of the trail.

The cat, apparently of common domestic ancestry, was caught in a trap set by Harold Berwick state custodian at Manawa Park and Ross Wiles, state park employe, Tuesday night.

A dark blue in color, the feline shaded almost to a black over the back. He was caught in a trap on the far south side of the lake where the men were trapping out predatory animals in the preserve there.

Puts Up Vicious Fight

The trapped animal fought so viciously that it was necessary to kill him before removing him from the trap: He measured just under three feet from the tip of his nose to the tip of his tail. He was only three inches short of standing a foot and one half high at the shoulder.

For quite some time, reports coming to state conservation officers of a black wildcat or panther observed along the Missouri River bottoms of Mills and Fremont counties.

Latest reports had placed him in the area just to the south of Lake Manawa in Mills County. Farmers in that area were awaiting a good snowfall in order to organize a hunt for the creature.

Whether the trapped cat of Lake Manawa is the "black panther" of the bottoms remains to be seen. But to quote Johnny, Borwick's 4-year-old son, the cat is "sure a big son-of-a-gun."

WAPELLO AREA JITTERY OVER JUNGLE BEASTS
Davenport, Iowa, *Morning Democrat*, October 15, 1955

Wild cries pierced the night. A coon hunter was attacked by a lion. A pastor saw a panther.

No, this was not in Africa. This was Louisa county, Ia., in December of 1946, when a wild animal scare was no joke.

As near as anyone could determine, a small circus released several wild animals not far from Wapello in the summer of '46. It seems that meat was especially short, and the show didn't have the money to feed the beasts, so they let them go in the woods.

Farmers really got the jitters when the animals started to get bold. One night, the Rev. R. J. Bloomquist, pastor of Wapello Presbyterian church, was driving on Highway 92 near Grandview. He thought he saw a dog on the road shoulder. When he got closer, he gulped to see that it was a black panther, five feet long with legs bigger around than a man's arm. Not long afterward, James Hartsock went coon hunting near Wapello with four dogs. He heard a commotion, and was puzzled when three of the hounds ran yelping like they'd seen a lion. (And whatta ya know—they had seen a lion.)

Hartsock flicked on his flashlight to outline one of his dogs, a 150-pounder named King, pinned against a barbed wire fence by a snarling lion.

"It was like a nightmare. I almost went out of my head," said Hartsock. The lion ignored the dog, and in one bound leaped toward Hartsock who jerked the butt of his gun up to protect himself. The gun smacked the lion in the face, and the animal bounded off into the darkness.

Frightened farmers of Louisa county quickly went into action. Elaine Hawkins, editor of the Wapello Republican newspaper, organized the farmers into groups and set up a telephone system so they could be mustered in different sectors whenever the animals were sighted.

But the critters continued to evade hunters; so on a cold New Year's Day in 1947 several hundred farmers fanned over the Louisa county countryside in a mass wild animal hunt. Their guns were loaded with heavy shot; they meant business.

However, the only thing the big game hunters shot were a half-dozen rabbits and a fox.

Throughout that winter, scattered reports continued to be received about wild panther-like cries and huge pawprints. By spring, the whole thing was forgotten and no one has ever again seen the "Wild Beasts of Wapello."

BLACK PANTHER SIGHTED AGAIN AT HAMBURG
Glenwood, Iowa, *Opinion-Tribune*, December 3, 1959

A large black animal, described variously as a black panther, a large bobcat, and a mountain lion, stirred up quite a flurry of excitement around Hamburg over the week end.

First spotted on the Missouri bottoms west of Hamburg Friday morning, the animal was seen at close range again Sunday night, northwest of town. Only one thing is certain: a fairly large black animal is now living on the bottoms west of Hamburg.

Clarence Hendrickson (Junior) first saw the animal and shot at it while stalking deer just over the Iowa line in Missouri southwest of Hamburg late Thursday afternoon. He had been with Dick Murdock earlier in the day when Murdock killed a large deer, and had returned to the same area hoping to spot another deer.

The animal came out of the willows without seeing Hendrickson who was standing quietly near a deer run. Hendrickson sighted in with his telescopic rifle, but missed him. The beast fled into nearby brush and disappeared.

A search of the area next morning revealed sets of tracks two and one-half inches across, but the animal was not sighted.

Several groups of hunters Saturday and Sunday failed to find a trace of anything.

When next seen it was at the Wayne Long home a block west of the Albert Propp place northwest of Hamburg. Long's son, Larry, was watching television while waiting for his parents and young sister to return from Hamburg.

He heard a scream similar to those of his exuberant younger sister, and thought his parents were back. Turning on the yard light he discovered the mysterious black beast standing 30 feet away in the drive. Larry rushed back inside for his gun, but the animal had disappeared when he came back outside.

The Albert Propp family, too, reported hearing the same child-like scream.

Speculation here is that the animal, whatever he is, is probably the same creature observed through field glasses for ten minutes a week ago by hunters at Pacific Junction.

Iowa's Cougar Mystery
Des Moines, Iowa, *Register*, April 28, 1974

Within an article otherwise dedicated to accounts of alleged mountain lions (of normal coloration), there is mention of a Swanson farmer who lost a calf to a black cougar.

Big Cat Sighting Raises Questions
Burlington, Iowa, *The Hawk Eye*, March 29, 2005
Cat Owner Speculates on Sighting
Burlington, Iowa, *The Hawk Eye*, April 9, 2005

A man photographed what he thought was a black panther, estimating it at one-and-a-half to two feet tall. It was sunning itself on some farmland just east of Mount Pleasant. Not long after, a man living in the same area noted it was probably one of his two over-sized black cats. They had the white throat markings noted in the witness' photographs.

KANSAS

That "Black Panther" is Shot, Proves to be Wildcat
Kansas City, Missouri, *Times*, October 25, 1950

Salina, Kas., Oct. 24. (AP)—Farmers hunting a mysterious black panther near here shot a 15-pound wildcat Monday night. It was on the same farm where the panther was seen.

Francis Roesner, who brought down the cat, said it was different from the animal he saw Sunday and believed to be a panther.

Strange Animal Seen at Taylor Ranch Early This Morning
Great Bend, Kansas, *Daily Tribune*, September 14, 1953

This morning about 6 a.m., Mr. and Mrs. Tom Taylor, who live on and are part owners of the Walnut Hill Hereford ranch northwest of Great Bend were awakened by the barking of their two dogs and the bawling of a number of Herefords in a field adjacent to the house.

Taylor got out of bed to see what the disturbance was and saw immediately behind the dog pen not 50 feet from the house, a large black animal with feline characteristics. He yelled at the dogs to quiet down and then called his wife, who also verified the fact that such an animal was there.

Taylor released the dogs who took out after the creature and he got into his car and gave chase down the entrance road and through a feed field. He said that the animal easily outdistanced the dogs, and he lost the black feline in a wooded patch.

Mrs. Taylor estimated that the creature was about two and one half feet tall, about five feet long and had a long tail that was ... bushy.

He had short legs, and a square ugly head. He was completely black except for a … patch on his shoulders, Mrs. Taylor said.

After the chase, the Taylors found several tracks around the house and driveway to prove that something unusual had been there. The tracks measured about five inches in diameter.

"BLACK PANTHER" SEEN AGAIN IN KANSAS AREA
Joplin, Missouri, *Globe*, October 16, 1954

Kansas City, Oct. 15 (AP)—Residents of neighboring Johnson County, Kan., had reason to be alarmed today after the third alert this month was out that a "black panther" was seen roving the area.

The latest persons reporting the animal were Mr. and Mrs. H. S. Fuller. She glanced out her back window this morning and was startled to see the awesome figure of what appeared to her to be a black panther. Her husband also verified seeing the "panther."

By the time six deputies had arrived at the Fuller home, the animal had disappeared. Officers who began a search of the area with dogs, reported seeing large tracks in some soft mud.

A member of the search party, Sgt. Edward Bowers, said he saw a black form that resembled a panther, but that it darted away before he had time to shoot.

A survey of the area failed to reveal the "black panther," which was first reported in March, 1951.

BLACK PANTHER ON TURNPIKE
Salina, Kansas, *Journal*, April 7, 1961

Wichita (AP)—A young black panther, skunks and deer have been reported seen by motorists traveling the Kansas Turnpike recently.

The Ray Goggin family of Derby reported the panther, apparently half grown, crossed leisurely in front of their car near Andover, between Wichita and El Dorado, Easter Sunday.

Hunters in the El Dorado area along the Whitewater River reportedly have seen panther tracks near their traps, turnpike officials were informed.

Deer and skunks have been sighted at several points along the tollway, as open weather starts the area's animal life roving again.

TESCOTT PANTHER OBJECT OF HUNT
Salina, Kansas, *Journal*, August 2, 1963

Tescott—The great Tescott panther hunt is on.

Harry Woody, Tescott farmer and sportsman, has been asked to use his dogs to track the animal reported to be prowling the Saline River and visiting Tescott at night.

Two women said they saw a big black cat in a Tescott alley Thursday.

It's believed to be a panther and it's still in the Tescott area. Bait left in a tree Thursday night was gone Friday morning.

The undersheriff said he had been told a panther had been caught in the vicinity several years ago.

Mrs. J. C. Murcdick and Mrs. Jessie Cains said the animal they saw Thursday was black and of the cat family. It was about 36 inches long with a 36-inch tail.

The fence the animal leaped when the women saw it is about five feet high.

Tescott residents reported hearing screams resembling those of a panther Thursday night and Enlow said it was reported sighted.

The screams have been awakening residents of the area for about two weeks. There also are reports that something has been stampeding cattle and horses.

HAS MYSTERY ANIMAL BEEN DRIVEN AWAY FROM TESCOTT?
RESIDENTS ARE HOPING SO
Salina, Kansas, *Journal*, August 5, 1963

Tescott—Tescott residents are hoping the animal which has been stampeding horses and cattle and frightening residents with its screams to the night has been scared away.

Sunday night residents slept without being awakened by any screams. This was the first such quiet night in about three weeks.

Dr. L. B. Eustace, Tescott said the animal was heard Saturday night when residents were out in force hunting it, but by the time men and dogs got to the area it was gone.

He raid there was no organized search Saturday night, but cars and people were all over the city waiting for the animal to show.

No one hunted the animal Sunday night.

The animal was first seen by two women, Mrs. J. C. Murdick and Mrs. Jessie Cains, both of Tescott, in a Tescott alley Thursday morning. They said it was a black creature of the cat family, about 36 inches long with a 36-inch tail

Whether it was a panther no one knows.

Dr. Eustace says there is nothing like its scream, "Once you hear it, you never forget it."

THAT WAS NO PANTHER
Salina, Kansas, *Journal*, August 12, 1963

The black panther scare hit Salina for a short time Sunday evening. A boy told police he had seen a black panther near Salina Concrete Products, Inc., 1102 W. Ash.

Tescott residents have reported seeing a black panther several times in past weeks.

Police checked the area the boy described, but found no black panther. They did locate a big black dog, however.

DEERFIELD: BLACK PANTHER HAS BEEN SIGHTED
Fort Scott, Kansas, *Tribune*, September 6, 1963

By Mrs. Orville Hillman

Norman Beeman, who lives near Milo, reported seeing a black panther while plowing in his field last summer. He said it was real brave. It sat and watched him for a while and then went off. He's been reported in Deerfield not long ago and one resident took plaster casts of his or her paw prints coming up from the Marmaton River. They also reported a smaller track which showed it has a cub. A blood curdling scream has been heard east of Nevada and east of Fort Scott in the winter.

'CAT' ROAMING AT GLEN ELDER?
Salina, Kansas, *Journal*, December 31, 1965

A Glen Elder resident reported a large black feline on the east edge of the town. Other residents had also seen the animal, closer to the sewer pond south of the town.

POLICE REPORT
Atchison, Kansas, *Daily Globe*, July 31, 1966

A deliveryman claimed to have seen a black panther cross the highway near Halls, while traveling in the morning to Atchison from St. Joseph.

LARGE BLACK CATS SEEN IN JEWELL COUNTY
Mankato, Kansas, *Jewell County Record*, August 9, 2007

While driving to work about 5:30 a.m., a construction worker, Joel Massey, hit a deer on Highway 36 between Montrose and Formosa. He turned off to the side of the road, got out with a flashlight, and went back to pull the deer off the road. Massey then realized that "two large black cats with long black tails" were feeding on the deer, jumped back in his truck, and alerted authorities. The Sheriff suggested that the felines were in pursuit of the deer when it ran across the road.

KENTUCKY

"American Leopard

"Frankfort, Kentucky, Aug. 21.—The following description of an animal lately killed in Ohio county, in this state, has been handed us by the sheriff of that county, who assures us that the statement contained therein is strictly correct:

"A leopard was killed on the 6ᵗʰ day of June, 1820, by John Six, living on the waters of Green River, ten miles south-east of Hartford, in Ohio county; length from the end of the nose to the buttock five feet, and a tail two feet long. Under the jaw the colour was white and black spots equally proportioned; the sides and back are yellow, with black spots curiously arranged; a row of black spots on its back, much larger than those on its sides, extending half way down the tail; small round ears, black outside, white inside; around its nose and mouth were long stiff bristles, some appear to grow out black half the length, then white, six inches long. The hair on the end of the tail is longer than elsewhere; tail slim; its legs were short, and its feet were like a cat's, only much larger; with large claws, large teeth; supposed to weigh about 150lbs." (Anon. 1820)

A TIGER IN KENTUCKY

Edinburgh, UK, *Advertiser*, September 5, 1823

We learn from Russellville, that a gentleman discovered an animal of an alarming appearance, a few miles from town, and hastened to the nearest house, where he was joined by three men, two of whom were armed with guns, and attended by a dog. The strange monster was again discovered, and while bayed by the dog, the two guns continued to fire on him at the distance of about fifty yards, without forcing him to move from his stand: a furious look and appalling brow, frightened the two men without guns, who fled to town. Experienced marksmen continued to fire, and on the twelfth shot the beast put off at full speed, marking his way by blood flowing from many wounds that it must have received. The dog was too much frightened to continue the pursuit, and the huntsmen dared not venture, although one of them was as fearless as Boone himself, and accustomed to the chase from early life.

When the news reached Russellville, about forty gentlemen repaired to the spot, and had a full view of the ground. The prints which the paws of this animal made in the earth correspond with the accounts given of his great bulk by those who had an opportunity of viewing him at a short distance for several minutes; he was of a brindle colour, with a most terrific front; his eyes were described as the largest ever seen in any animal. We are well acquainted with the party engaged in the attack, and give the fullest credit to the account we have received.

The conclusion drawn is, that the animal in question was a Tiger of the largest order from Mexico, and that it has, like the monsters of the deep, thought proper to wander into distant regions. There is nothing remarkable in his passing such a distance unobserved. Wolves have been seen of late years low down in the northern neck of Virginia, a distance of nearly 200 miles from the Blue Ridge, the supposed residence of those animals; they had to pass through a country of the thickest population, unprotected by large forests, until they arrived on the Potomac river, where cedar and pine thickets shelter them from all future danger.

The above Tiger was seen a few days after braving a dozen shots, making its way into the state of Tennessee, and there is still a prospect of its being taken, and the public gratified with a more minute description.—*Kentucky Gazette*, July 17.

A MONSTROUS BRUTE.
NOTHING LIKE IT KNOWN TO THE MOST EMINENT ZOOLOGISTS.
IT HAS A BROAD BODY, FLAT HEAD, BIG FIERY EYES, WOOLLY HIDE,
BUSHY TAIL, POWERFUL LIMBS AND BLEEDING MOUTH.
Monroe, Wisconsin, *Evening Times*, October 28, 1893

The "dog eater," panther, or whatever it is that has created consternation time and again throughout this section among the country folks, has again made its appearance, after an interval of something like a year, says a Danville (Ky.) dispatch to the Cincinnati *Enquirer*. The existence of this strange animal has been scouted at by the skeptics, but persons of undoubted veracity who claim to have seen the monster during its midnight prowling say they are willing to make oath to the statements concerning it.

About five years ago it made its appearance in this county, and several parties were organized in the vicinity of Perryville to hunt the strange beast down and exterminate it, but none were successful in their mission. From the fact that it seldom, if ever, attacked anything save dogs, the people gave it the name of the "dog eater," and by this it has been known for about seven years. Persons versed in natural history say they can recall nothing like it, and seem to think from the descriptions given by those who have caught glimpses of the animal, that it is a cross between a panther and a mastiff, though the descriptions vary so at times that such a conclusion cannot be relied upon.

Its last appearance was in Mercer County, a short distance from this city. James O'Connor and the colored the of R. E. Coleman's bus were returning from Burgin with several passengers aboard, and had just passed the old Walden farm and were coming down hill at a moderately rapid gait, when suddenly the team stopped, reared, snorted and plunged about, almost upsetting the bus and badly frightening the passengers, acting just as horses have been seen to do when scared by some strange beast.

In a moment the occupants of the vehicle were startled and almost paralyzed at seeing an animal of enormous size and ferocious looks spring out of the woodland into the road, glare at the conveyance a moment and then leisurely leave the scene without molesting anything. The animal was distinctly seen by Mr. O'Connor and the driver, who were sitting upon the front seat. They described it as being of a dark color, with a broad, flat-like body and head, large, fiery eyes, woolly hide, powerful limbs, bushy tail and a monstrous head and mouth. There can be no doubt of Mr. O'Connor having seen this animal, as he would not concoct such a strange story, and his testimony about the appearance of the beast is corroborated by others who have seen it.

The question asked by many is: what is this monster that comes and goes, and still molests nothing except the worthless curs of the country, except now and then destroying a fancy setter? It is no stranger in Mercer County. Several years back there was a current report that some strange animal had taken up its abode in Bonne's cave, and the people thereabout, especially the colored portion, were very much alarmed, and afraid to venture out after night. A few determined ones, however, explored the cave, but failed to find the monster, though they discovered strange-looking tracks in the moist earth on the floor of the cave.

Two other gentlemen, Mr. Phil Marks and Edward H. Fox, the artist, claim to have seen this remarkable beast one night as they were returning from a coon-hunting expedition. They were writing leisurely along the pike, engaged in conversation, their fine pack of hounds following behind, weary and worn out after the chase, when suddenly Marks' horse reared up and had it not been for Mr. Marks' expert horsemanship, he would have been thrown backward against the ground. Mr. Fox, who preserved his presence of mind, soon saw all the cause of the trouble. The dog eater had stepped out into the road ahead of the party and began drinking out of a small stream, and right here this animal's strange influence over dogs was illustrated. The hounds following along seemed to become paralyzed with fright. They huddled together, trembling with fear and whining piteously. Mr. Fox drew his revolver and shot the dog eater, which jumped over the fence and disappeared. The artist is confident that he hit the monster, but thinks that the thick coating of hair on it was too much for the small bullet used. After the animal had gotten out of the way the hounds struck for home at a 2:40 gait. Mr. Marks can be found at his place of business in the city at any time, and will cheerfully detail the story of his experience with the now noted animal. Mr. Fox, at the request of the reporter, made a rough sketch of the dog-eater as it appeared to him.

CATAWAMPUS . . . DISTURBER OF HUNTERS, BELIEVED FOUND

London, Kentucky, *Sentinel-Echo*, February 6, 1919

All the excitement about the strange wild animal that has been disturbing the peace of the citizenry in West London and thereabout, seems to have abated with nothing definitely learned about the cause and the wherefor of it all.

Last Thursday night, Mr. F. M. Tompkins shot at an animal in a tree near his home which he thought was the much sought for demon, and it was described as a most beautiful black and white spotted creature, that could jump further and run faster than any four legged animal ever seen by man, but the strange noise continued to be heard in various sections.

It is believed by most persons that it was all the work of a few boys, practical jokers, with what is called "dumbells," skins stretched tightly over the end of a bucket used as a drum, and a resined string attached. A demonstration of this sort of device was given on our streets Tuesday night.

But whatever it was, it furnished a mighty good story, kept little children and some grownups closer home than usual and was the basis of many a rude joke—it was a joke.

Moore (1977) collected a number of folk tales from Sara Cowan, who grew up in a small town at the foot of Black Mountain in Harlan County, Kentucky. Cowan mentioned that "big black panthers" were known to live in the mountains.

CAT SIGHTINGS BECOMING FOLKLORE
Bowling Green, Kentucky, *Daily News*, February 26, 1984

Joe Ford, of the League of Kentucky Sportsmen, has received "any number of almost identical sightings from all over the state" of black panthers. A state wildlife supervisor suggested they were just feral cats, noting that one Russell County man had shot a "black panther" and put it in his freezer. A state biologist took a look at it and identified it as a big house cat. One hunter saw a large cat, "as black as night," in Crittenden County about 15 years prior.

Several individuals in Scott County reported seeing a large gray cat over the last few years, some noting darker spots on its sides. One state police detective who saw it said it was the size of a German shepherd, with a head the size of soccer ball.

SIGHTINGS OF BLACK PANTHER ARE BEING TAKEN SERIOUSLY
Bowling Green, Kentucky, *Daily News*, June 22, 1993

Reports of a black panther near Dennis in Logan County were treated seriously by Sheriff's deputies, but doubted by the county conservation officer. The first sighting (June 7) came from a woman who said she saw a 3-foot-tall black feline sharpening its claws on a tree. All sightings were within a half-mile radius, and one deputy suggested that the feline may have been living in caves nearby until disturbed by road construction crews.

READERS REPORT BEAR, PANTHER SIGHTINGS
Portsmouth, Ohio, *Daily Times*, January 21, 2007

Residents in Body, Carter and Lawrence Counties reported seeing a black panther. One nurse was driving Route 3 to Ashland just before daylight when a large black long-tailed feline crossed no more than six feet in front of her. She braked quickly and could see the glow of its eyes in the lights. She said a neighbor had watched one take down a deer.

THE BLACK PANTHER OF SCHULTZ CREEK
Portsmouth, Ohio, *Daily Times*, March 1, 2009

A Maloneton woman looked out her kitchen window towards Shultz Creek and saw a large black feline walking through the field. She said it was at least three feet in length, not including tail and head, and was at least two feet at the shoulders. Other neighbors thought they had seen a similar animal.

LOUISIANA

FIND AN ANIMAL THAT EATS DOGS
CREATURE SHOT WHILE FEASTING ON
CARCASS OF MULE. RESEMBLES HYENA.

Trenton, New Jersey, *Evening Times*, June 9, 1910

Church Point, LA., June 9.—Newton Barousse found that his mule colt, which was born the day before, had been killed during the night. The carcass was partly devoured. All of the colt's head and neck, the left foreleg and part of the shoulder were eaten. Mr. Barousse was unable to account for such a depredation, but he suspected the much talked of "dog eater." He allowed the carcass to remain unmolested, and decided to lie in wait the following night, hoping the animal would return.

His anticipation was realized, for that night at about ten o'clock, Mr. Barousse saw the animal approaching. When within shooting distance he fired, and the animal was brought to the ground. After being wounded the animal did not yelp or howl as a dog usually does, but expressed its dissatisfaction at the reception by growls and snaps.

In the darkness Mr. Barousse hurriedly examined the animal, and left it for a well sized dog.

The following day, J. E. Daigle passed by the carcass, and after close observation he pronounced it a cross between a wolf and a dog. Since then a great number of people have examined the body, and nearly all pronounce it a wolf.

R. B. Boettcher and Edward Daigle drove out to make an examination of the carcass. The animal was found to be of a red grayish black color, about four feet long, one and a half feet high in front

303

and fully six inches lower in the hind quarters, giving the animal the appearance of sloping toward the back and the characteristic development of a hyena. The tail was only about six inches long and very thick. The animal's neck, shoulders and forelegs were of unusually large size and developed to double the proportion of the hind quarters. The head was bulky, with a long nose; very small, straight, erect ears and a mouth of unusually large size and ferocious appearance.

Covering the neck and shoulders was a growth of hair about three inches long, and running along the animal's back was a streak of stiff hair resembling hog bristles. From the appearance of the mane around the shoulders the animal showed traces of lion blood, but the conclusion was reached that it was a wolf.

R. B. Boettcher, who is familiar with the West Texas coyote, says the animal killed resembled that species of wolf except it is larger.

Some call the animal killed a wolf, others say it is a hyena, while some think it is a catamount, but all agree that it must be the famous "dog eater."

During the last sixty days there has been some wild beast running at large in this community, attacking dogs and eating them. In the neighborhood of Branch the beast had jumped at a young man riding along in his buggy and lately this animal was seen several times around Crowley. Owing to its fondness for eating dogs this beast became known as the "dog eater."

One early morning, around 2 a.m., in November 1945, Wanda Dillard reported seeing a large black panther cross the road in front of her as she drove along Highway 90 between Franklin and Morgan City. She turned around and drove back, and was able to locate it sitting by the edge of the woods. As there was no traffic, she sat and watched it for about ten minutes. She stated: "Having seen the mountain lion and the bobcat both in the wild and in captivity, I would relate this black panther more to the mountain lion in both looks and movement, but sleek and more graceful. In size he would have been a bit smaller than the mountain lion, but much larger than the bobcat, and with a much more dignified looking head, and the ears were not pointed." (Clark 1993)

Shreveport Girds To Continue Hunt For Big Panther
Albuquerque, New Mexico, *Journal*, January 14, 1952

Shreveport, La., Jan. 13 (AP)—Shreveport police are awaiting the arrival of larger hounds to continue their hunt for a "black panther" which has been seen several times near the city during the past ten days.

"Cat dogs," especially trained for hunting big game, will be brought from the Caddo Gap hunting lodge at Caddo Gap, Ark.

Det. Lt. L. E. Hall said several persons have reported seeing and hearing the animal within the past ten days. He said descriptions of the animal "ranged from a tan mountain lion to a black panther."

An estimated 200 volunteers and hunters of the Southern Coon Hunters' Assn. formed a posse several days ago, but no trace of the "panther" was found.

Experts have described the animal as a puma, a member of the cat family which includes the cougar, panther and mountain lion."

Some reports indicate that the animal has lived near Shreveport for the past ten years, living off rabbits and other small game. However, there have been no reports of dogs or other domestic animals killed during the ten-year period.

Lt. Hall said the animal is believed to be hiding in the Cross Lake section, a "well-settled" area north of the city.

Many residents in the vicinity have armed themselves with shotguns and locked themselves in their homes.

Panther on the Prowl?
Alexandria, Louisiana, *The Town Talk*, April 22, 2004

Residents of Gardner, Louisiana, have reported seeing a "black panther." One witness said "the panther was solid black and had a tail as long as the rest of its body. The animal was about the size of a large hound dog."

MAINE

Portsmouth, New Hampshire, *Journal of Literature & Politics*, April 25, 1835

Wild Animals.—The forests of Maine still abound in numerous species of wild animals, such as the Moose, Deer, Caribou, Loup-cervier, Lunkasoose and many others—most of them valuable for food and for their skins. The Lunkasoose, (the orthography is arbitrary) is an animal of which we have only heard recently; but tradition says that a ferocious animal of huge size, and with a mane like a Lion, has actually been seen to come to the borders of the river, and the Lumbermen say that they have heard him in the woods roaring most lustily. The Indians, too, talk about the "Lunkasoose,"—and they are conclusive authority on such matters.—*Bangor Whig.*

"Till lately I have strongly adhered to the opinion that a 'Black Lynx' was 'dyed in the wool'—*after death*. Recent researches have almost made me doubt. I have received assurances from men whom I think reliable, that there is, or has lately been, such an animal in existence. How it could have escaped the sharp eyes of our naturalists, I cannot imagine. It is represented as being of large size, almost as large as the black bear; in form and general habits resembling the ordinary Canada lynx; but is said to be as ferocious as the Canada lynx is timid. The hair is said to be thick, long and shaggy, and as black as Erebus. It is also said to have great local attachments, never leaving the impenetrable wilderness of swamp which it inhabits. The Indians have many wild and curious legends or traditions which perhaps refer to this animal. He is doubtless—if he exists—the 'Lunxus' or devil of the Indians of Maine. The 'Black Lynx' is said to be able to throw a full grown sheep across his shoulders and make off with ease." (F.R. 1894)

Milwaukee, Wisconsin, *Daily News*, February 2, 1864

An unknown animal has been shot in the vicinity of Moosehead lake, Maine. The body is from fifteen to eighteen inches in length, very slim, of a dark gray color, wide but short ears. The tail is nearly as long as the body, somewhat bushy, in alternate rings of black and white.

Camden, Maine, *Herald*, November 29, 1879

Some extravagant stories are being related about an animal that is said to infest the woods bordering our village, and just now our sportsmen are excited to such a pitch that to merely pass an opinion against there being such an animal, brings forth a look of scorn from one of the believers that would freeze a Behemoth. The prevailing idea however is (among the shooters) that the strange animal is a Lucifer, although nothing to indicate that such is the case has so far as we have learned been discovered. Some declare that they have heard howls and yells, that would put the king of the African forest in the shade, others have seen tracks as big as an elephant's, and more have seen where the "critter," had been indulging in ten feet jumps, and other yarns, that would take the conceit out of Peleg Pelter, of Peltersville. If now that young feller that used to slay lions with a club would put in an appearance, and rid us of this supposed monster, he would receive the everlasting thanks of an excited public.

STATE NEWS. YORK.

Bangor, Maine, *Daily Whig and Courier*, February 22, 1887.

It is reported that a wild animal, supposed to be the panther that has been seen in various towns, was seen in Wells Reach a few days ago. A little boy in the neighborhood set a trap on the bank of Little River near the shingle mill, and accordingly went to investigate the situation. The boy says "that as he was approaching the river he saw a strange looking creature come over the bridge and go down over the bank and behind the ice house and disappear." He also says "the creature was spotted and larger than a large Newfoundland dog." The boy was very much frightened and took refuge in a tree, but after

the creature disappeared he beat a hasty retreat for home. A party
of men armed with clubs made an investigation of the premises, but
the supposed panther could not be found.

A WILDCAT

OR SOME STRANGE BEAST SEEN ON A PHIPSBURG ROAD

Bath, Maine, *Independent*, September 17, 1898

Thursday Dr. and Mrs. W. G. Webber drove down by the back
road through Phipsburg and Parkers Head to Popham and while
riding up a high hill this side of Parkers Head, both Doctor and Mrs.
Webber saw, 300 ft. ahead in the road, an animal with a body about
a yard long, apparently close haired and striped with black and gray.
His tail was long and a black stripe extended its length. The animal's
tail was as big round as one's wrist and he looked like an overgrown
cat with a head that was a cross between that of a cat and that of a
bull terrier.

Hearing the team coming, the beast leaped into the woods and
disappeared. The doctor inclines to the belief that the animal was a
wild cat, but others think it may possibly have been a coon. The doctor says it looked like the wildcats he has seen mounted in muse-
ums.

STRANGE ANIMAL REPORTED TO BE SEEN

Kennebec, Maine, *Daily Kennebec Journal*, April 3, 1914

The report has reached the office of the Commission of Inland
Fisheries and Game of a strange animal seen in Township 7, Range
16, Piscataquia county, near the East branch waters. The report is
confirmed by several people in that section and the source of information is considered reliable by the commissioners. The tracks in
the snow made by the animal were twice the size of the tracks of
bobcats and places in the snow crust where the animal had broken
through proved that the animal was heavy.

One gentleman stated that he had followed the tracks for some
distance and found where the animal had killed a rabbit, and a little
further along had found where the animal had stretched. Hairs of
the animal found at that place were over an inch in length, were wiry,

and were grayish brown in color—not streaked like the hair of a bob-cat.

Some time ago the story was about that a Frenchman in Chessuncook region had been chased by a wolf, and it is thought that this animal in question might also be a wolf.

STRANGE BEAST SIGHTED IN EDGECOMB-BATH AREA
Portland, Maine, *Press Herald*, December 30, 1950

Another cat with a long tail, might be a loup-garou or a wahoo, if there's such an animal, has been spotted. This time it is roaming around Edgecomb near Bath and experienced hunters contend they've either seen it or its tracks.

Some time ago we received a report of a black panther roaming in the southwestern portion of York County. The latest panther or cat isn't black. It looks like a large bobcat only it has a long tail.

Although one attempt to track down the strange animal was made on a fresh snow, hounds being used in the chase switched tracks, eventually drove a medium sized bobcat in the open where it was shot down.

It is believed by hunters who have seen the tracks of the strange animal that it is of the wildcat or lynx family. The animal could be a freak. Only last year several hybrid wild dogs were killed in the Maine woods and before they showed up very few experienced woodsmen and hunters would have agreed that such animals existed in Maine.

Bruce Wright (1972) noted the report of Justice Randolph Weatherbee of the Maine Supreme Court, who was deer hunting in November of 1949 or 1950. Weatherbee and his hunting partner came across recent tracks in the snow, and followed them, thinking they were from a large bobcat. The came to a clearing, and saw the feline on the other side, realizing it was as large as the Judge's "fifty-five-pound Airdale dog but noticeably longer with a long slender tail. It appeared very dark, almost black, but it was wet from falling snow."

Researcher Matt Bille noted that his father had once seen an unusual feline (Arment 2000b):

"My father was on a rural road in eastern Maine about 1955 when a black cat jumped right across in front of his car. Dad describes it as much too big for a domestic cat, but it seemed smaller than black leopards he remembered from circuses. Dad was an experienced hunter and a very level-headed guy."

STRANGE CRITTER AT SCARBOROUGH, ME.
Lowell, Massachusetts, *Sun*, November 4, 1956

Scarborough, Me., Nov. 3 (AP)—Residents near Scarborough Downs racetrack would like to know if a strange animal, reported seen by several people, really is a black panther.

Charles B. Stone, office manager of the Downs, said today Donald Miner, a Standish painter, told him he was in a group of men working at the closed track when a strange animal crossed a nearby field. Miner said he recalled a black panther, shot by his father in Vermont when he was a boy.

Miner jumped into his truck and gave chase, but lost the animal when it went through a fence. He described it as twice as big as a bobcat, with a long tail and picket ears, Stone said.

Alfred E. Jackson, a former game warden and veteran bobcat hunter from Dry Mills, said he "wouldn't doubt it could be a panther," but that in this region it would be light brown or cream colored rather than black. He said the last panther killed in New England was in 1905 at South Paris.

Newton (2005) noted the May 1981 sighting of a black panther by a game warden's wife. She saw it near Harrington Lake in Baxter State Park.

SPORTSMEN SAY, BY GENE LETOURNEAU
Augusta, Maine, *Kennebec Journal*, February 17, 1983

This column noted the sighting by Frank Kendall of what was "definitely a black cougar" sunning itself on a Titcomb Hill ledge (west of Route 43). It was "about five feet long, had a long tail and was coal black."

SPORTSMEN SAY, BY GENE LETOURNEAU

Augusta, Maine, *Kennebec Journal*, December 5, 1983

This column included a couple sightings of black felines. Mrs. Flora Gehrke, several years prior, saw a "black panther cross the road on Route 148 in Madison near our home in a wooded area." ... "He displayed the sleek head, powerful forelegs, long curling tail and dark pelt." Mrs. Laurie Ann Vigue was driving with her husband to Telos Lake when she saw "a large black cat, with a very long tail, running across the road, from left to right, in front of us, back into the woods."

SPORTSMEN SAY, BY GENE LETOURNEAU

Augusta, Maine, *Kennebec Journal*, March 17, 1984

This column noted the sighting by Ben Rogers of a "quite dark" cat-like animal, estimated at "between 50 and 60 pounds," crossing Route 202 between Albion and Unity.

SPORTSMEN SAY, BY GENE LETOURNEAU

Augusta, Maine, *Kennebec Journal*, July 9, 1992

This column noted the sighting by Allen Linkletter and his father, both veteran loggers, of a "very large black cat with a long tail" that crossed Farrar Mountain Road in Township 1 Range 12. It was "very agile and larger than a big German Shepard Dog."

MANITOBA

Manitoba researcher John Warms (*pers. comm.*) has collected a wide range of stories about mystery animal from that province, including several carnivore-like creatures. These include:

Maned felines: A few stories scattered throughout central and northern Manitoba. There is even a small "Lion Lake" east of Lake Winnipeg.

Doglike animal: Around 2000, a 75-pound animal with shaggy brown fur was reported in southwestern Manitoba. It loped like a dog, but had a compact snout. One farmer thought it had lynx-like features, but a hunter who saw it didn't think it was a lynx.

Water dogs: An aquatic animal that is reported to have canine-like characteristics. Sleek, brown, agile, with a short snout, one witness described it as a "weiner dog with big ears and short legs." The traits sound musteline, though otters don't have big ears.

MYSTERY BEAST BLAMED FOR KILLING 3 DOGS
CBC News Manitoba, August 10, 2006

Four sled dogs disappeared in the Northlands First Nation in Lac Brochet, and three of them were found gutted on an island in nearby Reindeer Lake. What exactly killed them was unknown, but people were advised to avoid the lake for the time being.

MARYLAND

BLACK PANTHER SCARE SETTLED
Hagerstown, Maryland, *Morning Herald*, March 24, 1955

People were staying out of the way of the path of a black panther north of here on the Middleburg Pike for a while yesterday afternoon.

However, a farmer made everyone feel a lot better. He identified the animal which somebody had glimpsed as his friendly, but rather large, black Persian cat.

The black panther rumor spread in the area near the Colonial Motel when someone saw the cat out of the corner of his eye, and State Police were even asked to investigate.

HUNTING AND FISHING, BY GENE BROWN
Cumberland, Maryland, *Times*, December 11, 1955

A group of hunters from Hagerstown yesterday killed what they thought was a jet black wildcat on Green Ridge Mountain but it turned out to be a housecat which had gone wild.

The cat was considerably larger than the tabbies found in most homes and its fur coat was heavier than usual. The cat was in excellent condition and well fleshed, indicating it had been living very good on the small game in the woods. Cats which go wild kill much game, especially rabbits and birds, and some sportsmen believe they are a greater predator than dogs which roam the woodlands.

STRANGE ANIMAL REPORTED SEEN
Hagerstown, Maryland, *Morning Herald*, August 9, 1957

Despite a number of telephone calls received recently by game wardens of Allegheny County of persons having seen a strange animal of the cat family roaming the woods in the Little Orleans section, near the Washington County line, no one yet has got real close to the animal.

A few days ago William Price, a Herald correspondent at Little Orleans, reported that Raymond Creek and Frank Price had definitely seen the animal while working on a road. William Price is also a road worker.

Frank Price said he got a good look at the animal which was in no hurry and disappeared in nearby woods. He said it was spotted, had a long tail and was about the size of a Shepherd dog.

Game wardens have searched the woods in the vicinity of where the animal was last seen but no further trace has been found. It is thought it might have crossed the Potomac into West Virginia where some think it originated.

Being spotted some believe it might be an hyena, an African animal, which has a fairly long tail. Wild cats have stubby tails. The animal might have escaped from a circus, side show or zoo, according to some.

Wardens do not believe the animal to be a puma or mountain lion as they have not been seen in this part of the country for years, and are mainly in the West.

CATAMOUNT SHOT NEAR JUGTOWN; HAS BEEN KILLING CHICKENS
Hagerstown, Maryland, *Daily Mail*, June 30, 1961

Jugtown's chicken-killing animal, that has been keeping the people of the area awake by its piercing screams for the past several months, was killed last night.

Henry Holtzman, who has lost half a dozen chickens during the last month, shot the animal, and it has been identified as a catamount. The animal has a long tail that is marked by black rings circling it.

"When it screamed, it sent cold chills down your back," said Holtzman, who told how he had been trying to get a shot at the animal for weeks.

"Last night, after losing a chicken the previous night, I put the chickens in the coop. A neighbor, Albert Huntzberry, was over at my place when we heard the cries of the animal, which I thought might either be a wildcat or a bobcat. I got my shotgun and presently I saw two shining eyes in my garden near the chicken house. I drew a bead on the eyes and let go."

The fact that the animal, which weighed about 20 pounds, did not have a short tail, ruled out it being a bobcat. The long tail with black rings identified it as being a catamount.

Holtzman said he hadn't seen a catamount for fifteen or more years. Occasionally, several decades ago when a brother used to hunt raccoons in the mountains above Jugtown, the dogs would tree a catamount and would kill it. So he quickly was able to identify the species of wild animal that he shot last night.

Big Cat Seen Again in Area
Cumberland, Maryland, *Times*, February 6, 1966

A large black feline was spotted several times in eastern Allegany County. It was seen along Fifteen-Mile Creek, on Town Creek Mountain, and near Locust Grove. District Game Warden Eugene Law saw it cross Route 40 ahead of him, heading into the woods on Martin's Mountain. Law said it was about five feet long with a long curved tail. It was "black with some gray."

Baltimore Area Man Shoots at 'Big Cat'
Cumberland, Maryland, *Times*, June 4, 1966

Several individuals saw (and shot at) a black feline near the Eggers farm 10 miles south of Baltimore. One described it: "It stands about two feet tall at the hind legs. ... It's got a long slinky tail, beady eyes, a short nose, short ears and it's got whiskers about four inches long on each side of its jaw. He's a real slick black, about the prettiest thing I've ever seen. His coat shines as bright as silver. ... It doesn't run ... it takes a stride—a leopard's stride—sort of a swift lope."

Hunters Don't Find Panther
Annapolis, Maryland, *Capital*, June 6, 1966

Hunters, led by Walter Eggers, searched the Hanover area (near Dorsey Road) for the alleged black panther, without success.

POLICE FAIL TO FIND PANTHER
Annapolis, Maryland, *Capital*, June 9, 1966

After sightings and reports of strange cries near Linthicum, police set up all-night patrols for the "black panther," but did not find it.

PANTHER PROBE PRESSED
Annapolis, Maryland, *Capital*, June 10, 1966

County police focused on the "black panther" sightings, with a detective interviewing eyewitnesses to determine if there actually is a panther on the loose. The police chief was skeptical, as he hadn't heard of any circuses or zoos missing a big cat, and there still wasn't any conclusive physical evidence of the feline. A panther hunt with dogs was being organized. There was a report of a dog owned by one searcher, noted to be a Cherokee Indian, rushing into the woods only to emerge with an eight-inch gash on its neck. The dog died of its injuries.

POLICE SEE DANGER IN POSSE HUNT OF PROWLING PANTHER
Annapolis, Maryland, *Capital*, June 11, 1966

The county police chief urged citizens not to go hunting for the "black panther," but to stay out of the woods and let the wildlife officers do the searching.

PANTHER CASE 'CLOSED'
Annapolis, Maryland, *Capital*, June 13, 1966

The county police chief stated that the "black panther" case was closed as far as they were concerned, despite three more sightings coming in over the weekend. James Boraman, of Glen Burnie, said he was awakened by a cat-like cry and looked out his window to see a three-foot cat that ran off into the woods. Jerome Murphy of Linthicum said he saw it Saturday night on a neighbor's lawn. A woman nearby said she saw a "long, black animal with green eyes" earlier Saturday along the road.

COUNTY MAN STILL WARY OF PANTHER
Annapolis, Maryland, *Capital*, November 29, 1966

Walter Eggers was still keeping an eye out, and rifle ready, for the "black panther." He had seen it again three months earlier.

A man claimed to have seen a strange creature in the wooded hills of Shookstown, near Gambrill State Park (Chorvinsky and Opsasnick 1989). It drew his attention with a scream, and he saw it "crouched down flat on the ground about 30 feet away," backing away from him. It was "the size of a deer, dark brown, and had shaggy hair. Allen felt that it was too large to be a dog. The head was somewhat triangular with a pointed chin, pointed ears, and sharp teeth. According to Allen, the legs were most unusual—they were not vertical to the body like a deer or dog, 'but stuck out from the side of the trunk of the body making its movements appear almost spider-like as it backed away.'"

CATOCTIN RESIDENTS SAY BIG CATS PROWL PARK
Frederick, Maryland, *News*, July 10, 1991

Residents near Catoctin Mountain Park (home of Camp David) claimed that large felines (black and brown) prowled the woods. They "are knee-high, and weigh 100 or more pounds." A farming couple on Manahan Road often saw what they believe was a "powerful" black panther crossing the fields. Another resident, Larry Smith, off Manahan Road, said that he photographed a black cat feeding off dumped livestock entrails, but size couldn't be determined in the image. A wildlife biologist with the National Park Service noted that Catoctin is too small to support even a single adult cougar.

BIG CAT SPOTTED IN MAN'S BACKYARD
Washington, D.C., *WTOPNews.com*, July 28, 2006

A man mowing his back lawn in Brandywine, Maryland, saw a large black cat "walking slowly, about 20 feet away from him and into the woods." Compared to the man's 120-pound Rottweiler, he said the cat "was four times the size."

MASSACHUSETTS

A Circus Tiger Loose
He Kills Chickens and Calves
The Farmers Afraid of Him

Hartford, Connecticut, *Times*, September 6, 1888

Special to The Harford *Times*, Winsted, September 6, 1888

Information has been received here that the farmers and backwoodsmen in that vicinity of Massachusetts lying between the towns of Otis and Sandisfield, are in a state of great excitement, owing to the presence of an animal believed to be a young tiger. The information as yet is somewhat vague, but that a strange animal is exciting much terror in that neighborhood is an undisputed fact. The depredations of this roving beast are frequent. Last Sunday afternoon he entered the chicken-coop of William Watson and killed fourteen fowls, eating four and carrying away two. Monday, Andrew Mason, who owns considerable live stock, found a cow and a heifer lying dead in the pasture, while the rest of the drove were huddled in one corner of the lot, crouched down in abject terror. The hides of the dead animals were terribly mutilated, and at their throats were large and deep wounds, made evidently by very sharp claws. From appearances, it was judged that the attacking foe had sucked the blood, as, although the arteries were cut very little, blood could be found on the ground. A calf belonging to an Otis farmer has disappeared, and while theft was at first believed to be the cause, this murderous visitor is now suspected. That the animal is a tiger is not doubted. Mr. Cobb, of Wallace's circus, which was exhibiting in this vicinity last week, informed the Times representative that last Thursday night a tiger, 13-months old, had escaped while the circus was near Lee, Mass., and that although hunters were out after him, no trace of the beast had been discovered. This fact was naturally kept as quiet as possible. The missing animal was born in London, Ontario, last year, of the captive tigress, "Princess." The farmers near Otis seem afraid to hunt the animal.

THE WILD ANIMAL IN LEVERETT
Fitchburg, Massachusetts, *Daily Sentinel*, January 24, 1889

Leverett had an exciting day, Tuesday, caused by the wild animal having been seen in the morning near Rattlesnake gutter by a teamster going down the highway leading through that celebrated pass. This horses became frightened and almost unmanageable and started down the steep highway with a rush, giving the driver all he could attend to to keeping them in the track. Glancing to the side of the way to learn the cause of the strange action of his team, he caught sight of a strange striped animal which he thought was as high as a five-week-old calf and twice as long. It was but a glimpse as he rushed down the mountain and gave the alarm at the first house at the foot, and then to the scatter farmhouses until he arrived at the box factory at Gates & Henry. When he told his story again the firm decided to shut down the factory for the day and let the man go hunting the strange beast.

The news spread toward North Amherst and even to Amherst. Landlord Woods and [—] Messenger procured a few rifles from the militia boys and started for Rattlesnake gutter and the beast. Probably 40 men were on the ground and numerous tracts of a kind unknown to the experienced hunters were found in the snow, showing that the beast had been capering about the neighborhood in a lively way since the fall of snow, Sunday night. Nothing else was found except some fur or hair that had evidently been pulled from the body of some beast forcing itself through the thicket. The capture of the beast is still to be reported.

DEDHAM'S NONDESCRIPT WILD BEAST
Philadelphia, Pennsylvania, *Inquirer*, August 16, 1896

Dedham, Mass., is an attractive little town and the natives there are of the peaceful kind—people who like to live and let live. It is the ambition of the girls and boys of Dedham to have nothing to do except walk through the pretty promenades there, and they are legion. But lately the young folk have been confining their walks to the most thickly inhabited part of the town. The trouble is that there is a wild beast going around and no one can make out just what kind of an animal it is. Several months ago the people of Dedham were excited over the publication in the Boston Post of a story about a curious wild beast in the outskirts of the town, but the people scoffed and said the ones who had seen it had partaken of a trifle too much hard cider, that being the acme of dissipation around Dedham. In the spring this beast was termed a monster wildcat, and it was run across by a hunter in the Fox Hill woods of the West Dedham one day, and the sportsmen considered himself in luck to get away with his life, the member of the family of lynx rufus not only being willing to engage in battle, but expressing every evidence of being anxious to have its opponent come to the scratch. As its discoverer related, an encounter with it single-handed could only have been followed by disastrous defeat, and bearing in mind that discretion is the better part of valor, there was a prompt retirement, without close regard to the order of the going.

Frequently after this the cat was seen by different parties, and as its appearance indicated extreme ferocity, a good deal of alarm was inspired. After a time, however, it was killed by a well-directed shot and Dedhamites breathed more freely and did not make it an imperative duty to lock the chicken coop doors securely every night.

There is now something of a revival of the feeling of fear and apprehension by the advent of a newcomer with its nocturnal meanderings and maraudings. Just what it is no one seems to suppose to declare. Opinions differ. Some call it a wild dog, others incline to the belief that it is a wolf, and several have declared it a hyena. It is agreed that it is a bloodthirsty wretch, no matter how far men's minds may diverge as to its identity, as there is a very pretentious record of killed and eaten poultry to its credit. The ordinary safeguards of a valued flock that stand securely against cats, skunks and the regulation run of enemies of the hennery are as naught before the onslaught of the stranger. Wire screens are torn apart, slats clawed and bitten until entrance is possible, and all found the next morning to tell the tale of the violent and to the lives of several plump and tender pullets is a scattering of feathers and blood.

As it is now no one knows what will come next. A month ago it was common to meet berrying parties scouring the hills and woods in all directions. Children went alone far from the site of a habitation or sound of a human voice, giving themselves unreservedly to the enjoyment of facing nature in all its untenanted beauty. They do not do it now. The youngsters insist on company, and they keep along the open places and clearings. The women of the families in the vicinity of Fox Hill, where the homes are distant from each other, do not care to cut through the woods just now as they once found a convenient.

There is alarm in every crackling branch, falling twig and rustling leaf. In the dark shadows of the forest fancy pictures the gaunt form of a cruel monster, waiting with lips curling over its long glistening teeth to tear the flesh, drink the blood and crunch the bones of helpless victims.

The beast that has occasioned this state of feeling among a large number who live outside the most thickly settled part of Dedham was first met by a woman and several children who were blueberrying in the West Dedham woods at the base of Fox Hill looking to the southeast. They described it, so it is told, as the size of a large dog. In fact, they thought it was a dog. One thing made prominent mention of was that its hind legs were much shorter than the front. It was also asserted that it was striped. But it looks and actions were so decidedly ugly and threatening that they fled, not even looking behind to see if they were followed, but expecting every instant to feel themselves seized by the remorseless claws and white fangs.

The story went from mouth to mouth. It is an adage of long past days that "anything worth telling is worth adding to," and it may be that the description of the nondescript as now heard may be somewhat magnified from the reality. Several persons were found who claimed to have seen it at a distance, and to their minds the animal was as high at the head and shoulders as a table. Its color was black and brown, with a short tail. The eyes looked to be large, the face round, the mouth wide and the teeth showing plainly.

The Dedham Historical Society has discussed the matter, so it is understood here in Dedham, and concluded that the beast is a hyena that probably escaped from some menagerie. A trip of venturesome gunners have been out this past week with rifles, and while seeing it once skulking through the underbrush, did not get into position for a shot.

Information has been received from Hyde Park that an animal answering the description of Dedham's concern has been seen about there. There is an old graveyard between here and there, and a pedestrian over the road one evening is positive he sought about the graves. He stopped for some time and watched its movements. His description tallies with that of those who sought in the West Dedham woods.

It is supposed the animal has its lair among the rocks of Fox Hill, and takes the quiet of the night to explore the surrounding country for food. The place is especially favorable for a hiding place. The hill carries a heavy growth of oak and pine woods, with scattered open places. So far as could be gone over, it seemed in all respects and alluring resort for wild animals and game. West Dedham is at its foot.

SEARCH FOR PANTHER
Lowell, Massachusetts, *Sun*, December 4, 1941

Holden, Dec. 4 (UP)—Shotgun-armed hunters and state police searched woods in this area today for a "big black panther" which Joel L. Hodges, Jr., reported seeing at the rear of his home.

"BLACK PANTHER" SEEN IN HINSDALE
North Adams, Massachusetts, *Transcript*, December 9, 1948

The cat-like "what-is-it," seen most recently in the Plunkett lake area of Hinsdale, is a "black panther," according to Spencer Samuels of the Tully Mountain road, about a mile from the lake.

Mr. Samuels said the animal, five and a half feet long, stood in the roadway as he came out of his barn at night. He attacked it with a pitchfork from a distance of "two or three feet," but it escaped into the woods.

Mrs. Edward Reinhardt, a neighbor, reported seeing the panther carrying a smaller animal in its mouth.

Residents of the area are getting a thrill out of the apparent presence of dangerous wild animals in their vicinity. While not panicky, they believe discretion the better part of valor and are keeping their small children and pets indoors after dark, and panther hunts are planned.

'CANADIAN PANTHER' ON MOUNT GREYLOCK
North Adams, Massachusetts, *Transcript*, August 2, 1949

They can start seeing sea serpents off Cape Cod anytime now, because panthers are prowling the Berkshires again.

New Ashford residents reported today that wild screams have been heard nightly from the direction of Mount Greylock's slopes, and Mr. and Mrs. Henry Pierce of Williamstown, returning home Saturday night from New Ashford, said their car's headlights caught a panther-like animal in the road.

The Pierces said the creature was the size of a collie dog, tan in color with a striped body and a long tail. Mr. Pierce said it resembled a Canadian panther he once had seen.

CATS ARE GETTING BIGGER AND BIGGER
North Adams, Massachusetts, *Transcript*, August 5, 1949

That panther which a Williamstown couple thought they saw in New Ashford last Saturday night must have been of the freewheeling variety.

Either that or the hills are alive with the varmints, for a posse has been formed to track down a similar creature reported seeing in the Briar Hill section of Williamsburg, 60 miles away. Dr. Russell Burnett of Ashfield has rounded up a collection of cat-hunting dogs and has enlisted the services of Leon Pomeroy of Westfield, an expert wildcat hunter, in search for the creature, "six feet long and three feet high," whose nightly howls have been heard by summer campers in that area.

Outlandish screams were heard last week on the New Ashford slopes of Mt. Greylock and Mr. and Mrs. Henry Pierce of Williamstown said an animal resembling a Canadian panther appeared in front of their car on the New Ashford Road.

BAY STATE NOW HAS THAT BLACK PANTHER
Nashua, New Hampshire, *Telegraph*, September 2, 1949

Concord, Sept 2—Hold it now. They are claiming Massachusetts has got New Hampshire and Vermont's famed black panther.

For years editors of papers in the twin states have published reports of such an animal being seen. The Rutland (Vt) *Herald* has a long-standing reward offer for anybody bringing in the beast, "dead or alive." It never had a claimant.

George Proctor, NH game warden, in his column reported today western Massachusetts offering a $250 reward to anybody getting "that black panther."

"They say these big animals come from New Hampshire and Vermont," George wrote.

"You can have them," he advised the Bay State.

NOTES AND FOOTNOTES
Pittsfield, Massachusetts, *Berkshire Eagle*, April 25, 1973

Someone in Austerlitz reported seeing a "large black cat as big as a panther" near No Bottom Pond.

MICHIGAN

Posse Fails to Find Black Panther Claimed Near Leonard, Mich.

Saint Joseph, Michigan, *Herald Press*, April 11, 1952

Leonard, Mich.. April 11 (AP)—A posse yesterday found nothing but mud tracks in its search for a black panther reported seen here.

The panther report was made Wednesday by Carl Kuechenmeister, a 36-year-old barber here. But by the time the posse began its hunt all it could find were smeared tracks of some kind of animal—not necessarily a panther.

Kuechenmeister and his wife said they saw the animal skirting a field some 100 feet from their window.

Conservation Officer Cyril Adams investigated the Kuechenmeisters' story. He said he found the tracks but was unable to ascertain if they were those of a panther.

The reported appearance of the panther revived stories circulated near here last fall when a large black animal was reported near the Metamora Hunt club. Farmers in the area lost sheep and young pigs at the time. Some of the farmers thought the animal was a bear.

However Kuechenmeister said only a big cat could leap the way the "panther" did.

First of Mystery Animals Reported

Traverse City, Michigan, *Record-Eagle*, April 15, 1953

Pontiac, Mich., Apr. 15—(UP)—The spring rash of "mysterious animals" prowling the Michigan countryside got its start here this year.

A "black panther" is roaming the cedar bogs in Oxford township, according to Roby Snyder, 32, who maintained vigil early Tuesday armed with a deer rifle but the critter failed to show.

The "panther" was seen by Snyder's sister-in-law, Mrs. George Reynolds, 22, who spotted it both Sunday and Monday mornings while warming a bottle for her baby.

"It was black and it walked sort of slinky," she said.

The Snyders and Reynolds live together. Snyder said he caught a glimpse of the animal Monday. He armed himself with a rifle and kept watch Tuesday but the animal didn't put in its regular appearance. Reports of a panther circulated in this area last year.

LIKE BLACK PANTHER
SEE 'MYSTERY BEAST' AT MARQUETTE WOODS
Benton Harbor, Michigan, *News-Palladium*, October 15, 1959

A "mystery animal" that looked like a black panther was reported prowling around a Marquette Woods home Wednesday afternoon.

The beast, described as black with sleek fur, was reported sighted at 2:30 p. m. yesterday by 13-year-old Francis Crawford as she carried some rubbish to a trash pile behind the Crawford home.

Sheriff's Deputy Ed Bartz said Francis told him it was a large animal and looked like a panther.

She said it did not run, but loped into a nearby ravine when she came out of the house with the rubbish.

Tracks Resemble Dog's
Bartz said he found large tracks near the rubbish pile that could have been made by a dog.

The Crawford home is located between Hollywood and Scottdale roads. Undersheriff Joe Heward said deputies would continue their investigation of the mystery cat and turn the information over to the Conservation department.

BLACK PUMA SIGHTED
Ironwood, Michigan, *Daily Globe*, September 1, 1981

A Gogebic County road worker reported seeing a "black mountain lion" along Chaney Lake Road. He saw it in the afternoon, while paving the road, at a distance of 25 to 30 yards. He stated "the tail was as long as the body, and the cat looked scrawny but was big."

Black Panther Seen Near Michigan Village
Owosso, Michigan, *Argus-Press*, May 29, 1984

Residents of Manchester reported a large black panther, estimated at 100-150 pounds, with at least five sightings. It was first seen by a patrolman, who said it was the size of a German shepherd, and watched it cross in front of his parked car.

Black Panther Sightings Reported
Owosso, Michigan, *Argus-Press*, February 27, 1987

Residents of Muir, Ionia County, reported seeing "a hairy black animal" they thought was a panther.

Does 'Dogman' Haunt Luther?
Ironwood, Michigan, *Daily Globe*, July 23, 1987

Rumors of a "half-man, half-dog" creature started up after damage to a cabin door and window were determined by DNR to have been caused by a canine trying to get into the cabin. The animal itself was not seen.

Big Cat Sighted Within Ironwood
Ironwood, Michigan, *Daily Globe*, September 27, 1995

Police had received two reports of a big cat. Mark Ruppe, of Globe Concrete, reported that one of his employees had seen a "black panther" from about 75 yards away. The next morning some children reported a "mountain lion" that ran into the brush near their home.

Third Big Cat Sighting Reported
Ironwood, Michigan, *Daily Globe*, September 28, 1995

Police received a third report of a "mountain lion" when a witness stepped forward after the previous sightings. He said he saw it about a week earlier.

'LIGER' ON PROWL IN DETROIT SUBURB
Philadelphia, Pennsylvania, *Daily News*, September 25, 1997

A lion-tiger hybrid (wearing a red collar) was reported roaming the woods near Warren, a suburb of Detroit. (And *Napoleon Dynamite* didn't come out until 2004...)

Godfrey (2003; 2006) noted numerous cases of the Michigan Dogman, sometimes described as a wolf-like animal that can walk erect, or as a creature with a manlike body and a wolf-like head.

VAN BUREN TOWNSHIP STALKS MYSTERY CAT
Detroit, Michigan, *News*, September 19, 2003

Witnesses, including a police officer who was trying to trap it, reported a large tan feline, estimated at over 100 pounds, "with a lot of hair on the back if his neck, and it's kind of fluffy."

WOMAN SAYS SHE SAW BIG BLACK CAT
Oakland, Michigan, *Daily Press*, January 10, 2004

The general manager of a White Lake retirement community and four others at the facility reported sightings of a "solid black," "150-pound" long-tailed cat. Sightings of a "black panther" in the region go back at least to 1951 (with a sighting in Addison Township), and "at least 100 sightings were reported of a large black cat in western Oakland and eastern Livingston counties" during the 1980s.

CONSTRUCTION WORKERS REPORT PANTHER SIGHTING
Detroit, Michigan, *ClickonDetroit.com*, November 12, 2004

Two construction workers in Rochester Hills reported a large feline, "described as black, about 4 feet long with a long tail." A search by law enforcement officers did not come up with anything.

JULIE WROBLEWSKI: ELUSIVE COUGAR ALARMS RESIDENTS
Ironwood, Michigan, *Daily Globe*, July 20, 2006

This column first noted the recent sighting of a "black mountain lion" in Wakefield. The columnist then described her own sighting, also witnessed by her husband, of a black feline in Kimball, Michigan, in 2003. It was the size of a medium-sized black bear, with a tail two or three inches thick and as long as its body. She spoke with another Kimball resident whose dog was killed by a similar black feline.

RESIDENTS SAY BLACK PANTHER WAS SEEN
Hillsdale, Michigan, *Daily News*, November 2, 2006

A Hillsdale man and his friends saw a black panther in the wooded area behind his home, 200-300 feet from where they were sitting. He stated, "Whatever it was, it was bigger than a big dog."

MINNESOTA

FEAR PECULIAR ANIMAL
MEN AND DOGS OF MINNESOTA ARE
TERRORIZED BY FEROCIOUS WILD BEAST

Lincoln, Nebraska, *Evening State Journal and Lincoln Daily News*, November 8, 1919

Winona, Minn., Nov. 8.—Roaming in the wooded land where it evidently has a hidden place in which to hibernate, is a large, strange animal, so ferocious that it has caused men ordinarily frightened at nothing to flee in great fear at the sight of the beast, according to advices from Pickwick at the lower end of Winona county.

Reports of seeing the beast have persisted for several weeks. What it is none who has seen it will say.

The most reliable information thus far is said to have been gained from Carl Nelson, a farmer, residing on the edge of the infested woods. Nelson swears he saw the beast plainly and that it was light grey in color, striped and about as large as a yearling calf.

David Hoffers, a retired merchant, went to the woods with two good hunting dogs and a high powered rifle. Several miles below Pickwick his dogs picked up a trail. They followed it to a heavily wooded place which backs into a rocky draw.

The dogs began to bay, then suddenly broke and fled to their master, tails between their legs. Hoffers turned around and went home. He said he didn't see the animal—didn't even have a desire to see it. The fear of the dogs satisfied him, he said.

Farmers around Pickwick believe the animal escaped from a circus, has worked its way up the Mississippi river and is unable to cross it.

Those who have sent dogs on the trail of the beast declare they become greatly excited when the trail is first picked up, but after following it for some distance, break for home, displaying unusual fear.

'Black Cat Mystery' Puzzles Game Warden
Waukesha, Wisconsin, *Daily Freeman*, September 17, 1952

Fairbault, Minn.—AP—Veteran game warden Hugo Wicklund said today he wished people would stop hunting for a "five-foot long black cat" that has frightened residents of Monkey Valley.

Wicklund vowed to track down the beast himself, if it actually exists.

"I wish other people would quit trying to find the animal," he said. "The only way to hunt it, if it belongs to the cat family, is alone with two hounds."

"People are scared, whatever it is," the warden said.

Efforts to find and kill the big "black cat" have failed so far. Two large hunting parties searched Monkey Valley near here last week-end "without success."

Residents have described the beast as everything from a cougar to an ocelot—a South American jungle beast. However, neither of those animals are black.

Creature Lurking Near Hugo Butcher Shop isn't Your Average Cat
Minneapolis-St. Paul, Minnesota, *Star Tribune*, June 16, 1999

A Washington County deputy and a couple of motorists reported a large black feline near the Lutz Cuts butcher shop in Hugo.

Lots of Speculation Follows Cat Sighting
St. Paul, Minnesota, *Pioneer Press*, August 18, 1999

Photographs of a black feline taken on a farm near Hugo were examined by a zoo biologist, who said it looked like a young leopard, noting it did not have tufted ears and or look like a cougar. Witnesses described it as about three feet in length, two feet high at the shoulders, with a three-foot long tail. The photographs were not decisive, however, and some suggested it was a large barn cat.

MISSISSIPPI

Animal Killed on Pass Road Can Not be Identified
Animal Appears to be of Same Species as Catamount,
but is of Different Variety from Bob-Cat.
Whipped All Dogs That Undertook to Catch It.
Animal May Have Escaped from Circus or Wandered in from
Louisiana Marshes—Policeman Randolph Views Skin of Animal.
Biloxi, Mississippi, *Daily Herald*, February 3, 1915

John Astleford and others living on the Pass Christian road killed a wild animal of unknown species a few nights ago. The Astleford and Hurlburt families have been interrogated in regard to the animal which was killed in that section of the county, but neither they or those who have visited the place know what kind it is. The animal was skinned and it is being exhibited.

Those who have visited the place state that it was not a catamount, though, probably, of that species. Bob-cats are spotted, it is claimed, while this animal had black fur with a white spot on its breast. It was two or three times the size of an ordinary cat and was a fighter of much strength, chasing all of the dogs in the vicinity night after night.

The people living in that section set a steel trap for the animal, but it succeeded in breaking loose. It was later chased by fox terriers to a tree previous to this having beaten back all the dogs, scratching them up badly. Mr. Astleford came to the tree and killed it with a shot gun, using No. 6 shot.

Those who have examined the skin are puzzled to identify it. The animal had a long, bushy tail, while the catamount has a short tail.

It had long whiskers and eyes that were set closed together, with a pug-like head.

Policeman R. M. Randolph paid a visit to Mr. Hurlburt's to view the skin.

"I never saw anything like it," he said. "It is not a wild cat and just what kind of animal it is, I can not say."

Theories are numerous as to what the animal was, some believing that either this animal or its progenitors were lost by a circus, or that there is a cross-breed between animals of some kind making it. Others state that perhaps an animal that has not been classified has wandered in from unknown swamps in Louisiana or Mississippi marshes.

Hunters state that they have been endeavoring to kill this or a similar animal in the Forty-mile branch for years. They state that there are probably others of the same species in Harrison county.

Lions Spotted in Egypt?
Tupelo, Mississippi, *Northeast Mississippi Daily Journal*, March 24, 2006

After a woman in Egypt, Mississippi, reported seeing two lions in her driveway, residents in rural Monroe County were hearing rumors and taking precautions.

Woman Reports Black Panther in Amite County
McComb, Mississippi, *Enterprise-Journal*, September 29, 2007

A Gillsburg woman was driving to work one morning on Highway 584 just north of the Gillsburg crossroad when she saw a large black panther cross the road in front of her. She said, "It was not a fox or big dog or anything. It was a cat. It was just so fast. I was so stunned."

Are Panthers Fact or Fiction in Pine Belt?
Clarion, Mississippi, *Ledger*, June 18, 2008

This article discusses black panther stories and rumors in southern Mississippi. One man said he had taken a game trail camera photo of one in Forrest County, and that members of his family had seen black panthers over the years. A local taxidermist put out a $13,000 reward for three years, without having anyone provide a specimen.

MISSOURI

Blame "Panther" in Death of 276 Hens
Elusive North Missouri Animal is Reported Again in Grundy County
Chillicothe, Missouri, *Constitution-Tribune*, July 24, 1948

That old black panther is on the loose again, this time killing chickens, according to the Trenton Republican-Times. It has been gathering quite a reputation for its traveling, but seems to appear in Grundy County about two times for every time it appears in other places.

The last; time it was reported here, a fisherman claimed he saw it on the Thompson Fork early one morning.

The complete description of the latest incident follows:

None of the residents in the Trenton area who in the last two years have heard the cry of the panther or who have seen his tracks or perhaps caught a glimpse of the animal leaping off through the brush as they approached have had such an excellent opportunity to be close to observe the animal closely as have Mr. and Mrs. Jay Webster who live on the old Mary Webster farm about 18 miles northwest of Trenton.

The animal killed 276 of their chickens and returned a few nights later to awaken Mrs. Webster with its screams. Then they saw it.

276 Chickens Dead

It happened about three weeks ago. Mrs. Webster one morning went to the chicken house to find 276 of the young chickens dead. Some had been eaten, by what she did not know, and the remainder had beaten themselves to death in fright as they tried to escape. Tracks were found. They were five inches across.

And six nights later, Mrs. Webster was awakened one night by five or six screams, as if someone were being choked to death. She got out of bed and listened closely. She heard it again.

Carried Flashlight.

She awakened her husband. They went together outdoors and to the area near the chicken house. Webster flashed the flashlight around as they walked. And then green eyes caught the light and Webster flashed the light on the fence.

There on the gate sat the long black animal, his green eyes alight, his big head cocked and short ears perked up.

Look at Animal

For something like a minute, the Websters stood in amazement looking at the animal. It sat without making a sound, looking at them. Webster had no gun. He thought about going to the house to get his gun, but didn't want to leave Mrs. Webster alone holding the flashlight on the animal.

So they went together to the house, hurrying in order to get back before it got away. But when they returned, it was gone.

Reports of panthers being heard and seen have come from near north of Galt some time ago, and their cries have been heard from near the Livingston County area. One Livingston County man, and a Grundy county man some time ago in different spots came upon the animals they believed to be panthers, but the animals dashed away before they could shoot.

North Missouri Panther Bobs Up Again, This Time in Linn County

Chillicothe, Missouri, *Constitution-Tribune*, December 16, 1948

A long, sleek, black panther is lurking along the highways and by ways of Linn and Chariton counties, according to several residents of both counties who have assured a reporter for the Brookfield Argus that they have seen it.

"I saw the panther at two different times," said Mrs. Mary Lynn Cross, "while driving along Highway 36 near the vicinity of the old Bunker Hill coal mine. I noticed it in a ditch along the highway and turned and drove back. It was still there but sped away through the woods on my return. I saw panthers of the same variety while in the Panama zone and I could not be mistaken about this one."

Mrs. Cross said she saw the panther about three or four week ago, but it now seems to be haunting the borders of Linn and Chariton counties.

Several section men working on Santa Fe tracks near there say they have seen it and described it as being about the size of a large dog, with a long black tail that had that upward curl of the end of its tail characteristic of the feline family.

Jack Bridges, of Rothville, while flagging on Santa Fe tracks saw the panther on the right of way. One fanner near there found a calf that had been cut to pieces by some animal. None of the carcass had been removed. Lawrence Sportsman, Marceline, stated that he had talked to several who had seen it and they were all agreed that it was a panther.

It hasn't been too many years since wild animals roamed freely in the vicinity of Brookfield but times have changed and with the exception of this panther, the only wild animals now rampant are the reckless car drivers on the highways who daily take a toll of lives far greater than all the wild animals such as wolves, tigers or panthers.

To those who are alarmed by this report the Argus suggests that they follow the advice of Ogden Nash who is the author of the couplet:

"If by chance you are called by 'pantha', don't 'antha'."

'BLACK PANTHER' IS SPOTTED AGAIN IN NORTHERN MISSOURI
Mobery, Missouri, *Monitor-Index*, August 19, 1948

Princeton, Mo., Aug. 19—(AP)—A "black panther," frequently reported by residents in northern Missouri, has been seen again.

Earl Osborn of Half Rock in Mercer county, said he spotted the animal in a draw on his farm after noticing that his cattle had been frightened.

Osborn said he was "frozen" in his tracks, and that the panther fled after a few seconds.

Mr. and Mrs. Jay Webster, whose farm is 16 miles north of Half Rock, reported seeing the panther recently. An attack by the animal cost the Websters 265 chickens.

Black Panther is Fact, Not Rumor, Farm Couple Finds

Lawton, Oklahoma, *Constitution*, July 25, 1948

Spickard, Mo., July 24.—(UP)—The black panther of the Grand river bottoms was converted from rumor to fact Saturday by a flashlight wielding farm couple near here.

But the huge black animal was still on the prowl and as elusive as ever. For more than two years, residents of Grundy and Livingston counties had reported sketchy glimpses of a "big black cat" in the bottoms along the Grand river.

But the creature always darted into surrounding underbrush before witnesses could decide what nature of animal it was.

Recently, Mr. and Mrs. Jay Webster, who own a farm near here, lost 276 chickens in a single night

Webster, a former school teacher in Princeton, Mo., said some of the fowl were half eaten. Others, evidently, had killed themselves in panic.

Recently Mrs. Webster was awakened in the night by "terrible screams" from the farm yard.

She awakened her husband, and the couple dashed out toward their chicken house, armed only with a flashlight to investigate. As they walked toward the rear of their farm lot, Webster flashed the beam from the light back and forth across the yard.

Suddenly a pair of brilliant green eyes reflected the light.

Sitting on the gate in the barnyard was a "huge black animal."

Webster said it had "green eyes, short perked up ears and a long tail. It was a panther, all right."

Unarmed, the terrified man and wife stared at the creature. With huge black head cocked on one side, the panther stared back at the blinding flashlight.

"Then we ran as fast as we could back inside the house," Mrs. Webster said.

Queer Animal Seen, Heard Near Sedalia
Makes Lots of Hair-Raising Noise, Farmers Declare

Sedalia, Missouri, *Democrat*, April 14, 1950

A strange animal, which has roamed around the Walnut Grove vicinity, about nine miles northwest of Sedalia, for the past three years, was nearly captured last Tuesday night.

All the neighbors in that area, who live fairly close to a bluff, timberland and Muddy Creek, haven't definitely agreed upon what kind of an animal it is, but they all are sure of one thing, it makes lots of noise, that hair-raising type of noise, as one farmer explained it. He went on further to say, the noise sounded like a panther or a woman screaming

Clyde Shull, Sedalia, route 3, who hasn't seen the animal lately, but has seen it twice in the past three years, thinks it might be a wildcat, and also believes there is another type wild animal which hides out in the timber land alongside the creek and around a big bluff.

Shull's Description

The description Shull gave of the animal was, built like a bobcat, but unlike the bobcat, it had a long tail, shorter legs, and a pointed head. He said it was burnt brown in color and he imagined it weighs fifty pounds. He said the animal had long whiskers coming out from each side of its face, about four or five inches long.

Ralph H. Mosby, Sedalia, route 3, also is under the belief that there are two different type beasts running wild in that neighborhood. He said his son, Walter, shot nine or ten times at a wild beast last Tuesday night with an automatic rifle, but evidently had missed the animal, which had come up close to the barn that night, making a screaming noise. Mosby said he and his wife were in Sedalia that night.

Ran Into Oat Field

Mosby quoted his son as saying after he shot at the animal, it ran into the edge of an oat field into some buck brush. He related the son said the animal didn't seem to be afraid of him. The boy did not obtain a description of the animal, due to the darkness. He also did not feel like going into the brush after the beast.

Mosby said he has heard the wild screaming noise three times this spring, and has seen some type wild animal twice. The first time he saw it was when he took his boy to school one morning last winter and as the boy stepped out of the car, the animal jumped from a ditch, and ran off about a 100 yards, and then turned around and looked at them momentarily.

He gave a description of the animal as being yellow and was the size of a shepherd dog. Mosby said he believed one to be a panther, and the other a bobcat because one screamed a lot and the other didn't.

Missed Little Pigs

He said the farmers in that neighborhood have often missed little pigs. Mosby said the dogs have often chased the wild beast, but the dogs didn't seem to have much luck in capturing the animal

Other farmers in that vicinity who have heard the strange noises the unknown animal has made through all hours of the night are: Charlie Snow, Junior Riley and Howard Paige.

PANTHER WITHOUT A TAIL NOW IS REPORTED SEEN IN COUNTY

Chillicothe, Missouri, *Constitution-Tribune*, October 12, 1950

What seems to be that old black panther has put in another appearance, this time in Jackson township on the Grand river.

A trip into "red brush" this morning to check on an incident, which occurred two weeks ago Tuesday seems to add another story to the legendary black panther, except this one didn't have a tail.

Bookmobile driver Earl Willams brought the story into Chillicothe. A description he had heard of the animal's tracks, led him to believe it might be a bear.

But the description of Cecil Million and Ferd Russell, who saw the animal seems to bear out the panther yarn. They lived near Happy Hollow school.

They were fishing on the river about 3:45 o'clock in the afternoon, down stream from the Chula bridge, when the animal "jumped" Million's three dogs, gave them a good whipping, and then made off up a draw.

"The dogs were sleeping down by the sand bar when, it happened," Million relates. "There was an awful commotion and I thought the dogs had got into it.

"We ran over and saw this big black animal. It knocked one of my dogs from here to the middle of the road," he said, indicating about twenty feet.

"The animal ran up the draw for some distance before it disappeared. In a slough it left some tracks. They were six or seven inches long and had five toes, one of them set back a little bit. There was a definite heel to the track."

Million and Russell both thought the animal was of the cat family. It had no tail that they could see. It ran like a cat in a hopping fashion.

The animal was about eighteen inches high and from thirty-six to forty inches long.

Million said it could have been a bob cat, but didn't look too much like one. Some animals' pelts are dark this time of the year, however, he said. A true panther is supposed to have a tail.

"We killed a bob cat in our back yard about two winters ago," he said. "It was smaller, though. And had a short tail that stuck straight up in the back. The ears were long and pointed and it had whiskers."

In that instance, the same dogs treed the animal and weren't afraid.

"But they sure didn't want any part of that thing on the sand bar."

Million said the animal jumped them and the dogs never did bark. He said at one time one of the dogs was actually under the intruder.

The draw it made its escape up is a game trail and deer, coon and beaver tracks also dotted it. The tracks of the strange animal were washed away in a rain last week.

"This thing didn't look like a bob cat," Million says. "And if its that panther everyone talks about, someone has cut its tail off."

DIGS TRACK OUT OF MUD;
COULD IT BE THAT MYSTERY PANTHER?

Chillicothe, Missouri, *Constitution-Tribune*, April 26, 1950

That old panther is around again or at least Art Lile has found tracks he wonders about.

Sunday, Lile, who lives near Grahams Mill, dug a fresh track out of the mud, mounted it on a piece of cardboard and brought it into the Eagles Club rooms here.

Lile says there is a regular "run" cut by the animal, apparently leading to a sandbar in the river. He says tracks have apparently been made on different days.

The model he brought in now is dry and it is possible the track might have shrunk as it dried. As it is, it measures four inches from the tips of the claw mark to the heel. The width is about 3 ½ inches.

The "panther," usually identified as a black one, has been seen— or at least claimed to have been seen—in the Grand River bottoms off and on for several years. Conservation officials scoff at the idea a panther would have roamed this far from its native habitat.

In library books on tracking, this imprint resembles several different animals including a mountain lion, a wolf (not a coyote), or even a big dog. A mountain lion track, according to one book averages about 3 ½ inches. The book shows pictures of a lion's, or panther's, track sneaking and trotting. The claw marks are not exactly as pictured.

The nearest the tracking book comes to the Lile's track is to the forepaw of a Grey or Timber Wolf. The average track, according to the book, is five inches long. The track also resembles that of a Virginia Fox Hound, except that the pad is a little different. The average coyote track, according to the book, would be much smaller.

Lile says he saw a panther track in California while in service and this one, he says, is almost the same. He hopes John Madden, conservation agent, can identify it.

Another similar track is that of a wolverine. But the mud was so soft the wolverine's fifth toe, which sometimes registers, probably would have left a mark.

If Livingston County sportsmen wanted to get in on the publicity Oklahoma City recently had, they could claim Lile's track is that of a leopard. They are very similar, a leopard's track being 3 ½ inches long, but the claw does not usually show.

Another, and more similar track, would be that of a cheetah, a native of Africa.

BLACK PANTHER SEEN BY ST. CHARLES RESIDENTS

Edwardsville, Illinois, *Intelligencer*, December 23, 1950

St. Charles, Mo., (UP)—It could have been a big cat or even a mirage, but several residents of St. Charles swore Friday they had seen a black panther on the outskirts of town.

Contractor Herman Hindersman said he saw the animal running across a field Wednesday. He said it was about six feet long, coal black and had a long tail.

After hearing Hindersman's version, three city policemen spoke up. They said they had seen a similar beast recently, but were afraid to report it for fear of being laughed at.

Policeman Bill de Leal said the animal weighed at least 150 pounds. Officers Milton Schaeper and Lloyd Mungle said they caught it in the glare of their patrol car headlights a month ago, but it got away too fast for them to get a good look.

ARMED MEN ON HUNT FOR BLACK PANTHER IN MISSOURI TODAY
Greenville, Mississippi, *Delta Democrat*, January 1, 1951

St. Charles, Mo., (UP)—Almost 100 men armed with shotguns and rifles took to the woods Monday to hunt for a mysterious "thing" which residents believe may be a Black Panther.

The hunt was an official expedition of the State of Missouri. It was organized by State Conservation Agent Kenneth Amos who vowed to bring back the animal's hide as a trophy.

The volunteers began gathering before dawn, each with his shotgun or rifle ready for action.

They were carried in trucks to the wooded lowland where the "thing" is believed to have its lair.

Most of the men carried shotguns for close-in protection in case they were attacked by the beast.

Woodsmen Ready

But about 15 seasoned woodsmen and marksmen were selected to tote high-powered rifles. These snipers were posted at strategic spots where they could take pot shots at the "thing" if the safari's beaters managed to flush him from the brush.

The hunt was organized after a month of uneasiness among residents and farmers in the area.

Seven persons, including three police officers, reported seeing the beast. Most of them described it as a "jet Black Panther" about six feet long, weighing 150 pounds, with a "long black tail."

But some residents think its merely an out-sized coyote, wolf or bobcat.

'BLACK PANTHER' OF ST. CHARLES ELUDES 300 EAGER HUNTERS
Moberly, Missouri, *Monitor-Index*, January 2, 1951

St. Charles, Mo., Jan. 2(AP)—The "black panther" of St. Charles county is still on the loose despite a search yesterday by some 300 eager and heavily-armed hunters.

Net result of the hunt was discovery of some cat-like tracks each about four inches wide, at two locations south of here.

Don Schoene, who lives on a farm where some of the tracks were found, said he saw the varmint Friday night.

Reported Several Times

The tracks were examined by Raymond Hendrix, an experienced hunter, who expressed belief that they were made by an animal weighing 100 pounds or more. They were no more than four hours old, he said.

The beast had been reported seen by several persons, but their descriptions of it vary widely. Several agreed that it was coal-black, about six feet long, slightly bedraggled and with a long tail.

Three St. Charles policemen were among those who reported seeing the animal. Reports on it have been cropping up for more than a month. Missouri Conservation officers organized yesterday's hunt. Members of the party were armed with high-powered rifles, shotguns and even bows and arrows.

Paul Reeder of O'Fallon, Mo., gave out with some feline mating calls he learned from his grandfather. But if the critter was nearby it must be a female. Reeder explained he knows how to call only males.

CATLIKE TRAIL DEEP MYSTERY
Tucson, Arizona, *Daily Citizen*, January 2, 1951

St. Charles, Mo., Jan. 2. (UP)—Two sets of "catlike tracks" were found by more than 200 hunters who spent New Year's day trying to track down a "big, black thing" that has kept residents jumping at shadows for the past month.

CHAD ARMENT

The tracks were found within a two-square-mile area of dense timberland near here. But otherwise, the hunters—consisting of St. Charles county officers, sportsmen and farmers—found nothing.

At least eight persons, including three police officers, have reported spotting a big elusive creature on the outskirts of St. Charles. Some thought it was a black panther. Others said it looked like an over-sized coyote or bobcat.

The hunt was organized yesterday by State Conservation Agent Kenneth Amos and constituted an official state expedition to bring back the beast dead or alive.

The first set of tracks was found in a farm woodlot a mile south of St. Charles and a mile west of the Missouri river. Amos said the tracks were about four inches across and were spaced 18 inches apart.

The second set was found two miles away from the farm in a district where some unidentified animal recently killed a chicken. Other fowl and a calf also have been killed by the animal, officials believe.

Amos said the tracks indicated that they were made by a catlike beast, apparently weighing 100 to 125 pounds.

Yesterday's hunt included skilled trackers carrying high-powered rifles and three archers who hoped to loose an arrow at the "thing" if it were sighted. Most of the other hunters carried shotguns.

"PANTHER" HEADS SOUTH FOR WINTER
Mason City, Iowa, *Globe-Gazette*, January 5, 1951

St. Charles. Mo., (UP)— The "black panther" hunt here has gone to the dogs.

Reports that a large black beast was roaming the woods spurred a group of 300 hunters on an unsuccessful search for it earlier this week. Those who had seen the animal were sure it was a panther.

But Friday Morgan Bakewell, socialite farmer living two miles north of here, logically reasoned that the "panther" actually was his missing Labrador retriever.

'HUNTERS GREATER MEANCE'
Galveston, Texas, *Daily News*, January 7, 1951

St. Charles, Mo., Jan. 6. (INS)—Please tell us, what is "the thing?" chorus the residents of St. Charles county.

And Sheriff Lester Plackmeyer follows with the declaration that those bang-bang-bang hunters from St. Louis county are a greater menace to peace, life and property than the "thing," whatever it is.

Still at large is the mysterious black beast of St. Charles county today.

It was first reported earlier this week and described as something like a thick panther. It has also been explained as a renegade dog, or a wolf.

Something definitely is roaming the wooded area around Missouri's original capital. A farmer saw it slink across his moonlit barnyard, causing his dogs to bark frantically and his sheep to huddle together in terror. A woman "caught" it in her car headlights as it crossed the highway in front of her. Two coon hunters said their dogs treed the animal, but it jumped down and ran away when they shone their lights on it. Even two St. Charles policemen reported seeing "the thing" one night while cruising in a squad car, but they kept mum about it for a while because they thought they'd just be laughed at.

Pigs, hogs, a calf, many chickens, cats and dogs have been killed and partially devoured since the mysterious animal was first sighted. One farmer reported that 15 of his 21 pigs were carried off and eaten.

Witnesses report hearing weird, high-pitched cries, and "catlike" tracks have been found in the muddy Missouri river bottomlands about ten miles west of the city.

Two Lindenwood College girls, who may have been just getting in the spirit of "the thing," reported solemnly that they saw a large long-tailed black animal slinking across the campus as they went to breakfast.

George Vierheller, director of the St. Louis Zoo, says the animal's description matches that of a black jaguar or a black leopard—but he would surely have heard if such an animal had escaped and was at large in the area.

Coon dogs, fox hounds and wolf hounds have been used in unsuccessful attempts to track the animal down. And Sheriff Plackmeyer, who says he doesn't put any stock in this panther theory either, warned today:

"If some of these guys from St Louis county don't stop coming over here with guns, they're going to wind up in jail."

LIVINGSTON COUNTY PANTHER NOW SOUGHT NEAR ST. CHARLES

Chillicothe, Missouri, *Constitution-Tribune*, February 22, 1951

The same black panther which roamed Livingston County some four years ago, or perhaps a close relative, now is alarming residents of St. Charles County, Mo.

Charles Cornue of Omaha, formerly of Chillicothe, who keeps up to date on panther activities in the mid-west, sends clippings of St. Charles newspapers which relate that at least fifteen otherwise reputable citizens have observed a strange animal in the Missouri River area and which they say was not a lost dog or bobcat.

Since The Constitution-Tribune first publicized a strange varmint roaming North Missouri, as discovered by Mrs. Fred Ashlock, this animal has been reported in southern Nebraska, hunted around Des Moines and Rock Island, been sighted attending its business near its old stamping grounds adjacent to Chillicothe, and now is alleged to be cavorting in true panther style in the St. Charles area, says Mr. Cornue.

In view of its widespread travels it is interesting to note that the U. S. Fish and Wildlife Service states the catamount (or panther) may range as far as 300 to 400 miles, to spring a distance of nearly forty feet from a small elevation and to have the faculty of feeding where necessary upon small game only.

The charted course of the beast leads to Plattsburg, where a big-time hunt was sponsored for it; to Princeton, where irate farmers staged a nighttime hunt over Grand River bottom with a pack of dogs. At Princeton, the dogs were put to rout, a Trenton newspaper reported.

It is Cornue's belief that the large animal may have drifted to the Chillicothe area during a flood. Old-timers recall when early-day panthers and bobcats floated to high ground on logs and downed trees.

The Medicine Creek area was combed by local citizens after persistent reports had indicated it might be there. No panther was found, however, and the writer of this article came out of the hunt with a severe case of poison ivy.

At St. Charles, a crew aboard Wabash train No. 97 said they saw a beast answering the description of a panther. A report of the incident was dropped off the train and relayed to state highway patrolmen at Wentzville. A government fire patrol said he saw a strange animal and heard its eerie cry south of St. Charles. A 300-pound sow belonging to Arlie Reinwald was found dead in a field with its neck and back partly devoured. A St. Charles resident told of hearing a weird sound while coon hunting and finding animal tracks larger than any he had ever seen. One farmer reported the loss of fifteen young pigs. These were some of the reports.

Two More Witness What May Be Black Panther in Area
Chillicothe, Missouri, *Constitution-Tribune*, May 27, 1952

That animal, or one like it, has been seen again.

Two young men. Bill Beever, jr., and Bill Shipley told of working on the Bill Beever, sr., farm yesterday afternoon. As they started home, they spotted what they thought was a stray Black Angus calf lying down.

Upon their approach, the animal jumped and galloped off. They described it as black and with a long tail. The story corroborates reports of a panther seen in this area recently.

Beever and Shipley were working on a farm located about eight miles northwest of Avalon in the Grand River bottoms.

Varmint on Loose Has Folks Locking Doors at Diamond
Moberly, Missouri, *Monitor-Index*, December 3, 1952

Diamond, Mo. (AP)—A varmint on the loose in these parts has the folks locking their doors at night.

Rewards have been offered for the hunter who bags the beast, described as a black panther.

Martin D. Kelly and Don Heifner said they saw the animal, black with a yellow striped belly and a 30-inch tail.

Panther Returns
Chillicothe, Missouri, *Constitution-Tribune*, February 16, 1953

Rumors of another panther has traveled over the vicinity of Springhill. The protection of "Poosey" could possibly attract panthers as well as deer and other wildlife. John Volk and several of his neighbors tried to find some hounds to chase the panther, after reports drifted into the Springhill store that the "old black panther" had been sighted again. The efforts of a possible chase died out and as far as we know the cat or whatever it is may be 50 miles away by now. Wouldn't it be a relief if someone would catch the mysterious animal so we could identify it for sure?

CAMP CROWDER'S "BLACK PANTHER" IS A BOBCAT
Chillicothe, Missouri, *Constitution-Tribune*, June 13, 1955

Neosho—The "black panther" that has been stalking the country around Camp Crowder is probably a bobcat, according to Conservation Commission biologist Leroy Korschgen.

His opinion is based on hair samples sent in by Conservation Agt. Randolph Masson from the site of a recent "panther" kill of an Angus calf. They were bobcat hairs, Korschgen said.

"Some of their hair, collected from bushes in the vicinity of the kill, were dark," he said, "but some bobcat hair is quite dark, too. There is considerable variation in bobcats, ranging from ruddy to nearly black, and there is no doubt that these hairs were from a bobcat."

An animal described as a "black panther" has been reported in the Camp Crowder area by military personnel and the same beast is said to have killed calves and pigs. Korschgen pointed out that there just isn't such a thing as a black puma, the only other possibility.

MYSTERY SHROUDS REPORT OF PANTHER
Jefferson City, Missouri, *Daily Capital News*, April 22, 1955

St. Louis, April 21 (AP)—It's still a mystery today whether a shiny black cat-like animal seen prowling near Jefferson Barracks just south of St. Louis was a "black panther."

Bruce Dowling, wildlife biologist for the Missouri Conservation Commission, said tracks found in the area were too masked to make plaster casts.

"I can't say yes and I can't say no," Dowling said when asked if the animal was a panther.

The report by Irvin J. Wagner Tuesday revived talk that "varmints" still roam wooded areas of Missouri and Illinois. Many farmers have reported seeing panther-like cats.

This one was "four feet long, two feet high and had a long tail that stood out when it ran," Wagner said.

Wagner reported seeing the animal near the bank of a stream as he was working his beagle dogs. It leaped up the other bank and ran 1,000 yards toward a pile of rock before disappearing in the heavily wooded area, he related.

'Black Panther' Now Makes Appearance in St. Louis Area

Chillicothe, Missouri, *Constitution-Tribune*, April 29, 1955

St. Louis, April 29 (AP)—St. Louis County deputy sheriffs today renewed a search for a "big black panther" reportedly seen cavorting yesterday in a wooded area behind the Jefferson Barracks fire department station.

Deputy Sheriff Bird Sappington said tracks "much larger than a dog's" were found but lost in the underbrush by the search party, which also included about 25 residents and an officer of the Animal Protective Assn.

Joseph Trammel, a retired railroad man, said he saw a black, yellow-eyed cat, 3 feet long with a 2-foot tail and pointed ears leaping through woods behind the fire station.

A group of children playing in the woods said they fled when the animal growled at them.

About a week ago another "black panther" report was made by a man while working his Beagle dogs. No conclusion was reached by a Missouri Conservation Commission biologist who examined tracks.

Reports of a Black Panther Bob Up Again

Chillicothe, Missouri, *Constitution-Tribune*, March 23, 1956

Jackson, Mo. March 23 (AP)—Reports of a Black Panther in northern Cape Girardeau county have bobbed up again, this time with the partially devoured body of a calf to accentuate them.

The remains of the young calf were found on the farm of N. Burton Short near Fruitland. Short says he got a close look at the panther-like animal on his farm this week.

He described it as taller than a large dog and with a sleek black coat.

Earlier Roy Boren reported seeing the same type of critter in the same general area. Boren said his two dogs chased the animal into a wood.

Similar reports of panthers, black and other colors, in recent years from various sections of Missouri have stirred a controversy between skeptics and those who insist the panther has returned to Missouri with the increase in the state's deer population.

As yet, no one has been able to produce a panther, dead or alive, to settle the argument.

STRANGE ANIMAL SEEN NEAR BELL CITY SATURDAY
Cape Girardeau, Missouri, *Southeast Weekly Bulletin*, January 16, 1958

Bell City—Farmers in the Bell City area have been alerted and are on the look-out for a strange animal, reportedly a mountain lion or cougar, which was sighted among a herd of cattle on the farm of Herbert Bandy, four and one-half miles from Bell City on old Highway 25 Saturday morning about 10 o'clock.

The animal was first seen by Roby Kight, farmer and business man of Bell City, and was described as being considerably larger than a calf, striped, and having a long, shaggy tail.

Mr. Kight went immediately to the Bandy home to notify Mr. Bandy, who happened to be away from home at the time. Mrs. Bandy, however, got into the car and drove to the scene and Mr. Kight went on to his home on the adjoining farm and got his deer rifle. On returning with the gun and an employee, Gene Scisson, the animal was still there, but loped off into the distance before Mr. Kight could make a kill.

A hunt for the predatory beast was instigated and is continuing. Those initially engaged in the exciting hunt included Charles Rendleman, Leon Kielhofner, Herbert Bandy, Roby Kight, Gene Scisson and his brother, Lester Scisson.

A FEW REFLECTIONS ON WHAT'S GOING ON IN THE COMMUNITY
Chillicothe, Misouri, *Constitution-Tribune* (excerpt), January 23, 1960

That story of "the monster" reportedly seen in the Chillicothe vicinity has the skepticism of Conservation Agent Coval Gann insofar as its being a black panther. Mountain lion he might not be surprised at, but not a black panther.—Bill Plummer.

MEN AND CAT-LIKE ANIMAL GET LONG LOOK AT EACH OTHER
Chillicothe, Missouri, *Constitution-Tribune*, May 16, 1960

Gus Hogan and Joe Miller observed a strange animal in the bottoms area west of Farmersville yesterday and they are wondering if it might be the area's elusive "panther."

They aren't saying it is, but both admit they "never saw anything like it."

They were at Miller's cabin on a fishing expedition and saw the animal standing in a road watching them. It was there about 10 minutes and apparently was undisturbed by the men talking and moving about. It disappeared into brush and then reappeared for a short time.

Hogan said the animal was black, had a cat-like head but that it was bigger than any cat he had seen. It had a thick neck and long tail that pointed straight back. The body was approximately 14 inches long and the back was 12 inches or less high.

The men watched it from a distance of around 40 yards. "I wish we had had a rifle," Hogan said.

There have been occasional reports of a strange animal, possibly a mountain lion, roaming the area.

LEGENDARY BLACK PANTHER STRIKES AGAIN
Troy, New York, *Record*, April 13, 1963

Kansas City (AP)—The legendary "Panther of Noland Road" has struck again. This time it chewed up 12 little pigs.

For three years it has been rumored in the rural area southeast of Kansas City that a cat-like animal is hiding in a wooded hollow.

Four housewives reported sighting it 11 days ago. They said it was black, "larger than a tomcat, but its head was small in proportion to its body." Zoo Director William T. A. Cully said that accurately described a black leopard.

Wednesday night, Lewis Edward Montgomery found his pigs had been attacked. Eight are dead. Four survive.

Cully said they had been killed by "a powerful animal." He suggested children remain inside after dark, but said the animal would not likely attack adults unless it was desperate for food.

BIG BLACK MYSTERY CAT ON THE PROWL?
Moberly, Missouri, *Monitor-Index*, February 13, 1964

Over 100 individuals claimed to have seen a large black feline near Cameron, Missouri, in the six years prior to this 1964 article. A wildlife columnist, Gene Hills, from the Cameron, MO, *News-Observer*, managed to get a plaster cast of one of the tracks. It measured 3 ½ inches wide, and 3 ¾ inches long. A general description given of the animal is: "Color black; length of body about three feet; large, bushy tail about as long as the body; height about 18 to 20 inches. When seen, animal was 'creeping along the ground like a cat after a bird.'"

WAS IT OR WASN'T IT?
Moberly, Missouri, *Monitor-Index*, April 16, 1971

After a year of rumors of a black panther near Warrenton, a conservation officer heard that someone had shot it. The officer asked him about it, and was told: "Well, it was a housecat. ... But ... it was a pretty good sized one!"

I was contacted in 1998 by a man who reported seeing a black panther in June 1976 (Arment 1999). He stated:

"It was on a farm about 2 miles north of a town called Wellsville, Missouri. It was around mid-day and I was armed with a .22 single shot rifle with the intention of taking frogs from a small pond I knew of. I did not have permission to be there.

"The pond was heavily overgrown with brush and trees. I was being as quiet as possible so as not to scare the frogs away from the bank when I approached. Upon reaching the top of the dirt dam, I saw to the left of me something black. I stared at it for a moment before I realized it was a large black cat drinking from the pond. It was about fifteen yards away.

"It was perhaps four feet long with a long black tail. I would guess its weight at about 100 to 110 pounds. It was solid black, though I could not see the underbelly due to its position.

"It saw me and raised its head. We stared at each other for what seemed like ten minutes but in fact could not have been more than four or five seconds. I turned and ran back the way I had come until I got worried it might give chase and stopped. I could not see or hear the cat though I stood there for several minutes.

"The entire time I viewed the cat could not have lasted more than ten or fifteen seconds. I have seen panthers and cougars, (in zoos), and bobcats in the wild and I can say that this was without question a panther.

"I returned there the next morning, much better armed, in hopes of shooting the cat and getting my picture in all the papers in the state. I did find two prints near the water's edge where the panther was drinking but was unable to find any other tracks. After a couple of hours of searching I gave it up."

KILLER BEAST ROAMS WOODS—WHAT IS IT? TOWN ASKS
Chillicothe, Missouri, *Constitution-Tribune*, February 4, 1980

Residents of the Lone Jack area reported a shiny black dog-killing feline. One witness said it wasn't large enough to be a couger, as it was perhaps 50 pounds in weight, and about 2 feet in height.

PANTHER REPORTS SPARKED LIVINGSTONE COUNTY HUNTS
Chillicothe, Missouri, *Constitution-Tribune*, December 2, 1981

After a recap of the decades-old stories of a black panther in the Chillicothe region, the article writer mentioned that a farmer in the area claimed the animal still roamed near his farm. No details given, but reports were solicited.

BILL PLUMMER: REFLECTIONS
Chillicothe, Missouri, *Constitution-Tribune*, April 9, 1982

Mary Ashman Jensen, having read the recent discussion of whether black panthers were ever actually seen in the area, wrote in with her story, from the early 1920s. She stated that a circus train visited Trenton and lost a pair of black panthers which were never recovered. Her parents' farm was 1 ½ miles west of Chula and ½ mile east of Center school. About a month after the circus visit, they began hearing strange noises from the nearby woods and neighboring farmers began losing livestock to a predator. Hunters were unable to find the animal. One night, Mary and two others saw it as it passed near their home. They saw it flatten its ears and snarl at a passing car. It was "large, coal black and 3 ½ to 4 feet long, two feet tall at the shoulders, and had a tail about 18 inches long." A few nights later, it followed (and frightened) Mary's mother and a friend who were walking home from a neighbor's house.

BILL PLUMMER: REFLECTIONS
Chillicothe, Missouri, *Constitution-Tribune*, April 12, 1982

Mary Ashman Jensen continued her recollections of the Black Panthers of Livingston County in the 1920s. A few days after the panther was spotted trailing the two women, a neighboring farmer killed a black feline as it was killing several hogs in their pen. They decided it must be a male black panther that escaped from the circus. The farmer went off into the woods to hunt the panther's mate, but only found a pair of young. He shot them, and left them in the woods. That evening, the female found the bodies and screamed until about midnight. It disappeared from that area, but turned up killing livestock near Blue Mound and Dawn. It was eventually shot by a farmer there.

GETTIN' AROUND THE COUNTY, BY CHARLES CORNUE
Chillicothe, Missouri, *Constitution-Tribune*, February 11, 1988

The columnist noted that there were rumors again of a black panther in the community.

BLACK PANTHER ROAMING IN BRANSON ELUDES ANIMAL CONTROL
Chillicothe, Missouri, *Constitution-Tribune*, April 21, 2005

Over 100 reports of a "black panther" over a few months spurred Animal Control to try and trap it. One man was able to videotape it, showing what appeared to be a "60- to 80-pound panther."

A PANTHER ON THE PROWL? AREA WOMAN SEES WILD CAT
Neosho, Missouri, *Daily News*, August 15, 2006

An eastern Newton County resident, Kathy Newton, driving near her home, watched as a large black panther (she estimated at 100 pounds) jumped out of a field and crossed the road ahead of her. This was the second sighting for her, as she had seen one a few months earlier.

Godfrey (2006) noted stories from the Missouri Ozarks southeast of Springfield relating to hyena-like creatures. One woman had seen a pack of the animals, "muddy brown and black with weird heads and stove-pipe long heads," stalking cattle.

Deputy Kills Jaguar ... Or Was It Leopard?
Joplin, Missouri, *Globe*, May 20, 2008

Responding to a Newton County resident's call of a "black panther," a deputy shot and killed the cat after it lunged at him. It had been pawing at the door when he arrived, and examination showed that its claws were removed. Later articles confirmed its identity as a melanistic leopard.

MONTANA

Big Game in Great Plenty
Successful Bear Hunting in Montana Mountains
Portsmouth, Ohio, *Daily Times*, September 9, 1896

Charles Morell, who was formerly a Helenaite, but has for a few years been living on the Clearwater river, was in Helena the other day. He is in one of the wildest regions of Montana and his nearest neighbor is seven miles away, between the main divide, of the Rockies and the towering and barren Kootenai mountains. He says the country, where he has established a farm in the dense timber and ragged mountains, is alive with game. Elk, moose and deer are to be seen at all times and are easy to shoot, while bears and mountain lions are very numerous. A new species of bear has been seen there this season for the first time. It is an animal with a fur of a creamy white, nearly of the color of the polar bear, and none of the old hunters of the region ever saw that sort of bear before. The animal is smaller than the grizzly, black or silver-tip bear common in the country, and is very fierce. Of this kind three have been trapped and killed this season. ...

Woman Frightened by Black Panther
Anaconda, Montana, *Standard*, October 31, 1911

Half hysterical with fright, Mrs. Elsie Johnson, a washerwoman, who lives at the corner of the alley in the 400 block of East Seventh street, ran to her home last evening declaring that she had run into a black panther less than 200 yards south of the courthouse. Mrs.

Johnson thinks that the huge beast would have attacked her had it not been for the two dogs that were with her at the time.

As is her usual custom, late yesterday afternoon Mrs. Johnson took her hand cart and started for Sheep gulch, where she collects firewood for her home. At a point about 150 yards from the courthouse a black form, larger than a collie dog, came tearing down the hillside. The frightened woman gave a shriek and her two dogs, that were behind her, gave chase to the animal, which Mrs. Johnson says positively was about the size of a mountain lion, but jet black. The animal came down the hillside on the east, at a point near the water flume. When it saw the dogs it went up the other hill toward the upper corner of the Hill cemetery fence.

Mrs. Johnson at once returned home, leaving her push cart where she was stopped by the wild rush of the panther. In telling of her experience, she said that less than a week ago, at a point a few hundred yards farther up the gulch, she heard the animal give a blood-curdling yell, but at the time thought it was her 12-year-old son trying to frighten her.

E. J. Munger and others said last night that for the past several days they had seen the tracks of the animal around Sheep gulch, but thought that they were the tracks of a lynx or an extra large wildcat.

COYOTE-WOLF FREAK RAIDING HENROOSTS
Billings, Montana, *Gazette*, October 10, 1924

Melstone, Oct. 9—(Special)—Ranchers near McCalla have been hunting lately for a queer animal, thought to be a cross between a coyote and a wolf, which has been robbing henroosts and killing turkeys and small pigs in that vicinity. It is said, that the creature is larger than the ordinary coyote, and is almost black in color. It seems to be afflicted with the mange, for although its coat is thick and heavy there is no hair on its tail, which gives it an unusual and hideous appearance.

One rancher states that he had a chance to shoot the animal, but refrained when he saw the condition of its tail, thinking that he would have an opportunity to kill it later in the season when it had recovered from the mange, and when its coat was perfect. Two or three days after this, this same rancher lost a number of chickens, and on

watching discovered this strange coyote sneaking around his pigpen. He went to the house for his gun, but on his return the coyote
had disappeared, and has not been, seen since around this particular ranch.

However, he has boldly made his appearance at several other
ranches in the neighborhood in daylight and has robbed numerous
flocks of chickens. Several ranchers have tried to shoot him, but he
seems to have an uncanny knowledge of firearms, and keeps out of
sight when there is a gun in the neighborhood.

MOUNTAIN LION SIGHTED
Helena, Montana, *Independent Record*, October 23, 1958

East Glacier—(AP)—A large black mountain lion was reported
seen beneath the Two Medicine bridge near East Glacier by the crew
or a Great Northern railway train. Weight of the animal was estimated at about 200 pounds, considered unusually large.

I received an email (Arment and LaGrange 2000) from Valerie ReVander, who
stated that when she was a teenager, about 1980 or so, near Pablo, in northwest
Montana, her dog killed a small mammal that had been raiding the chickens. "The
animal hadn't eaten the pullets, but had just killed them and left them in a row. It
also had filled in a hole that had been dug to repair a well pump and shat upon the
top of the dirt. The feces was very musky smelling even from yards away." She
described it as mongoose-like, but with slightly longer legs, and somewhat feline
features. "It had soft fur that was ticked on the tips and rounded paws between the
shapes of a ferret's and a cat's. It was the size of a 6-month old house cat." The
coloration was grey, with lighter gray ticking on the hair tips. There were some
faint stripes on the tail. Her mother made her throw out the carcass. It had a musky
odor. (I wouldn't rule out an out-of-place ringtail.)

NEBRASKA

Newton (2005) noted the 1897 sighting of a black panther in Fairmont, Fillmore County.

Eyes Blazed in Moonlight.
Callaway's Wild Animal Becomes "Middle-of-Roader."
Lincoln, Nebraska, *State Journal*, December 31, 1908

Broken Bow, Neb., Dec. 30.— Another person claims to have seen the mysterious animal that is alleged to have created much excitement in the southwest part of the county. This time it is a commercial man, who drove over here Tuesday night from Oconto. He states he was about fifteen miles from town when he discovered the "mystery" lying in the middle of the road. Upon the approach of the carriage, the animal jumped to its feet and uttered a deep growl, its eyes fairly blazing in the moonlight. It then slunk away without hurrying in the least. The knight of commerce says he was armed, but lost no time in widening the distance between them. He describes the animal has resembling a mountain lion, but as large as a full-grown tiger. Incidentally, it may be mentioned that Oconto contains the only licensed someone in Custer county.

Jungle Beast At Large In State
Tiger Reported Roaming the Fields in the Vicinity of Norfolk
Lincoln, Nebraska, *Star*, September 27, 1923

Norfolk, Neb., Sept. 27.—The tiger which has been reported seen in several nearby counties during the last few months, and last week

was believed to be on a farm near Newman Grove, was seen standing on the banks of the Elkhorn River between here and Battle Creek
yesterday, according to a report brought to the city by Mr. and Mrs.
Ed Muffly of Battle Creek. They declare they got a good view of the
animal as they crossed the bridge over the river. Farmers have been
warned as it is feared the beast will make raids on livestock.

It is supposed the tiger escaped from a circus last spring.

SEARCH RENEWED

Pacific Stars and Stripes, November 23, 1951

Alexandria, Neb. (INS)—The search for the "lion of Ceresco" was
renewed near Alexandria Wednesday after a farm youth reported
seeing the lion like animal in a field.

Duane De Long, 15, reported he was bringing stock from the pasture on his father's farm when he saw the cat from a distance of 30
feet. He described it as about three feet high at the shoulders, light
brown and with a bushy tail. The youth reportedly fled when the cat
reared up and growled.

Three hogs were found slain in the area shortly before young De
Long saw the beast.

FARMERS HUNT 'AFRICAN LION' ON EASTERN NEBRASKA PRAIRIE

Sheboygan, Wisconsin, *Journal*, November 29, 1951

Waterloo, Neb. (UP)—Worried women and children stuck close
by their homes today while gun-toting farmers hunted what they
believe is an "African Lon" which somehow turned up in Eastern
Nebraska.

Townsfolk became convinced the animal was a lion—and not a
St. Bernard dog—after four deputy sheriffs and a rural mail carrier
reported spotting it in the headlights of their cars.

"It really looked like a lion," said Deputy Sheriff Dale Nelson. "It
had shiny eyes and a light red coat and just floated along with long
strides."

"You can bet your life we're worried," said Mrs. Nelson. "The
mothers didn't want to send their children to church doings alone
last night because the lion was believed nearby."

Her husband was hurriedly summoned at 2 a. m. yesterday by Carl Anderson, rural mail carrier. Anderson said he saw the lion in his car's headlights outside the town when he was delivering newspapers.

Nelson called in other deputies and the four staged their own private lion hunt

"I turned two spotlights on him." Nelson said, "and then started shooting. I think my first shot might have hit him but I'm not sure."

But the animal leaped over a six-foot wall at the local county fairgrounds and headed for Elkhorn River. The hunters lost its tracks there, and figured it had run alongside the river and disappeared in the underbrush.

The search was abandoned until daylight when farmers set out with shotguns and rifles. Planes flew overhead with other hunters searching for the lion's lair. But no trace was found.

A "lion" was first reported sighted by farmers a year ago near Ceresco, Neb., about 50 miles from Waterloo. It supposedly turned up again Nov. 12 when Farmer Joe Potacka said he saw an "African lion" stalking through his cornfield.

That report touched off a search in a 50-mile area. Farmers told authorities some of their livestock was missing.

But the hunters called off their bloodhounds and smugly concluded it was a missing St. Bernard dog.

NEVADA

"Grizzly" Adams sought out rumored "purple panthers" in the Humboldt Mountains, but only saw the tawny variety. (Hittell 1861)

THE WHOAHAW.
A HUNTING PARTY'S EXPERIENCE—THE STRANGE CREATURE THAT DOMINATES DEETH—STORIES OF HIS FEROCITY— ABUNDANCE OF GAME AND GOOD SPORT.
Reno, Nevada, *Evening Gazette*, August 26, 1879

Richard Smith, the express agent, and his brother H. R., got back last Sunday night from their hunting expedition. They shot over the country around about Deeth and Halleck. Game was abundant. During the trip they bagged 125 prairie chickens and one lynx, besides untold numbers of stagehands, rabbits, ducks, etc. The singular number of lynx shot by the party would appear to indicate a scarcity of large game in that section of the state explored by the expedition. One might suppose that where birds were so abundant, lynx and other of their four footed enemies would be more numerous. It is a striking fact that they killed only one of the carnivora during the trip. But there is an explanation for the scarcity of such animals that will be strange to many readers. The whole region round about Deeth is dominated by

A Mysterious Beast
known locally as the Whoahaw, an animal supposed to be a cross between the grizzly bear and the coyote. As the mule combines all the bad qualities of the horse and the ass, so does this hybrid display

362

the current and ferocity of the grizzly joined to the cunning and treachery of the coyote. The whoahaw has never been seen by daylight. He roams and ravages only at night. The beast has been known to carry off a horse. Cattle and sheep are often borne away by the monster. Mules he never attacks, for some unexplained reason. The brothers often set up far into the night listening to the ranchmen's tales of the strength and ferocity of the whoahaw. There is supposed to be only one of them in that section of the country. He has never been distinctly seen, but some of the ranchers have caught glimpses of him prowling about in the darkness. They always had business of importance to attend to somewhere else upon such occasions. The recital of tales about the fearful creature one night,

Curdled the Blood

of H. R. Smith, almost to coagulation. Richards says the ranchmen's stories often ran his own circulation down from seventy to forty beats a minute. Nothing but the speedy administration of tea or other stimulant could have enabled them to pull through those narratives. The brothers camped in the vicinity several nights, almost daring to hope that they might catch a glimpse of the dangerous hybrid. But they saw nothing of him, although one midnight they heard far off an echoing sound like "whoa—haw," which the ranchmen said was the cry of the monster, and from which they gave him his name. Should some adventurous hunter eventually cope with and kill the curious beast it is hoped for the sake of science, that his bones and skin will be carefully preserved. The brothers did not meet with any exciting adventures during their trip. They were much impressed with the exceeding swiftness of the coyote, when stimulated to exertion by the presence of fine shot under the skin. In order to observe his velocity they frequently introduced fine shot under the hide of straggling coyotes, always with the most gratifying results. They recommend the neighborhood of Deeth for hunting, but "Beware," they say, "of the whoahaw."

WAS IT A WAHHOO?
THE STRANGE BEAST THAT VISITED MR. ADAMS AT MIDNIGHT—
A CREATURE INSENSIBLE TO A STREAM OF WATER—ITS LONG
SNOUT AND EYES OF FLAME—A REMARKABLE STORY.
Reno, Nevada, *Evening Gazette*, September 2, 1879

A. A. Adams is a well-known and highly respected German citizen. His residence is near the corner of Fourth and Chestnut streets. Mr. Adams is a bachelor, and lives in a small cottage. He is not superstitious, and never clouds his mind with ardent spirits. His temperament is of the phlegmatic rather than the nervous kind. So when he heard some heavy animal walking slowly up and down his verandah last Friday night, he did not suppose that the noises were in any way supernatural. He simply wondered what kind of a beast could have got into his yard. He listened and could distinctly hear the tread of some four-footed creature, as it slowly strode to and fro. He could even hear the scratch of the creature's claws on the boards. His curiosity and length aroused, Mr. Adams determined to take a look at

The Pedestrian on His Portico.

He rose from his bed and, putting on an additional garment, he stepped forth into the rheumy and unpurged air of midnight. The moon was shining from a cloudless sky, and by its bright beams he saw what at first sight seemed a large black bull-dog. The animals stopped in its walk, and turned two brilliant, fiery eyes upon Mr. Adams. They glowed with an unnatural brightness; looking more like coals then visual organs. He noticed too, that it's "snoot" was very long, like a pig's; and its tail of surprising length, stuck straight out behind. Mr. Adams clearly saw that, whatever his strange visitor might be, it was no bulldog. After looking at him steadily for over a minute, the beast slowly retreated to the fence, which it climbed by means of its claws, after the manner of a cat. Perched upon the top of the fence, the creature sat, and resumed its survey of the astonished German. Mr. Adams thought he would direct a stream of water from the garden hose upon the animal, and thereby induce it to retire. He did so, but, to his amazement, the creature remained immovable, merely presenting its snout to the stream, its body enveloped in a cloud of spray. Mr. Adams persevered in the hydropathic treatment for about ten minutes, but still

The Beast Kept Its Position,

its eyes glaring at him like the red lights of a railway train. Mr. Adams owes to a feeling of dread at the creature's peculiar persistence under the circumstances, and began to think that ghost stories might be true after all. He retreated into the house and locked the door. He looked out of the window, and there are still sat the animal on the top of the fence, its baleful eyes throwing a lurid glare into the room. Mr. Adams now felt a great fall in the temperature. He was shivering with cold. He thought he would fire a charge of shot at the brute, and got out his gun for the purpose. He loaded the weapon hastily, but after he had put on the caps, concluded he wouldn't fire after all. He pulled down the blinds, and went back to bed. He listened for a long time, but heard no more footsteps. The cold continued, and Mr. Adams shivered nearly all night, and got very little rest in consequence. At daylight next morning he was out, and made a careful search of the yard, closely examined the fence and porch, but discovered no trace of the strange beast. He put his gun in careful order, and made up his mind that should the creature come again he would fire upon it at sight. He was surprised to find, on cleaning the weapon, that in loading it the night before, he had put the shot in first, and the powder afterward. That is the usual method of loading for ghosts, and is one of the principal reasons

Why So Few Spectres Are Shot.

Mr. Adams carefully loaded the gun that day, putting in the powder first to avoid the mistakes apt to attend that operation when performed in the dark. The following night he sat up late in company with a countryman of his, but saw nothing unusual, nor has he since been visited by the mysterious beast. No animal answering to the description is known to exist in this section of the country. In addition to the peculiarities already described, Mr. Adams says it had a long, slender neck. Could it have been one of those ferocious hybrids called wahhoos, which are said to prowl at night in the neighborhood of Halleck and Deeth, such as were recently described to a Gazette reporter by Richard Smith, the express agent? The mystery may yet be cleared up, but now the question is, among many hear the story, was it a wahhoo?

WAHHOO, OR WHAT?
Reno, Nevada, *Evening Gazette*, September 9, 1879

A strange looking animal was seen by two sportsmen on the road about two miles from Peavine last Sunday. This creature was not unlike a coyote but larger, yet too small for a bear. It was running on the side of a hill with wonderful speed, and disappeared in a moment. Could the beast have been a Wahhoo?

THE WILD WAHHOO!
A "MAN-EATER" THAT OPENS THE GRAVES OF THE DEAD.
A FULL DESCRIPTION OF THE STRANGE BEAST—ITS PEDAL PECULIARITIES—ITS HAUNTS IN THE HILLS—FOUND IN NEVADA, IDAHO AND MONTANA—A CREATURE WITH THE FORM OF A DOG AND THE VOICE OF A JACKASS.
Reno, Nevada, *Evening Gazette*, September 12, 1879

A recent number of the Gazette contained some account of an animal found in the neighborhood of Halleck and Deeth, Nevada, known in that section of country as the "Wahhoo." It appears that the creature is not known to naturalists, and finds no place in the catalogs of writers upon zoology. Some readers of the article referred to for this reason supposed the whole story to be a hoax. But it must be remembered that every day the researches of scientists are bringing to light hitherto unknown animals and plants, in every quarter of the globe. Animals well-known locally, in some remote localities of the earth, often prove entirely new and strange to the world of science. The narratives of travelers are often received with a great deal of incredulity because they frequently contain descriptions of things before unheard of. When Du Chaillu discovered the gorilla in Africa, scientists were slow to believe in the existence of the animal. To take animal stray should nearer home: there is a fish in Idaho waters called the "redfish," which is familiar to the people of that territory, yet has never been scientifically described, and of which no specimen has yet been placed in the Smithsonian Institute.

The Account of the Wahhoo,

which follows, is plain and unvarnished, and it may yet be found that this strange beast will possess a high scientific interest to workers in the wide field of natural history. In the brief and imperfect description of the Wahhoo to which reference has already been made, were given some data relating to the animal, which were furnished by Richard Smith, the agent of Wells, Fargo & Co.'s express, at Reno.

That gentleman, while hunting in the neighborhood of Halleck, heard from residents of that locality many stories of the Wahhoo and its peculiarities. He did not succeed in catching a glimpse of one, but his brother who was with him succeeded in obtaining the dressed hide of a Wahhoo, and took it with him to Los Angeles on his return from the expedition. Mr. Smith's Reno friends, to whom he repeated some of the stories he had heard of the strange creature, were skeptical of their truth. But the publication of the article in the Gazette brought to light some additional testimony concerning the curious beast. Before proceeding with an entirely new evidence, it will be well to state what the people about Deeth say of the animal. It is they are known both as

The Wahoo and the Man Eater.

The former appellation is supposed to have been given it in imitation of the peculiar noise it makes. The latter designation originated in the known propensity of the beast to dig up and devour bodies of the dead. Wahhoos which have been killed near Deeth exhibit a peculiar structure. The legs are short, and the paws very large proportionally, furnished with strong projecting claws of great length. This formation enables the creature to dig with ease and rapidity. The body is long and slender, the tail of medium length and usually curved over the back, the neck short, the head broad, and the jaws provided with formidable teeth. The skin is covered with long, fine hair. Its prevailing color is black, spotted with white. In weight it varies from fifty to seventy-five pounds. The creature is larger than a coyote, and in appearance, when seen at a distance, not unlike a large dog. In conversation with Mr. Smith, a young man said that he had shot a number of Wahhoos, that he carefully measured each specimen, and found that the left legs of each were somewhat shorter than the right legs. Although his informant persisted in the assertion, a statement must be

Regarded With Great Caution.

The probability is that he measured a malformed specimen and jumped to the conclusion that the others showed the same peculiarities. The young man stated, as an explanation for the inequality in the length of the creature's legs, that the Wahhoo was found only upon the hills, along the sides of which it was constantly traveling. The unequal length of its legs would be advantageous to the animal in traversing the hillsides. It would indeed be strange if nature had provided such a marvelous adaptation of structure to fit the creature for ranging upon the sides of hills. Incredible as such a statement certainly is, it might possibly be true. Knowing what strange forms of life have inhabited the earth in bygone epochs, nothing seems impossible in animated nature. In the presence of the fossil remains of the *Eohippus*, or five-toed horse, a laterally unequal wahhoo would not seem so strange a creation after all. Daniel Roberts, and express messenger on the Central Pacific Railroad, states that in Montana wahhoos are not uncommon.

He Saw and Heard the Creature

in Idaho, in 1867. He has a vivid recollection of his first sight of one of the beasts. He was approaching a station in Beaver canyon, one evening in the summer of that year, after a long journey on horseback. On the way up the canyon he heard what he supposed was the bray of a mule. He remarked to his companion, who was familiar with the country, that they must be near the camp. He was told that the noise was made by a wahhoo, and shortly afterward they saw the animal sitting on its haunches upon the edge of the cliff above them and giving utterance to the dismal cry, which had deceived Mr. Roberts. The creature walked along the edge of the precipice for some distance, at intervals sitting upon its haunches and sending forth

Its Prolonged, Wailing Bray.

A man named Thomas, well-known to Mr. Roberts, shortly after killed a wahhoo on the Montana road, near the Camas Creek station. It weighed about seventy pounds and fought wickedly after being wounded, until it was finally dispatched. Not long after that, Mr. Roberts, in company with Mr. Bassett, a superintendent in the employ of the Western Union telegraph company, saw two wahhoos together near a place called Summit Station. He states that the animal is well-known all over Montana. It is very shy, nocturnal in its habits, and abides in the wilderness away from the habitations of men. It is going to these reasons that so little is yet known of the wahhoo. The foregoing meager description is all that the Gazette has been able to learn of the mysterious beast. It is published in the hope that the attention of some zoologist may be drawn to the fact of the existence of such a creature, and that the animal may prove a subject for the study and investigation of some one qualified to classify and describe it. All communications designed to throw any light upon the nature and habits of the wahhoo will be gratefully acknowledged.

TOTALLY EXTINCT
WAYLAYING THE WAHOO ON A MOUNTAIN SIDE— THE LAST OF THESE STRANGE ANIMALS.
Reno, Nevada, *Evening Gazette*, January 8, 1880

The cold weather is said to have killed off all the wahoos about Halleck and Deeth. A Wells-Fargo messenger reports that on last Saturday morning a party of Deeth hunters found wahoo tracks in the snow five miles northeast of the station. The severity of the weather appeared to have forced the animals to band together, as the tracks could not have been made by less than six different animals. The circumstance excited the interest of the hunters, for anyone who knows what a wahoo is, knows full well that the creature is one of the most unsocial of all animated beings. The hunters determined to follow up the broad trail made by the wahoos, and at once started in pursuit. The tracks led them around the base of a high mountain and constantly ascended in a continuous spiral. The hunters followed the trail for hours, constantly climbing higher and higher, until at length the summit was reached and the trail began to wind back down the mountain. The hunters were now thoroughly exasperated, and pressed on with fierce determination. The chase went on for hours until they reached the plain again, and found that the wahoos' track had turned up the mountain once more. A consultation was held and some proposed to give up the pursuit, but one of the party said if they would follow him they would fix the wahoos. He reminded them that owing to the fact that the wahoos' left legs are shorter than its right, the creature always walks on the hillside from right to left. His proposal was that the hunters should ascend the hill in the opposite way, and thus head off the wahoos and take them at a disadvantage. So with fresh enthusiasm the party started off again, and after an hours' climb they met six wahoos face to face, half way up to the summit. Three of the creatures fell at the first fire. The others tried to turn and run, but owing to their legs being shorter on one side than on the other they immediately lost their balance and rolled helplessly to the bottom of the slope, where the hunters subsequently found their lifeless carcasses.

This little band of devoted wahoos probably was the last of these curious animals. The few others that roamed the hills in the vicinity of Deeth and Halleck are believed to have perished in the late cold snap, and thus the wahoo, like the dodo, may at last be considered extinct.

NEW BRUNSWICK

San Antonio, Texas, *Express and News*, January 31, 1954

What "Scare" stuff ala tabloid style has done to the young citizens of Birdton, New Brunswick, Canada, is typical of what happens when some novice outdoor writer unloads his imagination on the reading public. "Fear of the Black Devil panther, a mysterious and terrifying cougar once believed to be extinct held the isolated brush community of Birdton in terror. Men kept rifles close by and mothers comforted frightened children behind closed doors." Sheer sensationalism. Wildlife specialists say they haven't even seen tracks of the terrified area.

Bruce Wright (1972) reported several cases of black panthers along with "normal" eastern panther stories in New Brunswick. These include:

September 17, 1951: A driver along a "company road" about 16 miles from Bathurst saw what he thought was a young black bear standing alongside the road, then saw that it had a long tail.

October 1953: Children and adults reported seeing a large gray-brown feline in the Birdton area. Media reports called it a "'Black Devil' Panther." It does not appear to have been a truly black feline.

September 1955: The Camp Commandant of Base Gagetown was walking his dog along the foot of Blue Mountain when he spotted a black feline watching the dog from the brush. It was larger than the Commandant's Chesapeake Bay retriever. The cat ran off when it saw the Commandant.

May 1956: Two soldiers returning to Base Gagetown at night passed within 15 feet of a large black feline with a long tail.

March 1958: J. D. MacDonald was driving slowly along a gravel road a few miles from Chamcook in the middle of the afternoon. He spotted a large "quite definitely black" feline standing alongside the road, which then sprang off into the brush. He stated that it "may have been about three feet long, not counting the tail."

July to October 1959: Several people reported seeing a black panther in the region along the east bank of the St. John River, "from Sheffield to the Becaguimec Game Refuge."

1959: Location unknown, but somewhere between the St. John River and the Maine border (Wright's "Western Range"), someone reported a feline in their headlights at night, as "black as a bear" and "about forty pounds in weight."

1959: A black panther was reported crossing the International Paper Company Road six miles south of Dalhousie in the middle of the day. It was seen by three employees of the Mines Department.

October 29, 1960: Also in the Western Range, two men watched a black or very dark brown panther slowly cross a main road and woods road, in good light and a clear view, from about 75 yards. They said it was about three feet long with an additional 18 inch-long tail, and was about 16 inches high at the shoulder.

July 1961: Near Fredericton, three boys on bicycles, heading to school, saw a "police dog"-sized black feline. Wildlife officers found large cat tracks at the scene.

1962: A "jet black" feline was seen one afternoon by a Corporal as he was driving on Shirley Road, about a mile east of Base Gagetown. He was able to watch it for two to three minutes from about 75 feet.

June 1962: A black panther was watched through binoculars from 75 yards in broad daylight, near Devon.

June 15, 1962: Over a dozen people reported seeing a black panther near a Fredericton cemetery.

1962: A Corporal hunting deer in the Training Area near Blue Mountain saw two separate black felines just a few days apart. One was "half-grown," the other was reported to be four feet in length with a tail about as long.

1962: A woman was frightened by a large black feline near her home in Upham, Kings County.

August 16, 1967: A Public Works employee working in the morning on Hanwell road in York County watched a jet-black feline leap across the road. Head and body were four feet in length, and the tail almost that long itself. Five minutes later, a large bobcat crossed at the same area, giving the employee opportunity to note the differences. A man living in a camp about 200 yards from this spot reported seeing a large black animal two nights later.

October 1967: A hunter reported seeing a black panther ten miles from the last location.

April 12, 1970: Two persons watched a black feline for about five minutes near Fredericton.

MARITIME PANTHERS NOT MYTHICAL: RESIDENTS
Ottawa, Ontario, *Citizen*, July 6, 1985

The article reviews reports of both black and tawny panthers in New Brunswick. Kathy Kerton watched a jet black feline as large as a Labrador retriever on the outskirts of Fredericton. Jean-Paul Melanson saw a black panther behind his home in Moncton.

Michael Newton (2005) noted a black panther sighting near Miramichi, New Brunswick, in mid-December 2002. Two brothers on their way to work on a cattle ranch stopped to watch the animal for about 15 minutes before reporting it to wildlife officers.

NEWFOUNDLAND

BIG BLACK CAT SIGHTINGS CONTINUE AROUND REGION
Springdale, Newfoundland, *The Nor'wester*, October 17, 2008

This article reviews stories sent in by readers after an earlier report, when the journalist learned that wildlife officers had little to go on. Sightings have gone on for years, and one local legend was that American hunters brought in a group of big black cats to release and hunt in the area. One man said he and two others were driving on Salmon Pond Road when a 50-pound black cat crossed in front of them. "It was a cougar-style cat with a square nose. He was about 80 feet from us and we got a real good view of him because he he just walked slowly across the road and into the brush on the other side."

NEW HAMPSHIRE

Gunnin' and Fishin' with Bob Elliot
Portsmouth, New Hampshire, *Herald* (excerpt), August 11, 1945

A black panther has been roaming Proctor's district lately. He says:

"Well, folks, that black panther has been seen again and this time at Russell's station down near the beaver dam on upper Stoney brook. Guy Hollis of Antrim saw the animal. This is the second time Mr. Hollis has seen the animal or one like it. One day last week two women from Greenfield saw a bear cross the road somewhere between Dublin and Keene. Boy! What a lot of hunting we will do this fall."

Portsmouth, New Hampshire, *Herald*, October 25, 1945

Hunters take note!

The "black panther" is on the loose again.

It may or may not be a myth, but at least two men in Windsor have seen this long, black animal during the past few weeks and have even taken shots at him.

One of them aimed at the weird beast as it was molesting a couple of young heifers in his pasture.

Although sportsmen in authority on the subject say there's no such animal, several persons in Deering and Hillsborough reported seeing a creature meeting the same description last summer.

'BLACK PANTHER' BACK AGAIN—WITH A FRIEND

Nashua, New Hampshire, *Telegraph*, September 12, 1946

Greenfield, Sept. 12—Southern New Hampshire's black panther is back again, and this time he has a friend. At least, that's the word from this town where several residents have reported seeing the perennial—and footloose large animal bettered to be a "black panther."

The possibility of there being two of them is raised by the widely different descriptions given by people who claim to have seen the beast.

WATCH OUT, HINSDALE: N. H. 'THING' HUNTER WINGS A NEIGHBOR

Pittsfield, Massachusetts, *Berkshire County Eagle*, December 18, 1948

The reputation of the Hinsdale black panther, or "The Thing," if you are a skeptic, has reached clear to the Keene, N. H., woods.

A Keene dentist, Dr. Lorenzo Metivier, told police there he had heard tales of a black panther roaming the forest. He took aim at a dark and slinking object yesterday, and winged a fellow townsman, Almes Lower, in the left arm.

From today, Berkshire hunters intend adhering to the ancient marksman maxim: "Don't shoot until you see the whites of the black panther's eyes."

BLACK PANTHER, CHAPTER IV

Nashua, New Hampshire, *Telegraph*, December 22, 1948

George S. Proctor of Wilton, the "Pinup Warden," drops us a line along with a clipping from the Peterboro *Transcript* which reports that the Keene *Sentinel* is offering a $25 reward to the hunter who can produce the Black Panther, dead or alive but, of course, preferably dead.

The *Transcript* calls it the "Panther Sweepstakes" and lists yours truly as the last to positively see the Black Panther,

Frankly, if what we saw was the Black Panther, we're no little amazed.

Seriously, however, the black animal the writer saw and described to several persons has been identified as several different animals. One sportsman, is well known to many dog lovers, told us the black thing could have been a Canadian Fisher Cat and added that if it was, another was probably close by because this type never travels solo.

And then another hunter dropped in to say that what the writer saw was actually a bear, one of two this nimrod has been tracking for a considerable time without success. One, he said, was large. The other small.

Unfortunately the oversigned can be of little help in aiding the solving of the South Merrimack Black animal. The animal left no tracks in the pine needles and leaves. Even if he had they probably wouldn't have helped the writer much. Outside of a deer's and cow's tracks, yours truly has trouble recognizing his own footsteps in the woodlands.

PANTHERS ABOUT

Fitchburg, Massachusetts, *Sentinel*, December 13, 1948

For a long time, people have been wondering how many years would go by before some deer hunter knocked off one of those Vermont or New Hampshire panthers reported roaming about. The Vermont one almost became a trophy this season—but not quite.

Tony Gandin, a granite memorial manufacturer, came face to face with the big cat while hunting at West Grolon. Tony says it was as large as a young deer. He doesn't say he was a bit flustered by the sudden appearance of the big beast. He did finally fire a charge of buckshot at the critter and he believes that he connected, but he could find no blood trail. Tony says the animal was on a deer trail at the time and that "It looked awful big to me."

Dr. Earle Hall of Winchendon, Mass., who has a camp near Mountain pond in southern New Hampshire, reports seeing the mysterious White Mountain black panther twice. He didn't get a shot at the animal.

PANTHER RETURNS TO MONADNOCKS

Portsmouth, New Hampshire, *Herald*, December 29, 1949

Wilton, Dec. 29 (AP)—That elusive New Hampshire-Vermont black panther, occasionally "seen", never photoed and assuredly never killed has reportedly been visiting Wilton.

A Wilton news correspondent reports Mrs. H. F. MacMurray looked out her kitchen window recently and at a distance of 200 yards saw the panther eating on a deer carcass. When her husband, a doctor, returned home, he, too, is reported to have seen the animal.

PROCTOR INSISTS BLACK PANTHER IS IN THIS REGION
Nashua, New Hampshire, *Telegraph*, January 18, 1949

Wilton, Jan 18—In reply to a statement in a Sunday newspaper that the black panther was just "a figment of the imagination", former Game Warden George S Proctor said that "somebody's face is going to be red when they get the animal".

Ralph G. Carpenter, New Hampshire director of the Fish and Game department reportedly stated in the Sunday paper that he was tired of hearing stories about the black panther and did not believe such an animal existed. Carpenter is further quoted as saying that he was told by experts that no panther ever lived this far north, they all live in the tropics.

Proctor said that some 65 persons who have supposedly seen the animal could not all be wrong and he indicated that there would be a lot of red faces when it is killed or caught.

'MYSTERY ANIMAL' GROWLS AT NAVAL BASE COOK
Portsmouth, New Hampshire, *Herald*, November 22, 1949

York county sectors reporting the activities of the "Panthdeliarpard" were quiet for four days last week.

However, Saturday morning Ship's Cook Edwin Hibbert of the Portsmouth naval base reported that he saw a long, black shape, about the size of a lioness or leopard trotting along Route 1 near the bridge across the York river.

"It was about 5:30 in the morning," said Hibbert, "and I was on my way to work. I was driving slowly and I spotted this black shape trotting along the road near the York bridge between the Orchard farm and the McIntire estate.

"I pulled up alongside the animal and it was as big as a St. Bernard dog. My lights showed it to be sleek and it had a very long tail.

"When I pulled up alongside, it growled at me. And, brother, what a growl that was. I didn't have any gun in my car so I just kept right on going. The Panther, or whatever you call it, turned and went into the woods."

Cook Hibbert's story checks with that of Waldo Jones of the Round pond section who was hunting near Folly pond Sunday afternoon.

Jones said that he was in cover shortly after daybreak and he heard a twig snap. Glancing over his shoulder, he saw the long black shape of the "Panthdeliarpard" sulking in the shadows. He fired one shot but missed and the animal moved off toward the pond.

"Probably spoiled all my deer hunting for the day, too," said Jones. He failed to even see or hear a deer the rest of the day.

Two hunters in the North Conway region reported seeing the animal Sunday but their stories do not check with those of Hibbert and Jones.

The fact that it has been seen so often in the southern York county area leads the game officials to believe that the animal is sticking pretty close to the York county area.

There is plenty of food available for the wily creature as it hunts far deeper in the woods than hunters go.

GUNNIN' AND FISHIN'

Portsmouth, New Hampshire, *Herald*, January 12, 1952

By Hal Pierson

The "black panther" has been appearing in the news again. This time a couple over in Wilton reported seeing a large black feline, which they identified as a panther. With snow on the ground, it would have been easy to make a mould of the track for more positive identification.

A black panther (or puma) was bagged and identified several years ago in New Brunswick but none have been reported from this downeast province since that time. There, were rumors of a black panther in Rye and North Hampton a couple of years ago but that turned out to be Abbot Drake's laborador retriever. The big black dog was identified by several people as a "black panther."

Les Hall, Hampton sportsman, reiterated his story of spotting a large black cat last fall while bird hunting in the Coffin pond area. My two dogs flushed an average sized gray cat from under a juniper in the latter part of the season only a few hundred yards from where he spotted his animal.

It was a bit surprising to jump a cat out in the woods, but, there was no question about the one my dogs chased up a tree being a house cat. Les is an old enough hand around the woods so a house cat would not make him think of panthers, so there may be something to his story. Especially, if the "black panther" in Wilton can be positively identified.

BLACK PANTHERS UNMASKED, SIMPLY OVERGROWN WEASELS
Lowell, Massachusetts, *Sun*, November 13, 1957

Concord, N. H., Nov. 13 (INS)—Hunters in snow-topped White mountains were warned today that black panthers they have been seeking probably are Pennant Martens, members of the weasel family.

The conclusion was reached by John Brennan, one of the directors of the state planning and development commission, who offered a $1000 reward in October for any person producing the carcass of a black panther.

Dr. Harold Anthony, of the American Museum of Natural History in New York, told Brennan that there is no record of a black American mountain lion or a black Asiatic leopard escaping from a zoo. The latter is immune to snow and cold.

Incidentally. Brennan pointed out that a 40-inch Pennant Marten, or "fisher." with long black tail is mounted and on display at the New Hampshire capitol.

"There's a $50 penalty for shooting these Martens." said Brennan, adding:

"They apparently are increasing in numbers and getting more tame, and are coming down from their retreats along the Canadian-Northern New England border. We have warned hunters who hunt Canadian lynx with cat dogs not to shoot Martens. They are valuable in keeping down destructive animals."

Brennan said that for the past five years black panthers have been reported seen in New Hampshire, Vermont, Maine, Massachusetts, New York state and Pennsylvania, none in Connecticut and Rhode Island.

A black panther report, Brennan said, came this fall from Dalton, near Whitefield in the White Mountains, where paper mill workers saw a black animal drinking on the bank of the Connecticut river.

A Gilmanton woman reported a black panther escaping death after leaping from the road in front of her automobile. The town is in lower New Hampshire.

A New Hampshire man reported a black panther roving in Franconia Notch, near the Old Man of the Mountain ledge.

Brennan said he had scores of letters since he offered the $1000 reward. They came from persons in all walks of life, including a New York banker and Newport News, Va., financier.

Three writers protested the killing of black panthers on humanitarian grounds.

Lynx in Woods?

Nashua, New Hampshire, *Telegraph*, November 29, 1962

A large black animal believed to be a Canadian lynx was seen in the woods back of the Kenneth Scheibel home on Nashua rd. by Mrs. Scheibel and her daughter while picking Christmas greens.

Alerted by their German Shepherd dog who was with them they noticed a large black animal up in a tree not more than 15 or 20 feet from them. Startled by the dog's barking the animal came down from the tree, slithered through the tall grass, and climbed another.

According to Mrs. Scheibel the creature was a large, heavy, black, catlike beast with a long black tail, a face like a bobcat, short pointed ears and very sleek, weighing perhaps about 50 pounds.

A similar animal has been reported as having been seen in the Oak Hill section of the town. Neighboring towns have reported seeing like animals, some of which have been described as weighing considerably more and in colors ranging from black to grey or tawny. These have been called either cougars or mountain lions, all part of the lynx family.

In 2002, a large feline skull and some associated bones were discovered in the woods near Manchester, New Hampshire. They were later determined to belong to a female leopard (Lankalis 2006).

WOODS, WATER AND WILDLIFE, BY JOHN HARRIGAN
Manchester, New Hampshire, *Union Leader*, June 13, 2004

The columnist noted the sighting of a black panther by Conservation Officer Keith Kidder. It crossed in front of him on Route 3 in Pittsburg "in broad daylight."

NEW JERSEY

New Brunswick, New Jersey, *Daily Times*, March 21, 1872

The wild animal which has been the cause of so much alarm and frightened the young men of the upper part of Passaic county into keeping virtuous hours for a long while, has finally been plainly seen in broad daylight, and unmistakably identified as a large, striped hyena. Two farmers saw it the same day during the past week. The animal fled from them. Being unarmed, the farmers did not pursue at the time. Subsequently meeting and relating what they had seen, they procured rifles and started in pursuit of the beast, but a prolonged search failed to again bring the strange "critter" to bay. He is believed to have escaped from a menagerie which passed through the vicinity two or three years ago, one of the cages of which was smashed by accident, letting the animals out.

SEEING STRANGE THINGS
HUNTERS UNABLE TO CATCH UP WITH MYSTERIOUS ANIMAL
Philadelphia, Pennsylvania, *Inquirer*, January 9, 1906

Mount Holly, N. J., Jan. 8.—Hunters in the vicinity of Medford have not yet been able to capture the mysterious animal said to be in "bear swamp." It appeared at the residence of William Schreider, near Ballinger's Mills, a few nights ago, making a most unusual noise.

While Mrs. John Mingen and Mrs. Caleb Dudley, of Indian Mills, were driving along the Burnt House road, a strange animal came out of the swamp, passed the carriage, then disappeared in the thicket again. It was as large as an ordinary bird dog, black in color, short hair and had a long tail.

TIGER SCARE IN JERSEY
C. W. BROKAW AND FAMILY SURE THEY SAW BEAST—ON DUKE ESTATE ALSO
New York *Times*, July 1, 1921

Somerville, N.J., June 30.—Twice within a week an animal said to be either a panther or tiger has made its appearance here and disappeared as suddenly as it came. It is believed the beast escaped from a circus train.

The strange visitor was seen last Friday night by Mr. and Mrs. Charles W. Brokaw of 246 East Main Street who, with their son William and another boy, Wilson Vanderveer, were on the front steps of the Brokaw home. They thought it was a dog that ambled across East Main Street but when the animal came nearer they saw that it must be a denizen of the jungle. Young Brokaw and his friend ran into the house after rifles, but when they returned the animal had gone.

Since then small animals have been disappearing from this vicinity, one man reporting the loss of two calves, but not until last night was the animal seen again. Irving Schaeffer, Assistant Superintendent of the 2,000 acre estate of James B. Duke, tobacco magnate, said he saw an animal not unlike a panther or tiger outside of his house about 11 o'clock last night. The animal ran off, however, before he could shoot at it.

REAL "LION" HUNT STIRS SOMERVILLE
HUNDRED CITIZENS SEARCH FOR BEAST,
WHILE WOMEN AND CHILDREN HIDE
Trenton, New Jersey, *Evening Times*, July 26, 1921

Somerville, July 26.—the wooded section of this town, bordering on the Lottery residential section, was the scene last night of an aggressive "lion" hunt.

Expeditions of the townsmen, armed with as motley an assortment of firearms as the farmers of Lexington and Concord used in '76, were scouring the countryside seeking combat with the exotic marauder.

And while the men folks were out for big game, women and children kept strictly within doors.

Henry Squinger, manager of a store here, saw the lion—and he is a man of standing and veracity. Also, Richard Hoffman took a shot at it. Neither claims to be an expert on tropical zoology, but each has seen pictures of lions in the movies and lions themselves in menageries.

A month or more ago there was a vague report that some kind of a wild animal had escaped from a passing circus train. There were numerous reports of a trainer from the circus roaming the woods hereabout on a mysterious mission. He would not tell his quest. Inquiries at the railroad brought no confirmation of the escape.

The farmyard stock began to disappear. Chicken coops were robbed on a wholesale scale. Citizens traveling the roads of the Lottery section at night told of meeting a strange and ferocious looking animal that invariably fled at human approach.

The strange appearances ceased—only to have a recurrence last Saturday night when Mrs. Squinger and her daughter, on the way to their home on Davenport Street, saw a strange animal in the road. They reached the house hysterical and out of breath.

Then Squinger himself, on his way home early yesterday morning, saw the animal outlined in the moonlight. He identified as positively as a lion. He organized a searching party—and the searching party was augmented to nearly a hundred by night.

Among the hunters were three Cline brothers, two Hoffman brothers and Lamuel Hall. They searched the shore of Peters Brook, regarded as the favorite lair of the animal, and they say they sighted it. Richard Hoffman ran forward and, at a range of 50 yards, fired a load of buckshot at the animal. It turned and raced away, apparently unhurt. Hoffman is sure it was a lion.

The citizens are determined to track the "lion" to its death. It has confined its attacks thus far to poultry and other small animals—but they fear it will assert its man-eating proclivities unless dispatched.

Renew Hunt For Killer of Livestock
Seek Strange Animal in Hayesville Area
Mansfield, Ohio, *News Journal*, May 5, 1950

By Virginia Leed

Ashland—A strange animal which has been killing livestock in the vicinity of Hayesville for the past two years, is striking again in recent weeks, renewing the hunt by farmers and Ashland County dog warden Carl Koegler for the animal which is apparently not a dog.

A composite picture from descriptions offered Koegler by farmers in the area would look like something out of a nightmare, but it wouldn't resemble a dog, Koegler says.

According to information furnished Koegler, the marauder, which has successfully resisted every attempt to capture it, leaves claw marks on the livestock it kills, goes abroad only at night, has been heard making an eerie squawling noise, covers a wide area on its nocturnal ventures and attacks animals in a way different from that of sheep-killing dogs.

So far in recent weeks the animal has killed a ewe and two lambs at the Jones farm south of Hayesville, returning the following night to attack another lamb, then killed a ewe and lamb at the M. W. Shafer farm and attacked lambs at the Shafer and Dwight Donelson farms this week.

Koegler estimates damage caused by the animal in the past two years at $4000 in livestock.

MYSTERIOUS BEAST IS DOG (*sic*)
Indiana, Pennsylvania, *Evening Gazette*, September 27, 1950

Haddonfield, N. J., Sept. 27.—(AP)—The mysterious beast which has been sought by police for the last three days left some new tracks on a farm in nearby Delaware Township yesterday.

John A. Yowa reported he found the tracks on his farm—nearly three miles from the spot where the first traces of the mystery animal were discovered.

Yowa said the tracks were three inches in diameter with divisions indicating cat-like pads—the sort of tracks that might have been made by a leopard or similar beast.

Mrs. Yowa said her pet dog has been cowering in the house for the last two days.

Extra police patrols were ordered on Sunday by Delaware Township Director of Public Safety, Ernest R. Williams.

Williams himself reported he had found "strange, large" tracks across the lawn of his home two weeks ago.

The tracks, he said, "definitely were not made by a dog. Dogs don't come that large, but we're stumped when it comes to identifying the creature."

Panther Sightings Fascinate Vestigians
Passaic, New Jersey, *Herald-News*, January 14, 1979

The Fortean research group Vestigia collected sightings of black panthers in Sussex County. Two sightings in the last year were from Stanhope (where Vestigia was based). In one, a woman was walking her dog one morning and heard squirrels making a racket up in the trees. She looked and saw a large black feline. Vestigia co-founder Bob Jones stated, "The rear end of the cat was maybe 5 feet off the ground. It was clutching the tree with its paws. The woman was within 25 feet of it so she got a very good look at the animal. It was black and it was a panther. She described the body as between 4 and 5 feet long and she said the tail was about 3 feet long, which is a typical panther. The thing that stood out in her mind—aside from the fact that she was seeing an unusual sight—was the extreme healthiness of the animal. It had a beautiful coat, there was nothing scraggly about it at all." Later that year, Vestigia (Jones 1979) reported in their newsletter that a black panther was seen within 100 yards of their headquarters the past April. Two sisters saw a large black cat, about the size of an adult Irish setter that lived nearby, walk up to the woods behind their house. A large feline print was found later by the group members. I measured 4 x 3 3/8 inches. No claws appear present on the track, as published in the newsletter.

Panther in N.J.? It Ain't Garfield!
New York, New York, *Post*, November 5, 1983

Lacey Township Police and Humane Society officers chased a black animal "which stands 2 feet high and lopes like a wild feline."

Mysterious Beast Baffles Rural Burlco Town
Trenton, New Jersey, *Times*, March 20, 1988

A Burlington County horse boarding farm owner reported sighting a large black feline, "bigger than the Dobermans" that she kept. A game warden and an executive from the Safari wild animal exhibit visited to look at the prints, and were intrigued but uncertain as to whether a large feline was responsible.

PANTHER IN S.J.? DEP SAYS NO
Cherry Hill, New Jersey, *Courier-Post*, May 1, 2007

Some residents were worried by sightings of a large black cat, while others were unconcerned. Wildlife officials could find no physical evidence, and a photo was examined by animal control and suggested to be a very large domestic feline.

NEW MEXICO

THE MOUNTAIN DEVIL.
SPORTSMEN SEEKS INFORMATION ABOUT
VARMINT OLD HUNTERS BELIEVE IN.
Washington *Post*, January 5, 1914

From *Forest and Stream*.

It is a trite saying that fools can ask questions but it takes wise men to answer them. For example: Does the panther scream? Do snakes spit? &c.

Now as a member of the class that asks questions, I desire to propound an interrogatory to the members of the second class; a question possibly not of such vital importance as the scream, probably not so pleasant to consider as the spit question, but nevertheless, one of considerable interest to us timid fellows who frequently sleep alone in the deep mountains, with the pale light of the moon for a blanket, the shadows for sheets, and the soft side of the rock for a pillow.

What is a mountain devil? To satisfy yourself that such a creature really exists, ask almost any one of our Western mountaineers. Of course no two of them agree about the size, shape, or general characteristics of the "cuss," but on the general proposition that there is such an animal, and that he is a "holy terror" you will find remarkably unanimity of opinion, and refreshing volubility of expression. I was talking last evening with an old hunter who had seen one by moonlight in the mountains of New Mexico. He undertook to describe it for my edification, but the poverty of the English language forced him to give up the job, and he finally contented himself with the declaration that it more nearly resembled a jackass than any other animal he knew of. He also informed me that the Mexicans and Indians pointed out to him next day the home of the creature among the crags and caves of an adjacent mountain, with a gentle admonition to avoid said mountain on general principles, which they did, and for the further and separate reason that it was the home of the devil.

For my own part I have seen but one mountain devil. It was dead. Billy Newman and myself, while at Ilwaco, near Shoalwater Bay, were invited to call at Jawbone's saloon and inspect the carcass and give our opinion as to the "kind of varmint it might be." The proprietor informed us that it had been shot out of a tree about 70 feet from the ground in the mountains back of the bay; that it was an active tree climber, but absolutely fearless, and when shot was preparing to descend to attack its enemies. It was a plantigrade animal, and its hair much resembled that of the cinnamon bear, but here all resemblance to the bear family ceased. Its legs were long and straight; its back straight as a wolf's, while the body and head much resembled the hyena. Will some sportsman kindly inform us just what a mountain devil looks like, so that we may know his excellency when we meet him?

TIGER MAY BE LOOSE IN AREA NEAR GALLUP
Albuquerque, New Mexico, *Tribune*, November 22, 1955

Gallup, Nov. 22 (AP)—A tiger may be roaming the Navajo reservation in the Western New Mexico and Eastern Arizona.

Drivers for the General Services Administration and Navajo Indians say they have seen large cat tracks which a professional government hunter says are not those of a mountain lion.

One of the drivers says he has seen a large, striped animal jump across the highway.

Indians who have seen the tracks say they believe the cat may be holed up in one of the abandoned mine shafts in the Ft. Defiance area.

The cause of the speculation goes back four years when a circus wagon train was wrecked about 50 miles north of Gallup. A lion and a tiger escaped and never were recaptured. The tiger may have survived, Indians believe.

Journalist Jerry A. Padilla kindly shared a fascinating story he was told by a close and respected relative, Mack:

"As best as I remember, when I was about 11-13 years old in the mid 1960s, I was playing with plastic toy dinosaurs and prehistoric animals, and my favorite was a sabretooth cat. Mack would often engage in conversation with me and tell true stories about experiences he'd had during his life in New Mexico and other places he'd been during his military service overseas.

"That particular day he said, very matter of fact, 'I saw one of those kind of cats once.' To which I replied, how could that be, they have been extinct a very long time? He just said, 'Well, I saw one.' My next question was, where?

"He replied, 'On the old road on the back way from Cimarron to Ute Park, one night I was driving a coal truck from Dawson to Taos. It was at night, and I saw it in the headlights when it crossed the road, very clear.'

"I had heard him tell of his truck driving days in 1946, among other jobs before he re-enlisted in the Air Force. With this particular account, I insisted that sabertooths were supposed to be extinct.

"He continued, 'Well, I know what I saw, and it crossed the road as I came around a curve. It was a rough mountain road, the old route, a rough dirt road through some very wild country.'

"Insistingly, I questioned, 'Maybe you saw a mountain lion with deformed fangs, or it had something in its mouth?'

"At that point, somewhat annoyed that I questioned him, Mack replied, 'I didn't see any deformed anything, I know what mountain lions look like, I've seen them up close before; I know what I saw, and this sabertoothed cat was bigger and I could tell the difference even in the headlights. It wasn't any kind of mountain lion or puma or anything like that. I know what I saw. It was crossing the road and when it turned its head to look at the truck, I could see those real long teeth pointing down from its mouth, and then just like that it was gone into the timber.' He left and went outside to work on his old 53 Chevy. End of conversation.

"He never talked about it again to me, or anyone else I know of.

"I didn't pursue it anymore, but as I got older, remembering that story, and as an adult working, driving through and hunting wild remote country in Colfax and Mora Counties, I often wondered if there could still be a breeding population of sabertooths.

"The current paved two-lane road between Ute Park and Cimarron still runs through very rough remote mountainous land somewhat parallel to the old road.

"Mack died very young, in his early 40's from a service connected disability. He was not prone to telling untrue stories, often entertaining us young people on long rides with true stories about old times, and interesting things he experienced over seas and as a young person herding goats for his father in the wilds of neighboring Taos County. He would rarely speak of his combat experiences, and then only with his brothers or other male relatives who were also WWII vets. Mack had lived an exciting, interesting life. He was a strict father and uncle who always encouraged truth and honesty, even if it meant consequences, and very respected by his peers."

Nick Sucik (2002) interviewed a man in New Mexico who reported seeing a large shaggy white wolf near Dulce, New Mexico, in the late 1970s. When the witness mentioned this to friends in the Navaho community, he was told he had seen a "medicine wolf."

NEW YORK

"About this time reigned the King Atotarho IV. At the fort Ke-dau-yer-ko-wau, (now Tonewanta plains) a party went to hunt and were attacked by the Ottauwahs, which created differences between the two nations as they entered on no terms but to commence hostilities; the To-hoo-nyo-hent sends a band of warriors to attack some of the hunters as to retaliate the vengeance upon their enemies. The warriors advanced above the lake named Geattahgweah (now Chatauque) and made encampment and agreed to hunt two days, after which to proceed towards the enemies country; the warriors went in various directions; a certain warrior passed a small brook, he discovered a strange animal resembling a dog, but could not discover the head; the creature was a greyish color, and was laying asleep exposed to the rays of the sun; and also discovered a den, supposed the place of his residence; the warrior returned to the camp at evening and related the kind of animal, and informed them, as he imagined was a very poisinous animal, and he was afraid to approach it again, but one of the jokers laughed at him and he was called a cowardly fellow; the joker determined to go himself and kill the creature without trouble, but wished some of the warriors to be spectators in the time of the engagement; accordingly the warrior went, accompanied by a number of warriors: he was directed to the spot and discovered the animal. After beating it short time with his club, he seized the animal and tied it with a tumline; but while he was lifting it the creature immediately moved to the den. With all his might, he held the tumline, but he could not stop it, he was compelled to let go the tumline when the creature went beyond his reach; the warrior was confused at not being able to kill the animal; he hastened to retire from the spot, but when a few paces he was taken with the pestilence which was influenced by the creature, and suddenly died; another warrior was at sight and directly fled to carry the intelligence, but also died at a short distance, and the others returned to the camp; but the pestilence soon prevailed among the warriors, and many of them died in the same manner; a few of them escaped by leaving the camp before the plaigue appeared, and thus ended the expedition. The Ottauwahs continued their hostilities and attacked the hunters; the Senecas sent out a small party and fought—drove the enemy off, but their engagements were small and continued many winters." (Cusick 1848)

The Westchester County Panther

New York, New York, *Times*, June 23, 1880

The animal, supposed to be a panther, which has been making such havoc among the live stock of the Chappaqua farmers in Westchester County, by killing calves, chickens, dogs, and anything that came in its way, made its appearance yesterday in the Village of Pleasantville two miles south of Chappaqua, and almost caused a panic among the pupils attending School No. 9 of the Town of Mount Pleasant, which is about half a mile east of the Pleasantville rail-road station. Just back of the school-house is a dense patch of woods. A boy saw a large, striped animal, with a long tail, bushy at the end, jump from a limb in a large chestnut tree to the ground, and then proceed through a lot toward another piece of woods further east. The boy scampered to the school and reported what he had seen. The pupils looked out of the window, saw the animal, and very soon the whole school wanted to go home. Dr. J. M. Brown, on Broadway, who was reported to have had a personal encounter with the panther, said yesterday to a *Times* reporter that a few days ago he was going along a path in the woods with a small boy named John Doody, when he came within 10 feet of a large animal lying in a crouching position, watching him suspiciously. At first he did not know what to do, he was so frightened, but, keeping his eye directly on those of the beast, he picked up a large stone lying in the path and hurled it with all his strength at the animal's head, which it struck. He looked for another stone, and as soon as he found one the animal was gone. He said it was a panther beyond doubt. It had a cat-like face, though many times larger and very broad and stubby. He particularly no-ticed the stripes on its body. The eyes glared like two coals of fire, and the immense mouth was open, as though it was lying there in the shade to rest and cool off. Dr. Brown is a thoroughly conscien-tious man, and people who have heretofore scouted the idea of a panther or any wild animal being in the neighborhood, when they heard him tell his experience, at once believed "there was something in it after all."

ATTACKED BY BLACK PANTHER

WESTERN NEW YORK HUNTER HAS FIERCE
BATTLE WITH ENRAGED BEAST.

New York *Times*, February 11, 1900

Special to *The New York Times*.

Rochester. Feb. 10.—Fred Emerson of Bolivia, Alleghany County, while hunting squirrels in the woods back of his home, came across the tracks of a strange animal. Following them up he came to a dark, shallow cave out of which peered a pair of gleaming eyes. Without stopping to take aim the thoroughly frightened hunter raised his gun and fired the contents, a charge of small shot used for squirrels, into the cavern. A snarl, followed by the sudden appearance of a wounded black panther, came from the cave as the result, and while the hunter endeavored to reload his gun the animal attacked him. In less than two minutes his dog was killed, being literally torn to pieces, and the hunter was obliged to defend himself with his clubbed gun. Emerson struck the beast again and again with his gun, nearly shattering the weapon to pieces. He finally dispatched the beast, but not until he was badly cut up and perhaps fatally injured by the sharp claps of the panther.

The animal measured nine feet from the tip of the nose to the tip of the tail. He was a full-grown male specimen of the black panther, and it is believed escaped from a circus train that was wrecked in this vicinity last Summer.

STRANGE BEAST WITH BLOOD CURDLING CRIES RUN DOWN AND SHOT

Syracuse, New York, *Post-Standard*, August 7, 1906

Eaton, Aug. 6.—A wild animal resembling a lynx, believed to be the so-called "catawaubus," which has been terrorizing the adjoining county for some time, was killed in the vicinity of this place last week by Rupert Taynter, a young farmer. Taynter was hunting woodchucks when his attention was attracted by the cries of a strange animal from a neighboring wood. He located the beast at the root of a large tree and dispatched him with a rifle shot.

The dead animal was pronounced a lynx. It was of a grayish brown color, about fifteen inches in height, with a life, supple body. Old age had apparently enfeebled it as nearly all of its teeth had dropped out and its movements were slow. The head resembled that of a monster cat.

For several days previous to the death of this animal the presence of a strange wildcat in the vicinity had been reported from various sources. Lynn Dyer, a boy of 13 years, was twice frightened by the intruder on the premises of H. B. Leach. A week ago a lynx-like animal, dubbed a "catawaubus," was reported near Cortland, thirty-five miles south of here. A posse of hunters failed to discover the trespasser.

The lynx is believed to be the one that is alleged to have escaped from Sig Sautelle's circus some time ago.

SHOT A STRANGE ANIMAL

Gettysburg, Pennsylvania, *Adams County News*, December 10, 1910

For several months an animal, that it was believed was a panther, although no panthers have been seen near Middletown, N. Y., for twenty years, had been terrorizing the people of White's Pond, carrying off young sheep and pigs and making itself generally disagreeable, has been killed.

Although hunted for months by scores of good shots, it remained for C. M. Wisner, of White's Pond, to land the prize and the $50 reward that was offered for it, dead or alive. Mr. Wisner shot the animal in a swamp two miles from his home.

Following the shooting the animal was inspected by many people, including Game Warden Coy, and no one had yet named it. It was about seven feet long and had a small brown tail, with a black streak the full length. Some believed it to be a lynx and others that it is some animal that escaped from the menagerie of a circus that visited the vicinity early last summer.

Black Panther Seen in Hills

Kingston, New York, *Daily Freeman*, June 16, 1923

Clarence R. Peters, who lives at Lake Hill, recently saw what appeared to be a black panther near Wittenberg Lake, one of the wildest sections of the Catskills. Mr. Peters is manager of the Catskill-Shawangunk Mountain Association, which among other things secures as much publicity as possible calculated to call the attention of tourists and vacationists to the unrivalled possibilities offered them in this section, but he is a truthful man, so known to hundreds of Ulster county citizens. Besides, his wife and son saw the animal, and, besides again, Mr. Peters admits that any publicity secured thus might back fire; timid mothers hesitating to bring their children to a wilderness infested by panthers.

The animal when first seen, was, about 50 yards distant, was sitting on its haunches and resembled in every way save color a puma, panther or mountain lion. It was as large as a calf and jet black. Mr. Peters called the attention of his wife and son to the beast and then walked toward it. At first it showed no inclination to move, regarding him steadily out of its big yellow eyes, but finally it arose, stretched itself, switched its long, club-like tail and walked into the thick brush.

It is several years since a panther has been reported in the Catskills and the black variety is very rare.

Troopers Hunt Lion

Middletown, New York, *Orange County Independent*, July 11, 1929

Deputy sheriffs, State Troopers and volunteers are searching the hills northeast of Binghamton for an African lion which it is believed escaped from a circus which passed through Binghamton when one of the motor trucks was wrecked. Four people reported seeing the lion crossing open fields while several farmers reported unusual commotion among their cattle during the night caused by the howling of a strange animal.

Believes Dog Killed by Claws of Black Panther

Olean, New York, *Times Herald*, December 6, 1940

Belmont—Has the town of New Hudson a black panther?

During the deer season it came to the attention of one of the game protectors in Allegany County that Frank Cooper, a farmer residing in that town, lost a dog while coon hunting in that section of the county. While dressing the coon Mr. Cooper sits a black object by the dim night light, heard his dog cry out from a distance of fifty feet and going where the dog had been lying found him dead on the ground, disemboweled by some animal. It is believed some wild animal, attracted by the scent of blood from the dressing of the coon was attacked by the dog or came onto the dog, and quickly killed him.

The possibility that the animal might have been a black panther is believed by those familiar with the work of that animal and by the fact that what is believed was such a wild animal was seen last summer on a lonely road near the border of the towns of Clarksville and Wirt. The animal seen had three small animals of its own kind with it. Black panthers are known to have been seen in lonely areas across the Pennsylvania border. Residents of the town of New Hudson have heard strange cries in the night that they think may have come from such a wild animal.

Au Sable Forks, New York, *Record-Post*, November 25, 1943

Local hunters and woodsmen are still uncertain as to identity of the strange animal caught by Archie Brooks on his trap line last week. The creature has a foxlike face, while the ears resemble those of a wildcat. The short, stubby tail is about four inches long, while the paws are typical of a woods fox, having two distinct toemarks. The coloring, too, is peculiar: grey on the back, with reddish flanks, shading into white. The animal weighed eleven pounds.

BIG GAME HUNT MAY BE DUE
Dunkirk, New York, *Evening Observer*, August 17, 1944

There're more rumors flying around about this panther business than you can pick up at a barber shop.

If one was to believe even one-sixteenth of them, he'd barricade his home, post hound dogs at every corner of the lot, build a 10-ft high tight-board stockade around the place, arm himself with

Abercrombie and Fitch's best tiger rifle and sit down and wait with an edition of "Livingston in Darkest Africa."

Just what there is to all the stuff flying around, we can't—or haven't been able to as yet—learn. There does seem to be something to it alright—that is, the existence of some sort of a wildcat, whether it be a black panther, a dark grey one or what.

Responsible people claim to have seen "it". Tales come through every now and then of such an animal having been seen as close to Dunkirk as the Laona area. Other stories place the animal nearer Stockton.

And it wasn't very many hours ago that someone even spun a tale to us that some man who chose to sleep out on his country-area lawn because of the night's heat, was awakened only to see such an animal slink right across in front of where he was lying on the lawn.

There's little doubt as to the actual existence of some sort of a wildcat. Long-experienced hunters whose word can be depended upon claim they have seen the footprint of the animal and say it measured at least four and a half inches across. They add that it definitely was not a dog's footprint because no dog in these parts leaves a footprint that large. Nor is there a cat that large.

Trolley-car crewmen claim to have actually seen the cat. So have farmers in the field. Another story has it that a young calf was found dead in the field, his hide torn to ribbons and ripped as though some great tiger had gone to work on him.

Sporting goods store owners are interested in organizing a "posse" to go out with rifles and hunt the alleged panther down. Many of those who took part in recent fox hunts, would be only too willing to join and something may came of the idea in the near future.

"We must first find out everything possible about the existence of the thing," explained one store owner. "Otherwise we wouldn't know just when to start on the hunt. We would have to talk to those people who claim to have seen the animal and get everything on it we can in that district. There's no use going out half-baked on the idea. Whether there is just one running about or several, we don't know, but certainly by virtue of the number of different stories I've heard , one would be led to believe there are more than just one.

"Another thing I can't understand is: where are the carcasses of other wild animals the cat or cats live on? Perhaps they kill small game only, but a panther usually goes after a cow, a calf or a deer or something on that order. No one yet, to my knowledge—other than that story about a farmer finding a slashed body of a calf—has found any real evidence of domestic animals having been killed. I've heard tales of one having been shot by a state trooper over near Stockton. Gosh, if that was true, we would all have heard about it—and would have seen the trophy—long before this. The papers certainly would have gotten some word on it."

PANTHER REPOTTED NEAR STOLP GULLY
Syracuse, New York, *Herald Journal*, September 6, 1945

That panther's in again.

Latest report on its movements was received by the sheriff's department yesterday afternoon and placed the animal in the vicinity of Stolp gully and adjoining woodlands.

Deputy Sheriff Joseph Seeley, dispatched to the scene, organized a panther hunt, with eight members of the Elmwood Fish and Game Club participating. Armed with rifles and shotguns the men searched the vicinity for hours without success.

The animal, described as jet black and shiny, was seen yesterday by Lois Clayton, 16, of 126 Fairfield av., and Joseph Weston, 12, of 130 Fairfield av. Fred Keefus, Jr., of Young rd., three miles southeast of Onondaga Hill, reported he saw the beast last week.

"PANTHER" IN CITY!
Syracuse, New York, *Herald Journal*, September 6, 1945

Three boys, breathless from pedaling their wheels from Elmwood Park to police headquarters reported that elusive black panther was in the park at 11:15 A.M. today. So out of breath were the boys that they could hardly tell Sgt. George O'Connell about it.

"It was a great big black cat—so long—" indicating "on a limb 50 feet from us."

The boys who reported the panther today were Raymond Brown, 12, and Paul McConnell, 10, both of 409 Niagara st. and Harry Ryder, 11, of 218 Gifford st.

Panther Gets in Hair of City's Police

Syracuse, New York, *Herald Journal*, September 7, 1945

Panther, panther, who's got the panther? A case of panther-nerves sent Police, S.P.C.A., and city health officials on several "wild panther chases" today.

Dr. Maurice J. Dooling, director of the city bureau of food and sanitation, joined S.P.C.A, attendants Bert Hunt and Peter Bushey on a fruitless dash to "137 LaForte av.," to see a family named Wilson who reported a panther cornered in the garage about 11 A.M.

It turned out that the 100 block of LaForte only runs to 111 and that there is no family named Wilson on the street.

The trail of the panther led up several other blind alleys. Reports were received from Midland av. and Tallman St., and along the Onondaga Creek. Police reported finding a footprint. The animal was reported seen yesterday Elmwood Park.

Black Beast Mystery Should Be Solved Soon

Syracuse, New York, *Herald Journal*, September 13, 1945

At latest count the majority of Syracusans had not seen the peripatetic panther that is reported ranging our parks and the Onondaga fields and woodlands. Most of us don't want to see it. It is enough to have the vicarious thrill of reading that someone has seen it or heard it purr or give vent to the curdling panther-cry in the dark of night.

Some of us are sceptical but some have been cleaning up the old musket and preparing for the eventuality that the black thing may snarl at our open window some night. Some a-hunting have gone. Wherefore the sheriff's office has warned that an organized hunt will be conducted and that persons firing high-powered rifles, scaring folks in the environs, will be stopped.

This is as it should be. Residents should be permitted to sleep and cows should enjoy, without anxiety, the last days of Summer pasturage. We don't know whether there's a real panther or two in our alarmed midst or not. If there isn't, Syracuse should be permitted something of the kind. Our geographical situation makes a sea serpent wholly out of the question and a panther is far more logical. Perhaps a panther or some bugaboo may be helpful: it keeps us from worrying about mosquitoes. Anyway, we'd like this mystery solved before election and at the latest before the holidays. It shouldn't be a political issue and we want to do our Christmas shopping without toting a gun around.

DID PANTHER SCREAM?

Syracuse, New York, *Herald Journal*, September 21, 1945

Those panthers are screaming again, in North Syracuse this time.

Deputy Sheriffs Leo Beebe and Charles Post answered a call from John Dicamelli, 211 Shaver av., and found the neighborhood in wild excitement. Dicamelli stated that he was listening to the radio when the program was drowned by terrific screams, "like a woman, only louder." Three neighbors affirmed the story of the screams.

Beebe and Post set out on foot to explore three acres of woodland in the rear of the houses and came up with exactly nothing, not even a scream.

NO PANTHER, DOGGONE IT

Syracuse, New York, *Herald Journal*, October 4, 1945

The prowling panther preyed on the peace of mind of the populace at Amboy yesterday, doggone it.

A woman living near the Syracuse Municipal Airport claimed she saw the animal near her home. So a posse of men of the neighborhood was alerted, including Gordon Chapman, James Walsh, Edward McGlynn and William Gooley, "prepared for any eventuality," it was said.

The posse found the culprit: A large black dog, innocently roaming about the airfield.

BLACK CAT KILLED: WAS HE 'PANTHER'?

Syracuse, New York, *Herald Journal*, January 6, 1946

Alfred Norris of Taunton is one of Onondaga County's most relentless enemies of the fox. He is a chap who loves to hunt and likes nothing better than to put on one-day extermination drives against the fox any time he can. The latest trophy to fall to his skill was a big black cat weighing about 20 pounds. He and friends wonder if he has solved the mystery of Split Rock's black panther.

Norris, was out hunting fox, when his dog picked up a trail. He followed as closely as he could over the rough Split Rock terrain and eventually he sighted, up ahead, a black animal, slinking along the ground.

His dog was on a fox trail and Norris had to turn him and start chasing the black animal himself, before the dog became interested. Finally the dog treed the animal and Norris got a good look. It was the biggest black cat he has ever seen.

Norris took a shot, but the cat jumped out of the tree and the dog was quick in pursuit. Soon the cat was treed again, this time high in a tree at the edge of a ravine. Norris took another shot. This time he hit his prey, right in the head. The body toppled into the ravine.

The Hunter took a good look, marveled at the size of the cat, and not being interested in cat fur, traveled on, hunting foxes.

Telling friends about it later, he commented on the sleekness and blackness of the cat, which he felt had been running wild a long time. Some asked if it could be Split Rock's black panther.

Stockton Watches for Black Panther
Dunkirk, New York, *Evening Observer*, June 19, 1946

Stockton, June 19—Area residents were on a cautious watch to-day for a black panther reported to have disappeared into woods near here.

The animal, which has menaced Chautauqua county dairy herds for more than a year, was spotted Monday night by Mr. and Mrs. Fred Conk of Casselman Corners. They notified neighbors, one of whom, Kenneth Arnold, shot at the animal but missed as it raced into the woods.

A paw mark seven inches in diameter was found in a roadway.

That Phantom Panther Feared on Prowl Again
Syracuse, New York, *Post Standard*, June 30, 1946

Onondaga's panther that kept the western part of the county in a nervous state last summer, but unheard of since early fall, may be back on the scene again.

Yesterday morning Frank Smith, 2608 Bellevue ave., just beyond the city line, told Deputy Sheriffs Leo Beebe and Charles Post he was awakened by a commotion under his window at 1230 a.m. He said he slipped out of bed, grabbed a flashlight and went to the window

Turning the light on the spot from which angry snarls and growls were emanating he said he saw a huge black animal mauling the family cat. It did not look like a dog and it might be a panther he told the deputies. When he went to the spot after sunup he said he found the cat dead, its body cut and slashed by the teeth of its assailant.

PANTHER OR WHAT HAVE YOU SEEN IN FREDONIA TODAY
Dunkirk, New York, *Evening Observer*, August 22, 1946

The panther comes to Fredonia, or more probably, a bob cat.

Anyway, Frank W. Horton of Porter avenue, former Fredonia village trustee, and Mrs. Horton both saw the big cat early this morning.

Mr. Horton says it was 5:15 when he was awakened by a weird wailing which he thought at first was some night bird, but as he became fully awake, he realized it had a distinctive sound: a sound not like a bird or any cat he had ever heard, but much heavier, a sort of wailing.

He got out of bed and ran to the window overlooking the lawn toward the gas station and saw a long cat-like creature slinking toward the road. He saw it plainly on account of the street light throwing rays across the yard. The animal was dark on the back and lighter underneath and he estimated it to be 30 to 40 inches long.

His dog remained discreetly in the kennel. A neighbor's dog was noticed soon after acting as if he had seen something unusual. The strange animal disappeared across the highway.

While not taking much stock in the panther stories circulated about the county, Mr. Horton does believe the animal might have been a bob cat. He is positive it was too large to be an ordinary cat and its cry was definitely not like any cat he had ever heard.

Mrs. Horton also saw the animal and heard the wailing again after the creature had disappeared from sight.

Boys Think They Saw Panther Today

Dunkirk, New York, *Evening Observer*, August 26, 1946

Another story of an alleged panther prowling in the vicinity of Fredonia comes from some boys who claim to have seen a long black creature stalking a cow and a calf early this morning.

The boys had gone to the field to feed a cow and calf tied there and saw a long black animal crouching and working its way toward the cow and calf. The boys went to the house and obtained a .22 rifle and then returned and took a shot at the animal. Police Chief Floyd Thompson was called, but no trace of the prowler could be found.

Black Panther and Cub Seen by Jamestown Men

Dunkirk, New York, *Evening Observer*, August 31, 1946

Jamestown, Aug. 31—Two Jamestown men, one of who claims he knows panthers, insist that the animal they saw on Big Tree road near the city was a black panther. It had a cub panther with it, the two declare.

The two men saw the animals once as they crossed a bridge. The men stopped and gave the animals as good an inspection as they could at a respectable distance at 4 a.m. When the panthers disappeared they followed along a nearby road and saw them again passing through a farm yard in the direction of this city. It all happened Friday morning. The larger panther was about four feet long, had a long tail that it carried straight out behind it.

Armed with Riot Guns, Officers Search in Vain for 'Big Panther'

Salamanca, New York, *Republican-Press*, October 23, 1946

Police and firemen spent several hours early this morning in a fruitless search in heavy fog for a panther which a truck driver reported hitting in Wildwood avenue near the city line.

The driver, Jack Hawkins, operator of a Jamestown Furniture Express truck, reported at police headquarters at 12:45 that he had just hit a black panther a short distance east of the B. & O. underpass in Wildwood avenue.

CHAD ARMENT

The animal, he explained to Capt. William Schaefer, on duty at the desk, was dragging itself off into an adjacent nursery field on the north side of the highway.

"It was a big panther," police were told by Mr. Hawkins, who said he had seen them before and was sure it was a panther.

Armed with riot guns and rifles, Officers K. C. McClune, John Griffith and John Kowalski sped to the nursery lot, but their search was in vain.

Unable to see any distance in the heavy fog with the patrol cars' small spotlights, they obtained permission from Fire Commissioner Joseph Zafron for use of a fire truck equipped with a large spotlight.

The truck, the fire department's small emergency vehicle, was driven into the field by Fireman Mike Vincosk. Again the area was searched with the aid of the more powerful spotlight, but again no animal was found.

The officers reported they did find a spot by the highway where "something" apparently had dragged itself through the grass.

They also searched fields in Ellicott street and the back road from Killbuck to the Great Valley highway without success.

FIRST CATCH A COON
CALLED BEST PANTHER HUNT LURE
Syracuse, New York, *Herald Journal*, March 25, 1949

The Onondaga County black panther hunt became a little more picturesque today and it may turn into a doggone 'coon hunt. That is, to catch a panther one first must catch a 'coon, as a lure of course. This is on expert advice.

Meanwhile, Sheriff Robert G. Wasmer and Deputy Cliff Black believe the panther may be just a dog or four dogs, since one family reported four panthers together. Deputies went, today to check up on John Hinerwadel's retrievers at Hinerwadel's Grove, north of Syracuse. There was an idea that perhaps the retrievers had been mistaken for panthers, which would relieve the minds of the not very merry huntsmen of the Sheriffs Department.

At Hinerwadels they found he has four big Labrador retrievers, black as panthers, and they've been running loose while the grove proprietor is in Florida.

Sheriff Wasmer just wouldn't commit himself as to whether or no these dogs are guilty of being black cats. If he asked them they'd be mad as panthers and so what then? But it is possible these may solve the mystery. The sheriff and the sheriff's men hope so.

Meanwhile, Justice of the Peace William Oley of Pompey, who is in the treasurer's office at the Court House, advised the Sheriff that the way to catch a panther is to go out at night, carrying something bloody. Oley recalled that when an alleged panther had been seen in Pompey two winters ago three hunters were returning from a 'coon hunt, carrying dead 'coons. These huntsmen, James and John Hartnett and Charles Welch, said they saw a panther in a cornfield, stalking along. They didn't shoot him because they had only three shells and if they missed the panthers might get their 'coons and possibly others, or "et al." as the lawyers say.

Yesterday's search for the panther reported, by residents in the Buckley and Wetzel rds. areas brought no results. Deputies did not find the panther or any trace of him.

Onondaga's Panther Takes Spring Stroll in Kelley's-Cards Corners Farm Field

Syracuse, New York, *Post Standard*, April 28, 1949

Making an earlier appearance than in the past three years, Onondaga county's panther has been sighted roaming its old stamping ground in the western part of the county and deputy sheriffs said they feel certain that it's the "McCoy" this time.

Miss Jean Stoddard, who resides in the Kelly's Corners-Cards Corners road, about midway between the two intersections, reported to Deputy Sheriff Fred Armstrong at the courthouse jail that she sighted the animal in a field back of her home about 6 p.m. yesterday. She told Armstrong that she had a look at the animal last week, but that it was farther away and she said nothing of it.

"I was afraid I would be kidded with all these panthers being sighted, so I said nothing," she told the deputy. "But I had a better look at it and closer up this time, so decided to report it."

Miss Stoddard said that, her father returned home a short time after she saw the animal and that he loaded a rifle and went into the field in search of it.

She assured Armstrong that she was not frightened and that she was not given to flights of imagination. She described the animal as the size of a mountain lion and nearly black in color.

"It appeared slick and was a beautiful animal," she said.

Armstrong radioed Deputies William Behan and Walter Foote on duty in that area and they sped to the scene, but the big cat had disappeared, probably into a swampy woodland not far from where it was sighted, they said.

GRAY PANTHER SEEN AT ONONDAGA HILL
Syracuse, New York, *Post Standard*, May 1, 1949

Does the black panther have a double?

Deputy sheriffs reported yesterday having seen a strange animal, but it was gray not black. They saw it near the county gravel bed off the Onondaga Hill rd., and one of them shot at it but missed, the distance being too great.

A few hours later Mrs. William Evans telephoned the sheriff's office that a strange animal was howling in the barnyard at the home of her parents, Mr. and Mrs. Orrin Danforth of Bussey rd. The parents said they could not see the animal and it was gone when deputies arrived.

RADIO DOTH OF WHODUNITHS
MAY BE ANTHER FOR PANTHER
Syracuse, New York, *Post Standard*, August 31, 1949

Onondaga county's panther hunt moved from the southern and western towns to the town of Clay yesterday, when a strange black, sleek animal with the grace of a cat, but larger and more powerful, was discovered in a barn on the Warren J. Moore farm, Clay road, town of Clay.

Like the other hunts, the black shadow eluded those who sought to slay him. C. William Hounder, who discovered the animal Monday night, reported if it shows no more fight than it did when he saw it, there should be no fear of attack anywhere.

Houhder and Mrs. Hounder, daughter of Moore, occupy an apartment in the Moore home. He said the livestock in the barn totals—

two small dogs, a Spitz and a cocker spaniel. He said that it was grow-ing dark when he went to the barn to feed the dogs. They had walked in the barn and turned on the lights, when a sleek, black animal made two jumps, one of about 10 feet and a second even longer, to a lad-der leading to a trap door into a haymow.

"I saw it disappear thru the trap door like a black flash," said Hounder. "I don't know which was the most scared, the animal or myself. Mr. Moore was not at home and I had no notion of going to the upper part of the barn filled with baled hay to look for the ani-mal, so I telephoned the sheriff's office and Deputies Charles Post and John Bateman were out here in a few minutes."

The deputies reported that with Hounder, they went to the mow and found evidence that the animal may have lived there for two or three days. A search failed to uncover him. They believe that he left the barn by an open door after being scared by Hounder. Neighbors and Hounder searched the barn yesterday and last night, but felt sure that the animal was not hiding among the bales.

"It was about the size of a medium-sized dog, but slender and fast," said Hounder. "It was not a dog and it was much too large for a cat. I never saw anything jump so high and with such speed. It was not a coon. I think it was a panther or some such animal."

The Moore farm is not far from Cicero swamp.

Git Yer Gun, That Panther's Back
Deputies Go A'Huntin' Traveling Black Beast
Syracuse, New York, *Herald Journal*, March 24, 1949

The black panther is back again, in all his ebony color, unfaded by time. Deputy Sheriffs Clarence Walker and Arthur Kasson, armed with high-powered rifles, were sent out hunting for the beast this morning by Sheriff Robert G. Wasmer. His instructions were defi-nite—not to try to catch the black fellow but to shoot to kill, without quarter.

It was reported the sheriff remarked they were not to shoot till they see the whites of his eyes, which, of course, complicates mat-ters. The deputies can't play the black which is bigger, but must play the whites, if a panther's eyes are white. They had no warrant for arrest.

But all fooling aside, Harry Kilian of Laurel av., off Buckley rd., reported it is a panther in fact. He called up Deputy Sheriff Fred Armstrong at 9:45 A.M. He said he knows cats and that this animal is a panther and no question about it.

He and his mother in their yard saw the panther and watched it for five minutes. Kilian said he was about 1,400 feet distant and is about four feet long from head to rump, exclusive of his lashing tail. He went toward the swamp.

This is the first time this wandering zoological specimen has been sighted in months. A year or two ago, he was reported in other parts of the county and within three months others have reported seeing him. So Walker and Kasson are on the hunt and Sheriff Wasmer is waiting tensely for their report: "Sighted panther and sank same."

DOG'S ACTIONS AND 'LONG TAIL' PLACE 'PANTHER' AT MATTYDALE
Syracuse, New York, *Post-Standard*, July 3, 1950

Having changed his haunt again, the "black panther" supposedly was seen again in the Mattydale airbase area over the weekend.

Exercising his dog in Avenue B about midnight Saturday, Thomas Sunkeimer of Building 854, saw a "dark moving object" at the intersection of Avenue B and 10th st., as he told Deputies Andrew Hoffman and Henry Coughlin.

Sunkeimer told the deputies he walked in the direction of the object, and, as he did so, his dog "acted strange" and "pulled away," eventually escaping to make his way home.

The master, however, now walked toward the object and noticed it had a "long tail" and was moving toward him. Sunkeimer then followed his dog home as fast as he could, he told the deputies.

YOUTHFUL HUNTERS TAKE TO FIELD
HOT ON TRAIL OF MOUNTAIN LION
Syracuse, New York, *Post-Standard*, January 28, 1950

A report that Onondaga county's panther is on the prowl again and was observed in a field off Grand ave., near the city line Tuesday night, sent a small army of young nimrods armed with .22 caliber rifles and BB guns into the area yesterday afternoon.

Residents of that section of the county said the reported sighting of the panther or mountain lion did not cause them much concern, but that groups of young hunters were creating a real danger.

Alert and set to shoot the animal on sight, residents of the area in which the animal was reported seen fear that the 1950 Daniel Boones will go "trigger happy" and in their excitement shoot each other or other persons, dogs, cats, kitchen or other windows, housewives, children, or other objects, moving or still.

Animal Not Seen

At sundown none of the young hunters had reported sighting the animal and no casualties of the hunt had been reported With no school today, the army of young hunters is expected to be larger and deployed over a much wider area.

Deputy sheriffs will be keeping a sharp eye on the lion hunters and last night issued a warning to all boys going afield with rifles or shotguns to be careful in handling their arms, to look well before they shoot and be sure they are shooting at a lion.

Carl Putzer of 807 Butternut st. stirred up the panther excitement when he reported sighting what he said he is sure was a mountain lion in a field only a short distance from the road and near the city line in Grand ave. at about 9:30 p.m. Tuesday.

'Froze Dead Still'

Putzer said that he was accompanied by his wife and two small children when he spied the form of the animal and stopped the car to watch it. As it moved slowly across the field, he said he moved his car to the shoulder of the road and turned it so he could keep the headlights on the big cat which he described as black in color and about six feet in length with a tail nearly as long as the body.

Putzer said the animal "froze dead still" when the light beams fell on it. He said he later returned to the spot with a gun and camera, but that the animal had left.

THAT PANTHER AFIELD AGAIN
Syracuse, New York, *Herald Journal*, April 14, 1950

Onondaga County's black panther is on the march again.

He was reported seen yesterday, slinking about the Onondaga Hill area. Miss Mary Ann

Doyle, W. Seneca tnpk., was the witness.

Miss Doyle said she sighted the beast stealing across a field about 300 yards from her home. She said it climbed a tree to escape from three dogs in pursuit.

In a hurry to get to work, Miss Doyle did not have time to capture the animal and the reward money awaiting a black panther slayer. When her father, A. T. Doyle, went out to the tree later in the day he said he found claw marks on the tree but the tracks were made indistinct by freshly fallen snow and the panther was gone.

'BLACK PANTHER' FRIGHTENS RESIDENT OF SOUTH SIDE
Syracuse, New York, *Post-Standard*, April 6, 1950

Police were called to 133 Kenmore ave. on the south side about 7:30 p.m. yesterday by a black panther alarm.

Miss Sally Dodge, alone in the house, reported she was talking on her telephone to a friend when she heard a noise on the front porch. Thinking it was her boy friend, she said, she looked out and saw a strange animal.

She described it as about four feet long, 18 inches high, black and with a long tail. It turned and looked at her, then waved its long tail and departed from the porch. It had a face like a cat.

Police investigated briefly and then departed.

Investigation revealed tracks in the film of fresh snow on the porch, but they appeared to have been made by Miss Dodge's little dog, Pudgie.

A neighbor said his black setter was touring the neighborhood and he advanced the theory the setter had ventured on the porch. He was afraid an excited neighbor or police officer might shoot his dog.

Miss Dodge, who was quite excited about the proceedings, declared the animal wasn't a dog.

However, none of the evidence appeared to have aided in the solution of the black panther mystery.

SAW MOUNTAIN LION IN CITY, DRIVER SAYS

Syracuse, New York, *Herald Journal*, January 27, 1950

Carl Putzer, a city dweller who enjoys the time he spends in the woods, saw what he believes was a mountain lion, and he's pretty unhappy about the whole thing.

"First time I didn't have a gun in the car since hunting season started last fall," he complained. "The least you'd want to have at a time like that is a camera, anyhow."

A resident of 807 Butternut st., Putzer saw his lion Tuesday night in a field off Grand av., near the city line. It was about 9 p.m. Tuesday, and Putzer was driving by with his wife and children, Judy Ann, 18 months, and Carl, six months.

His eyes sharpened by many years of hunting in Central New York, Putzer spotted the animal some 500 feet from the road as his headlights pierced the darkness. "Could have been that black panther that they've been seeing around North Syracuse," he said, "but the minute the lights hit him I figured it was a mountain lion."

He said the cat-like animal was slinking through the field, moving around the area for about 15 minutes before finally disappearing over a distant ridge. Putzer swung his automobile around on the shoulder of the road so that he could train the headlights on his discovery.

"He froze dead still the minute the lights swung around to him," the local hunter reported. "Probably thought we couldn't see him. I've heard about fellows seeing lions up north, but this was the first for me even though I've done a lot of hunting.

Spreading his arms to indicate the animal's size, Putzer said the animal was six or seven feet long, fairly low to the ground, jet black and waving a tail approximately as long as his body. "He had cat features," Putzer maintained. "I've seen pictures of mountain lions and I've seen them in the zoo. This one looked just like them.

"I don't know about it being a panther. They say that panthers can live in the tropics but not here. You'd think that they could live here all right if tropical plants and animals at the zoo live. But I think it was a mountain lion."

Pointing out that there's open season on mountain lions and black panthers all year long in Onondaga County, Putzer said that he has returned properly equipped to the site of his discovery. He was armed and ready for action, but the mountain lion failed to cooperate. There wasn't a sign of him on the return visits.

Putzer wasn't able to do anything the time he did see the lion, but throw a few stones toward it. The animal didn't seem to mind that, Putzer observed.

The Syracuse hunter hopes that he'll see the animal again when he has a gun or camera in order to establish proof that big-game stalks Central New York.

DEPUTIES NOW MIGHTY HUNTERS
THEY'RE STALKING PANTHER, BOBCAT AND OTTER
Syracuse, New York, *Herald Journal*, February 8, 1950

Those Versatile sheriff's deputies—now they're stalking a panther, bobcat and otter.

The latest adventure of the black panther was reported last night by Elmer Morin, of Buffalo, a government employe.

While, driving near Card's Corners on W. Seneca tnpk. at about 1:20 A.M. today, on route to Albany with his wife to see an ill relative, Morin reported a black, strange looking animal crossed in front of his car. He stopped and so did the creature, which had a smaller animal in its mouth.

Morin told Jailer Patrick Rooney at Cedar st. jail he got out for a good look; but the animal did not scare easily and stood defiantly in the glare of the headlights. Getting back into his car, he started down the highway and the animal followed alongside the car until it became frightened by the horn and took off through the fields. Deputies Robert Ashmore and James Roche made a thorough search, but failed to see the strange animal.

At 10:23 last night, Lloyd Gearhart, of 312 Bronson rd., Westvale, called the Cedar st. jail and told Fred Armstrong a bobcat was perched on a fence across the street from his home. He insisted he knows a bobcat when he sees one. Deputy Bert Reardon investigated, but all he could find was prints of a dog.

Meanwhile, Deputy Frank Rice still was convinced it was an otter that had been reported in the Split Rock area and he was out to get, not his man, but the mysterious beast.

Black Panther Worth $25, Dead or Alive

Syracuse, New York, *Herald Journal*, February 24, 1950

Syracuse and Onondaga County residents today launched an all out effort to capture the "black panther" dead or alive.

There's now a price on the head of the beast, reported to be an ebony animal with a menacing snarl.

Syracuse newsmen, members of Sigma Delta Chi fraternity, are offering a $25 reward to the person who catches a real black panther inside the county,

Burnet Park Zookeeper Dan Hanley has a cage all ready for he animal if it is brought back alive.

Meantime a wildlife expert, Andrew Buff, suggests that the beast is an otter, a bob cat, or perhaps a black fox.

It's closed season on otters, Buff points out. So be careful if you spot a black, long-tailed animal which makes no noise.

Buff, who runs a sports store at 1516 Erie blvd, E., says some otters are from four to five feet long.

Another guess is that there's a bobcat at large in the county.

"When one of them yowls it makes you shiver in your boots," Buff reports. It's just possible the animals seen may be a fox, tom cat or large dog, Buff thinks.

Sigma Delta Chi members pointed out their reward is only for the capture of a genuine black panther.

Black Panther is $75

Outfox Mystery Animal and You'll Claw Onto Rewards

Syracuse, New York, *Herald Journal*, February 26, 1950

What is the black panther?

A bobcat, an otter, a fox, a tom cat, a dog, a figment of imagination—or an honest-to-goodness real, live, black panther?

Seventy-five dollars says, in effect, that the animal scaring Syracuse and Onondaga County residents is no panther.

The money has been put up as a reward to the hunter who produces a county-caught, genuine panther. Those who offer the reward frankly say their dough is safer than it would be in a Brink's vault.

Members of the Elmwood Fish and Game Club are offering $25 of the total. They promise none of their own members will compete for the prize. Andy Buff, widely-known sportsman, is to be their judge of whether any animal is or is not a black panther.

Buff, by the way, has some panther theories of his own. He names the otter as a leading candidate. Otters are sometimes five feet long, he says, and can look pretty fierce slinking through the dark. They live near water, but have been seen far from trout streams and lakes.

To give would-be hunters the lowdown, Buff shows visitors to his trading post at 1516 E. Genesee st., his collection of mounted animals.

Those who object that an otter doesn't yowl while the reported black panther does, are shown a mounted bobcat. Bobcats make a terrifying noise. Add to one bobcat a little imagination and you may have the black panther, he says. He thinks it's also possible that foxes, yipping occasionally, could be the disturbance of the countryside.

Sigma Delta Chis, Syracuse newspapermen, also offering a $25 reward, make no claims as animal experts. "If by long chance a black panther is found we'd get enough news stories out of it to make it worth $25," says Gene Cohen, spokesman for the group.

Earl Budd. 243 Oneida st., who offers the third $25, is quite scornful of residents' tales of black panthers stalking about. "The only panther I ever saw was in the zoo," is Budd's retort.

While posses are being organized for the wild animal hunt, dog and cat owners are quaking in fear. "Those trigger-happy guys will be killing our pets," one moaned. "I'm thinking about offering a counter-reward for the hunters."

Considered apt prey for the hunters, since they are large and black in color, are the Labrador retrievers owned by John Hinderwadel of Fay rd., North Syracuse. Hinderwadel, however, is keeping his dogs under close surveillance until the panther scare has passed.

In spite of opposition, however, the hunt is still on for the panther, dead or alive If some valiant citizens catch him alive, the panther will go to the Burnet Park Zoo. If caught after mortal combat, the panther will be given the full S.P.C.A. funeral.

A note of encouragement comes for panther hunters from a Syracusan who saw them once in Idaho.

"Real panthers are cowards," he says. "Don't be a bit afraid."

He reports there's no reason at all for people to be upset by a panther, because even a hungry panther is scared of his own shadow.

"Now, if people were seeing dinosaurs or pink elephants," he concluded, "that might be serious."

PANTHER'S IN TOWN AGAIN
Syracuse, New York, *Herald Journal*, February 27, 1950

The mystery of the elusive black panther, allegedly on the prowl in the Syracuse area for several weeks, continued today when tracks of an animal were found in the 3900 block of S. Salina st.

Edmund Stearns, operator of the Salina Motor Co., 3927 S. Salina st., called for police today to investigate large paw prints in the snow at the rear of the motor company building.

Stearns described them as "larger than a St. Bernard's." He said the marks were first noticed last Friday and appeared again today.

"We're convinced the prints are too large to be made by a dog, even a St. Bernard," Stearns said, explaining that he once owned a St. Bernard and was familiar with its paw prints.

Deputy Sheriff Fred Armstrong answered the police call, but turned it over to the Syracuse, police when he found it was in the city. Patrolman Frank Pleklik was dispatched to the scene.

"I don't know," Pleklik said. "Let's call in the S.P.C.A."

PANTHER'S IN AGAIN, OUT AGAIN, GONE AGAIN
Syracuse, New York, *Herald Journal*, February 28, 1950

Sheriff's Deputies were hot on the trail of the "black panther" again today after a Southwood resident, off E. Seneca tnpk., reported the elusive animal was trapped for a time under an automobile by six dogs.

The newest report on the much-sought-after panther, dog, bobcat, otter, or whatever it may be, came on the heels on the discovery of large animal tracks yesterday, about a mile from the Southwood residence.

Deputy Sheriff Fred Armstrong said the Southwood report was telephoned in by a woman who identified herself only as "Mrs. Brown."

She said an animal, believed to be the panther, was cornered under a van by the dogs shortly before midnight, but it broke away, heading through backyards into nearby woods.

Deputy Sheriffs Bert Reardon and Harry Ginsberg were assigned to investigate.

Yesterday Edmund Stearns, 3927 S. Salina st., discovered large tracks right in the middle of a populated area within the city limits and near his place of business.

Depth of the prints indicate they were made by a heavy animal with claws. They led off into an unpopulated area which extends beyond the city limits towards Jamesville.

On the basis of Stearns' find, hunters are expected to comb the vicinity. Some may venture into the woods, with the $75 reward for a genuine black panther in mind.

The reward has been put up by the Elmwood Fish and Game Club; Sigma Delta Chi. Syracuse newspapermen's organization, and one civic-minded individual, Earl Budd. All are skeptical that the animal, reported seen many times, is a real black panther.

Yesterday, someone here remembered that a mountain lion escaped from a circus in Watertown in 1939.

The lion was a mangy old animal, but imaginative Central New Yorkers think his escape may have some connection with the black panther.

They suggest the current "panther" may be the offspring of the mountain lion and some animal he met in the Adirondacks.

Another Syracusan recalled that Slate Police conducted a successful hunt in 1945. The "panther"' of that chase turned out to be a dog—a Labrador retriever. It was shot down in a farmer's chicken coop by Trooper Clifford Christensen.

Bait Loaded With 'Mickey' For Panther
Syracuse, New York, *Post-Standard*, March 2, 1950

For an animal that appears to be having so difficult a time in finding something to eat that its hunger has driven it to the fringes of the downtown business district, that town and country panther has the strength to cover a lot of territory.

While horsemeat bait loaded with sleeping potions were being placed in the vicinity of St. Anthony of Padua church in Midland ave., where tracks of a large animal, differing from those of a dog, were reported found Tuesday night, Joseph Borrow of Baldwinsville, RD 3, reported sighting the animal on the Liverpool-Baldwinsville rd.

Borrow telephoned Deputy Sheriff John Schuyler at the courthouse jail early last evening that he was given a scare when he saw the animal cross the road ahead of his car, about 300 feet north of the entrance to Onondaga Lake park picnic area.

Borrow described the animal as being dark, about two feet in height, three to four feet in length, and as having a long tail which curled upward. He said the animal crossed the road from west to east, moving at a long, easy lope.

Hunter Tells About Cat Family

Syracuse, New York, *Herald Journal*, March 5, 1950

To the *Herald-American*:

I am not after any reward but maybe I can open your eyes to the cat family. I lived in Maine in camps, hunting and fishing, and also in New Hampshire. Warden Lowe of Jefferson, N. H., is a hunter of the black lynx.

While hiking along the route to Maine I met Warden Lowe and his dogs. The stories are true. The black lynx of the White Mountains really exists today as it did years ago. A telephone call to Warden Lowe will convince you. My experience in Maine and that of Warden Smith of Ellsworth, Me., will prove to you that the "woman scream" of the cat family comes from the male cat and is the hunting call of the bobcat family. These are still plentiful in Maine, Vermont and New Hampshire.

The last panther near us here was traced over three hills and was lost in the hills near Binghamton a few years ago. Also a couple of boys neat Middlebury, Pa, a few years back saw a black object in a hollow tree along a creek and they pulled up a single barreled shotgun No. 4 and shot and killed a black panther.

I believe the Catskill and Pocono Mountains near Blossburg and Roaring Branch still have bobcats.

The area is also good territory for black panthers or brown panthers or lynxes. Any coon hunter in your vicinity with good young dogs will tail your cat or panther for you provided you find a track or spoor. They will soon put him in a tree or rocks. If he screams, believe me it is a male bobcat or wild cat.

Joseph A. Warren.

Corning, Box 367

SAW PANTHER IN 1930
APPEARANCE NOW COULD BE REAL THING
Syracuse, New York, *Herald Journal*, March 6, 1950

Mrs. Jennie Lombard says people are not just "seeing things" when they spot the black panther.

The Cazenovia woman is sure she saw one 20 years ago. By now, she reasons, there must be lots of them in Central New York area.

In a letter to the Herald-Journal she writes she was on her back porch shelling peas back in 1930. All of a sudden, Mrs. Lombard says, she noticed some crows were cutting up a fuss.

"They were striking at a black thing that looked like a bull dog with a long body and a tail longer than the body. The animal just walked along. He kept rolling up his tail and then whipping it out," Mrs. Lombard writes.

She says the animal had long legs, slender ears, and was as black as coal.

Mrs. Lombard says she took a dog and gun and went after the animal, but lost the trail. Later she said she heard the animal in a swamp near her home.

"The animal looked like a black lynx and sounded like one," she says, "and the difference between a black lynx and a black panther is hardly worth mentioning."

BOBCAT SEEN NEAR WATERTOWN
Syracuse, New York, *Herald Journal*, March 19, 1950

Watertown—A big black wildcat is being sought by the authorities in the woods south of Cold Creek Inn on the Watertown-Burnville rd., just on the outskirts of the city.

The elusive animal, declared to be more than six feet long by former State Trooper Gilbert C. Blackmer, a resident of that area, so far has not been found.

Friday afternoon and yesterday, police armed with rifles, went to the area but failed to get a shot at the animal, in fact they failed to get even a glimpse. Only tracks in the snow, declared to be large ones, were found.

Blackmer told police that he first spotted the "big cat" 20 feet from the residence of Manfoord Corey, of the Burnville rd., Friday afternoon. He said that suddenly the animal disappeared in a small woods nearby.

Here is a description of the cat as given by Blackmer: Overall length, ranging from five to six; feet, 60 to 70 pounds weight, two and one-half foot long tail, pointed head, fur short, black and sleek.

Panther Stalking Again in Greenwich Vicinity

Troy, New York, *Record*, October 11, 1952

A Greenwich newspaper has offered a reward for the capture, dead or alive, of a black panther which has been stalking the area and also has come forth with specifications for the panther. The panther was last reported near Sherman's Rocks just outside the village limits last Sunday. Willis Johnson of the North Greenwich Road reported seeing the animal.

The Greenwich Journal wants the critter but also wants to make sure the right one is captured. The specifications call for a "cat-like dark-furred animal weighing at least forty pounds with a body 3 feet long and a tail 2 feet long." The animal must be killed within a seven mile radius of the village. The reward is $25.

Reports of a panther created quite a stir in Greenwich more than a year ago.

Johnson and his daughter were returning from church Sunday when he saw the panther crossing a field near the Greenwich Cemetery. He stopped his car, he said and watched the animal go into the cemetery. Then he took his daughter home and returned to the field where he found large footprints made by the animal. He walked toward Sherman's Rocks, he said, and as he took a turn in the trail he suddenly saw the black cat again.

The panther was coming toward him, about 25 feet away. Johnson said that both he and the panther stopped for a second and then the animal turned and disappeared in the bushes. The animal, he said, stood from 18 inches to 2 feet at shoulders and probably weighed about 100 hundred pounds. The face was definitely cat-like and at least seven inches broad. The animal was entirely black.

The panther was reportedly seen in the came field a year ago by Dr. C. D. Williams of Greenwich and the footprints are similar to those seen by William L. Sharp of Greenwich at what is known as the "upper reservoir" a year and a half ago. About two weeks ago, a motorist reported seeing a black panther cross at the Old Schuylerville Road.

'BLACK PANTHER' ONLY ALLEY CAT
Syracuse, New York, *Herald Journal*, October 22, 1953

Canastota —The panther scare in the southeastern part of the village turned out to be nothing more than a large black cat.

Patrolman Roger New reported that he received a call from a woman in that part of the village on Tuesday claiming that a strange animal was in her backyard. The officer responded to the call and discovered an old cross black cat which he shot.

The residents of the area had been somewhat terrified after Saturday's report of an alleged black panther seen crossing Route 5 near the residence of County Highway Superintendent and Mrs. Lynn E. Johnson.

MOTORISTS SPOTS STRANGE BEAST AROUND AMBER
Syracuse, New York, *Post-Standard*, October 29, 1953

To the Editor of *The Post-Standard*:

I have read several articles in your paper about the panther being around Marcellus and Navarino and thought it just another story.

Last Sunday night I was driving through Amber and saw something by the roadside. I thought it was a dog.

I blew my horn. It jumped into the road in front of my car. I saw it was no dog. I blew my horn again. It gave a loud shriek and ran towards the lake.

It had a round head, small ears, long slim body, short legs, large feet and a long tail that curled around at the end.

I live in the western part of the state and passed through Amber quite often, but never saw an animal like that before.

I thought the people in Amber should know about it.

C. J. Shulins.

Black Panther Reported Spotted
Syracuse, New York, *Post-Standard*, September 6, 1954

A man and wife hunting woodchucks in the Onondaga Hill area yesterday reported they had spotted a large animal, about the size of a police dog, resembling a black panther.

Mr. and Mrs. Edgar Morrow of 128 W. Bissell st., reported they spied the animal about 5 p. m. near the Hitching orchard. The long-tailed animal was seen near a hedge. Morrow fired his shotgun, but said it was out of range.

With the hunting couple were Mr. and Mrs. Gerald Hayden, relatives, of Buffalo. Mrs. Morrow said the cat studied the hunters for a moment at a distance, then jumped the hedge and disappeared.

Wary Eye Kept for What-is-it
Strange Animal Suspected as Panther or Cougar
Syracuse, New York, *Herald Journal*, January 9, 1955

Auburn—A strange animal is on the loose in Cayuga and Wayne Counties, it was learned yesterday. If it is identified as a panther, as suspected, there may be a systematic hunt for the animal in the near future.

Ross Humphrey, who resides northwest of Red Creek in Wayne County and not far from the Cayuga County line, no longer has to convince his neighbors that there is a strange animal lurking in his neighborhood. He has a footprint of the beast to show that it really was in his pasture lot.

He discovered the big track, which measures 4 ¾ inches across, and gently lifted it with its cast of snow and ice and placed it in his freezer, where he is holding it for a photographer to take a picture of it. He plans to send the picture to the State Conservation Department, Albany, to get official identification of the animal.

And that is not all. For behind the supposedly mother's tracks were those of a baby animal, apparently of the same species.

After discovering the tracks, Humphrey went on a sleuthing trip of his own. He found where the animal had taken refuge in a hollow tree and there the trail ended.

Ernest Penneman, who lives near the Humphrey home, recently saw strange animal as he was going home after work. He described it as black in color, having a very long, sleek tail and resembling a panther or cougar in appearance.

The cry of some animal resembling a panther's has been heard over a period of several years in that neighborhood. The late Arthur Humphrey was close enough to it in the spring of 1951 to get a shot at it but missed.

The tracks of the animal have been seen by hunters and land owners in the area many times.

Panthers or cougars hunt by night. They are especially dangerous, sometimes leaping great distances to catch their prey.

PROSPECTING FAMILY SEEKS URANIUM IN DELAWARE COUNTY
Oneonta, New York, *Star*, July 20, 1955

Delhi—Whether or not there will be a wild scramble to acquire real estate on Scotch Mountain, an extensive acreage between the Little Delaware River near Delhi and the Cabin Hill road connecting DeLancey with Andes, remains to be seen.

Carlton R. Hinaman Sr., and his wife and son Carlton Jr., live on a 35-acre tract of land on the northernmost of three Scotch mountain roads. The road starts at the County Home flat near Delhi and the eastern extremity is at the George Thompson place on the Cabin Hill road.

In the last few weeks, Mr. Hinaman and his son have been operating a geiger counter all over the mountain.

"In certain locations", said Mr. Hinaman, "the instrument clicks like mad. The instrument is adjusted to test for uranium, and we sure are keeping our fingers crossed."

"Of course," said Mr. Hinaman, "many people might not relish tramping through these woods, and climbing around the rock ledges up here, because it is right in this area that a little over a year ago my wife and son and I saw on more than one occasion, a large black mountain lion roaming the terrain.

"And we aren't the only persons who saw the animal. Mr. and Mrs. Howard Graham of Delhi, were driving up through here one day, and watched the big black beast roaming around, for some 20 minutes."

'CAT' STALKS GOAT
BLACK BEAST ROAMS FARMS
Syracuse, New York, *Herald Journal*, June 15, 1957

A "black panther" scare is circulating through the sparsely-populated country north of Plainville in the Town of Lysander.

For several weeks, rumors of a "large, black animal" roaming the section have been reported.

Yesterday, the mysterious animal was seen by a farm boy as it stood over a goat it had killed, sheriff deputies said.

The boy, James Snow, told his father, Arlyn, who lives on Prine Rd., that the animal was "large and black."

The goat was owned by Snow, who called deputies and the Lysander dog warden, Roswell Homer.

Deputies reported that the tracks, "like a large cat's print," were found near the goat's body.

The prints, they said, were about four inches wide.

Snow told deputies that he was worried that the animal might attack small children in the area,

Mrs. Homer, wife of the dog warden, said today that her husband believed that the animal might have been a large, wild dog, one of a pack in the area.

WOOD, FIELD AND STREAM
FAST CHORUS OF THAT OLD BLACK PANTHER
THAT THEY THINK THEY SEE SO WELL
New York Times, October 30, 1957

By John W. Randolph

Here come a couple of letters that bring up that old black panther again, that old black panther that keeps going round and round.

This one is from Gabriel Aiello of 5905 Twenty-third Street, Brooklyn:

"On Friday, Oct. 11, at about 8:30 P. M., while driving my truck south on route 9W near Highlands, N. Y., with headlights on because it was dusk, my helper and I saw two black panthers right in front of us as we turned one of the sharp corners at the top of the Storm King Highway north of High Tor blinker.

"We were flabbergasted because we thought at first they were bears, but tails a yard long and sleek, velvety fur, were giveaways.

"I've been hunting bear and deer in New York State for over six years, and never heard of anyone seeing a black panther before. Were we seeing things? Why should black panthers be prowling around North America, anyhow?"

The second letter comes from John Brennan, publicity man of the New Hampshire Planning and Development Commission, who never sleeps by night until New Hampshire has been widely venerated by day as a paradise.

It advises me that on Oct. 22, Mr. John Brennan had offered a reward of $1,000 of his own scratch, not state money, for anyone who could produce a New England black panther, dead or alive.

He means any authentic black panther, "with reasonable proof that the beast was killed or caught in a. wild state in New England woodlands, preferably New Hampshire."

There is a second stipulation: That the black panther must be "a mutation of a species native to the United States—the puma, cougar or American mountain lion." It may *not* be the black leopard mutation called black panther that is found in zoos and animal farms and is not native to the United States."

Brennan is a man whose faith is built on the solid rock that if any New England state can produce a black panther, New Hampshire can. But he is a state employe who cannot throw $1,000 of his own scratch around as heedlessly as a TV jollyboy throws sickening grimaces. He is one man who does not believe there are authentic United States black panthers in these parts.

He was inspired to make his offer by reports that a black panther had been seen by several people at Dalton, N. H. It was described as four feet long, with a three-foot tail, and glossy black. It was said to have come down to the Connecticut River to drink.

Brennan adds that "black panthers" have been "seen" up his way for about fifteen years. He still feels safe.

It may be added that black panthers have been reported pretty nearly everywhere in the East in just about every year since the truth was young. I haven't heard of one being killed or caught, but then I haven't heard everything, yet.

Mr. Aiello, permit me to introduce Mr. Brennan. Mr. Brennan, Mr. Aiello.

HUNTERS KEEPING POWER DRY FOR BLACK PANTHER

Syracuse, New York, *Post-Standard*, September 13, 1958

By Howard Merritt

Panther hunting which engaged pioneers as a matter of self defense rather than a sport, will provide an exciting diversion for fox and coon hunters this fall if one of the big cats reported on the prowl in the county remains on the scene.

Inspired by two reports of persons claiming to have sighted the animal in the Town of Onondaga Thursday and yesterday, gunners who hunt with hounds are anxious to give them a try at panther trailing when conditions become more favorable.

Idle during the summer, the dogs are "soft" and with foliage still in full bloom and some crops unharvested, the chances of picking up the animal's scent and giving it a worthwhile chase would be poor, hunters said.

Because mountain lions, the species which hunters believe the animal reported sighted to be, take to a tree when pressed, coon hunters are hoping that it remains in the area for another month. They are convinced their dogs have a better chance than foxhounds of treeing and holding the animal until a shot can be fired.

Reappearance of the "black panther" was first reported Thursday when Carl Christensen of McDonald, road told deputy sheriffs he sighted it in a large fallow field across the road from his home.

Miss Jane Harding, 32, reported that about 8 a.m. yesterday morning she looked out of a window of her trailer home in Tanner road, a half mile off the Cedarvale-Navarino road, and saw a panther standing only a few feet away.

She told Deputy John Checkowsky that she hurriedly picked up a rifle and loaded it, but when she opened the door to take a shot at the animal it was moving away across a field to a woodlot. The animal stopped and turned for a look, but it was already out of range of her gun.

Miss Harding told Deputy Checkowsky that she was certain the animal was a panther. She became familiar with panthers while working for several seasons with a carnival. She recalled that a panther with one of the shows escaped from its cage and killed a horse.

Deputy Erenst Monica said a man living near East Syracuse in the town of DeWitt reported sighting a panther in a field near his home last week.

SLAYING OF RARE ANIMAL CLIMAXES AREA MYSTERY
Schenectady, New York, *Gazette*, February 25, 1961

"Eerie reports of a black leopard lurking on the outskirts of Glenville are expected to come to a halt as the result of an investigation yesterday upon the death of a 'strange animal.'" A large fisher was killed after it attacked a dog. It was four feet in length and weighed between 15 and 20 pounds. The local game warden suggested it was responsible for black panther sightings in the area the previous two years.

MYSTERY ANIMAL TERRIFIES HAMDEN AREA
Oneonta, New York, *Star*, October 11, 1961

By Glen Harper

Hamden—Is it a panther, a mountain lion, a lynx, a coydog, bobcat or possibly a rabid fox which is attacking and killing dogs and terrifying farm families and others living in the Town of Hamden?

Paul Moody, whose farm is located three miles north of DeLancey on the Covert Hollow

Road, said yesterday:

"Our little two year old shepherd dog, Jinx, which was just beginning to get good with the cows, was attacked and killed. He was torn all to pieces.

"Jinx, who weighed 40 pounds, never went off the place, so we can't believe that he had been in a fight with another dog. Our two boys, Bill, 7, and Bobby, 4, thought the world of the dog, and the dog loved the boys. We have never seen the animal which is preying on dogs in the area, but I heard it last night.

"It let out a shrill blood-curdling shriek."

Scared Dog

Mrs. Robert Kelly, a neighbor of the Moody's, said:

"I was blackberrying about the time the Moodys' dog was killed. My dog, Henry, was with me and usually she runs off chasing rabbits. That day, however, she stayed right near me. She just kept crying and howling.

"But when I got back to the house I looked on the hill and saw an animal much larger than a woodchuck of a sort of reddish color. It stayed for quite a while. At night, we are keeping our windows closed, the dog in the house, and are leaving no calves out of the barn."

"We've been hearing an animal sounding off like a big tom cat since the middle of September," Clyde Miller, whose farm is located a mile north of Hamden Village on the Launt Hollow road, said.

"It would sound-off just once, and then again in less than a minute, usually right after dark. I think probably it's a bobcat, but I don't know."

Weird Noises

"We hear it, again between 5 and 6 a.m., just before we get up, but as soon as we turn the lights on, it takes off. My son-in-law, Raymond Jones, thought he saw the animal entering the woods, but it was almost dark and he couldn't see its color.

"Something, also has scared the Burt Aikens family, which lives over the hill on the West Launt Hollow road. It chased their cows, driving them to another field, Mrs. Aikens told my wife," Mr. Miller said. "We are definitely quite concerned.

"The last time I heard the animal was Friday morning, when I went to the barn. I thought it might be after our calves. I can't see why a bobcat, mountain lion or panther would need to be concerned about food now, as there are lots of rabbits and wild game, but they claim a bobcat will even kill a deer."

Mervin Bryden, whose family lives further up Launt Hollow, said Mrs. Bryden saw the animal on the hill just before sundown on September 10, She described it as being dark brown, bobtailed and big as a large coon.

Calves Protected

"We're kind of careful and not too anxious to go out at night. We shut our calves in at night, and have over since we first heard the animal shriek. I haven't seen it," Mr. Bryden said, "But I'd like to get my eye on it. If it's what I think it is, it's a bobcat. If it were a panther or mountain lion, it would attack cattle. I once saw a Canadian lynx, and it was bigger than a dog."

"My wife heard something out near our woods, and was so scared she felt like taking off," Dock Choate, a neighbor of the Brydens declared.

"The noise started off like a tomcat, but ended up sounding much coarser. I've heard the same noise different mornings. My black and white shepherd dog, Butch, came home this morning, with a great big gash on the top of his nose and one right in the front of his left leg.

"A wild cat doesn't scare me, I've seen too many of them. They won't bother you unless you get too near their young. It might be a fox making a funny noise. They'll scare you to death."

Whatever the animal is, it is surely the chief topic of conversation in the Hamden-DeLancey area right now.

MYSTERY BEAST DEAD?

Syracuse, New York, *Herald-Journal*, July 28, 1965

This article noted that 1965 hadn't yet had any "mystery beast" sightings from Onondaga County, and recapped several reports from the previous few years. The black green-eyed animal was reported from the Cedarvale area in 1962 (a woman fired at it, and missed), and 1963 (it clawed a horse). Dogs treed the animal in 1963 on Onondaga Hill, but a deputy just watched until it descended and escaped.

'PANTHER' LIVES AGAIN

Syracuse, New York, *Herald-Journal*, August 2, 1965

An Oneida family reported to the newspaper that they had seen a black panther-like creature in the pasture during the spring. They had hunted for it without success. It had a "blood-curdling eerie cry."

BLACK PANTHER, ANYONE?
Syracuse, New York, *Herald-American*, January 30, 1966

This article reminisces about the alleged black panther. A conservation officer said a man reported to him (several months prior) that the animal had chased sheep across Van Buren Road, between Lakeland and Baldwinsville.

PANTHER'S BACK; OR IS IT?
Syracuse, New York, *Post-Standard*, March 31, 1966

A pair of eyes were seen reflecting in the headlights along Pleasant Valley Road, near Marcellus. They were apparently too high off the ground to be a rabbit or fox, but the animal itself was not seen.

IS THE BLACK PANTHER BACK IN CIRCULATION?
Syracuse, New York, *Herald-American*, May 1, 1966

A Nedrow resident called conservation officers to say he had the black feline in his sights, and wasn't sure what to do. The officer told him to shoot the animal, and he would come over to look at it. When he arrived, the conservation officer found that the resident hadn't taken the shot, as he didn't have a hunting license and didn't want to get in trouble with the law. A policeman later told the same conservation officer about his own sighting of the animal. He saw the large black catlike animal walking slowly away from the Nedrow area past his Sentinel Heights property. Searching the dense high weeds with a friend later, he was unable to find any corroborating tracks.

'PANTHER' BACK AGAIN?
Syracuse, New York, *Herald-Journal*, June 2, 1967

An Onandaga couple reported that a "large dark animal" killed 11 hens and a rooster within the space of 10 minutes, without making a sound. Their dog remained quiet throughout. The animal fled with a chicken in its mouth.

MYSTERY REVIVES TALES
Syracuse, New York, *Post-Standard*, March 9, 1973

The strange death of a horse (apparently frightened to death, though not attacked) made some people think of the old panther stories. A sighting of a large black animal in Otisco was reported the previous week.

HUNTER'S 'BIG HOUSE CAT' COULD BE BLACK PANTHER
Syracuse, New York, *Post-Standard*, November 16, 1974

A bowhunter, Walt Pulaski, was sitting one morning on a log in northwestern Cortland County, along a deer trail. He watched as "out of nowhere came this thing that looked like a big house cat. It was close to six feet long and was almost black. It looked like mottled gray or maybe black with a sort of frosting. It just loped along as if it was sniffing that fresh deer trail. Then it was gone." Pulaski was unsure of the game laws, so didn't attempt to shoot the animal. Waiting half an hour, he then paced off the distance to where he saw the animal, at 30 yards. He found a fresh deer carcass nearby with only a shred of meat left on it. He continued his hunt, but later in the day met up with another archer and told him what he saw. The second hunter, Jimmy Root, said he had seen a similar animal about 12 years prior. Jimmy was 16 at the time, and in the woods helping out a boy scout group. He was walking along a road in the woods, when he "heard a screech and this big, black cat ran across the road. It was about four feet long not counting the tail." The state game officers suggest that a fisher might have been confused for a panther.

Karl Shuker (1989) noted the 1977 sightings of a black panther in the swamp regions near Van Etten, New York. A large clawless track was found, but there was debate over its identity by local scientists.

PANTHER SIGHTINGS IN NORTH BAFFLE DEC
Syracuse, New York, *Post Standard*, February 11, 1984

This article notes several sightings of both black and brown large felines in Lewis and St. Laurence counties. Robert Fager claimed to have seen a black cat along the banks of the Black River in January. He believed it raided his traplines. Another outdoorsman claimed to have seen a black panther along the Grasse River near Clare. A DEC animal specialist said that black mountain lions don't exist, and the only recorded specimen ever of such was given to a fur trader by an Indian in the 1600s.

BOB PEEL'S EMPIRE STATE
Syracuse, New York, *Herald Journal*, January 13, 1985

State police and the SPCA checked out a report of a black panther at a sand pit on Long Island. They found tracks indicating a 30-40 lb. "cat-like creature." A few weeks later, a Suffolk County police officer reported seeing a similar animal about three miles from the sand pit area.

'Cat' Outside Window Turns Man into Panther Believer
Syracuse, New York, *Post-Standard*, March 16, 1985

Jim Pitcher reported seeing a big black cat prowling around his Hammond area farmhouse. He looked out the window on a Sunday morning and thought he saw his black labrador, but realized it was bigger and moved differently. He watched it for a while, then grabbed a camera and took a picture. Then, he and his wife spent about 45 minutes watching it through binoculars. A DEC officer later looked at a track, and said it was definately "cat-like," without claw marks, but was smaller than would be expected from a mountain lion (being about 3 inches wide). The photo and a hair sample were taken for analysis.

Mystery Animal Puzzling to Expert
Syracuse, New York, *Herald Journal*, March 19, 1985

DEC animal specialist Gary Jackson was puzzled by Jim Pitcher's photograph of the strange catlike animal seen on his Hammond area farm. The animal's body seemed to have too much heft, especially in the upper leg, to be a cat. Also, the tail seemed too "clubby" and short. But, Jackson wasn't convinced it was a canine, otter, or fisher, due to its movements and tracks. The hair was still waiting analysis.

Animal Hairs Cast Doubt on Panther Theory
Syracuse, New York, *Post-Standard*, March 25, 1985

Preliminary examination of the Pitcher farm hairs by the DEC showed they were probably not feline. The photograph was too blurry to be of any use in analysis. A fisher was suggested as the best candidate by another DEC officer.

Pathologist Says Panther Can't Be Mystery Animal
Syracuse, New York, *Post-Standard*, March 27, 1985

The state wildlife pathologist ruled out several felines (mountain lion, lynx, bobcat, and leopard) as candidates for the Pitcher farm animal, through microscopic analysis. He was still comparing a reference sample from a fisher. (This appears to be the last published news item on this particular incident, with the animal left undetermined.)

DRYDEN WOMAN SAYS PANTHER GNAWED CALVES
Syracuse, New York, *Post-Standard*, November 3, 1988

Frightened horses and dead calves in Tompkins County were the work of a black panther, according to Donna Rogers. Several locals reported seeing the animal, including Gordon Gabaree, a state certified nuisance wildlife remover. Ron Trofenof saw the animal in a field on a Dryden Road farm. The black cat "was three to four feet long and two to three feet high."

DICK CASE: A BIG CAT PROWLS AROUND CAMILLUS
Syracuse, New York, *Herald Journal*, July 12, 1993

The columnist noted that Ralph Tolone, a long-time hunter, discovered large fist-sized feline tracks in the mud along an old right-of-way. He tracked them for a while, but never saw the animal.

DICK CASE (column)
Syracuse, New York, *Herald Journal*, July 21, 1993

The columnist received information on another large set of cat-like tracks from Amboy Center. Also, he spoke with a witness, Mary ... Licardo, who saw a "black animal that moved like a cat," "too tall for a dog," behind her home in Syracuse.

DICK CASE (column)
Syracuse, New York, *Herald Journal*, March 22, 1995

The columnist related a story from Elmwood Park in the 1940s, passed along by reader Bruce Seeley. Seeley's father was playing in a ballfield with other kids at the time, when they heard a loud screech. They looked up into a tree to see a large black feline. They ran to get adults, but when they returned the animal was gone, leaving only tracks behind. Casts were made of the animal's footprints and sent to the state, who described it as similar to a mountain lion. Further sightings followed in the area.

Dick Case: What Prowls the Fields
Near DeRuyter? It's Anyone's Guess
Syracuse, New York, *Herald Journal*, May 12, 1995

A Madison County woman, Thelma Bandorsky, told the columnist she and members of her family had seen an unusual black creature multiple times in the fields around her home near the Otselic Valley. They watched it through binoculars. She stated: "If it's a house cat, it's a huge house cat. Its tail is as long as the body. It curls up over the body." Her brother-in-law said: "My wife thought it might be a bear. ... I watched it walk. I definitely was seeing a panther. It wasn't a cougar; they have long, slim tails. This is a furry cat's tail. And it's not a bobcat, either. I know bobcats; this would make three bobcats." He estimated it as 25 inches tall at the shoulders, and 4 feet in length. It left a 2 ½ inch track in the snow.

Dick Case: Fishing Around, We Find
an Angle on the 'Black Cat' Mystery
Syracuse, New York, *Herald Journal*, May 19, 1995

The columnist reviews the theory that the Bandorsky sighting was of a fisher, noting a letter from a reader who reported seeing a very large black fisher on a mountaintop just east of the Bandorsky sighting.

Dick Case: Here Are Several Central
New York Cats With Very Tall Tales
Syracuse, New York, *Herald Journal*, July 5, 1995

In a follow-up article on black panther stories, the columnist prints the story of a fly-fisherman in the Otselic River who saw a large sleek black cat-like animal on the opposite bank. It was about 3 ½ feet in length, not including tail, and probably over 40 pounds. The legs were shorter than similar sized dog's would have been.

Could There Be a Panther Prowling the Neighborhood?
Buffalo, New York, *News*, March 23, 2001

Residents of Hamburg reported seeing a black panther-like animal, about 2 ½ feet tall at the shoulder, weighing about 45 pounds. One witness, a hunter, said he was sure it wasn't a house or alley cat.

MISSING YOUR LEOPARD?
Lake Placid, New York, *News*, April 24, 2008

Hearing a noise outside, and thinking a bear or fisher was raiding their compost, a family in Keene looked out to see a "huge black cat." Dan Plumley stated, "It had to be four and a half to five feet long from its head to the tail. I judged its length in comparison to the driveway." It had a long, thick tail and moved its head side to side as it walked.

'LARGE CAT-LIKE CREATURES' MAY BE PROWLING AROUND PALISADES
White Plains, New York, *Journal News*, March 19, 2009

A few residents of Palisades have reported seeing black panthers. Dorian Tunell and his son were biking on a trail in Tallman Mountain State Park when they saw two of them from 25 to 30 yards away, estimating they were larger than German shepherds and weighing as much as 150 pounds. An Orangetown woman watched one for about ten minutes in her backyard.

NORTH CAROLINA

Plattsburgh, New York, *Sentinel*, June 17, 1870

A strange animal that roars like a lion, roams the woods of Mecklinberg, N. C., and proves of valuable service to the farmers in keeping their hands at home in the evening.

Indianapolis, Indiana, *Sentinel*, March 25, 1873

A Strange Monster.—Persons who have any regard for their lives and general safety will avoid the vicinity of Jonesboro, Tennessee, to say nothing of that of Shelton Laurel, North Carolina. A strange unknown animal, a monster of hideous mien, is terrifying the inhabitants of those districts, and a Laurel gentleman furnishes the following account of what he knows about the "What Is It:" "I was out in the jungle hunting up some lost hogs, when all of a sudden there came into my path a beast, the appearance of which, I must confess, caused me to quake for the first time in many years. Aside from its strange and unusual appearance, the unearthly yell it uttered on perceiving me, which reverberated and re-reverberated through the forest, was enough to shake the senses of the most daring adventurer. The animal was some hundred yards distant from me, and appeared to be a huge black bear, with mane and head like a lion, but had horns like an elk upon it. Its tail was long and bushy, with dark and light rings around it to its very extremity. Its eyes gleamed like a panther, and its size was that of an ox, but somewhat longer. Just previous to making its appearance, I had shot off my gun at a squirrel, and felt little prepared to meet such a ferocious beast without any weapon of defense. I immediately set about reloading my rifle, but had scarcely begun when it started toward me. I retreated in as good order as possible, and must say I did some good running—not looking back until I had reached an open spot—when I found the animal had disappeared in the laurel thicket. This is no story gotten up to scare not be children, I am not the only one who has seen the monster. Several have seen it since I have, and as sheep and calves are lately missing, it is presumed to be a carnivorous brute. Many have fortified their homes to prevent a night attack from the strange monster, the like of which was never seen in these mountains before. Some think it has escaped from some ambling menagerie, while others superstitiously think it is sent to warn people of some approaching danger."

Statesville, North Carolina, *Landmark*, October 3, 1884

As John Robinson's circus went into Portsmouth, Va., last Saturday morning, two black tigers broke out of their cages, and, after killing two trained goats, escaped in the direction of the Dismal Swamp. A throng of hunters with packs of hounds started in pursuit, but so far the animals have not been captured. The tigers are valued at $2,500, and the goats at $100 each.

Reno, Nevada, *Evening Gazette*, March 14, 1885

Some men hunting on Black river, in the Great Swamp near the South Carolina line, killed a black tiger six feet long. The animal died hard, after killing two dogs, and one of the hunter's had a narrow escape. Another tiger of the same species of equal size has been killed in Sampson County. It had killed a cow, several sheep and a colt. It is believed that these tigers are those which escaped from Robinson's circus near Portsmouth, Va., last summer.

THE ANTELOPE, GLUTTON, SANTER OR COUGER

Statesville, North Carolina, *Landmark*, September 4, 1890

The excitement among our colored friends, concerning the remarkable beast which is said to be ranging the woods near town, as told, of in these columns last week, continues unabated, and we have great pleasure in presenting herewith an accurate picture of him, drawn by our special artist, who was so fortunate as to get a view of him a few days ago.

With the idea of securing the latest and most reliable testimony concerning this brute and his operations, we sought an interview, yesterday afternoon, with William Newland and Curt Chambers, and William having been asked to recite what he knew and had heard of it within the past week, deposed and said:

Mr. Fettle, the policeman, shot at it last Thursday night just below the engine house; it was going down the branch and was after a dog and kept right on after it without appearing to be disturbed at all. Its tracks were seen Friday morning in the branch below Mr. Jo. Young's blacksmith shop. Saturday Major Pendleton, his folks got some fish from Mr. Phifer and that night were cooking them, and the Glutton smelt the fish frying and came out of the woods, close up to the house, and whined and then they say it jest roared right out. Saturday night Tom Houston heard it in the woods close to his house. Addison Poe and another fellow heard it make a quare fuss in Miss Cely Alexander's swamp Sunday evening. Sunday night it scratched on the door of the black folks's Baptist church while preaching was going on and flustrated some of the women folks. That same night it was seen between Major Pendleton's and George Weaver's. It has et up seven pigs for Mr. Clint Summers this week and Sam Allison says it has et up fifteen of Dr. Mott's cows and run his bull offen the hill. Don't know for certain that this is so but suppose it is as they say. Mr. Sherf Allison says it has been out on they side this week.

William and 34 other colored men were out hunting the Antelope, Glutton, or whatever it is, all day Monday. They did not see it nor hear it, but saw its track, which measured 8 inches long and 4 wide. Its hind track is like a bear's track. William understands that Col. Sharpe will give $50 for it, and also that the county commissioners at their meeting Monday offered a reward for it. He thinks that decided and systematic efforts ought to be made at once to kill it as the chinquepin season is coming on and the woods will soon be full of chaps. He is in favor of a town meeting being held and the proper steps taken in the matter. Doesn't know certainly of any chaps being et up by it this week but it is canted around that it has et severals.

Uncle Curt confirmed in every particular the report in *The Landmark* last week about his estimable wife's adventure with the strange beast, and made this addition to it: that after she had jabbed it the fifth time with the white oak stick it ran out from in under the house, walled its eyes at her and tuck down toward the branch. He says it may be a lion, as many suppose, but its track features a bears track. Has seen bears in Tennessee.

From a stranger from out West who was in the barber-shop a few days ago Rich Sherrill gathered that this animal, from the description, must, be a Couger, but the drift of opinion in town this week is that it is a Santer.

THE "VARMINT" AROUND AMITY HILL

Statesville, North Carolina, *Landmark*, September 11, 1890

We have ... no news of consequence except the excitement here the last few days over the Santer, Glutton, Couger, Catamount, Antelope, or whatever it is. The same outlandish varmint that has been roaming around Statesville has undoubtedly moved south, as there is something in the woods and on the branches near here.

Tom Foster, colored, heard it plain the night Dr. Cowan dipped Blind Sam Allison in Withrow's creek, which was Sunday night, and

Mr. Jo. Goodman saw and heard it near his house same night. Sherman Goforth heard it cracking bones on Monday morning at daybreak, near Mr. J. T. Goosman's. Mr. J. W. Brown and two other young men heard it Monday night as they going home from the post office. It was also heard by Blind Sam and by J. A. B. Goodman and J. G. Orbison and his family the same night in a body of woods south of Amity Hill, and they say Mr. J. T. Goodman's hogs at the distillery ran around the lot and made a considerable noise, as if something was after them; and a large and very curious looking track was seen on a branch, running through the above-mentioned woods, by J. W. Brown, John Colbert and others, who were out squirrel hunting. Brown says the track measures about three by six inches, and looks more like a catamount's track than anything else. Colbert says he has been a hunter all his life and has seen all sorts of tracks that ever were in this part of the country, but he never saw any such track anywhere. Blind Sam says he never heard any noise like it made outside of a show canvas. Joe Allison says it is so about the thing eating Dr. Mott's cows, for his brother Sam never was known to tell a lie.

They say Mr. Long, the depot agent at Elmwood, got a telegram from Greensboro to offer a reward of $200 for it; that it had broke out of a show; and Mr. Mell Martin says he heard the same thing and also heard that if you would call its name it would run right to you, but the man wouldn't tell its name. ...

Statesville, North Carolina, *Landmark*, October 9, 1890

The *Progressive Farmer* wants Maj. Robbins and Dr. Anderson to have a festival to raise a fund with which to employ a force to hunt the Santer.

It Wasn't the Santer, It Can't Be Killed

Statesville, North Carolina, *Landmark*, December 10, 1891

Gastonia *Gazette*.

Something like the *Landmark's* Santer played havoc with the poultry and kittens of the Steel Creek neighborhood. The varmint was as large as a dog and would cut the throats of his unfortunate victims and leave their lifeless bodies scattered about over the yard.

Some of the neighbors were so much afraid the thing would break into houses and eat the children that they stretched wire gauze across the windows at night. Mr. Brem Campbell went a hunting for the terror and when he shot it he found it to be a large wild cat with teeth an inch long.

FROM NORTHWESTERN CONCORD
Statesville, North Carolina, *Landmark*, April 28, 1892

One of our young men, whose best girl's home is in Sharpesburg, had quite a combat with the Santer the other night while on the road to his girl's home. The Santer roared like a horse snorting, the young man's mule refused to travel in that direction, and the more he would holler at the Santer the more it would roar. But after much difficulty he bolted his way through and landed safe after retiring time instead of suppertime. It is much safer to travel in day time while the Santer is sleeping.

THINKS THE SANTER IS IN WILKES
Statesville, North Carolina, *Landmark*, January 18, 1894

Wilkes correspondence, Charlotte *Observer*.

The natives in Wilkes seemed very much disturbed over what I think to be Landmark's "santer," of a year or so ago. The monster reappears (so the natives say) in all his former hideousness, and just where he comes from and goes to, no one yet has been able to find out. It first appeared in a luminous body, shedding its radiant light for hundreds of yards around, keeping the natives at a safe distance, but the last we heard of it, lo, it had changed from a bright, beautiful light into a black dog.

When pursued the other day, it ran into a hollow stump and defied capture for the present, but next morning our people met en masse and proceeded to dig the black dog out, and one Mr. Dula (not Col. Tom Dula), after digging for some time, discovered a large piece of coal (anthracite, I suppose), and taking out his knife in order to test the material, the blade and backspring flew out and left him lifeless. This was enough for Mr. Dula and all the rest. They left the place in a hurry.

Please let us know if the *Landmark* man has missed his santer, and if he has, tell him to send up to Mr. Dula's, in Wilkes County, and get him.

[The santer has not been about Statesville for quite awhile and we had begun to think it was lost. Some months ago we heard of it in Caldwell County and have no doubt that it went from there to Wilkes. Bill Newland, of color, so it many times when it was running about Statesville, and if the Landmark can persuade him to go to Wilkes we will send him up to see if the varmint herein before described is our santer, and if so to bring it home. Bill will know the moment he lays eyes on it whether it is genuine or a base imitation; and we warn the Wilkes folks in advance that they can't put off a bogus santer on us.—*Landmark*.]

THE NEWS FROM OLIN AND ABOUT
Statesville, North Carolina, *Landmark*, March 28, 1895

The "Santer" or some other wild animal is exciting the colored population of the town. On last Wednesday and Thursday nights it visited Jack Dobbin's and Andy Click's houses. They both alarmed the town by their distressing cries for help. Several of the neighbors went down to the cabin and gave the unhappy d— a few words of consolation and sent him to bed. The same thing ran a tobacco peddler's driver all over the town, causing him to lose his watch in the time of the race. Some one offered to help him hunt for it but he decided not to look for it until morning.

THE NEWS OF OLIN AND THE NEIGHBORHOOD
Statesville, North Carolina, *Landmark*, December 17, 1895

The "Santer" or some other wild varmint was seen by Mr. Jim Parks on last Sunday night. When the Mr. Parks first saw the animal he says it was standing on its hind feet with its fore feet extended forward. When he turned near (if he ever got close) the thing crouched down flat on the ground and then disappeared. We would be glad to know whether it is the "Santer" or not.

Statesville, North Carolina, *Landmark*, January 31, 1896

Mr. G. W. L. Cavin, who lives near here [Troutman's], caught recently what was first thought to be the *Landmark's* Santer. It filled the description of the Santer exactly. It was three feet long, short legged and had a hatred of dogs. But when shown to some persons versed in bruit-ology, they said it was an otter. These animals, it is said, never leave the river or large water courses, but this one was caught six miles from the river in a small branch.

A STRANGE "VARMINT."

Statesville, North Carolina, *Landmark*, February 7, 1896

Mr. Clifton Vinson, who lives near Connelly's chapel, reports that he was riding along the road leading from the Wilkesboro to the Taylorsville road, a few nights ago, when an animal having the appearance of a big black dog came out into the road near to him. Vinson had his dog along and endeavored to set it on the stranger but the dog refused to go near it. Some people in the neighborhood thinks the animal Vinson saw is the Santer, just returned from Guilford County.

THE SANTER HAS BROKEN LOOSE AGAIN

Statesville, North Carolina, *Landmark*, January 3, 1896

Brownsboro *Record*.

There is great excitement in the Ready Fork settlement. A strange animal has turned up out that way and is playing havoc with the dogs. Perhaps not so much objection would be raised if residents could only discover exactly what kind of a varmint it is. It has been seen several times and taken for a black dog but it is not much of a dog for it eats up other dogs.

The first notice of its presence was a month ago when it was seen above Ready Fork bridge, loping off into the woods. A week after this three hounds attacked something a mile north of the spot where it was first seen. There was a traffic fight in which one of the dogs was so badly torn and bitten that he died in next day.

During the week following this two miles north of Beaver swamp, a strange animal attacked a Mr. Apple's dog at night. He heard the fight, got up and drove the animal away, which he took to be a black dog, but says it loped off into the woods. Later the same night he heard his dog again fighting, but did not get up. Next morning he found his dog dead in the yard, with one shoulder eaten off. On the same night, half a mile above Mr. Apple's, some animal tore to pieces a neighbor's dog.

It has never been known to tackle anything but dogs, and we submit that as long as it confines itself to them that it should not be interfered with.

[This is of the *Landmark's* Santer, sure as fate. We hadn't heard of him in a long time and had begun to mourn him as dead. But we are glad to see that he is in good health. Yes, this can be none other than the Santer, for the public will remember when he used to be around Statesville a few years ago his special diet was dogs; and William Newland, of color, to whom was granted the privilege of seeing him on one occasion, said he resembled a big black dog and that he loped away when he saw him. Yes, this must be the Santer. He was fond of dog meat and he was a great "loper." We want the Guilford folks to send him home. He has been away a long time and we need him about Statesville again.]

WILLIAM NEWLAND
Statesville, North Carolina, *Landmark*, January 12, 1897

Bill Newland, a well-known citizen of color, died yesterday of pneumonia. Bill was the real discoverer and promoter of the *Landmark's* santer and told many interesting stories to further its career when it used in this section several years ago.

He was a well disposed toward man and was very well-liked by his white acquaintances.

Statesville, North Carolina, *Landmark*, August 3, 1897

Please allow me a little space in your valuable paper to let the people of Iredell and adjoining counties know of some of the dangers to which they are now exposed.

It appears beyond all doubt that there is some kind of a wild animal "using" about this section. It has been seen by two or three reliable parties within the past week. It is a little larger than a dog and is white spotted.

Almost every night for the last week it has been heard—some nights at one place and the next night probably two miles away. It makes a growling, blowing sound that can be heard quite a distance. A party of citizens made a close search on last Tuesday about Union Grove Church but failed to find it.

From all information the writer can gather, this apartment completely outstrips the *Landmark's* Santer of two years ago.

Statesville, North Carolina, *Landmark*, October 22, 1897

The *Landmark's* Santer is roaming through this part of the country. It appeared at two tobacco barns and its track resembles a bear track. It was seen several times and some say that it is spotted others that it is white and yet others say it is brown and about the size of a goat. It seems to be very quiet. The *Landmark* ought to take its Santer home for it has terrorized the country long enough.

THE NEWS OF THE TROUTMAN NEIGHBORHOOD.
Statesville, North Carolina, *Landmark*, March 1, 1898

Mr. Thomas Kerr, who lives near Sherrill's Ford, tells me that The *Landmark's* Santer is again in that community. It certainly has grown a great deal, especially its feet. Its track will measure at least 12 inches long. It looks like a bear's track except that there is a great claw exactly in the center of it. It is causing no little uneasiness amongst those who try to make their "nightly" bread, and they are anxious, if it is the *Landmark's* Santer, that you call it in or send it to some other community, for they have been tormented with enough varmints, without the Santer. They have had measles, mad dogs, smallpox scare, Spain war, and now the Santer. So I think it should be called home.

Statesville, North Carolina, *Landmark* (excerpt), October 19, 1900

For several decades the people of Iredell county have been pestered by a santer. It is invisible, but very distinctive. The Statesville

Landmark has offered big rewards for its capture and at one time went so far as to send out special hunters, but all in vain. The dreadful beast returns after each hunt is over to carry a way more dogs, cats, hogs, sheep and negro babies at will. After having read Lawson's history, I am convinced that he heard of the same animal on the Neuse river, over 175 years ago. Read what he said: "I have been informed by the Indians that on a lake of water towards the head of the Neuse river, there haunts a creature which frightens them all from hunting thereabouts. They say he is the color of a panther, but cannot run up trees; and that there abides with him a creature like an Englishman's dog, which runs faster than he can, and gets his prey for him. They add that there is no other of that kind that they ever met withal, and that they have no other way to avoid him by running up a tree. The certainty of this I cannot affirm of my own knowledge, yet they all agree in this story."

The animal in the Iredell is of the same character as the one that used [to be] on the headwaters of the Neuse in the time of Lawson.

The Santer in Union County

Statesville, North Carolina, *Landmark*, October 22, 1901

Monroe *Journal.*

That terrible wild animal which has been committing such depredations among the dogs and other domestic animals of Lancaster county and which the Lancaster papers designate the "dog killer," has made his dreaded appearance in this section. At Mr. H. J. Belk's, in Buford township, the varmint appeared one night last week. Before it, or he or she came, Mr. Belk had three dogs but now he has only one and he is so badly dilapidated that the other two wouldn't know him on his own premises. The thing came among them one night, there was a terrific fight, and the three dogs disappeared but the next day one of them came back all cut up and bleeding and torn. The next appearance that we hear of is in the Helmsville community. It attacked Mr. W. E. Helms' dogs a few nights later. Mr. Helms' son took his gun and went out to shoot the varmint. He shot four times and still the animal came towards him. He saved a fifth ball for emergencies and retired into the house. It has been suggested that perhaps the Statesville Landmark's "Santer" has wandered afar from home and is now prowling around in this community. If so the Landmark is respectfully asked to take notice of its whereabouts.

[This is doubtless the Santer. When he began his career in Iredell he was monstrous fond of dog meat and we notice that he still has his appetite with him.—*The Landmark*]

Statesville, North Carolina, *Landmark*, December 16, 1904

The *Landmark* has told of the strange varmint that is using about Greensboro, killing dogs by the dozen and frightening the colored people, and some whites too, almost out of their wits. The Salisbury *Sun* says "the Greensboro folks are attempting to appropriate the Iredell County Santer as their own" and it refuses to believe a word of it "until the Statesville *Landmark* is heard from." The actions of the Greensboro varmint, if truthfully described, are exactly similar to those of the Iredell Santer, but we can't think the Greensboro varmint is the original Santer for two reasons: first, because the Iredell Santer used about here something like 18 years ago, and it is reasonable to suppose that if alive is hardly so active as the Greensboro animal is alleged to be; second, the Iredell Santer, as shown by a picture which Mr. J. F. Anderson secured and which was printed in the *Landmark*, sported a tremendous long tail, which curled over its back and went clear past his head. Now the Greensboro *Record* says that the animal down bear has a very short tail; and in view of these matters we are inclined to agree with the Charlotte *Observer* that this is a pup of the original Santer, the Santer family having probably adopted a shorter tail as more stylish, or this particular one may have lost his tail by accident. In any event we are inclined to believe that the Greensboro beast is of the Santer family.

THE SANTER IN WILMINGTON
Lumberton, North Carolina, *Robesonian*, February 1, 1909

Wherever the first pinch of prohibition abounds, there does the Robeson County santer much more abound. Last summer a year ago, while off duty a while, he roamed about meditating in Wishart's township, this county, in Hog swamp, hard by Cyprus pond, but since then he has been active in other places, and he has had business in other States. Just now he is devoting his attention to Wilmington. Read:

"Half the colored population and not a few of the whites in the eastern and southern suburbs of the city are in a state of alarm, bordering closely upon terror, especially at night, over the appearance in their midst of some wild animal which is reported to have slain several dogs and committed numerous other depredations of startling nature."

Sure. That from The Morning Star, and the news item from which the above is taken further says that none can say what the nature of the beast is. Not on your life you can't. No man was ever able to say what its nature was when it roamed these coasts. Mr. W. F. Willoughby knew more about it than anybody else, and he never could tell. Sometimes it would cry like a woman in distress, sometimes it would moan soft and low, sometimes it would give wails like a lost soul— however that wail is—and sometimes—Oh, it's an accomplished monster, all right, and so terrifying are the various sounds it makes that whoever hears them straightway as urgent business in the opposite direction: no man yet has been hardy enough to come close enough to the santer to investigate. The Star says that "several who have seen the varmint solemnly avow that they are not hankering for the experience again." Quite so; nobody ever did want to see that santer twice.

This particular beast terrible is described as a large brown beast with claws like a catamount, but of much larger size. He is said to have killed several dogs in Wilmington, partially devouring at least one, and when one owner went to his dog's rescue and tried to kill the santer with a baseball bat he missed the santer and killed the dog. Do tell! What do you think of that? The Star wants to know if it is too much to presume that it is the Robeson county santer which is now causing all that howdy-do in Wilmington, and it further deposes and says, "Certain it is, it is not a blind tiger, for they are unknown here."

Well. Bless his wild heart, we bet anything it is our Robeson santer. Wherever the first pinches of prohibition abounds, there he does much more abound.

UNNAMED MONSTER SLAIN
MR. ANDREW BRITT THINKS HE HAS KILLED DEAD THE MONSTER
THAT HAS SPREAD TERROR THROUGHOUT THESE COASTS.
Lumberton, North Carolina, *Robesonian*, June 29, 1914

It will be remembered that several years ago a great howl was made down in the Globe Swamp section about a "monster" that was seen and heard in that and other nearby sections. This "monster" would kill a hog, a dog and a cat occasionally, and, judging from reports, the folks thought he was capable of devouring a man or most anything else that might come across his path. While there has not been so much talk of him recently, he could be heard from occasionally.

This "monster" may still be alive but Mr. Andrew Britt, who lives in Columbus county, near Boardman, thinks he killed him dead one day recently. Mr. Britt's dogs treed a sure enough "monster," one that nobody was able to give a name, up in a high tree. He went to the tree and saw him perched up in the top, and after looking into his frightful eyes was almost persuaded to leave him alone, but "let loose" with his shot gun, and down came the "monster," but he was not too dead to slap the face off of two or three dogs before Mr. Britt "let loose" again. Mr. Britt says he has not got a name for the "varmint," but rather thinks he was a European jackal. However, he is dead, and the people who saw him would be delighted to know that there was no more like him in these coasts. He was larger than a dog and the hair on his head hung down over his dangerous-looking eyes. Here's hoping he was the "monster" and the only one.

SANTER LOOSE ON BROAD RIDGE AGAIN, BY GEORGE!
Lumberton, North Carolina, *Robesonian*, March 20, 1916

Mr. A. H. Bissell of the Broad Ridge section was in town Saturday. Mr. Bissell says a certain animal that must be some relation to the Globe swamp monster, has recently been prowling around on the "Ridge." This "thing," which looks bad and makes a worse-looking noise, seems to play about a certain cemetery o' nights, says Mr. Bissell. It has caused some few boys of that section to break the speed limit, too. What it is and what it wants are two things the folks have not been able to learn.

Varmint Frightening Folks

Statesville, North Carolina, *Landmark*, January 31, 1921

Monroe *Journal.*

Some two or three months ago the southwestern section of Gaston county and the upper part of York county were much disturbed by reports of a "wild varmint" at large in that section of the country. Some said it was a catamount, others said it was nothing but a dog running at large. Of late there has been nothing heard of it on that side of the river. Something like it has, however, appeared in the Providence section of Mecklenburg county, according to the following from the Charlotte news:

A group of Charlotte hunters may form a party and go to Providence township to run down the strange animal that has been causing much talk among the people residing on both sides of Four Mile Creek, in Providence Township. It is likely the idea will be carried out Wednesday night or Thursday.

Referring to the report that the dogs of possum hunters have refused to run the "varmint" but leave the brush and come back to crouch in fear at the hunter's feet, one hunter has made the suggestion that a pack of airedales be taken and turned in the woods.

"If two or three airedales don't bring that beast of the woods and swamps of Four Mile Creek, I will pay all the expenses of the expedition out there," Jim Houston, hunter and admirer of the airedale strain of dogs.

The description given of the animal's appearance by Providence citizens in the city answered in a general way to the description of a panther, although most of the people of the neighborhood, realizing how long it has been since a panther, which is a native of North Carolina, has been seen in this section, believe it must be a wildcat or a catamount.

The two close-up views of the beast, as reported by reliable and trustworthy people, indicate that instead of having the stumpy tail of the catamount or bobcat, it has the long tail of the panther, the tiger or the wild cat and turns up at the and in true cat fashion, or else is carried straight out behind.

The report that Ed Sussel shot at the beast and had an especially good opportunity to see it was confirmed by people in Charlotte Wednesday. He is quoted as saying that he was within ten or twelve feet of the animal and fired at it three times with a pistol, which he had been carrying in the hope that he might run across it. He is quoted as saying that he hit the animal twice because at each of the first two shots it jumped high in the air and bounded over a hedge and fence at the third shot and disappeared in the underbrush. Luther Squires is another reliable man who has seen the animal. He came upon it at a turn of the road and had a good view of it before it took fright and ran into the bushes. His description also indicates that the animal is either a wildcat or a panther. If it is a wildcat it must be the largest seen in this part of North Carolina, Mr. Squires thinks.

IREDELL SANTER BELIEVED IN GASTON COUNTY
Statesville, North Carolina, *Landmark*, February 10, 1930

Monroe *Enquirer*.

The Gastonia *Gazette* recites that "residents of West Gastonia are hot on the trail of a blood-thirsty animal which murders by night and feasts on rabbits, pigs and other domesticated creatures. A score or more of men gathered last night at the home of Harley Redding, West Fifth avenue, to keep watch near the scene where the unknown creature Tuesday night killed and partly devoured a fifty-pound pig after having killed and eaten three tame rabbits penned up in Redding's back yard."

No doubt the varmint which is causing so much excitement around Gastonia is a Santer. Several years ago a Saner kept n— indoors for months in Iredell county. At the time the Statesville *Landmark* kept the public well-informed as to the maraudings of the beast and if it has ever been kilt no on has heard tell.

ANNIVERSARY OF THE WAMPUS HERE
Statesville, North Carolina, *Landmark*, July 26, 1932

See in the Mooresville *Enterprise* where "that d— wampus" has come to life in south Iredell. Let's see, wasn't it about watermelon time last year when they "shoo-ed" that bloodcurdling creature with the keen "holler" from the thickets?

ANIMAL CHARGES AT AUTO OF SALESMAN
WILD HYENA-LIKE CREATURE IN MCDOWELL COUNTY
MAKES OPEN ATTACK ON DRIVER OF CAR.

Statesville, North Carolina, *Landmark*, August 30, 1932

Marion, Aug. 26.—A wild hyena-like animal said to be roving through the neighborhood countryside charged at a shoe salesman between here and Spruce Pine yesterday, but failed to get within striking distance of his car, it was said.

Hearing something snarl beside the road as he drove this way, he said, he looked up in time to see a vicious animal leaping toward his car, its ugly fangs showing plainly between drawn lips.

The animal seemed to be leaping at his elbow, which stuck out through an open window of the car, and would have been within, easy reach of the beast had it been able to get closer to him.

He was not positive what the animal was, but said that it looked more like a hyena than anything else he had ever heard of. He declined to give his name.

This is the first time that the animal has been reported seen since Ben Price saw it assault his bull dog Sunday afternoon.

Statesville, North Carolina, *Landmark* (excerpt), April 11, 1933

The Santer is an animal of vast proportions, if seen after dark. All Santers, and sich like, look twice as large after dark than they do in daylight. This Santer paid us a visit last winter, but outside of keeping some of our folks in the house after dark, did very little, if any, damage. Being as how he left these parts sometime last winter, we thought it was gone for good. But not so. 'Tother night he raided Mr. Lester Rumple's hog lot, and carried off one of Lester's pigs, which was found about a mile down the road, partly devoured. Strict search has been made hereabouts but no trace of the Santer has been found. Some think it was a "Wampus" but others say a darned Wampus couldn't carry a pig as big is that. His tracks were about the size of a small dog when first discovered, but they have increased in size till they look like a bear's, or sumpin'.

Statesville, North Carolina, *Landmark*, January 31, 1949

This, dear friends and lovely ladies, is the Santer, known to a later and more profane generation as the Wampus.

Note his feet. Would they, you think, leave tracks like the tracks recently found in Bethany Township? Could it, do you think, be the progenitor of the strange animal that is presumably prowling these parts today?

Who knows. The story, as recorded in the Landmark files a half a century back, relates how a strange animal heard, but never seen, frequented the outskirts of Statesville in the days when the town was scarce more than a watering stop. Weird sounds were heard in the nighttime. Strange footprints were seen in the day and here and there animals disappeared, chickens squawked and died and folks, home-ward bound after dark, glanced warily about them and made haste.

No one ever saw the creature. Its picture, portrayed above, is the brain child of some earlier employee of this paper, the artist's name is lost in the shadows of the years and we term him here, "The man who drew the Santer." He drew the strange prowler in accordance with the best descriptions given by the many who told of shadowy suggestions of a "long, thin thing, bristled like a pig (or maybe a coonskin coat), tailed like a lion with a mule's twist in it and braided neatly like a sukey cow's that is put to graze in rough country. Hoofed like the chimpanzee, the creature was said to have toes like men that walk, to be human-eyed with the drooping ears of a faithful hound and nostrils not unlike the end of an elephant's tusk. The mouth, with the whiskers, was drawn in keeping with the description of the low rumbling sounds that rose into a screech like whistle that the animal made on occasion when nights were darkest or fog most dense.

The generations past. New events brought new interests and the Santer remained but as a tale that is told. Then some 18-20 years back, from the low lands along the Free Nancy and from the area about Fourth Creek there came rumors of an animal with strange, long feet, toes almost human in their imprint that was walking alone and lifting a mournful note in the darkness of the night. Old folks remembered, and remembering said, "another Santer." But the younger folks, brave in their ignorance, laughed derisively and made jokes about the animal terming it a "Wampus." None saw that Wampus either. No one ever caught a glimpse even of his thin shadow traced against the ground on moonlight nights—for no one, so they say, used to stay around the places where he prowled after dark had come. But there were likenesses in the stories. So many likenesses that the idea grew that perhaps the Santer's wraith was wandering or maybe a younger Santer, offspring of the old Santer's mating with who knows what strange denizen of woods or marsh.

And now we have another day—and another creature. Santer, Wampus or the third generation of the strain, bred in silence, come to life to haunt the places where his kind had their borning. Who knows? Anyhow out there in Bethany Township there are strange tracks. Here above you is the creature, the original Santer, who left strange tracks in his own day and who was followed in the space of a lifetime by another who made his tracks in the nighttime and went unseen, unidentified. They talk about a bear out there—but no one has seen a bear. They talked bear in the days when the Santer walked, when the Wampus roamed, but no one ever saw a bear—no one ever saw anything in fact. And so which may be now. Anyhow, if there be those that would try it, they might take the little picture above, camp around Bethany where the strange footprints go, and some night when the moon is low and the woods are dark, maybe this strange beast will come, moving slinkily among the brush and the watcher can compare him, find what kinship, if any there be, in the Santer of green memory, his later kind, the Wampus, and this last, and as yet untitled four-footed, strange tracked Iredell County wanderer.

ARMED HUNTING PARTY TO SEEK BLADENBORO'S 'VAMPIRE BEAST'
Lumberton, North Carolina, *Robesonian*, January 5, 1954

Bladenboro, N. C. (AP)—A "vampire beast" that sucks blood from its victims had Bladenboro citizens up in arms today.

Armed posses roamed the town after the discovery of three mutilated dog bodies recently. Police Chief Roy Fores said the body of the latest victim was opened yesterday and it contained only a few drops of blood.

He said the three dogs all had their bottom lips broken open and their jawbones smashed back. He said the ear of one dog was chewed off and the tongues of the others chewed out.

Fores said the vampire probably is a mad wolf.

By John Gause
Robesonian Staff Writer
Bladenboro—Bladenboro Police Chief Roy Fores, working around the clock with a group of armed citizens in an effort to bring to bay the mysterious 'vampire beast,' indicated this morning that the animal could be a huge wildcat or mountain lion.

Tracks leading away from the latest dog killing obviously were not those of a dog. Observers say they were spores evidently left by a huge mountain lion, probably weighing between 80-100 pounds.

The tracks, deeply imprinted even in a hard grain field, led away from the latest dog killing in the community. Last night a puppy owned by Johnny Vause was killed, the seventh dog in that section to be a victim of the mysterious killer.

The puppy last night, although not killed vampire fashion, had it's nose chewed off.

Tracked to Swamp

Chief Fores and his men last night tracked the animal to a swampy section near town by using a pack of dogs brought in from Wilmington. They were led to the spot by Lloyd Clemons, who first saw the animal out near the Mill Section. He told Chief Fores 'the animal was around three feet long, and low to the ground, probably 18-24 inches high.' Fores said this description sounded like a cat.

Bladenboro citizens, meanwhile, aren't taking any chances. Fores said this morning that the town was quieter now than he has ever known it. "People stay off the streets and lock their doors now," he said.

Seven Dogs Killed

The toll in dogs through this morning was seven. Other than the puppy lost by Vause last night, he has lost another dog. Also losing pets to the mysterious, killer were Woody Storms, Harry Waystrom, and Ray Callahan,

Fores said he was 'definitely worried' about the animal, whatever it is, running loose in the community. "An animal of that size could very easily attack a child ... even a grown man."

An armed hunting party with dogs was to be formed again today. Two or three people have seen the animal. All describe it nearer to being cat-like. Two carpenters noted the animal late yesterday afternoon but couldn't offer any description.

'VAMPIRE' STRIKES AT WOMAN; POLICE CHIEF WARNS PARENTS
Lumberton, North Carolina, *Robesonian*, January 6, 1954

Bladenboro, N.C. (AP)—Worried parents kept a close eye on their children today as a strange "vampire" beast continued to roam the countryside.

The beast, which police say sucks blood from its victims, attacked its first human last night. Previously more than seven dogs were reported killed.

The vampire apparently is some species of the Family that is a bloodsucker. Lloyd Clemmons, who claims to have seen the animal, said it looked "like a cat." He said it was about three feet long, 20 inches high, and had a tail about 14 inches long. He said it was dark in color.

Two sets of tracks were found which authorities could not identify. The extra set of tracks led them to believe the "vampire" has a mate.

Mrs. C. E. Kinlaw said the beast attacked her when she went on her front porch to investigate a noise. It fled, she said, when her husband came out.

At least seven dogs have been killed and their blood drained by the strange beast. While a posse of nearly 500 men and dogs searched last night, the animal struck, dragging a dog into the swamps within 100 feet of the searchers.

Police Chief Roy Fores warned parents to keep a close eye on their children.

By John Gause

Robesonian Staff Writer

Bladenboro Police Chief Roy Fores said this morning that he and his posse of some 20 armed men would work "from now on if necessary" to put a stop to the cat menace which has put citizens of the Bladen community under arms.

Plans were underway this morning to surround a bay area about a mile from town, where the cat-like 'vampire beast' last struck. A climax was reached in the 5-day-old reign of terror at last night when the beast, described by some to be a huge mountain lion, struck at its first human victim.

Although the animal inflicted no injury to its intended victim, Mrs. C. E. Kinlaw, it was the first indication that it would not stop at merely attacking dogs.

Since the animal was first reported in the area, more than seven dogs have been found drained of blood, leading hunters to believe that the beast is a blood sucker.

Chief Fores and his men spent the night searching for trace of the mysterious dog killer following the news that Mrs. Kinlaw had been attacked by the beast. She told newsmen this morning that "it looked like a dog from the rear." Understandably excited, she didn't observe the animal close enough to offer definite description.

She did add, however, that "it appeared to be gray."

The posse went ahead with plans today to bring the search to a climax. Fores said that they were surrounding the bay area where the search was concentrated last night and would set a trap in hopes of enticing the beast near enough for the men to shoot.

According to the Bladenboro officer, the trap will be baited with a dog.

Meanwhile, people living in the Mill Section, where the beast has seemingly picked as a base of operation, are indoors while the posse combs the bay section nearby.

Vampire' Hunters Outnumber the Residents of Bladenboro

Lumberton, North Carolina, *Robesonian*, January 7, 1954

Bladenboro, N. C. (AP)—Tension mounted in this terrorized village today as an unidentified "vampire" beast continued to roam the countryside.

No one knows exactly what the strange animal is that has mangled six dogs and sucked their bodies dry of blood. A seventh was dragged screaming into a swamp Tuesday night by the marauder.

Only one human, a pretty 21-year-old mother, Mrs. Charles Kinlaw, has been attacked. She escaped unharmed.

Police were forced to call off the search for the beast last night as a safety measure when the number of armed hunters exceeded the town's population.

Police said between 800 and 1,000 hunters tramped through the mill town with a 786 population. The hunters were armed with shotguns, rifles and pistols.

The plan had been to stake out live dogs as decoys, hoping to attract the beast.

At least one veteran hunter pressed into service said he thinks the "vampire" is a maddened panther. S. W. Garrett of Wilmington made the guess after hearing the animal scream.

Lloyd Clemmons, a mill worker, said he saw the cat-like beast Monday night. He said it is about three feet long, 20 inches high and has a tail 14 inches long. Hunters estimate the animal weighs at least 90 pounds.

Previously hunters discovered two sets of tracks leading them to believe the "vampire" has a mate.

By John Gause
Robesonian Staff Writer

A party of hunters was concentrating on the swampy area about two miles from Bladenboro late this morning in an attempt to keep the mysterious 'vampire' beast located until an experienced hunter with trained cat dogs can come in from Wadesboro tonight.

D. G. Pait, one of the party of some 800 men who remained up all night in an effort to bring to bay the strange blood-sucking animal, last reported in a swampy area near Bladenboro, said this morning that nothing new has been discovered about how the beast might look.

Whatever, it is that has terrorized citizens of that town no one can say definitely. Only one person has offered any description, leading hunters to believe they are on the trail of a huge mountain lion, or panther, that possibly has a mate.

Pait said this morning, however, that the beast was heard by a resident of the Mill Section last night before the hunt began. That person, who is unidentified told hunters 'it sounded like a baby crying.'

The search last night was greatly hampered by the thronging group of hunters and spectators. A plan to trap the animal by using dogs as decoys had been cancelled when it became obvious that the crowd would keep the beast in hiding.

Wilmington hunters, using fine Blue Tic hounds, called the dogs off the trail when they grew apprehensive about the safety of their animals.

Mr. Pait said the hunt was called off late last night but armed groups patrolled the streets all during the night in an effort to locate the beast.

KITTY KORNER

Lumberton, North Carolina, *Robesonian*, January 8, 1954

Bladenboro, N. C. (AP)—A mysterious monster that "crys like a baby" and drinks blood continued to elude hunters today.

The catlike beast, possibly a maddened panther accompanied by a mate, has killed at least six dogs in the Cotton Mill Hill area. Another dead dog was found yesterday but Police Chief Roy Fores said it had not been established that the dog fell victim to the maddened beast.

Still another dog was spirited into the swamps as helpless residents listened to its death screams.

Hundreds of gun-bristling volunteer hunters swarmed into the tiny mill town of about 800 population, making it difficult for professional hunters and their dogs to trap the killer. Police feared the volunteer help would shoot each other.

The beast came close to a woman on her front porch Tuesday night but slumped off when she screamed. She described it as a big catlike animal, with a body about three feet long, and a long tail.

Residents have described the beast's cries as sounding "like a woman in pain," "barking like a coyote" and "crying like a baby."

***

Fayetteville—Two dogs found mysteriously dead and a Plum St. yard Thursday morning apparently were poisoned and not victims of the mysterious Bladen County "vampire," but strange footprints of some kind of "varmint" were discovered this morning in the yard of a home on Oak Ridge Ave.

No signs of external violence or discovered on the bodies of the dead Plum St. dogs. They were lying about 20 feet apart and on their sides. They had last been known alive about 3 AM, when they were heard barking at the milkman.

They were found in the yard of Sgt. John S. Jones at 306 Plum St. One of them was owned by Thomas McLaughlin, 308 Plum, and the other was a stray.

Unusually large footprints of some animal were discovered early this morning in the front yard of E. B. Martin at 522 Oakridge Ave. Mr. Martin noticed the tracks as he went out to his car this morning.

The most distinctive footprint measured about four inches in length and was about three inches wide. It clearly showed the outline of four toes.

No one has reported seeing an animal of unusual size in the Oakridge Ave. area.

***

Fresh Shave

Lumberton—The "varmint" came to Robeson county Thursday afternoon, according to a report received by Capt. Wilbur Lovette of Lumberton Police department, and was located near Robeson County Health department.

Capt. Lovette said no investigation would be made by the department because it was outside the city limits, even with its extended boundaries.

Capt. Lovette said the report was made by a man named Mills who telephoned to say that the "thing" was seen crossing a field from Red Springs Road toward the health center, carrying a dog in its mouth. He said Roy Shipman was the person who actually saw the beast, and described it as having much bushy hair about the head and neck, but seemed to be clean shaven from the neck back, and had a long tail.

***

Feline Fantasy

Lumberton—An unidentified (he prefers it that way) man who told newsmen last night he saw "a strange looking, shaggy brown cat-like creature" in the vicinity of the Robeson County Memorial Hospital, called this morning to retract his statement.

He said "he must have been dreaming."

Continuing further, the mysterious phone caller said he "had read so much (about Bladen's 'vampire' beast) that anything moving at night looked like a demented varmint.

"I could believe all this about 'vampire,'" he said, "but when I heard this morning he et up 1000 feet of timber in South Robeson, I gave up."

Plus the timber, the mysterious wampus cat has set a pretty enviable record to date: eight dogs, including blood; a spare tire from a parked automobile, including tube; a family of kittens in Robeson County, including milk and bedding; one lamb, too innocent to protect itself.

One famous animal trainer, who also prefers to remain anonymous, said he would capture the beast, train it to do right, patent his accomplishment, and sell it to Robeson and Bladen towns for Chamber of Commerce use.

BLADENBORO BEAST HUNT IS CALLED OFF

Florence, South Carolina, *Morning News*, January 9, 1954

Bladenboro, N. C., Jan. 8 (AP)—The vampire beast apparently has won the "Battle of Bladenboro."

Mayor W. G. Fussels said late today the hunt is off—unless the mystery animal strikes again, or unless it is actually sighted.

The hordes that swarmed this small mill city with rifles, pistols and shotguns for the past week were absent tonight.

The mystery beast is a blood-lusting killer that has left seven mangled and lifeless dogs in the area and is known to have vanished with an eighth dog. Other reports have been received, but unconfirmed, of other missing animals.

The town remained in a state of apprehension today.

The mayor himself admits he is scared:

"When I go home at night from my theater, I look around plenty before I get out of my car."

Mystery Beast Believed Gone Back Into Woods
Panama City, Florida, *News*, January 9, 1954

Bladenboro, N.C., Jan. 8—(UP)—Fears of this jittery community were calmed today by belief that a strange, catlike beast has fled back into the deep swamps from which it emerged a week ago to prey on dogs.

The "mystery animal," believed by most authorities to be a panther, has not been seen for the past three days nor has there been a sign of it since before dawn Thursday. The beast was believed to have been heard prowling around a stockyards east of here, but fled into the swamps when packs of dogs were put on its trail.

Since early last week at least eight dogs have been slain by the beast's night-time attacks. The throats of the dogs were ripped and their bodies clawed.

Veteran hunters believed the animal was a panther from the wilds of the Cape Fear River swampland which surround this town. Wildlife officials said panthers, though extremely rare, still are present in the deeper forests and swamps.

Other theories included those that the beast was a large wolf or wild dog, but hunters said a panther would more likely be responsible for attacking dogs and killing them in such a manner.

For the past three days Bladenboro has been like an armed camp, with hundreds of hunters and packs of dogs converging on the town to try to hunt down the animal. Experienced hunters said the swarms of people trampling the countryside probably frightened the animal and believed it would not be seen again.

MYSTERY BEAST BELIEVED GONE BACK INTO WOODS
Panama City, Florida, *News*, January 9, 1954

Bladenboro, N. C., Jan. 3—(UP)—Fears of this jittery community were calmed today by belief that a strange, cat-like beast has fled back into the deep swamps from which it emerged a week ago to prey on dogs.

The "mystery animal," believed by most authorities to be a panther has not been seen for the past three days nor has there been a sign of it since before dawn Thursday.

The beast was believed to have been heard prowling around a stockyards east of here, but fled into the swamps when packs of dogs were put on its trail.

Since early last week at least eight dogs have been slain by the beast's night-time attacks. The throats of the dogs were ripped and their bodies clawed.

Veteran hunters believed the animal was a panther from the wilds of the Cape Fear River swampland which surround this town. Wildlife officials said panthers, though extremely rare, still are present on the deeper forest and swamps.

Other theories included those that the beast was a large wolf or wild dog, but hunters said a panther would more likely be responsible for attacking dogs and killing them in such a manner.

For the past three days Bladenboro has been like an armed camp, with hundreds of hunters and packs of dogs converging on the town to try to hunt down the animal. Experienced hunters said the swarms of people trampling the countryside probably frightened the animal and believed it would not be seen again.

TWO MORE SUSPECTS ENTER BLADEN 'VAMPIRE' GALLERY
Lumberton, North Carolina, *Robesonian*, January 14, 1954

Bladenboro, N. C. (AP)—Two more suspects, both dead, have entered the Beast of Bladenboro rogues gallery.

Both animals were killed yesterday near Bladenboro. One, a bobcat, was trapped and shot in the Big Swamp about four miles from here.

The second, a spotted "leopard cat," was run over and killed by a service station operator near Bladenboro last night.

The widespread search for the strange "vampire" beast was set off when seven dogs were found killed recently, their bodies drained of blood. The search was called off last week after officers theorized that the animal had been frightened away from the district by the mobs of armed hunters.

The bobcat was trapped and killed by Luther Davis. The "leopard-like" animal was killed on the highway by Bruce Soles. This possibly is an ocelot.

Both animals generally fit the description of the "vampire" beast.

The bobcat was gray, about 30 inches long and had a stumpy tail. The bobcat was lighter than the "vampire" which was estimated to weigh about 90 to 100 pounds.

The animal killed on the highway was spotted like a leopard, was about 20 to 24 inches high and weighed between 75 to 90 pounds. Its tail was about eight inches long.

The bobcat and the "leopard cat" join such other nefarious suspects as a black panther that has never been seen and a half-wild German police-hound dog.

NEW 'VAMPIRE' RUMORS FAIL TO EXCITE BLADENBORO CHIEF
Lumberton, North Carolina, *Robesonian*, January 15, 1954

Bladenboro—Contrary to wide-spread reports that the Bladenboro Beast had struck again, Bladenboro Police Chief Roy Fores said this morning "he wasn't getting excited" over reports that three more dogs had been killed by the mysterious 'vampire'.

Chief Fores indicated that rumors were flying too fast and frequent for accurate appraisal and said he had orders from Bladenboro Mayor W. G. Fussels not to take any more action until the mysterious animal actually struck again.

Not Substantiated

The Bladen officer said the latest reports that the beast had killed three more dogs in the Bladenboro section had not been substantiated.

The three dogs, owned by Spurgeon Little of nearby Pine Ridge, were reported missing Monday by Little's boy. Fores indicated that the dogs could well still be alive, and not victims of the blood-sucking 'vampire' monster.

The report that a giant 90-pound "leopard cat" was killed in the Bladenboro section, an animal that possibly fits early descriptions of the dog-killer of Bladenboro was spiked this morning by Chief Fores. He said the animal was actually-a small bob-cat run over near the Bladenboro school by Bunn Soles of. Tabor City.'

Bladenboro hunters, although not active and organized, still are trying to trap the beast, which at last count had mangled and killed seven dogs.

The theory that the mysterious Beast of Bladenboro might be a vicious, blood-hungry, escaped watchdog remains unproved. However, the opinion that the mysterious beast is a huge panther or mountain lion also remains pure speculation.

Chief Fores said this morning he "still couldn't convince himself the beast is a dog." He added however, that if it did prove to be a dog "he wouldn't be too surprised."

May Be Dog

Dr. N. G. Baird, a Lumberton veterinarian, advanced the theory that the Bladen killer could well be the cross-bred watch dog formerly owned by Zeke Stanton of Lumberton, who gave the dog away when it became blood-hungry and killed some neighborhood chickens. Stanton gave the dog to an Indian boy who lived near the Big Swamp.

Although seven dogs have definitely been established as victims of the 'vampire', none of the animals were examined by a veterinarian.

Chief Fores said, however, that four of the slain dogs were definitely drained of all blood.

He indicated that if the beast struck again he would call in a veterinarian to perform an autopsy in an effort to establish the identity of the killer.

Meanwhile, he said, he would sit tight until the beast, whatever it may prove to be, strikes again.

Lumberton, North Carolina, *Robesonian*, January 20, 1954

The Bladen Vampire has changed from its usual diet of dog blood to something better. He's now dining high on the hog, more specifically, the hog that once belonged to Berry Lewis of near Bladenboro, who lost the animal Monday night to the mysterious Beast of Bladenboro. Here J. C. (Tatter) Shaw, who runs a service station in Bladenboro, holds up all that remains of the pig that was killed and devoured on the Lewis farm. Although no one saw the animal that ate up the 60-pound pig, tracks around the remains pointed to the 'vampire' killer that terrorized Bladenboro two weeks ago. All that was left of the pig was the two hams, hung together by skin. All the animals' bones had been chewed off clean, evidence of a powerful killer. Mr. Lewis, whose farm fringes Big Swamp along the Bladen-Robeson County line, said he thought whatever killed his pig was a cat, but didn't say it was the 'vampire.' He said he didn't think it could have possibly have been a dog that did the killing, squashing one theory that a wild dog could possibly be responsible for the Bladen dog slayings. Almost any type animal could survive the Big Swamp, he pointed out, and he was going along with the belief the pig was killed by a huge cat. The pig was the first animal he had lost since the vampire scare arose. Tatter Shaw, who said he had followed the Bladen killings closely, also believes the dogs around Bladenboro were killed by some type of cat. "Whatever it is," he said, "it must be mighty powerful to chew up that pig's bones like that."

'BEAST OF BLADENBORO' TYPE KILLER STRIKES IN ROBESON
Lumberton, North Carolina, *Robesonian*, December 15, 1954

A mysterious: animal struck last night within shouting distance of Robeson Memorial hospital, killing five medium-sized-pigs and three chickens, on a K. M. Biggs tenant farm, giving rise to the belief that the famed 'Beast of Bladenboro' is once again on the prowl.

The pigs, averaging around 60-75 pounds, were discovered this morning by Marvin McLamb, strewn around a sty approximately 10 by 15 feet in area. The animals were mutilated and four had crushed skulls. Three of the pigs had legs torn apart from their bodies.

Strangely, enough, no blood was evident, indicating, the killer employed the same blood-sucking traits as the Bladenboro beast.

Tracks were evident many places on the farm, and, according to McLamb, led off to a small bay area just behind Meadowbrook cemetery. McLamb said he started to follow in that direction last night when his dog awakened him around midnight and started trailing in that direction.

At that time, Marvin and his father, Perry McLamb, got out of bed when chickens and the barking of the dog caused some concern.

The younger McLamb said he followed his dog a short distance, but the dog stopped. When McLamb came up to the point the dog had trailed, the dog was whimpering. It then bolted for the house. McLamb said he too returned to the house.

At that time nothing had been killed.

This morning a chicken was found on the back porch its head chewed off. Another was found a short way from the house

The pigs were discovered dead at feeding time this morning.

The many tracks around the house and pig pen indicate the animal is as big, or bigger, than the 'Beast of Bladenboro' which last year terrorized Bladen County.

McLamb couldn't identify the tracks, but said, "I don't think they are dog tracks." Worth Pittman of 7th Street Road, who is working nearby at a corn mill, said also that he didn't believe they were tracks made by a dog.

The spoor was four inches from heel to toe. Many could be observed around the farm.

The pigs were enclosed in a regular board type sty with the board nailed close enough together to prevent the passage of an animal indicated by the size of the track. The fence was approximately four feet high.

During the reign of the infamous 'Beast of Bladenboro', many reports of the beast's appearance in Robeson arose. Some killings were reported in lower Robeson around the Fairmont area, and a few reports of chicken killings were reported in Lumberton.

At that time it was believed that the beast was either of two things: (1) a wild dog, or (2) a cat-type animal that came out of Big Swamp near the Robeson-Bladen county line.

A few bobcats were killed at that time and eventually the scare passed without the animal being definitely identified.

McLamb has reported the incident to the Robeson Sheriff's Department. Officers were to investigate this morning.

'BEAST OF BLADENBORO' SCARE ENDS IN DEATH OF LARGE DOG

Lumberton, North Carolina, *Robesonian*, December 16, 1954

A short lived 'Beast of Bladenboro' scare in Robeson County apparently ended yesterday afternoon with the death of a large stray mongrel dog which was killed about one-half mile from the scene where five hogs were mysteriously mutilated and killed Tuesday night.

County Dog Warden Carol Freeman, called in to investigate the strange killing of five hogs on a K. M. Biggs tenant farm Tuesday night, reported later Tuesday that the dog found in the neighborhood 'most probably" was responsible for killing the shoats.

The dog's owner could not immediately be located since the animal carried no identification or inoculation tags. It weighed 63 pounds and apparently was a cross between shepherd and collie.

The animal's feet were not compared to 4-inch spoors found around the farm, but Freeman said he thought they were large enough to be the same. Tracks leading away from the farm were in the direction the dog was slain.

Marvin McLamb, who lives on the farm where the pigs were killed, also reported the death of three chickens. He said they were taken from roost in a tree and then killed. The Dog Warden didn't attempt to explain how a dog could be responsible.

Freeman, explaining killings, said: "It's entirely possible for a dog, or pack of dogs to do this type thing, although it doesn't happen often."

The report of the hogs being killed reminded people of this section of the time a killer 'beast' roamed the Bladen County area in search for animal prey. There was one report that the 'beast' (dog or bobcat) attempted to attack a woman.

In each case of an animal death the Bladen section, there was also the report that the mysterious killer had "sucked blood."

This type animal most probably was responsible for the death of the hogs Tuesday since no blood was observed around the small sty in which they were kept, although all the pigs were badly mutilated and chewed.

Freeman said a blood-thirsty dog would lap up warm blood.

County Dog Wardens are of the belief a pack of dogs was responsible for killing the hogs. In the general area where the stray was killed, two other dogs belonging to homes across town were located.

Warden Raymond Kinlaw predicts there will be further cases of such killings as long as some people feed raw meat to their dogs. "It definitely would cause a dog to become blood-thirsty," commented Kinlaw.

Asked whether a dog of the type killed could mutilate hogs as badly as the ones on McLamb's place, Freeman said, "I think so."

He pointed to the dead dog's 'holding' teeth, better known as fangs.

They were approximately three-quarters of an inch in length.

The dog was exceptionally thin, although big-bodied and broad-chested.

STRANGE ANIMAL REPORTED BY SCHOOL CHILDREN AND PRINCIPAL
Lumberton, North Carolina, *Robesonian*, December 17, 1954

An elementary school principal and one of his teachers said yesterday afternoon a mysterious animal which "looked like a cross between a big monkey and a dog" was chased off the Panthersford school yard Wednesday by a group of students and a janitor armed with an axe.

The principal was H. E. Williams and the teacher was Christine McMillan.

The animal was first spotted by a group of youngsters during a chapel program and, according to the teacher, "at least 25 of the kids ran outdoors to see what it was."

Panthersford is a Negro elementary school located between Lumberton and Red Springs on highway 211. The school is approximately 10 miles from the site of the recent mutilation of five young hogs on a farm near Lumberton by what is thought to have been a big dog.

Attracted By Yell

The teacher said she was first attracted to the animal's presence in the school yard by student Beula May Sanders' yell: "Look, a monkey!" There was an immediate scramble by the students to get a closer look.

The time was approximately 3 o'clock in the afternoon, according to Miss McMillan.

When the school's janitor, Clifton Williams, saw the commotion and spied the animal, he picked up an axe.

Principal Williams said he didn't see the action take place, but from what he could learn from spectators, it didn't appear the animal showed any fear of the janitor.

"It would run when they ran," he explained, "but when they (the children and janitor) stopped running, it stopped, also." He said he understood the janitor got close enough to the strange beast to swing the axe but the animal dodged the blow and ran off.

Miss McMillan said she saw the animal plainly at a distance of about 50 feet. "It looked just like a big monkey and a dog, I don't know which. It didn't run, it loped."

She said the animal appeared to be colored "a dark, reddish brown." It was about "three feet" high and had an "extremely long tail."

Supt. Williams said he went out to examine the animal's tracks and found them unlike any he had ever seen. "But they definitely were not dog tracks," he exclaimed. He added, "and I certainly know what a dog track looks like."

Description Differs

The children, said Williams, differed on a concrete description of the animal but agreed it wasn't a dog. Some said it was a monkey.

A beast scare in Robeson began Tuesday with the strange death of five medium sized hogs which were ripped and torn apart in a pigpen on a K. M. Biggs farm near Robeson Memorial hospital.

A 65-pound mongrel dog was killed in the vicinity the following day by Dog Warden Carroll Freeman and the case was closed.

The report of the monkey-like animal is reminiscent of the recent 'Beast of Bladenboro' scare which produced many varied and weird reports of strange beasts roaming the country-side seeking prey.

The infamous 'Bladenboro Beast' was never positively identified but it was presumed to be either a cat-like swamp animal or a blood-thirsty dog.

Whatever the animal, or animals, it apparently doesn't like to fool around with humans. It operates under the cover of darkness and operates silently.

A woman emailed me (Arment 2000b) regarding her sighting of a black panther in 1959 or 1960. She was a young teenager, and was playing with her sister on a dirt road near her house, south of Concord, North Carolina. They saw a large black "slick haired" feline from about 50 feet away. She thought it was about three times larger than a bobcat.

UPPER RANDOLPH BOASTS A 'BEAST'
High Point, North Carolina, *Enterprise*, August 24, 1965

An insurance salesman claimed to see a long black feline crossing Route 1 in the Trinity area. He caught it in his headlights, and described it as: "It looked like a black panther. It was about four feet long with big legs. His eyes looked like they were green." It turned to face him as he approached. Dogs had been killed recently in the area by an unknown animal.

IT HAPPENED IN NORTH CAROLINA, BY WILLIAM A. SHIRES
Gastonia, North Carolina, *Gazette*, February 7, 1966

Southern Railway trainmen on the line south of Lexington saw a black panther along the tracks. Some residents say it has prowled Davidson County for at least ten years.

MYSTERIOUS PROWLING ANIMAL KEEPS COUNTY RESIDENTS EDGY
Lexington, North Carolina, *Dispatch*, June 10, 1966

Miss Annie Musgrave on Rock Crusher Road and other nearby residents reported seeing a large black animal they thought was a panther. "It's jet black and glossy ... with a rounded head and pointed ears like a cat. It has a very long tail which would drag on the ground if it didn't curl upward. It keeps its head down low—not like a dog—and it twitches its tail. It seems totally unafraid, like it might have been in captivity."

MYSTERY BEAST KILLS HOG
Lexington, North Carolina, *Dispatch*, January 29, 1969

A 175-pound hog was killed on a Randolph County farm by an unidentified animal. The strange animal was described as jet black, "a head and body resembling a German Shepherd dog," with a long tail that curled up over the back. The hog was "found with claw marks on its back and a gash on its neck behind the ear where apparently the killer had bitten." No blood could be found on the ground, and the flesh wasn't eaten, leading them to conclude the animal "had only sucked the blood of its victim." Earlier in the month, half a dozen hogs had been killed on a farm near Franklinville in the same way. The animal there was described as "black and somewhat larger than a fox, with a tail like a fox and a long snout. Claw marks were found on the backs of the dead hogs, and the animal had cut their throats with its teeth and drunk their blood but had not eaten the flesh." In another incident near Millboro, a calf was killed and partially eaten.

Mystery Beast Strikes Again
Lexington, North Carolina, *Dispatch*, February 13, 1969

The Randolph County mystery animal killed another hog, this one weighing 100 pounds, in the same blood-sucking manner. A local woman reported seeing a strange animal in the vicinity the next day: "Red with a little black, and having a long nose which was flat on the end, and a bushy tail. She said it ran with its hind quarters held higher than the front."

Mystery Beast Killed in Randolph
Lexington, North Carolina, *Dispatch*, December 18, 1969

A Randolph County man, Miles Moffitt, shot and killed what appeared to be a jaguarundi as it left his hen house. It was healthy, and measured 41 inches from nose to tail. It had a slender build, a small head with rounded ears, and kinky gray fur sprinkled with lighter gray. One local avid hunter suggested that jaguarundis had "slowly and unnoticeably increased its range and reached this area," "flourishing in a new area unknown to local people ... because it is timid and seldom seen." Moffitt planned to have the cat mounted. The articles accompanying photo (not reproducible, unfortunately) of the animal does show a profile consistent with a jaguarundi.

VAMPIRE KILLER STRIKES AGAIN
Gastonia, North Carolina, *Gazette*, May 16, 1971

A "blood-sucking" monster was blamed for at least two livestock attacks. In one case, a cow was found hanging on a barbed-wire fence, "dead with a round hole in its heart and all its blood gone." Seven months prior, a horse was found in a similar position, with "a round hole where the heart would be. The heart was gone and something had sucked all the blood out." A "big black bobcat" had been seen in the area. One policeman said: "A logical explanation will be found ... but until it is I can certainly see where people in the area might be alarmed."

Newton (2005) noted that on January 27, 2004, a witness, Denise Williams, videotaped a black panther behind her Asheboro residence. Police and zoo employees were unable to locate the animal.

PANTHER AT THE BEACH? OFFICIAL DOUBTS IT
Wilmington, North Carolina, *Star News*, August 17, 2007

A Virginia man vacationing at Sunset Beach claimed he saw a black panther in the brush. The local police chief said the only other local report she had heard was of a sighting on Bird Island five years prior. A NC wildlife officer didn't consider the report likely.

MYSTERY PREDATOR RETURNS, KILLS RURAL ANIMALS
High Point, North Carolina, *WGHP Fox 8 News*, November 28, 2007

A Davidson County goat breeder lost 13 goats in a month, 4 in a single night, to an unknown predator. NC Zoo staff suggested an exotic feline or a wolf hybrid may have been responsible.

Mystery Animal Roaming Iredell

Statesville, North Carolina, *Record & Landmark*, May 6, 2009

A couple driving "on Wilkesboro Highway near Taylor Springs Road and Rupard Road" saw a large light gray feline loping across the road in front of them in northern Iredell County. It was six feet in length, including its long curling tail.

Where's the Lion?

Burlington, North Carolina, *Times News*, June 13, 2009

A 76-year-old Pleasant Grove farmer reported seeing (three times in a short period of time) a "female lion" on his 175-acre farm. It was brown with a long dragging tail, and appeared malnourished. Another individual on a neighboring farm phoned in a sighting to authorities also. Why it was described as a "female lion," rather than a mountain lion, was not discussed.

NORTHWEST TERRITORIES

THE VALLEY WITHOUT A HEAD
By Frank Graves, as told to Ivan T. Sanderson (n.d.)

Starting when I was a kid in junior high, I became ever more bugged by the story of what some writer way back in the 30s named "Headless Valley." My first introduction to this crazy story was in some pulp mag and what impressed me was that it was listed as a fact article and just a fiction tale. It avoided giving the exact location of this horrifying place except to say that it was in western Canada and way up by the Arctic circle. Neither my parents, who were very well read, nor my teachers, nor any library in Philadelphia where I lived, could tell me anything about the place. But over the years I kept stumbling across stories about it.

I am a mechanic by trade, specializing in auto and truck testing, but in 1962 I took on a second job as I wanted to put aside enough cash to take a year off to travel—anywhere as long as I could see for myself what life is all about outside a modern city. This job was as a stock-boy in the publishing house of Chilton Books, in downtown Philly. One of the first books I had to tote around was your Abominable Snowmen: Legend Come to Life, which I read while baby-sitting with the xerox machine. The subject had always interested me anyway but I found in it the first exact reference to the location of this valley, which you gave as the Nahanni in the Canadian Northwest Territories.

Since you were an editor at Chilton at the time, you will remember that five of us came to you to seek advice on this year-off trip that we were planning; and of course I led off with Headless Valley. [Note.—To make a really rather long story as proverbially short as possible, let me just say that all Frank's companions dropped out for one valid reason or another—as almost invariably happens in plans such as his—but he joined up with another gang I had been helping, out in Minnesota, led by one Michael Eliseuson and who were also aiming for the Nahanni. Frank took over the transport and got them up to the Mackenzie River in a fourth-hand school bus; and thence, by a specially built boat, up the Liard River and then on into the Nahanni. He also got them all out and in one piece, which was quite a feat in itself! I.T.S.]

The stories about this "Headless Valley" were really weird. It was the number one legend of the northlands, and had as its background, stories of tropical growth, hot springs, headhunting mountain men, caves, pre-historic monsters, wailing winds and lost gold mines. Actual fact certifies the hot springs, the wailing winds, and some persons or animals who delight in lopping off prospectors' heads. As for the pre-historic monsters, Indians have returned from the Nahanni country with fairly accurate drawings of mammoths burned on raw hide. The more recent history began some sixty years ago when the two MacLeod brothers of Fort Simpson were found dead in the valley, reportedly decapitated. Even then the Indians shunned the place because of its "mammoth grizzlies" and "evil spirits wailing in the canyons."

Canadian police records show that Joe Mulholland of Minnesota, Bill Espler of Winnipeg, Phil Powers and the MacLeod brothers of Ft. Simpson, Martin Jorgenson, Yukon Fischer, Annie LaFerte, Ed. O'Brien, Edwin Hall, Andy Hays, an unidentified prospector and Ernest Savard had perished in the strange valley since 1910. In 1945 the body of Savard was found in his sleeping bag, head nearly severed from his shoulders. Savard had previously brought rich ore samples out of the valley. In 1946 prospector John Patterson disappeared in the valley. His partner, Frank Henderson, was to have met him there, but never found him. Two fellows told an even wilder one back in 1948. They said that some years before they had been chased into a cave by a Mastodon though they insisted it had enormous spirally curved tusks and very long dark hair so that it sounded much more like a Mammoth! I remember thinking when I read that one that people have great imaginations; I was to learn just how lousy imagination can be when compared to reality.

I must explain that my purpose in going on that first trip was not just for adventure or to 'explore.' I had always planned to have a serious-minded objective, and I had looked around for something I could tackle that did not entail hauling great piles of equipment in with me before I knew what sort of environment I was going to encounter; especially as I had never been in any wilderness area before. Perhaps you'll laugh when I tell you that I had always had an interest in plant life, so I went to Dr. William C. Steere, Director of the New York Botanical Gardens, to seek advice. I did not know, when I did, that he was the leading expert on arctic and subarctic plants and especially the little ones—that you call "lowly plants"—such as lichens, fungi, mosses, and what are called liverworts. These had always seemed to me to be "fun," so I planned to collect them. [Frank brought back a rather fine collection that is still being classified and described under Dr. Steere's direction. I.T.S.]

Only later did I come to realize that it was this somewhat unglamorous pursuit that opened the local doors to me. You see, there are not only indigenous Indians in the area but a number of long-resident white families as well; and transients or comparative transients like resident priests, Hudson's Bay Company agents, trappers, prospectors, and even some geologists and surveyors. Most of these prove to be a confounded nuisance to the authorities and a pest to the residents, while the Indians take their traditionally dim view of all palefaces. Then also, there have been several so-called "expeditions" to the valley, starting back in the early 30s; and almost all of these have been pretty wild. The earlier ones came looking for lost gold mines and so on; the latter-day ones arrived announcing their intention of catching a Sasquatch, or Bigfoot, or what those who don't know what they are talking about call a Yeti or "Abominable Snowman"! None of these intrepid explorers and hunters had the foggiest notion what the terrain was like, what equipment to take, or even how to get there. Most of them failed even to get into the lower end of the valley; and almost all of those who did, had to be hauled out physically. Even the old sourdoughs and surveyors met such rough going that they often had to give up or stay and (yes, literally) be decapitated. And if the records are true, this is a fact. When the locals opened up to me a bit, they told me really hilarious stories about some of these unprofessional 'northmen'; how they had got plain scared of the world up this river, and how they had just given up and crawled back as best they could to the settlement at South Nahanni, where the North and South Nahanni rivers junction with the Liard that runs east to the Mackenzie. (Incidentally, the South Nahanni River flows south from the north!)

This is the place where there are numerous hot springs that are ice-free all winter and in which the Amerinds bathe regularly. Like many volcanic thermal springs they are encrusted with crystalline and amorphous deposits of sulphur, but they also support most marvelous beds and rim-fringes of algae and some extraordinary fungus growths. It was water piped from these hot springs to a greenhouse built by one of the Catholic Missions years ago that gave rise to the story—almost a legend—of bananas growing up there, because somebody did ship some banana-stools up from Florida and they thrived in the intensely hot, though short, summer sun, and then through the long winter under artificial light and in the high temperature and moisture content of the air induced by these hot springs.

At first I took the stories of the old-timers about these scared trappers and prospectors as being nothing much more than the sort of snide accounts that permanent residents of far out of the way places relate about the behavior of outsiders. But this attitude of mine seems to have gotten under the skins of the locals, and the resident Amerinds, and even those other Indians who wander in from the outlands from time to time. It then dawned on me that I was giving offense by not believing what they told me, so I sort of indicated that I did want to believe what they said but that I had thought they had just been pulling my leg as a greenhorn from outside. That did it.

They led me to a number of places where I was shown burned-out shacks and, what impressed me most, was that I personally found the remains of some very old rifles and other things of the utmost value to such people as hunters and prospectors in these overgrown camps. Later, I talked with members of the Canadian Mounties who now patrol the NWT regularly by air, and they confirmed the long string of disappearances of experienced prospectors and trappers in this area. Also, they told me that the decapitations reported were actually on the police blotters!

As a result of all of this I was led to another discovery: namely, that there are a heck of a lot of Indian Bands, or seemingly whole small tribes, up north of the place called South Nahanni. This is a positively vast country composed of three roughly parallel mountain ranges running from northwest to southwest, called (from the west) the Selwyn Mountains, the Backbone Range, and the Canyon Range of the Mackenzie Mountains complex. The Nahanni runs down from the north between the two former and originates somewhere around the southern slopes of the 7500-ft high mountain called Mt. Christie, which is at 64° North. Nobody has ever reached the head of this river on land and because of the heavily forested and terrifyingly steep gorges that chop this country up, no clear map has become available from aerial surveying. And, as a matter of actual fact, the most part of it has never been surveyed. Thus, the Nahanni is "headless" in more ways than one.

The Indians and their (presumed) "relatives" who live in and around South Nahanni, keep just turning up around the Liard River. Nobody can talk to the "outlanders," but they don't show any surprise at passing aircraft. They are intrigued by motorboats, and they offer anything, even their young women, in trade for metal knives. They don't seem to possess anything store-bought. The Canadian authorities don't know who they are or where they come from. And this brought up another fascinating 'discover.'

Several people—and notably a highly educated white man who has lived in the valley most of his life—remarked to me quite casually one day that enormous airplanes quite often came down from the north and sometimes fly so low in good weather that he could read their large markings even without his binoculars. Besides numbers, they bear "names" or identifications in the Cyrillic alphabet. Some of these he had tried to copy down, and on showing them to another educated old-timer who had also seen these planes for some years, he learned that a priest from Ontario had had them translated, and that they were Russian, and standard markings for certain series of overfly planes known to the Canadian authorities. I have asked around, but I never heard of such planes being spotted anywhere else; so why are the Russians so all-fired interested in this crazy valley? And crazy it is, and in all kinds of ways as I later found out.

First off, there is a really great waterfall about a hundred miles up the river, called Virginia Falls. This is the stopping point for the greenhorn invaders, and the starting point for the real explorers. Yet, the trip up the river even to that point is rough going. We were luck in that, when our boat sank in one set of rapids, we saved all ourselves and most of our gear. Most greenhorns lose one or the other, or both. I got above the falls because I wanted to collect those plants, and also because I had heard of a cave that one old fellow had told me somebody else had penetrated for no less than nine miles! I don't know if this is true, but an Indian showed me the entrance to several caves, and any one of them could have gone on into those mountains forever. I had no lights or proper caving equipment so I did not try my luck. And it was at about this point that I got sidetracked once again.

One of the Indians who was sort of guiding me, but only because he and his family were headed the way I wanted to go, became very much interested in my search for funny little plants, and then proceeded not only to tell me about but led me to and pointed out what seemed to be an endless lot of plants that his people use for medical purposes. I had heard of the subarctic peoples having discovered things like aspirin in the bark of certain willows centuries ago, and I had heard an awful lot of tales about our North American Indians being old-time herbalists, but I was quite stunned by the number of such medicinal plants that this man showed me, and by the equally endless list of the alleged uses that he said his people made of them. And then another thing happened.

I needed some game for food, and the whole area literally crawls with both game birds and mammals. However, once again, a greenhorn can wander about in the woods for months without seeing much more than an out-of-range grouse, so I asked another of my Indian companions to lead me to some game as quickly as possible, and I told him I would just follow him and shut up. I even gave him my gun so that we might get some meat as quickly as possible. We made several successful forays up side valleys and canyons during the days while we moved up river, and we got our food quite fast. But then one misty day we set off up a canyon that the Indian said he did not know personally but which was "not lucky." And in truth we did not spot a living thing in three hours; so we started back down to the river. Then suddenly my pal stopped, and pointed down at the soft wet ground in a little clearing and actually gave one of those grunts

that movie-makers love to have their "Red Indians" make. He was a
bit rattled and so was I, for there, most clearly marked in the mud,
were three footprints of what appeared to be a barefoot man who
would have had to take a shoe with an internal measurement of at
least sixteen inches! My friend gave this thing a name, but I never
really did catch up with that as we went down that valley at no dog-
trot, I can tell you.

Later, when the locals had finally decided that I was neither a
nut nor some kind of government agent, they really opened up on
this business of the "old-time hairy woodsmen." They'd never heard
of California's so-called Bigfoot, or B.C.'s Sasquatch, or even of the
Yeti or Abominable Snowmen of Asia; and they were quite amazed
and not a little skeptical when I told them what other people had
said about these. In fact, they obviously didn't believe it any more
than I originally had believed their stories about decapitated prospectors in their country. To them—and this went for the white-men
even more than the reticent Indians—their "big-footed-ones" were
just perfectly natural, normal people, but living a more "primitive"
life than they did. Their whole attitude was, very simply, "What do
you mean? Fakes?"

I didn't want to bring this up, but you know it was in a way this
business of what you call ABSMs that really started me on that trip;
and let me tell you [And Frank got quite hot about this. Ed.] I never
believed one single word of your book, and least of all about the
Canadian NWT. But that's the way it happened, and I have to tell
you this because it makes what comes next a bit easier.

While out hunting for food one morning with an Indian friend
who was a pretty fair woodsman, I ran into another real horror. These
were definitely bear tracks but if I told you their size—and I measured them with my collecting tape—you would just laugh at me. They
were bear tracks all right, and the Indian said that they were made
by giant brown bears that, as far as I could make out from his pacing
things off between trees, would make any Kodiak Brown look like a
Black Bear pup.

Then, three days later we were out not too far from the river,
and my friend went ahead with one of his dogs that he had brought
along, since we wanted only a few birds to eat. They saw something
over a bank down in a thick patch of pine woods. I waited up the rise

in open stuff with bushes. This Indian was a pretty good tracker and moved without making a sound, but his dog was not a 'hunter.' So, when I heard a noise and saw some brush moving about at the edge of the trees, I thought it was the dog coming back, and I did not raise my gun.

But then an enormous white thing that I at first thought must be a Polar bear just sort of wandered out of the trees. It wasn't a bear; it looked more like a gigantic dog. It stood straight up on rather long legs, more like a dog or a wolf. I had seen plenty of wolves and some of them are enormous enough up there; but this thing was twenty times the size of any wolf I had ever heard of. By a sort of reflex action I fired at it—and it was less than twenty paces away and only partly screened by little bushes. I hit it with two barrels of ball-shot. It didn't even jump, but turned away from me, and just walked back into the forest. I reloaded and fired again, and I know I hit it in the rear, but it just kept on walking. Shortly afterwards, my Indian friend bobbed up, asking what I had got. I didn't know what to say for a bit but, when I told him, we did another of our famous disappearing acts, and this time we loaded the boats and pushed off up river—real fast.

I tell you, weirdies and mysteries just keep coming in that fabulous country. Super-giant bear tracks, and wolves the size of giant bears were bad enough but—and don't get me wrong—if the ravens and eagles I saw every day were really the size I estimated they were, we'd have had every zoo keeper in the world in an uproar. But then, I didn't shoot any, and so I never measured one of either. And you know how silly one can be about estimating size at a distance.

Most unfortunately I never got far enough to see what is to me the most fantastic story of all about this "headless valley." This is the "frozen lake." Now, they have snow and ice up there all their long winter, and up in the mountains it never melts; but, down in the valleys that far south, the rivers run free from June, and the lakes are completely ice-free. However, everybody there told me that there is a large lake up near the headwaters of the Nahanni that never melts. It is in a ten-mile-wide basin between steep mountains, and its surface is said to be absolutely smooth, bright blue, crystalline ice. But what is much more incredible—and it will remain incredible to me until I see it personally—is that it is alleged that the whole of that valley is permanently frozen! By this I mean that there is a dense forest of spruce all around it and up the slopes of the mountains on each side, and all the trees are clothed in pure ice, just like after an ice-storm down our way. But this ice never melts. One old Indian told me that he had been there, and that he had found little animals also encased in clear ice along the shores of the lake. What is this? And I ask the geologists. Could it be true; and, if so, how come? Of course, this bit may well be nothing more than a pipe-dream but, having seen that dog-thing, and those human-like foot-tracks, and the bear tracks and the caves, I'm not saying anything more for now.

I'm going back again—this year, and on my own. If other whitemen can live there the year round, and year after year, I can; even if I am a city boy. I've been there, and I know; and I've asked the friends I made up there for a job and I've told them I am going to stay all winter; and if I make the grade and don't get in their hair, I just might ask to be allowed to stay on. People are always going off to Africa, or South America, or the Antarctic. I don't get it. Here is this fabulous place right in our own backyard and it's full of all the damned 'adventure' anybody could ask—plus a lifetime of wondrous things to look into, and decent people, clean air, and a real chance to discover something worthwhile.

NOVA SCOTIA

Bruce Wright (1972) reported a few sightings of black panthers from Nova Scotia:

September 1, 1952: A three to four foot long black panther with a long tail, estimated weight from 80 to 100 pounds, stood alongside a road snarling at the witness for several minutes from about ten feet away. This was apparently in the Cobequid Hills. Wright stated that a similar animal was seen in the same area in December 1952 and February 1953.

OHIO

Ohio *Statesman*, February 13, 1846

Singular Wild Beast Killed.—Several hunters in Lorain county recently gave chase to an animal prowling about Elyria, and after a pursuit of seven days, succeeded in killing him in Huron county, about 50 miles distant. The chase was continued about 150 miles, and the Sentinel says, it is singular that in the whole distance the animal never crossed a fence. The animal is described as a big dork, brindle color, three feet high, six feet eight inches from the end of his nose to the end of his tail. Old hunters give it as their opinion that he must have been a stray inhabitant of some other region.— Cleveland *Herald*.

A TERRIBLE BEAST

Columbus, Georgia, *Ledger-Enquirer*, November 8, 1867

The Dayton *Journal* of the 29th has the following relative to a frightful looking beast, discovered among the denizens of Preble County, Ohio. When we saw the item we were looking for the announcement of the discovery of a gold mine, an oil well, or a yellow ochre bed:

The people in the northeast part of Preble County are just now having a sensation which is not at all pleasant. Rumors were current for several days last week that a wild animal of large size was prowling about in the woods between El Dorado and New Paris. It was alleged that he had been seen by several persons, who described him as about 2 feet high, and from 5 to 8 feet long. Whether the animal was a panther, catamount, or lynx, could not be determined from the description given, as no one who was near enough to see the varmint was disposed to remain long enough to make a very close inspection. One man cleared that coming home late he was followed by some large animal, and hurrying to his house and closing the door, the stranger reared up and looked in at the window, thus affording a full view of the head, which resembled that of a bull dog.

Another person while driving along the road about eight o'clock last Sunday morning, saw the creature in the woods about 50 yards off rearing up against a tree, and the brief inspection then made makes the length of the animal from nose to tip of tail no less than 8 feet. Of course the party in the buggy was not disposed to tarry long in such a neighborhood, and is not able to give a more definite description then is afforded by the apparent length of the beast. The tracks made by it, resembled those of a dog, but are larger and more deeply indented in the ground.

So well satisfied were the people of the truth of the reports of the animal and the locality named, that on Sunday last about five hundred men and boys, armed with rifles, were in the woods hunting for the unwelcome visitor. Up to Sunday night the hunters were unsuccessful in their search.

We understand that the schools in the section of country where the animal has been seen have been all temporarily suspended, parents being unwilling to expose their children to the danger of meeting it while going to or returning from school.

We have not learned that any depredations upon stock have been committed by this unauthenticated beast, which is satisfactory proof that the woods afford it abundant rations without trespassing upon the herds and flocks of the farmers.

ANOTHER WILD BEAST AT LARGE
Marion, Ohio, *Daily Star*, December 18, 1877

Ben Bigford reports another wild beast at large near Pucky Huddle. Sunday the woods in the vicinity was full of men armed with shotguns and accompanied with "Bird" dogs. Considerable shooting was done, but that of course was to frighten the wild beast from his or her (as the case may be) lair. Some large yellow animal was seen. Some say it was a lioness, others said it was a tiger. But most of them say it was the "yaller dorg" that Dumbie and Christian had such a controversy about.

The Leopard Captured and Killed
Marion, Ohio, *Daily Star*, December 3, 1877

Probably no one person can give an accurate and complete history of the great wild beast excitement that has of late pervaded this part of the State, and culminated last Saturday in the killing of the animal. Various rumors are afloat as to when, where and how it was first let loose on the country, and the report we today give to our patrons we believe is as near correct as can be compiled from the multiplicity of statements:

The leopard escaped last June, from a cage of animals near Mansfield, Ohio, while being shipped over the P. F. & C. R. R to some Geological Garden. Strange as it may seem, from the time of escape until within the past few weeks but little was heard of it. A few weeks since, it was seen in Union county by Malcolm Stamates, who got on a log to get a better view to shoot it, but it leaped upon him before he could fire and would have torn him to pieces had not his own unearthly yells frightened it away. Mr. Stamates says "it could jump like the devil, and was in size between an elephant and a cat."

The next man to encounter it was Joseph Cathell, and a few days later Albert Wallace, both of whom came near losing their lives. It was next attacked by some half dozen persons, who went out specially to hunt it down and when they found it, it sprang upon one of the number by the name of John Sterling and they would have been powerless to prevent it from tearing him to shreds had it not been for the assistance of the dogs they had with them. A young man by the name of Johnson in company with several others then had the next encounter with the animal, and Johnson was severely bitten on the shoulder and elbow, but got one shot at it after which it made a retreat. A number of other persons have caught glimpses of the beast at different times and places, but above named are the only ones injured by it, and none of the wounds have proved fatal.

Last Saturday Mr. Burnalson, an engineer on the C. C. C. & I. R. R., left his post for a bird hunt; he got on the train at Galion and rode to Gurley station, seven miles west of Marion, where he got off and started for the woods north of the railroad track; in a short time he got on a flock of quails and jumped up on a log to get a whack at them, when to his amazement, he saw a fearful large leopard only a few rods of looking directly at him and waving his tail. Mr. Engineer's hair stood on end as he backed gracefully out, forgot the quails and made for the station, and gave the alarm. Eight or ten men joined him and they started in pursuit of it, but the leopard retreated on their approach. Three of the party, Johnson, Kesler and a German boy, after following the leopard apparently in a circular direction for about 9 miles when they overtook the leopard and one of the men fired at it but missed his mark; the leopard sprang upon Johnson and had him down in a twinkling. Johnson yelled and screamed for dear life, and Kesler, though terribly frightened boldly went to the assistance of Johnson, his brother-in-law, and pounded the leopard over the back so hard with his gun that he broke the stock from the barrel. He then took the barrel and continued to beat the animal over the head and shoulders until Johnson was released. (The injuries of Johnson are not a bit serious.) The leopard releasing Johnson was about subdued, and sneaked off a short distance, being followed up by Kesler who continued to beat it furiously. It finally backed up to a tree and showed signs of defense against the dogs that had come to molest it. At this stage of the drama the German boy put two shots of small balls in the head of the animal and that was the end of the poor leopard.

This was about one o'clock in the afternoon, and the party being near the station put the carcass on the train and came to Marion, and went direct to the City Hall where a large and excited crowd soon congregated and were permitted to gaze upon the wild beast that had gained such a notoriety, free of charge by paying 10 cents admission fee. The party started for their home at Gurley station about nine o'clock but before leaving there were about four or five hundred Marionites took a peep at the beast.

The leopard is being exhibited in Richwood to-day. It was a fine large animal and thought to be very old as its teeth were quite dull. The party now think they could have captured the leopard alive if they had not been so excited, as it at times seemed quite tame, and did not show vicious fight but its leaping and roaring in the woods made it appear more frightful than it really was.

THE ANIMAL IS DEAD.
IT PROVES TO BE A LEOPARD.
ON EXHIBITION—HUNDREDS OF PEOPLE GO TO SEE IT.

Richwood, Ohio, *Gazette*, December 6, 1877

Word reached this place on Sunday morning last that the wild animal that played such sad havoc in this section a few weeks ago, had been killed in Marion County, near New Bloomington and that it had been on exhibition at Marion Saturday afternoon and evening and that it would be in Richwood on Monday. A great many people— doubted the story, cracked their jokes as usual, in regard to the matter, but on Monday the animal was brought here on the 11:30 train and was placed on exhibition at the Engine house hall. Hundreds of our people rushed there to see it. We entered the hall and found the animal, in the center of the hall, just as it had been killed, with the exception of the removal of the entrails, and we found it to be the largest leopard we ever saw, rigid in death, and it looked peaceable enough in that condition. Mr. Samuel Johnson, one of the captors, readily gave us the following information in regard to its capture: on Saturday morning last as Mr. Burnison, an engineer on the C. C. C. & I. R. R., left his train at New Bloomington for the purpose of hunting for quail, and while so engaged discovered the animal in a little thicket about one mile east of that place. Of course he did not attack the animal, but immediately struck out for town, where he gave the alarm. Quite a number of men, with dogs, started in pursuit, and after hunting some four hours, came upon the beast, the dogs having it bayed in some brush. Mr. Johnson being nearest to the beast, was at once attacked, by the animal leaping for his head, striking him in the face with one of its paws, cutting a deep, ugly gash, and catching with its mouth his left arm, between the shoulder and elbow, both went over together. At this time there were only three persons together: Samuel Johnson, Edward Kesler, and George Hagerman. As the animal and Johnson rolled over on the ground, one of the parties gave it a terrible blow on the back of the neck, when it let go of Johnson and the third party struck across the back with his gun, breaking off the stock; picking up the barrel he struck again, a terrible blow over the small of the back, it is supposed breaking the bone, as the animal sunk to the ground and tried to creep away. At this time a boy came running up, from the crowd that had just arrived, and holding the muzzle of a shotgun against the animal and fired a load of shot and slugs into its head. It was found to have had an old wound on the right side of its head immediately under the eye, from a rifle ball from which it had lost sight of that member. This is supposed to be the one given it by Elias Johnson on the first night of its discovery and encounter at Summersville, and in regard to it being the same animal there is not the slightest doubt.

As the animal is now dead, and it was a dangerous and ferocious one to be at large, all conflicting stories as to its species ought to end. We confess we were mistaken last week in stating it was a lioness, but from the description given us, although as varying as the winds of Heaven, by the parties who had the encounter with it, we could not call it anything else. But is this to be wondered at, when even now, men who have seen it, claim it is a tiger. However we should have credit for the assumption or theory that it had left this part of the country, when every day men were reporting that they had seen it, found its nesting place, &c. There is proof unmistakable that the animal had been at the place where it was killed for nearly two weeks.

Johnson, Kesler and Hagerman will exhibit the animal as long as they can, and will then have it stuffed. They left here on Tuesday evening for Marysville, where they also placed it on exhibition.

Newton (2005) noted the 1890 sighting of a black panther from Lamartine.

THE MADISON TOWNSHIP TERROR
NOW AT FOSTORIA—A FULL DESCRIPTION OF THE "CRITTER"—
ITS RAVAGES IN GRAVEYARDS
Fort Wayne, Indiana, *Daily Gazette*, January 25, 1884

It will be remembered that the *Gazette* some weeks ago, mentioned the presence in Madison township of an animal then supposed to be a panther or catamount and its chase by an armed band of citizens. The animal has worked its way into Ohio and is now in Perry township, Wood county, near Fostoria on the line of the Nickel Plate.

James Bates, an old hunter and a gentleman, writes to Herman Krohne of this city, giving this description of the novel graveyard ghoul: "Its neck and breast are white, and the rest of the body black; its front tracks are about eight inches long and three wide, making tracks in the snow with its claws about twice the length of a man's finger; the tracks made by the hind feet are nearly round, and about the size of a large dog's, except the claws, which are longer and sharper. The animal is about three feet long and eighteen inches high. It burrows into the ground in the graveyard, and penetrating the coffins therein contained, devours the contents thereof. It travels with much rapidity and all attempts thus far to kill it have proved futile. The gentleman who last saw the animal says it was in the middle of the road, having gone from a farm by literally tearing the fences to pieces. His dog gave chase to the animal, but soon returned scared almost to death. The people living in the vicinity have frequently heard strange noises, which are now supposed to have emanated from this peculiar, unnamed, unknown beast. The animal is said to be slowly working its way toward Toledo."

The Graveyard Ghoul
The Nondescript Animal Caged at Fosteria—
An Interview With a Citizen of That Place—
W. W. Cole Secures the Creature for Exhibition.
Fort Wayne, Indiana, *Daily Gazette*, March 7, 1884

Mr. E. P. Nichols, of Fostoria, O., is in the city, and last night a *Gazette* reporter met him. Mr. Nichols informed our person that the terrible animal that has been ravaging graveyards in Allen county and along the line of the Nickel Plate as far as Fostoria, O., was captured last Saturday five miles north of that place.

J. W. Branderberry, a young farmer, set a wolf-trap for the creature which has an appetite for decaying human flesh, and burrowing into the graves has maybe licked its chops over the remains of the ancestors of many citizens of this and adjoining counties. Saturday morning Branderberry found the four-footed ghoul in the trap, struggling to be free. Just as the farmer came up the beast extricated itself from the sharp steel jaws, leaving a portion of its right foot, and hobbled off with unexpected agility. Mr. Branderberry, however, brought it down by a well directed shot from his rifle, penetrating the animal's head, but not killing it. Mr. Branderberry secured the strange thing and brought it to Fostoria, where it is now in a cage and being exhibited to thousands of people. Mr. Nichols learned from Branderberry that the animal, while in the wolf-trap, burrowed into the ground with its free fore claws and was nearly covered over when he came upon it.

The ghoul is about the size of a large terrier, and resembles in form a badger. It is of a reddish-brown color, and its claws are very long and sharp. Its genus has not been yet determined. It refused to eat until yesterday when it devoured a ground-hog thrown in the cage. Al Richards, the general agent and brother-in-law of W. W. Cole, the showman, lives in Fostoria, and is negotiating with Branderberry for the purchase of the prize.

"The American Badger in Ohio.—Last summer there appeared in our local papers several sensational articles about a so-called "ghoul," which was robbing the graves and devouring the dead in some village or villages in Wood county.

"I did not, of course, pay much attention to these wonderful stories; but some time after noticing them I saw outside a beer saloon on one of our principal streets a placard to the effect that the "Wood county ghoul or grave-robber," was to be seen within. The man inside informed me that many people had been to see the strange creature, but no one knew exactly what it was, but the best informed pronounced it a nondescript animal hitherto unknown. A single glance sufficed to show that it was a find and unusually large specimen of the American badger, *Taxidea americana*.

"The poor beast had been caught in a large steel bear-trap, and had lost one of his fore feet. He seemed quite docile and contented except when the keeper stirred him up with a stout club, when he snarled viciously, and displayed enormous strength.

"I thought that perhaps the capture of a badger in Ohio, where it has long been believed to be extinct, might be a matter of some interest. His burrowing among the soft earth of the new-made graves, is in accordance with its well-known subterranean habits; and the story of its devouring the dead is in strict accordance with the mental characteristics of a rural population, though of course destitute of any foundation in fact. The rarity of the animal hereabouts is evident enough from the failure of the numerous persons, hunters and others, who saw it to identify it.—J. H. Pooley, M.D., Toledo, Ohio, Oct. 22, 1884." (Pooley 1884)

THE CAT DESTROYER OF SCIOTOVILLE
Bryson, Quebec, *Equity*, May 17, 1888

The strange animal which for some time has proved a menace to the belated travelers, promenaders, and cats up at Sciotoville, Ohio, has evidently either exhausted the feline creation up there or concluded that food from other sources would prove more relishable, for he has ruthlessly invaded the precincts of Portsmouth and commenced a war of extermination upon the city felines. The strange creature was seen by Dr. James P. Bing, a gentleman of great veracity, on the back porch of his residence, and the doctor relates a thrilling experience with his strange visitor. The doctor was awakened about 2:30 o'clock at night by his wife who protested that something was wrong about the premises, she having reached that conclusion by the furious barking of a dog in the yard between the porch and the alleyway. The doctor arose and parting the curtains of the window, was astonished to see a strange wild animal occupying the porch not four feet from him. The dog had the animal at bay, and while he seemed afraid to attack, it was evident that the other animal had no desire to force the fight. The electric light shone dimly through the yard, but still it was strong enough for the doctor to get a good look at his unknown visitor. It was of a dark brown color, and of a strange species, without doubt. The face was small and the nose pointed, somewhat like the head of a fox. It was too large for a catamount and too small for a panther, and unlike either in color. When it raised itself to a sitting posture, with nose erect, it seemed to be about three and a half feet high. The doctor on turning on the light in the room so it shone full upon the animal and dog, unfortunately caused the latter to withdraw, when the strange animal quickly disappeared in the direction of the hills to the east of the city. There is no doubt but that this is the identical beast that has been terrorizing the people of Sciotoville.

THIS MAY BE A LIE
LEOPARD HUNTERS FIND MANGLED REMAINS OF THREE HOGS
Marion, Ohio, *Star*, November 26, 1895

The following was sent out to the city papers from Upper Sandusky, under date of November 25.

Yesterday being a bright, pleasant day, a great number of farmers, armed with guns and all other conceivable weapons, turned out
in quest of the escaped leopard, which is terrorizing the denizens of
southern Wyandot and northern Marion counties. Every bit of woods
in the vicinity of LaRue was scoured. Dogs were urged through hollow logs and into suspicious holes and caves and all sorts of underbrush were thoroughly and scrupulously searched. The only evidence
of the leopard was the discovery of three dead hogs, which, according to their mangled condition, were killed by the leopard.

Two hunters, Frank Gear and Kay Demarest, who have just returned from a week's hunt between LaRue and Marseilles, report
having seen the animal Saturday. They claim that it attacked their
dog and would have killed the canine had they not scared the animal
away by shooting at it. They believe they wounded it, for when the
gun was discharged the animal gave a terrific howl, and when gone
they found blood on the ground and by further blood drippings traced
it through the woods and across the Kenton highway fully a mile.

A farmer in town this morning claimed that the leopard is again
in the county, having been seen twice yesterday in Marseilles township.

Not A Myth.

Miners Insist That They Have Seen Wild Animals.

Salem, Ohio, *Daily News*, January 25, 1896

The Mapleton correspondent of the Canton *Record* furnishes
another version to the story of a wild animal that is said to inhabit
the coal mines in Osnaburg township. Many acres of coal have been
taken from under the farms in the vicinity of the Eli Brown mine,
and the tunnels are miles in extent. Some of these mines have been
worked for forty or fifty years. Three miners, Will Blackburn, John
E. White and J. Grant Brown, affirm that they have seen this wild
animal, and say that there are other miners in that vicinity who can
vouch for its presence. They describe the animal as being about 2 ½
feet high and about 4 feet long. It has a small head about the size of
a big dog's, with a long, whiskered nose, and with eyes as large as a
silver dollar. The footprints made by this animal are as large as those
made by a mastiff, with six claws on each foot. It utters fierce cries.

It appears that there are two of these animals in the neighborhood, as often an answering cry can be heard, as if the animal were calling to its mate. The miners claim to have discovered in the mines the bones of animals that had been devoured by these wild beasts. That the animals are fierce and vicious is insisted upon.

Recently one of the men was pushing a car of coal to the opening of the mine, when one of the beasts jumped on his back, and with horrid growls, commenced clawing him. In the struggles to free himself the man had the clothes almost torn from his back. Another miner hearing his cries for help came to his assistance. The light of his cap frightened the beast and it disappeared in the darkness of the mine before firearms could be used with effect and without endangering the life of the man attacked.

Last week a man going home from Osnaburg on horseback saw one of the animals at the foot of John Sausser's hill. The beast leaped over the fence, scaring the horse and almost unseating the writer. The man put spurs to his horse and escaped. A few nights since the beast was seen in the yard of Eli Brown, near the coal mine. Mr. Brown owns a large dog, and he and his sons were awakened by the dog bounding on the porch and leaping against the door, whining and howling. Mr. Brown and his son went down to investigate.

When they opened the door the thoroughly frightened dog took refuge in the house. The men saw the wild beast bound away towards the mouth of the mine and disappear in the darkness. The miners say they can often hear the animals howling in the mines, but they will not approach the lights on the caps of the miners. The community is getting excited over the affair and hope that the mystery will soon be solved. Very little credence has been placed in the wild animal story, but the Record is informed through its correspondent that the three men whose names are mentioned in the foregoing insist that an animal of some sort inhabits the mines. It may be an overgrown cat that has grown wild. At any rate, the populace is excited.

EXCITEMENT AT CONANT

Delphos, Ohio, *Daily Herald*, October 17, 1898

A Delphos gentleman, who visited at Conant, Sunday, brought home with him a story that was verified by some of the best people in the neighborhood.

Conant lies about ten miles southeast of Delphos, and near there is a large swamp. Last Friday four young men, one of them armed with a shot gun, were out hunting for squirrel. When they reached the swamp and were about to climb the fence skirting a neck of the swail they heard a peculiar sound. On the opposite side of the fence, only a short distance from them they saw a strange looking beast that is described as being about 18 inches in height, 3 feet long, with a tail 2 feet long that is of equal thickness. In color it is described as being a dark red, with yellowish stripes around the body, most prominent on the back.

The young men, whose names were not disclosed on account of their extreme fright, which would make them laughing stock, stood transfixed as the beast walked back and forth, emitting low growls, and seemingly desirous, of getting across the fence at them. The man with the gun hadn't the presence of mind to shoot the animal, but turned and fled with the other fellows, yelling like they were pursued by Indians.

Such well known men as David Diltz and Clell Ditto verify the story, and also report that a man named Sawmiller met the animal, and that another man on horseback chased it. A movement is on foot to organize a posse to surround the swamps and with dogs hunt the strange beast to its death.

Strange Beast in a Mine

Canton, New York, *Commercial-Advertiser*, January 2, 1901

Warren, Ohio, Dec. 31.—The mining village of Coalburg, near Hubbard, is excited over the presence of a strange animal in Phillips Brothers' cold mine, a quarter of a mile from there. The miners describe the animal as being 3 feet high and 4 feet long, with a small head, a long whiskered nose, and eyes as large as silver dollars. Its tracks show a good-sized foot, with six claws. While pushing a car in the mines, recently, a miner was attacked by the animal, which jumped on his back with fierce growls. The miner's clothing was torn from his back before other miners, rushing to his assistance, frightened the beast away.

Massillon, Ohio, *Independent*, November 18, 1901

West Brookfield, Nov. 16.—History shows that at regular intervals West Brookfield has discovered a wild beast, a wild man or something else of the awe-inspiring kind to be around in its fields and forests. Just now it is a leopard. Harry Miller was the first person to see it. He came across it one evening in the vicinity of a straw stack on his father's farm. Friday night Peter Cornbaner excitedly tore into the corner grocery to notify the public assembled there that he had met the leopard face to face in the very heart of the village. The villagers, particularly the younger element, flew to arms. No section of country hereabouts was left uncovered, but traces of the animal could not be found. That it is here, however, no one doubts, for it has been seen by many.

DID LEOPARD ESCAPE FROM BOSTOCK SHOW?
RESIDENTS OF STONY RUN THINK SO.—
SAY STRANGE ANIMAL IS RUNNING AMUCK.

Portsmouth, Ohio, *Times*, December 15, 1906

Did "Pasha" the untamed leopard, break away from the Bostock wild animal show?

That is the question that is agitating the minds of the people of Stony Run.

Fred Klefford is willing to swear that an animal answering the description of "Pasha" attacked him and his helper while hauling logs at Stony Run. The animal was beaten off and Klefford and his hired man made tracks for the house.

This, at least, is the story brought to Portsmouth by Joe McDermott, who organized a party and with guns and dogs attempted to capture the animal.

It has been seen by several parties and some claim it is as large as "Lord Baltimore" the ferocious lion, while others say it resembles "Pasha."

RESIDENTS SOUTH OF CITY JOIN IN PURSUIT OF SUPPOSED LEOPARD

Coshocton, Ohio, *Daily Times*, July 17, 1909

The vicinity of the Wilson farm, four miles below the city, was thrown into a furor of excitement Saturday shortly before noon, by the appearance of a large and strange animal supposed for a while to have been a leopard, escaped from a circus that exhibited at Zanesville the early part of the week. It was said to have followed children, who were playing along the canal, and a searching party was finally sent in pursuit. Shortly after noon, the animal was discovered crossing the fields on the Wilson farm and was shot by two men. The mysterious animal proved to be a huge dog, the like of which has never been seen in this vicinity. It is described as having a very massive body striped with dark and light brown.

Followed Little Boys

Brian Jacobs, while walking along the river Saturday morning, learned that an odd looking animal was following his small sons, who had been straying behind their father. He obtained only a fleeting view of the animal

Animal is Shot

Harry Norris at noon saw the animal on the Wilson farm, near the canal. The alarm was raised and a number of men obtained guns and started in search of the animal. About an hour afterward, Mr. and Mrs. Fred Ohlinger were standing in their yard on the Wilson place, when they saw the grasses part a short distance away and the dog appeared. About the same time two shots rang out and the animal gave a scream of pain. The men ran to the scene and found not a leopard, as some persons had suspected, but only a huge dog of unknown breed. It was probably a cross between a Great Dane and a hound. Before the animal could be dispatched, it ran into the underbrush and has not yet been found, that it may be put out of its misery.

HYENA IN WOODS NEAR TO DELPHOS
THE BEAST IS THOUGHT TO BE ONE THAT
ESCAPED FROM SHOW—YET AT LARGE.
Fort Wayne, Indiana, *Sentinel*, July 27, 1910

Delphos. O., July 27.—Considerable excitement was created here yesterday when Isaac Good, who lives on the Noah Miller farm In Marion township, about a mile east of Delphos, came into town and reported that he had seen in the Pohlman woods, near the W. C. Baxter farm, between 8 and 9 a. m., a strange looking wild animal that had all the appearances of being a hyena.

Mr. Good first saw the animal in the woods. It was sitting back on its haunches and licking its front paws much in the same manner as a cat or dog would. Mr. Good was quite near the animal before he saw it or before it saw him. After staring at him for a few moments, Mr. Good says the animal ran rapidly toward a cornfield and disappeared. His description of the strange beast left no doubt that it was a wild animal, that it was a hyena. Mr. Good remained about the vicinity for some time, hoping to get another glimpse of the animal, but was unable to do so.

He was on his way to Delphos to work, but after seeing the strange intruder in the woods, he gave up the idea of work for the day, and came to Delphos and endeavored to organize a number of armed men and procure dogs to chase the animal down. No reports had previously been heard concerning such an animal being seen in this part of the country, but Mr. Good's word is not doubted. The animal may have escaped from a circus.

CANADIAN LYNX IS KILLED BY FARMER AFTER HOT CHASE
STRANGE ANIMAL SHOT NEAR BUCKLAND
AFTER KILLING MANY GEESE AND HOGS.
Lima, Ohio, *Daily News*, January 6, 1918

A Canadian lynx whose presence in the neighborhood of Buckland was a mystery almost as big as any of the unsolved In that section, is no more.

Mr. Lynx made himself too well acquainted not only with Buckland farmers but with their geese and chickens. Loafing around the general store of Noah Silien in Buckland where daily losses of farm valuables were recorded in indignant tones by farmers of that section, it was decided to start a hunt for the creature doing the work of destruction. Two hound dogs, property of Theodore Ramsey were included.

The hunt was started early in the morning on Wesley Broerein's farm, three and a half miles northwest of Buckland where two geese had just been killed by the blood thirsty lynx. An entire day's hunt proved fruitless. It was resumed yesterday morning. The hunting party had determined not to be losers and allow the lynx continue to have the fruits of their labor. The hounds scented the tracks of the animal and at noon it was shot by Albert Adams.

Five shots were fired at the lynx before it climbed a tree to which it fled when the hounds took chase. Members of the party say that the animal had been in the neighborhood for some time, as it was accustomed to climbing trees, poles and fences and jumping when at a safe distance.

The lynx measured five feet, two inches in length and 27 inches in height. It was sold to Charles Wheeler for mounting for $15. Those who have seen Canadian lynx before say it is the largest they ever saw. It is believed the animal had been in the neighborhood of Buckland for a long time and probably had wandered there from the north. Residents recall that 26 years ago another lynx was killed in that locality. Not far from where the lynx was killed it was discovered it had buried a rabbit, half of which it had eaten and was evidently keeping the rest for another meal. The long fur of the animal is a peculiar steel gray.

SEE WILDCAT

Richmond, Ohio, *Gazette*, August 4, 1921

Excitement has been at full stage the past few days in the vicinity of the Harris woods three miles east of Washington C. H. on the Waterloo and Circleville Pike all because some berry-pickers reported to Washington police that they had seen a wild-cat in that section.

Milton Woods, John Selmon and Artie Selmon, who made the report, were picking black-berries in the Harris woods a few days ago when an animal, three times as large as a cat, gray in color, with cropped ears and a long tail, walked out from the woods and confronted the berry-pickers.

Before the men could make any attempt to kill the animal, it had disappeared in the thicket. When the tracks of the beast were examined, it was found that they were large and showed plainly that the animal was some species of the cat family.

Patrolmen Bull and Haggard of the Washington C. H. police force, armed with shotguns, spent an entire day this week in an unsuccessful hunt for the "cat." The tracks of the animal were plainly visible around several mud holes where it had been drinking. The dogs, which were taken to track the beast, for some reason flatly refused to take up the chase.

The last wildcat reported in this section of the state was killed several years ago in the Smith woods, just across the road from where the berry-pickers claim to have seen the strange animal.—Madison *Press*.

PANTHER IN MORROW COUNTY
Richwood, Ohio, *Gazette*, October 5, 1922

Howls and blood chilling cries in the woods and door yards of some of the farmers near North Woodbury for the past few nights have sent the men folks to the windows with shot guns, the children crying to their mother's laps, and the mother to the far corner of the room. Lovers slowly driving along by-roads are brought out of their dreams by the unearthly screech of the mysterious something of the woods of northern Morrow county, and stepping on the gas speed out of the range of the devil-like yell. ...

It's a panther. The animal is about the size of a collie dog, it's color is black and white. It is thought that it is living in a straw stack on the rear of Zolman's farm. A large nest was found in the stack. ...

Some of the huntsmen of the northern part of the county now fees sure that it must have been the tracks of the panther that they saw in the snow early this spring and wondered about.

LION HUNTERS GIVE UP CHASE
Reno, Nevada, *Evening Gazette*, January 4, 1936

Cincinnati, O. Jan. 4.—(AP)—County Police Captain Charles Coddington called off his organized lion-hunting safari today.

When and if, however, the beast actually is sighted, officers will be dispatched. He saw the "lion" against a patch of snow. Miller fired his shotgun from a window of his home, he said, but the animal fled.

Officers found tracks in the snow and soggy ground to back up Miller's claim.

Final Chapter in Wildcat Serial
Written on Road Near Wills Creek

Coshocton, Ohio, *Tribune*, June 1, 1944

Coshocton county's mystery wildcat made one appearance too many sometime last week-end and was run down and killed by a motorist along State Route 76 near Wills creek.

R. E Holskey, Coshocton Route 2, walking with his dog Tuesday evening, was returning to his home when he found the body of an animal more than 40 inches long lying dead at the side of the road. It was estimated that the animal would stand at least 20 inches high. The fur was tawny in color just above the paws and shaded to a heavy brownish gray on the back of the pelt and tail, which was "too long to be the tail of a bobcat," according to veteran hunters who viewed the remains yesterday.

Cuspid teeth on the animal were more than three-quarters of an inch long, its ears were erect and covered with long heavy hair.

Game Protector Everett Bailey, who was called to the scene, was frank in admitting that he didn't know the animal's identity, but did say it was definitely a wild animal and a member of the cat family. John Lennon, who has been in the fur buying business for many years, said the fur was the coat of a wild animal and not "house cat fur" as one skeptic suggested.

Rev. M. S. Kanaga, who lives near the place where the animal was found, said he believed it to be a young lynx, as the body closely resembled ones he had come in contact with in Canada.

However, as one farm woman of the neighborhood put it, "I can't say I rightly know what it is, but I do know it's not a barnyard cat— I've got 12 of my own at home."

Wildcat Is Shot Near Allentown

Lima, Ohio, *News*, December 9, 1946

Fred Laman was wondering Monday just what kind of a cat he shot Sunday evening on his farm near Allentown.

He described the animal as weighing more than 11 pounds and measuring; 31 ½ inches in length. The color was dark gray, with a darker stripe on the legs. It had a long tail.

Cornered by Clarence Louth, in a tile, the cat had a broad head, long teeth and long claws.

General belief was that the animal was a sister to one which was shot a month ago in the same neighborhood.

BLACK PANTHER BEING HUNTED
Lima, Ohio, *News*, July 10, 1947

Warren, July 10—(INS)—Searching parties were organized today in nearby Bloomfield tp to track down a reported "black panther" that was terrorizing residents of the area.

State highway patrolmen combed the wooded area in the Bloomfield sector yesterday but found nothing but sets of unidentified tracks.

Orville Earl told the police that the animal was a panther. He said he had "seen them before." Another resident of the area who reported spotted the "panther" was Alfred Carlson, who said he fired a shot at the "cat."

Many others have insisted they heard the animal scream in the night. Dogs were being pressed into service to aid in the organized hunt of the woods.

DEPUTIES HUNT 'BLACK PANTHER'
Massillon, Ohio, *Evening Independent*, September 12, 1947

Warren.—Sheriff Ralph R. Millikin and his deputies Thursday were out after that "black panther" in the Parkman area of Trumbull country.

Mrs. Ida Bronson, who lives just off route 422, said she saw a "coal black beast jumping over a fence in a field near my home." The sheriff said that "the beast killed a calf Wednesday night on a neighboring farm and carried most of the carcass away."

WOMEN FRIGHTENED INTO OHIO HOME BY "ROARING" ANIMAL
Richmond, Indiana, *Palladium-Item*, August 19, 1948

Reports of the "varmint" came from the New Madison area again Thursday after a hunt Wednesday proved a fizzle.

Mrs. Omah Stowe, living on a farm west of New Madison, said she and her mother and a third woman were frightened into their homes about 10:30 p.m. Wednesday by what she described as a "roaring animal."

Mrs. Stowe said she and her mother, Mrs. W. M. Hubbard of Knoxville, Tenn., and a friend, Mrs. Nannie Garrison, of Crane's Nest, Ky., who are visiting with her, heard the animal near their grape arbor. The grape arbor is near a cornfield.

"We screamed and ran into the house. It didn't bark or howl. It actually roared like the lions we've seen in circuses and it the zoos," Mrs. Stowe declared. "We thought that whenever it was was going to attack us."

Men of the household seized a gun and started a search but found no animal. Tracks were found, however, which Mrs. Stowe said were large.

Wiegand Says Black Panther Probably Loose
Richmond, Indiana, *Palladium-Item*, August 22, 1948

Greenville—Robert Wiegand, Darke county game protector, said Friday night that probably a black panther was loose in the area accounting for the reports that a large wild animal has been running the area.

Mr. Wiegand said the tracks he has been called to see were definitely not made by any dog. He said they were too large to be made by any domestic animal.

Several reports have been received by Mr. Wiegand on the "varmint." The latest was Thursday afternoon when George Royer said he saw a black panther near the Slagle gravel pit at noon. He told Wiegand that he was about 20 feet from the animal and "could not have mistaken it for something else. It was a black panther."

According to Mr. Wiegand, no farm animals have been reported killed by the "varmint."

He pointed out Friday night that it was possible the animal, whatever it may be, had escaped from a circus or carnival. He said if it was a mountain lion coming close to people was unlikely, as it would be frightened by humans. However, a circus animal would be used to people passing its cage and probably would venture more closely to them if turned loose.

Mrs. Daisy Mills, living west of New Madison, said she saw the animal and that it "definitely was not a dog."

Mr. Wiegand said he never had seen the animal and was "skeptical to a certain degree" but added he had investigated all reports.

OHIO POSSE HUNTS 'BLACK PANTHER'
Lima, Ohio, *News*, July 13, 1949

Hamilton, July 13—(AP)—A black panther was hunted today in Fairfield-tp, just south of Hamilton.

A group of men, headed by Constable Steve Gay, beat the woods last night in an effort to find the animal. Gay said he did not know who started the story, but that the housewives were so frightened they were keeping their children indoors.

One man in the posse, whose name was not learned, told Gay he had seen an animal about four feet long, and he believed it might be a panther.

There have been several circuses in the vicinity this year, but none has reported loss of an animal.

RANDOM SHOTS... WITH DON MILLER
Mansfield, Ohio, *News* (excerpt), July 17, 1949

A lighter note of the week was the announcement that a "black panther" has been sighted near Hamilton. Residents of the Fairfield township area claim to have seen the animal, and say it's about four feet long and has black fur.

The report has been the cause of numerous chuckles this week, but I'm wondering if the creature doesn't exist. I don't think it's a panther, but it could be a dark lynx or wildcat.

Remember all these dogs which have been slashed up on coon hunts this past year? One in particular was almost disemboweled, a typical cat trick.

Think about that one before you toss the story off as fiction.

HUGE BLACK BOBCAT HUNTED FOR WEEK, FOUND DEAD TODAY
Coshocton, Ohio, *Tribune*, July 18, 1949

Hamilton, O.—(AP)— The wild animal which frightened the country-side in Fairfield township, southeast of here, was found dead today. It was a large, nearly black bobcat. Constable Steve Gay and a posse had been hunting a wild animal for nearly a week, and mothers had kept their children indoors, or close by.

Frank Yauger reported finding the bobcat beside the Tylersville road. It had been shot. Constable Gay and others said they shot at an animal last Thursday night.

Yauger said the animal was four feet long, weighed about 150 pounds, and had fangs and claws as long as a man's fingers.

Mansfield, Ohio, *News Journal*, November 24, 1950

Strange Animal—H. F. "Bud" Byrd and his brother Melvin, who live at the Ken-Mar trailer court, report that while rabbit hunting in the vicinity of Walker Lake Rd. west of Mansfield yesterday they spotted a panther.

Bud said the animal looked like a giant black tomcat about four feet long. Panthers (leopards) are not native of the U. S.; but bobcats, cougars and pumas are often referred to by the name "panther." Whatever the animal was the Byrds say they gave up hunting for the day after spotting it.

BLACK PANTHER 'SEEN' IN OHIO
Van Wert, Ohio, *Times-Bulletin*, July 31, 1953

Millersburg (AP)—Hunters in this Holmes County area are hoping to get their sights on a black panther which several residents claim to have seen roaming the countryside.

Bert Giaque said he first saw the beast three weeks ago as it walked across a road about 100 feet from where he was doing farm work.

'CAT-LIKE' CREATURE IS SIGHTED
Elyria, Ohio, *Chronicle Telegram*, August 7, 1956

Birmingham—Residents of the area north of Rt. 113, between Florence and Birmingham, today sought a tracking-specialist to determine identity of a strange "cat-like" creature seen last night.

According to Mrs. Gene Graham, near whose place the animal was spotted, it left tracks which were "awfully big" and sighting of the animal set off a minor-scale tracking expedition.

Unlike many reports where just tracks are found, this creature actually was sighted.

Dorothy and George Schiebe, who are staying at the Lee Dean farm while the Deans are in Florida, watched the strange animal through binoculars about dusk last night.

They said it as catlike in appearance, had a long tail, and was black. The Schiebes watched it slink along across a field as if it were in pursuit of something, and promptly went to the Graham home to enable the latter to see the "visitor."

The Grahams, however, missed the animal but found the tracks.

Today Mrs. Graham said she would contact a game protector to see if he could identify the tracks.

Some weeks ago, a similarly described animal was reported sighted at Oberlin.

MYSTERY BEAST SPOTTED TWICE
Lima, Ohio, *News*, December 23, 1958

An unidentified animal was seen bounding across Rt. 25, east of Wolf Road, by an out-of-state motorist early this morning.

The motorist, who was not identified, told his story to an attendant at Arnold's Plaza, south of Beaverdam.

He said that he saw a large black animal bound across the road, leap a fence, and disappear in the woods on the south side of the highway.

The wooded area is near Rt. 25 close to the Ohio Power substation.

The attendant at Arnold's Plaza notified the Highway Patrol at 9:08 a.m.

Patrolman Don McComb and Game Warden Lewis Dawson searched the woods and surrounding area.

Even though the woods were snow-covered, no fresh tracks were found. Dog tracks were discovered, but they were not new.

At 11:03 a.m., Mrs. Katherine Early, Rt. 5, reported that she saw a strange animal near the substation a few minutes before.

Mrs. Early described the animal as being a "large black cat."

"There is definitely an animal in the area which people are not used to seeing," commented Game Warden Dawson.

"What it is, I do not know," he added.

Dawson refused to make a statement on the type of animal involved until the facts are more definite.

Elmer Griffith sighted a strange animal Sunday just three miles north of this latest observation.

Between Us

Van Wert, Ohio, *Times-Bulletin*, January 9, 1959

By I. Van Wert

Just Between Us, one thing they've always said about journalism was that it was a romantic vocation, even though notoriously it's a low-paying one. ... Certainly our staffer Ray Miller is a living symbol of how romantic the profession can become. ... He gets himself into some of the most intriguing escapades. ... Like the time a year or so ago when we all stood laughing in the newsroom window while Ray went hustling down the alley across the street, camera in hand and his coat-tails flying. ... He had just spotted a pheasant wandering down that way and he wanted a picture. ...

Then about a month ago, there came a report into the newsroom that a group of deer were in the fields south of Van Wert along state route 118. ... Ray grabbed his camera and set off. ... We had visions of Ray sneaking through the underbrush, trying to get close enough to photograph the deer. ... An hour or so he was back. ... He'd driven all over kingdom come. ... And seen no trace of deer. ...

This morning, Ray set out on the trail of that mysterious black beast which has been reported seen several times around Convoy and Wiltshire. ... Ray was gone for hours on the prowl for the wild beastie But he came home empty-handed. ... The story we'd been hearing all morning wasn't true. But Ray can console himself with the knowledge that, after all, journalism is really a romantic vocation. ... Ah, yes. ...

The story around town this morning went like this: A farmer near Convoy spotted a beast attacking his dog in the barnyard. ... The farmer shot the animal, dead in its tracks. ... He skinned it and took the hide to Albert Pancake, a Convoy fur dealer, selling it for $75. ... Turned out that animal was a black cougar. ...

That was the story circulating the streets. ... This was the quote Ray brought back from Clarence Pancake, son of Albert: "They may have ketched it, but so far we ain't seen nothin' of the hide." Said the elder Pancake: "I've been buying furs for the past 50 years and I wouldn't know a black cougar if I saw one." ... Ray writes: "So, at the time of going to press today, the mystery beast is still at large." ... Anyway, Ray, it's a romantic profession. ...

PHANTOM 'BEAST' ELUSIVE
Lima, Ohio, *News*, January 19, 1959

Van Wert—As snow pelted down on the Van Wert area to-day, jittery farm residents southwest of here were wondering whether the mystery beast can be tracked and killed.

The "beast," described by at least a dozen persons as a massive black animal of the cat family, has proved too elusive for hunters so far, and officers say best chances for finding the animal occur when the snow is heavy enough to show tracks clearly.

The unidentified animal, first sighted more than a month ago near Convoy, was spotted again Saturday by Mel Brown, an employee of the state highway department.

Brown, whose brother Sam keeps a .22 caliber rifle close, by as he works on his unfinished house two miles south of here, saw the animal behind the home of his mother at 3:30 p.m. Saturday.

Brown said the beast was "crawling on its stomach" as if it were stalking prey. As he approached, it ran away.

The large animal now has been seen jumping a high fence, galloping like a horse and crawling on its stomach.

The beast first was seen by a farmer near Convoy, whose dog was so frightened that it ran into the house and hid under a washing machine. The farmer saw the big animal scurrying across a pasture. It leaped a ditch, jumped over a four-foot fence and disappeared.

Report of the animal sent Van Wert County officers into action early the next morning. Deputies, Game Warden Joseph Wernert and three members of the Van Wert Outdoors Club, all armed with shotguns, patrolled a two-mile square area searching for the animal, without success.

At that time Wernert said he did not think the animal, described as "bigger than a panther," would be a threat to human beings unless it was cornered. However, the best may be a danger to livestock, he said.

Two weeks later, the phantom beast was sighted by Fred H. Grossman while he was hunting on the Jake Pulman farm near Willshire. Grossman said the animal held a rabbit in its jaw, and when it saw him, it turned and galloped, like a horse, into the woods.

This incident occurred at almost the same time that Charles Saams told neighbors a dead calf he had left in the woods behind his home, located southeast of here, was found stripped to bones and hide.

Several other area residents had fleeting glances at the animal from time to time, but the next definite report was received last Wednesday from Mrs. Bob Johnson, who saw the animal on the farm of Mrs. Jack Brown, mother of Mel and Sam.

Mrs. Johnson said she was within 100 yards of the animal. A busy phone line prevented her from contacting the sheriff's office immediately, and the animal was gone when deputies arrived.

Mrs. Johnson, who was rocking her small child to sleep when she noticed the animal in her neighbor's woods, said the beast "jumped up and down" as if it had something cornered.

"I saw it three different times, know it wasn't a cat and it wasn't a dog. It was dark in color and had a tail."

Mrs. Jack Brown said she heard noises in her barn, but thought they were made by her domestic cat.

Marvin Putman, who lives immediately north of the Brown residence, also spotted the animal Wednesday. He was within 175 feet of the animal and said it definitely was of the cat family.

Putman, when he saw the beast, picked up his shotgun from the house but was afraid to shot the animal with small shot with which the gun was loaded. He said he feared the animal would be dangerous if it were only wounded.

He said the animal was dark in color and about four feet long.

The beast has variously been sported as a black panther, a bear, a large dog and a wolf. One man even suggested that what one witness saw was a kangaroo.

Whatever the animal, residents in the area are on the alert for the "phantom beast" They are asking themselves, Will he be caught and killed? Where is he now?

MYSTERY ANIMAL REPORTED AGAIN;
PAWPRINT IDENTIFIED AS DOG'S

Elyria, Ohio, *Chronicle Telegram*, May 26, 1959

By Don Miller

"Mystery animal" hysteria continued to hold sway in the Avon-Ridgeville area today amid growing indications that the animal apparently is a house cat which has averted to the wild state.

Best description of the catlike animal came from Mrs. Anne Kozel, Stoney Ridge Rd., who saw it crawl from under her chicken coop.

The animal, she said, had a long tail and body, was a little larger than a tomcat, and had a long, bushy tail.

In color it was dark, with a grayish cast, and had a brownish stripe on it, Mrs. Kozel said. The catlike animal, she said, was the "fastest thing" she ever saw, and it was the "biggest cat head" she ever saw.

House cats, reverting to the wild state, grow exceptionally large and easily can match the description given by Mrs. Kozel.

Meanwhile, however, George Hricovec, Stoney Ridge Rd., Avon senior, reported seeing a large animal with a long tail and whiskers. It was, he said, on a back porch. The tail was smooth, with a tuft on the end.

This sighting, too, was at 9:30 p.m.

One youngster claimed to have seen it, and said it was larger than his large dog.

According to George Zilka, brother-in-law of Hricovec, the creature was also sighted by Veronica Ondros as she drove past the cat down a farm path.

The car was doing 40 miles per hour, she said, but she couldn't catch the cat which ran in big leaps and jumps.

A hunting party of some 20 members looked around the area last night, some remaining in the field until 3:30 a.m. Nothing was spotted.

Others was reported to be going out today.

Harold Dalrymple, county game protector, expressed concern today that "drug store cowboys" out with guns seeking the phantom animal are more dangerous than any wild creature ever would be, and feared someone would be shot in the excitement.

"It's been seen from North Eaton to Columbia to Ridgeville," he said wearily, "and I imagine by now they've seen it a couple of times down in Wellington."

Dalrymple left little doubt that he feels the excitement is unmerited.

A paw print he took from near a killed dog was identified by game management officials as that of another and large dog. For kittens were reported killed, but Dalrymple believes this to have been the work of a tomcat.

"One thing," the game protector added, "cats are learning it isn't safe to roam the fields!"

Meanwhile every flicker of movement in the woods or fields is bringing new calls to the newspapers, law enforcement officials and to the game protector, reporting "sightings" over a broad area—and many at the same time.

Dog Attacked
Cat-Like Animal Spotted in Eaton
Elyria, Ohio, *Chronicle Telegram*, May 29, 1961

Mystery animal scares, like municipal elections, soon to crop up every two years. One sighting of a "large animal" was reported today by a family, and two other families reported incidents which could have a bearing on such an animal.

It is two years since the last reports in the Elyria area.

The sighting, and other incidents, apparently occurred last Wednesday and Thursday, a few days after a similar report from the area.

Latest report in the "mystery animal" category came from Mrs. Kathryn Roll, Royalton Rd. near Rt. 82 on Eaton. She said she and a daughter watched the animal run across the road with movements resembling those of a cat, and disappear into woods on the William Sobol property. She said the large animal was so black it had a shiny appearance.

Dog Bitten

Thursday morning of last week the Charles Zelenka family, also of the Eaton area, found their beagle dog had been bitten on the head and body, and had an injury to the neck. The latter may have resulted from the dog chain and collar as the dog attempted to get away from whatever attacked it.

A German shepherd dog belonging to the Earl Ashby family, next door to Zelenka's also was discovered to have broken his chain, but was uninjured.

Previously, on May 19, the sighting of a large black, cat-like animal was reported by Mrs. Robert Penfound and by Mr. and Mrs. William Porter, in the area near Rt. 20 and Rt. 10.

They said it was black, with a long tail, and larger than a cat.

Porter and sheriff's deputies searched the area, but could find no evidence of the animal.

No Reports

A sheriff's department aide this morning said the office has received no reports on the incidents in the Eaton area.

Two years ago the sighting of a "mystery animal" in the North Ridgeville-Eaton area kept deputies, other law officials, sportsmen and private citizens in an uproar for most of May.

Descriptions ran from those which indicated the animal resembled a small lion, to those suggesting it may have been a large domestic cat which had gone wild.

Volunteer hunters scoured the area on several occasions. They found nothing which resembled a mystery "beast."

But—the searchers did mark up two victims—a large pet dog which in no way resembled the alleged mystery animal, and a domestic cat.

The search finally was abandoned after Sheriff Vernon Smith obtained a paw print at the scene of one of the animal's visits.

The print was identified as that of a large dog.

PANTHER BELIEVED TO BE IN AREA AROUND LONDON
Marysville, Ohio, *Journal-Tribune*, January 8, 1963

(London)—A large black panther is reportedly roaming the London area.

Max Allen, Prairie Pike Rd., South Solon, said he saw the large black cat New Years Day. He said he spotted the animal, which leaves five inch wide foot tracks, about 5 p.m. He chased the animal in a truck and shot at it about nine times before losing the race.

Several other persons have also said they have seen the cat.

Fear Panther Near Canton

Mansfield, Ohio, *News Journal*, August 8, 1964

Two thoroughbreds were attacked on a Minerva area horse farm, reviving rumors of a "black panther" sporadically seen over the last five years. A veterinarian treated the horses for large claw marks on their flanks.

Don Miller's Afloat and Afield

Elyria, Ohio, *Chronicle Telegram*, January 14, 1966

The columnist reports on an unusual feline, a "long-tailed wild cat," killed by a local man. Some local hunters were familiar with the type, but the columnist had never heard of such an animal and deferred to the local wildlife officer's suggestion to "mark it down as a feral domesticated cat." It was described as a young cat, furry, with an 8-inch raccoon-like tail, big heavy feet, sharp claws, and stripes along the gray body. It had the start of tufts on its pointed ears.

Village Policemen Search for Big Cat on the Loose

Coshocton, Ohio, *Tribune*, September 5, 1975

An auto mechanic told police that he was working in a garage when he saw a lion come out of the woods nearby. He said it was three feet tall and weighed about 250 pounds. He picked up a wrench and went towards it until it put up a paw and made a noise, at which point both mechanic and lion disappeared in different directions. A police search didn't come up with anything. No mention of a mane is made in the article, however, so whether he meant an "African lion" or a mountain lion is uncertain.

'Panther Like' Cat Sought Near Lima

Toledo, Ohio, *Blade*, May 21, 1977

A "black panther" like animal killed 140 sheep near Lima, Ohio. It was described as "about 3 feet long and 24 to 30 inches in height, weighing 100 to 125 pounds." Identification was based on "recent sightings and examination of animal droppings and plaster casts of its tracks by Dr. Wayne Kaufman, Lima veterinarian." A hunting party hoped to tree and shoot the animal with tranquilizer darts.

Panther in Ottawa County?

Toledo, Ohio, *Blade*, October 21, 1989

Earlier in the month, Howard Appling, a Carroll Township farmer, saw a large black feline next to a four-foot-high brush pile. He drove within 50 feet before it leapt over the brush and disappeared. Another witness said it was about two feet high, four feet long, and had a long tail that curled at the end. A cast of tracks was examined at local zoos, where it was speculated it might be a large dog.

Panther Sighting Reported in Portage

Port Clinton, Ohio, *News Herald*, May 4, 2004

A man living on East Harbor Road heard his dog barking one night, and discovered a 50-pound "pure black" cat that took off into the marshy brush nearby. It "had catlike facial features and pointed ears that stuck straight up off its furry head, though he recalled few other details."

U.S. Town Launches Lion Hunt

Yahoo! News (Reuters), May 4, 2004

A police officer first reported a lion as it crossed a roadway in Gahanna, Ohio. It was seen later by a courier, crouched under a parked semi-trailer. Police called in a helicopter and big-cat experts to help track it. It was spotted a third time near the Port Columbus International Airport (NBC4, Columbus). Later, it was reported near a lumberyard west of Columbus.

Residents Claim Lion Prowls Licking County

Newark, Ohio, *Advocate*, May 14, 2004

What may have been the same animal seen in Gahanna earlier in the month, was reported near the Kirkersville Elementary School. It was "large, brown and cat-like." One teacher described it as "large, butterscotch colored, with a large head and a very muscular body." Another witness noted it as "a tan, stocky animal with a short tail and hanging ears. It appeared to have an orange tint to the short hair. The animal made no sounds and appeared to be keeping its distance from the persons in the area." The animal warden suggested it was more likely to be a cougar than an African lion, if it was an escaped captive.

More Panther Sightings

Port Clinton, Ohio, *News Herald*, June 16, 2004

Several more witnesses in southern Danbury Township, along the Bayshore Road corridor, reported seeing a large black feline. One witness said it had a tail that was "long and fat and curled at the end."

Gahanna Lion May Be Moving West

Newark, Ohio, *Advocate*, July 9, 2004

A Madison Township firefighter says he saw the lion along U.S. 33, as it disappeared into a cornfield.

Rumors, Sightings of Big Cat Resurface

Newark, Ohio, *Advocate*, July 30, 2004

A large brown cat, suggested to be a cougar, was reported by multiple witnesses in Muskingum County. A pony was treated for wounds from an attack, and later news accounts noted paw tracks the size of a man's hand as well as other livestock attacks. Interestingly, there didn't appear to be any immediate attempts to connect this to the Gahanna lion sightings, despite the similar "big brown cat" descriptions. Later in the year, as sightings of a large lion-like cat were reported from the Granville area, the papers did start to connect the dots.

OKLAHOMA

NIGHT PROWLING MONSTER TERROR TO OKLAHOMANS

Fort Worth, Texas, *Star-Telegram*, August 19, 1921

Slick, Okla., Aug. 19.—Residents of this little oil town and vicinity for miles about our more nonplussed than ever following another night of failure in an attempt to capture a heavy mysterious animal which has been invading pastures and barn lots and slaughtering cattle and horses.

Each night the searching parties, heavily armed, are reinforced in numbers by determined and curious-minded individuals of various occupations, set upon bringing the monster to death. But each night the animal strikes in an unguarded spot.

Vacationists from other Oklahoma cities hearing of the mystery have decided to combine adventure with pleasure and have joined in the hunt.

A few persons have had a fleeting glimpse of the prowler, and describe it as black, weighing more than 400 pounds and peculiarly marked on its head. Natives declare that the animal is neither bear nor wolf, and are completely mystified.

PANTHER'S SCREAMS JUST TOO MUCH

Dunkirk, New York, *Evening Observer*, August 24, 1950

Cushing, Okla.,—(UP)—Pinky Mullins was behind bars again today because "I'd rather be close to a cop than a panther."

Mullins, 19, referred to a mysterious black panther reported seen by farmers in the vicinity of Cushing. A former Tulsa zoo official has offered a $100 reward for the jungle cat.

Young Mullins told Payne county officers he had been hiding in a cave and farmhouse since he escaped from Granite reformatory Aug. 17.

"The cave was a lot better than the farmhouse," he said. "I bedded down there for three nights, and every night I could hear that panther screaming. Funny thing, he seemed to be getting closer. I decided I'd rather be closer to a cop than a panther."

Mullins was returned to the reformatory to serve 10 more months of a term assessed him for stealing cotton.

LEOPARD OR BOBCAT, IT'S DIFFERENT
SMALL CRITTERS FOUND NEAR JESSE KIND
GAME RANGER HADN'T SEEN BEFORE
Ada, Oklahoma, *Evening News*, February 12, 1951

A group of Jesse community residents, are confident they have found a den of leopards, but their stories are being discounted by some authorities, who say there are no leopards In this section of the world.

Those authorities contend that the animals are bobcats because of the terrain where it was found.

Tom Winton and W. W. Whittington of the Jesse community were burning grass on a 40-acre plot Sunday afternoon when they saw a large animal come out of a hole in the side of a bank, and run.

Seven Small Ones

On further investigation, the men found seven small animals that were a light yellow with dark brown stripes running from their back to their belly. The dark lines were broken.

On each leg, the lines went around the leg to the foot, where they stopped.

One of the small animals brought to Ada Monday morning weighed three-quarters of a pound, measured 13 inches long, had a paw span of three-quarters of an inch and had a long fat tail.

Whittington said a state game ranger from McAlester identified the small animals as leopards.

"I don't know what it is, but I do know that I never saw anything like it before," Whittington said Monday morning.

Hunt Starts

The older animal that was first seen was reported to be the size of a good sized dog, orange in color with dark brown or black spots and streaks about its body.

Whittington said the animals were found in Boggy bottom about three-fourths of a mile north and two and a half miles east of Jesse.

It was reported late Monday that a group of Stonewall and Jesse men got their guns and started a "leopard" hunt in the Boggy bottom area where the animals were found.

Whittington said the ranger told him that the animals were about four days old when they were killed by the fire.

HUNTERS, DOGS AFTER CAT-LIKE ANIMAL IN AREA AROUND JESSE
NOT LEOPARD, MAY BE OCELOT; TIED IN WITH MISSING HOGS

Ada, Oklahoma, *Evening News*, February 13, 1951

A crowd estimated at 100 persons gathered at the Stevenville farm in the Jesse community again Monday night to make an extensive search for a cat-like animal that caused some excitement Sunday afternoon and night and left farmers and ranchers guessing as to the identity of the creature.

Dick Truitt, nationally known in major rodeo circles for the past 20 years and a rancher in the Jesse community, is heading the hunt for the animal.

Truitt said that there must be about 20 of the cat-like animals in the Jesse community. He pointed out that the animals were first noticed some three years ago, but that nothing was done at that time.

Hogs Disappear

A farmer in the area where Tom Winton and W. W. Whittington found seven small kittens Sunday afternoon reported that he lost eight good sized hogs from a pen last week and that since that time two others have disappeared.

Winton told hunters Monday night when they gathered in Boggy bottom that they were looking for an ocelot, which he said met the description of the cats being sought.

The farmers discounted a story that was reported Monday that the cats were leopards.

An hour-long race was heard Sunday night by those gathered around a camp fire near the Odie Sanger farm, which is located just west of the Stevenville farm.

Cat Eludes Pursuers

Truitt reported that the race ended on his ranch when the cat ran across some freshly burned grass-land.

Mrs. Sanger said she and other women in the community are scared to get out of the house by themselves at night for fear the cat might come along.

Mr. and Mrs. Sanger's daughter refused to go to school Monday morning as did some other children in the community. However, most of them were back in school Tuesday morning only because they were taken by their parents.

Hunters Come In

For the past two nights, coffee and sandwiches have been served to hunters who came to take part in a search for the feared animals.

An unidentified Tupelo resident is scheduled to take some new dogs to the area tonight (Tuesday) to by their luck at finding and catching the animal.

The dogs from Tupelo have hunted mountain lions and big cats in Colorado and California and their owner told hunters around the camp fire that his dogs will hunt nothing else.

Willie Hall and Riley Jennings, who are the leading hunters, have a large number of hounds they say will find the animals.

Truitt is laying odds on the dogs owned by Jennings and Hall. He points out that the two men have hunted the area before and that they know how to hunt the cats.

The place where the baby cats were found Sunday afternoon is located about 20 yards off a county road. A big bed of grass and underbrush was their bed. At the bottom of the bed was some dirt that had been piled loosely.

Winton told hunters that he got a glimpse of the animal several times last fall, but couldn't tell just how it was built.

Larger Then Bobcats?

He said the cats are definitely not bobcats—they are much larger and more vicious.

Bill Johnson of Woolley has been on hand with his dogs for the past two nights. He reported that he had some dogs torn up badly when he hunted near the Stevenville farm a few weeks ago.

Residents of the community say the animal makes a noise something like a panther's, but cuts its scream shorter.

Another hunt is planned for tonight with new tactics planned. Hunters say they will be on hand despite the disagreeable weather.

HUNTERS TO RESUME WHEN SLEET GONE; LIVESTOCK ATTACHED

Ada, Oklahoma, *Evening News*, February 14, 1951

A group of prospective hunters in the Jesse community quickly disbanded Tuesday night when sleet started falling heavily, so third night of the hunt for an ocelot was not successful.

Numerous queries from all section of Oklahoma were received by Stonewall and Jesse residents Tuesday. Many were seeking information about the cat-like animals that are now doing damage in the Jesse community.

Hunters called Wednesday morning and indicated they will be on hand for a hunt when the weather clears.

In the meantime, Dick Truitt, who is pressing the hunt, said that several animals have been injured during the past two nights. In all instances, the injuries are attributed to wild animals that are the objects of two unsuccessful nights of hunting.

A nest of seven baby cat-like animals were discovered on the Stevenville farm, northeast of Jesse, Sunday afternoon.

The news spread rapidly and some 100 persons were on hand to take part in a hunt Sunday night. A similar group was on hand for another hunt Monday night.

Truitt said Wednesday morning that one of his prize Brahma bulls suffered severe injuries Tuesday night.

"The bull was scratched all over and was bleeding badly when it entered the lot Wednesday morning," Truitt reported.

"I first thought the bull had been shot, but on further examination found it had been scratched about the head, shoulders and back," Truitt said.

Truitt told hunters Sunday night that the unidentified animal would soon start attacking domestic animals and Wednesday morning had proof.

While checking activities of Tuesday night, Truitt learned that a farmer had to shoot a mule because it was so badly injured.

The farmer told Truitt that the mule had one hip practically torn off and was clawed all over.

Jesse community residents are more determined than ever to find the animals that are doing the damage.

Truitt estimated that there are probably 20 ocelots or similar animals in the area at the present time.

'CRITTERS' ARE STILL THERE AND PEOPLE ARE STILL CURIOUS
HUNTERS ARE STILL DETERMINED TO TAKE, IDENTIFY ANIMALS
Ada, Oklahoma, *Evening News*, February 18, 1951

Those quaint little creatures that were found northeast of Jesse just a week ago have yet to be positively identified, but they are still being talked about.

Since those first seven cat-like cubs were discovered an the Stevenville farm, hundreds of stories have been told about them, but as yet an adult of the same species has not been found.

People Curious
Only Friday 151 person! visited the home of Emory Manuel, who caught two 'kittens' in the Jesse area, to see them and, said Manual, "They called them everything from Mickey Mouse to an elephant."

Some 100 hunters gathered at the farm last Sunday night so listen to dogs chase the older animals, but nothing was discovered.

Livestock Attacked
A similar group of hunters gathered on Monday night for the same purpose, but again nothing turned up.

Dick Truitt, prominent rancher of the Jesse community and nationally known for his rodeo performances, reported one of his prize bulls was attacked by one of the animals.

On the same day, a mule was so badly torn up that it had to be killed and nine pigs were reported missing from their pen.

Weather Slows Hunters
When the animals started attacking domestic animals, hunters

became more determined than ever to find the older animals and destroy them.

However, bad weather slowed hunters considerably until Emery Manuel caught a pair of two-months old kittens about a quarter mile from where the first group vas found.

A local veterinarian declined to say just what member of the cat family they belonged to, but said they were not house or alley cats.

The hunt was reported resumed Saturday afternoon when a group of Ada and Stonewall men returned to the area in an effort to capture or kill one or more of the older animals.

Some dogs, especially trained in hunting big cats, will be brought into the area this week as the search is expanded.

Women and children in the Jesse community don't feel safe away from their homes alone. "I will feel much better when the animals are all killed," one woman said.

Two More Kittens Found and Hunters After 'Old One'
Dick Truitt, Convinced is Ocelot, Ready to Try Roping Animal From Horse; Dogs Turn Back After Getting Close

Ada, Oklahoma, *Evening News*, February 25, 1951

Those kitty cats at Jesse jumped back into the news Saturday following an all night hunt Friday and that fact that two more young cats were found Friday afternoon.

Judging by an estimate made by an Ada veterinarian on the age of two cats found about 10 days ago. Jesse men estimated the age of the two latest "finds"' at six weeks old.

The two kittens were about the same as others that have been caught. However, the latest additions were darker with fewer spots than cats previously found.

Surprising enough, the kittens were found in the same general area as other cats that have been found.

A group of Ada, Stonewall and Jesse hunters gathered northeast of Stonewall Friday night for another hunt that was reported to have produced nothing.

Dick Truitt, who contends that the sought after animal is an ocelot, brought his horse and his roping ability into the picture Friday.

He has been hunting the area horseback and says he will rope the "old" cat if he gets a chance.

His brother, Charley "Fat" Truitt, is reported to have roped a lion once and Dick would be happy to turn in such a performance with a cat on the other end of the rope.

A pack of dogs, hunting on the Stevenville farm, found something they didn't like Thursday afternoon.

The dogs carried on a chase for more than an hour before apparently getting too close to their object.

After chasing a cat into the open, the dogs got close enough that they saw more than they wanted to fight and returned to a group of men who were following some distance behind on horses.

The dogs didn't even offer to hunt any more Thursday afternoon.

Tom Winton, who first found the baby cats on a plot of land that was being burned off, said Friday night that he would pay $10 just to get a look at the older animals.

'Critter' Sends Dog to Hospital
Hunters Out Tuesday Night Report One Dog Missing, One Injured.
Ada, Oklahoma, *Evening News*, March 7, 1951

One dog is missing and another had to be taken to a veterinarian Wednesday morning following an encounter with what hunters believe is an ocelot in the Jesse community.

Hunters spent Tuesday night and until sunup Wednesday morning chasing the animal, which completely eluded hunters.

A dog owned by Jess Corbin was scratched and torn during the fight and other dogs had scratches on them, but were not injured.

Emory Manuel said one of his best dogs is missing following the night-long hunt.

The chase lasted about eight hours after the animal was found by the dogs until about 11 p.m. Tuesday.

Hunters said the animal was traveling toward Stonewall and was within a mile of the town when the chase ended.

Ada hunters reported Wednesday morning that they plan to return to the Jesse area tonight (Wednesday) in an effort to capture the critter.

Mary Sue Sure of 'Critter'

Everett Shaw's Daughter Saw It—Mounts and Rides Away From Big Cat

Ada, Oklahoma, *Evening News*, March 26, 1951

Stonewall—(Special)—

Many hunters have just about given up hopes of catching what they call an ocelot, but Mary Sue Shaw, daughter of Mr. and Mrs. Everett Shaw, is sure that the animal is still in the Jesse-Stonewall vicinity.

The other day while rounding up a herd of cattle in Shaw's pasture, Mary Sue had an interesting experience with the cat.

She had got off her horse and was proceeding to whip her dog for chasing cattle when she looked up and saw a big cat running toward her.

She got on her horse and outran the bounding animal.

Mary Sue described the cat as being about the size of shepherd dog, but not fast in a race.

A number of other people have reported seeing the cat, but hunters have been unable to catch the critter.

Dogs have cornered one of them on different occasions, but after one encounter with the animal hunting dogs have been reluctant to hunt for it again.

Dick Truitt and other Jesse residents are determined to find and catch one of the grown critters.

Fitzhugh Hunters and Dogs Bag 110-Pound 'Wild Kitty'

Ada, Oklahoma, *Evening News*, June 3, 1951

It has been several months since cat hunting at Jesse has been in the news, but it jumps back now in a new area.

Hunters have been visiting the Jesse area almost every week this year in an effort to locate what Dick Truitt of that community called an ocelot. The hunts have been fruitless, for no game has been returned.

However, a group of Fitzhugh hunters made the catch of the season several miles southeast of Fitzhugh and about eight miles west of Jesse one night last week.

They reported catching a big cat that weighed 110 pounds and measured six feet from the tip or its tail to its nose.

The cat was big as a dog and had spots and rings about its body. It was an orange color with dark markings.

Making the hunt were Leonard Young, Jess Moss, Morris Moss, Jack Hudson, Leo Denham and Ben Newton.

A group of hounds started chasing the cat about 3 a. m. and continued until sun-up, when the cat ran up a tree. The dogs stayed with the cat until hunters arrived.

Jess Moss started up the tree to get the cat down so the hunters could hear another race, but the cat wouldn't budge.

Morris Moss threw a rock at the animal and hit it hard enough to knock it out of the tree. The dogs wrestled with it over an acre plot before the cat was killed.

The hunters agreed that the cat "looked like a wildcat," but they were not sure.

Jess Moss said the race was the best their dogs have had in several nights and he is planning another hunt.

The Jesse cat was reported to meet all the descriptions of this cat, but a grown one has never been caught in the Jesse area. Hunters started spending many days and nights hunting for the Jesse cats after it was reported that pigs were missing in that area.

More hunters visited the area when Truitt admitted that a cat had injured ore of his top brahma bulls and cut up a horse.

For weeks hunters reported seeing the Jesse cat. Now they will be wanting to see what the cat caught by the Fitzhugh hunters looks like.

In 1951, a Calumet, Oklahoma, woman saw an animal she said "looked like a cross between a wolf and a deer." It had thin "deer" legs and huge paws, a thin body and head, long hair, a bushy tail, and pointed ears. Her husband said he had seen a similar animal two years prior (Sucik 2002).

Elusive Black Beast Kills Oklahoma Stock
Bismarck, North Dakota, *Tribune*, February 13, 1952

Kremlin, Okla.—(AP)—A black beast, believed to be a panther, has played an elusive game of hide-and-seek with ranchers in the Kremlin area the past few weeks.

It vanishes when searching parties comb the area. Then, when least expected, it turns up to kill a young steer or lamb.

Authorities believe the animal escaped from a small circus whose owner failed to report his loss.

Tuesday night it was spotted again, crossing the road in front of a car driven by 19-year-old Darrol Hayes of Kremlin.

After patrolling the area thoroughly, searchers found only huge footprints where the beast had ducked under a bridge.

Hundreds of Grant county ranchers, angered by the loss of their stock, have attempted to track down the animal many times. They were aided at one time by five small planes.

PHANTOM PANTHER IN GARFIELD COUNTY
AND THERE ARE TWO OF THEM, SAYS FARMER WHO SAW 'EM
Ada, Oklahoma, *Evening News*, February 15, 1952

Enid, Okla. (AP)—Grand County's phantom panther: has moved its operations into Garfield County and now there are two of them.

Farmer Alvin Nichols reports he saw the two animals eight miles north of Enid Wednesday. He summoned Bill Cross and Lucius Nichols who confirmed his description of "a large black panther and smaller one."

Fred Wehrman said yesterday he saw a panther near Lahoma.

Johnny Reynolds, hunting in the vicinity, reported his dogs also were frightened by a large beast.

The panther first was hunted by farmers near Jefferson in Grant County. The beast had been killing livestock.

FAIL TO LOCATE WANDERING LION
Jefferson City, Missouri, *Post-Tribune*, March 9, 1961

Big Cabin, Okla. (AP)—First efforts by a safari of armed volunteers, led by Oklahoma zoo officials, failed to locate an African lion reported to be roaming the northeastern countryside.

Tulsa zoo director Hugh Davis, who led an all-day search Wednesday, said he is convinced the lion exists and plans to resume the search.

The lion was reported seen by several residents of Mayes and counties, the last time by Fred Campbell of the Big Cabin area. He said he saw the animal sleeping early Wednesday in a hay barn on his farm.

It remained a mystery how such an animal got loose, but Davis noted that last December a circus truck was wrecked at Adair. He said although an investigation by the Oklahoma Highway Patrol indicated no animal escaped, the lion may have been from the circus.

The reports persisted the past several weeks that a lion described as a full-grown male has been devouring livestock, sun bathing along the Will Rogers Turnpike, and raiding chicken pens.

Mrs. Joyce Propp said the headlights of her car spotlighted the lion the night of Feb. 9. She said the lion was only a few feet from her home and was eating a chicken.

She made photographs of the cat's tracks the next day because she said, "I knew my husband wouldn't believe me if I didn't."

Davis, said the photographs and other evidence convinced him the lion exists.

A woman also telephoned Davis that she saw a lion along the turnpike, but she rode 20 miles before she could bring herself to tell her husband what she had seen.

Two large paw prints which could possibly have been made by such an animal were found

Wednesday near the Propp home, and searchers also found four cattle carcasses.

Hunters Find Crows, But No Animal King
Syracuse, New York, *Post-Standard*, March 21, 1965

Hunters were in search of a large lion near Adair, Oklahoma. While some thought it could be a mountain lion, others were convinced it was an African lion, as several witnesses (going back four years) described it as having a mane.

'Cat' Sighted in OKC Sector
Ada, Oklahoma, *Evening News*, October 26, 1977

Several residents of north Oklahoma City reported seeing a large black feline. Police and Animal Welfare officers found tracks measuring 3 ½ wide by 4 inches long "that definitely were not made by a dog," but looked more like cougar tracks.

Newton (2005) noted a strange canid sighting along a highway in 1982 by two cattlemen driving near Cheyenne. One stated, "It was bigger and broader than a dog would be. Its head pretty much sat down on its shoulders, and it walked on four legs." The other said, "It was not fuzzy or furry, but slick-haired, like a pig. It was kind of smooth moving. It didn't bounce any. It was pretty heavy, and had a pretty big body."

OKLAHOMANS SPOT LEOPARD
Syracuse, New York, *Post-Standard*, April 18, 2002

A melanistic leopard of unknown origin was spotted roaming Mayville, Oklahoma. In one case, it walked up to a car and planted its paws up on the hood, staring at the driver before wandering off. An animal park director confirmed it was a melanistic leopard after catching a glimpse of it, and set out to try and trap the feline. He estimated it weighed between 150 and 170 pounds.

ED GODFREY: BLACK CATS NO MYTH, READERS SAY
Oklahoma City, Oklahoma, *The Oklahoman*, March 28, 2004

After having written a column apparently dismissing black panther sightings, the columnist received a barrage of letters from readers who disagreed. Sighting locations included near Edmond (just six weeks prior); Bromide, south of Ada (approx. 1984); Atoka County (approx. 1997); between Glencoe and Morrison (2003); Rush Springs area (1990, killed and played with a litter of pigs); south of Lake Dahlgren (1969, family saw it and reported it to head ranger, who said he had seen it before). In the latter case, "the cat was only 35 to 40 yards in front of us. We had time to get a good look at this animal, and this cat was 4 to 5 feet long from head to tail and 2 ½ feet tall."

BLACK COUGAR MORE TALK THAN FACT
Tahlequah, Oklahoma, *Daily Press*, February 1, 2006

Oklahoma Game Warden Brady May said the Wildlife Department gets a number of reports of black mountain lions from northeastern Oklahoma, but thinks they are likely misidentified black bears.

Man Thinks Black Leopard is Killing His Livestock

Oklahoma City, Oklahoma, *The Oklahoman*, August 18, 2006

A rancher by the name of Atkison claimed that a large black feline was responsible for killing 200 animals (poultry, goats, and a llama) on his ranch over the last two years. "It's a black cat. A big black cat. It's about 3½ feet tall and about 4½ feet long, plus a four-foot tail. This is a big cat. I'm not talking no little bobcat." Game wardens suspected feral dogs.

ONTARIO

Newton (2005) noted the 1960 sighting of a maned feline with a tufted tail by a farmer (Leo Dallaire) on his farm near Kapuskasing. The feline was five feet in length, three feet tall, and had a four-foot long tail.

Brown's Beat: Mother's Panther
Ottawa, Ontario, *Citizen*, March 19, 1981

The columnist notes a black panther sighting in the Gananoque area two weeks earlier, and a sighting a year earlier near Delta.

Panthers Spotted in Region; MNR Has Yet to Confirm Reports
North Bay, Ontario, *Nugget*, July 13, 2007

Several witnesses reported black panthers near Bonfield and Rutherglen. One rumor gave their origin in the closure of an exotic animal farm years ago that had leftover cubs.

Possible Panther Sighting
North Bay, Ontario, *BayToday.ca*, July 27, 2007

A North Bay woman walking her dog near McKeown Avenue saw a large black feline, larger than a dog, with a long "turned up" tail.

Tourist Spots Panther in Nipissing; No Sightings Confirmed by MNR
North Bay, Ontario, *Nugget*, August 7, 2007

A vacationer on Lake Nipissing's South Shore was jogging when she encountered a large black cat that bolted across the road.

Port Franks Woman Spots Cougar-Like Animal

London, Ontario, *Free Press*, November 16, 2007

A few more individuals reported seeing large black felines. One woman saw it sitting high up in a tree one afternoon, before it stretched, climbed down and wandered off.

Witness Says Cougar Left Paw Prints

London, Ontario, *Free Press*, December 15, 2007

A London apartment manager, Charles Snyder, reported a black cougar-like feline that left 2 ½ inch long tracks. It was in the brush behind a Kipps Land apartment complex.

Possible Cougar Seen on Farm

Milton, Ontario, *Canadian Champion*, August 27, 2008

While checking on her neighbor's hobby farm, west of Campbellville, Jill Rogers discovered a large black feline (estimated at 100 pounds) moving toward the barn. She said the cat's body was eight to ten inches longer than that of her golden retriever's.

Paw Prints May or May Not Belong
to Port Franks Cougar: Expert

London, Ontario, *Free Press*, November 3, 2008

Port Franks resident Bob Rutledge reported seeing normal cougars and a large black feline on several occasions over the last few months. A cast he had from one print was shown to a wildlife official, but was indeterminate.

OREGON

"Grizzly" Adams reported the following: "We also, while near Klamath Lake, saw a strange beast, which resembled a hedgehog with the head and feet of a bear. We made all the endeavors in our power to catch it, but in vain; and now, in looking back and harrowing my memory of this curious animal, I am unable to describe it more particularly. It was entirely unknown to me, and I had, very unwillingly, to leave it as one of the nondescript wonders of the Pacific coast" (Hittell 1861).

A Huge What Is It Monster Killed
San Francisco, California, *Bulletin*, February 12, 1870

For several years the farmers and Soap Creek Valley, in the southern part of this county, and the northern part of Benton, have been annoyed by a large animal, which destroyed all kinds of young stock, and was particularly destructive on sheep. During the last year it grew very bold, and was frequently seen in daylight. Some thought it was an American lion, and it was called a cougar, a panther, a tiger, etc. It was often hunted with dogs, but always succeeded in whipping all the dogs that went after it, killing and wounding a great many of them. About a year ago, as two boys about 15 years old, were passing through the hills at evening twilight, it attacked them, rearing upon the breast of one, who frightened it away with some lucifer matches. A large reward was offered for its scalp. It was hunted and hunted by all the neighborhood, and by parties who came from a distance with trained bear dogs, but to no purpose; the hunters came off with wounded dogs, and sometimes fewer dogs than they started with. It would frequently approach the farmhouse in the night and kill any dog that had the temerity to attack it; then help itself to a sheep, a hog, a calf, a colt or anything it happened to fancy. It could be sometimes heard around the dwellings, making a noise much like a hoarse purring of a huge cat; at other times it exposed itself in a growling roar, something like a lion. Some time ago John Miller procured a heavy bear trap, and set it in a path near a drift, where they had observed this animal sometimes crossed Soap Creek. One day last week, in going to the trap in the morning with a neighbor, who carried a gun, he found the enraged monster in the iron tolls. In leaping from the drift to the bank, he set one of his hind feet in the trap and it closed upon him just above the hock joint. He had gnawed his leg off, except the hamstring, and would have gnawed that off and got away if he had not broken all of his teeth out on the trap. When Mr. Trip came up with the gun, the animal sprang at him the full length of the chain, uttering a frightful roar of rage and pain; the gun snapped. Just then it became painfully apparent that the leg was only held in the trap by a slender string. The next time the gun fired; and the huge animal, measuring over three feet high, and seven long, lay dead at the feet of his captors. His foot measured over five inches long and nearly as broad; his leg measured 18 inches in circumference, and he was nearly as large in the breast is a horse; with thick round ears, a head very much like a lion, and a bushy tail about 14 inches long. It made great use of his paws in striking like a bear, and is about the color of a grizzly. It seems to partake of the character of the bear, lion, cat and wolf, and yet it is neither. The destruction of this monster is a great relief to this neighborhood, for the people have lost much stock by its ravages, and have been kept in fear of their lives for a long time; and it seemed to increasing boldness, yet, was very shy of anyone carrying firearms—it seemed to smell the powder and keep out of their way.

PENNSYLVANIA

Before 1840, Burrel Lyman of Roulet, Pennsylvania, shot two "bobcats" with long tails (Lyman 1973).

A WILD BEAST AT LARGE

Lebanon, Pennsylvania, *Daily News*, November 2, 1883

Huntingdon, Nov. 1.—The citizens of Cass township, this county, are greatly excited over the appearance of a strange wild animal supposed to be a tiger, that has already commuted serious depredations among the poultry and stock of the formers in that locality. Although several hunters have pursued it, all efforts to capture or kill it have proved futile, as it invariably escapes into the fastness of Terrace mountain. A number of persons have been confronted by the animal while journeying over the mountain, and serious apprehension is felt by the people of the entire neighborhood. The supposition is that it has escaped from some traveling menagerie.

Connellsville, Pennsylvania, *Keystone Courier*, July 8, 1887

Wm. Howell, when in the woods near Ebensburg, started up a strange animal, about two feet high and five long, of reddish brown color, with stripes on the sides, which he thinks is an American lion that has escaped from some menagerie.

Harmonsburg Folk are Scared by Strange Beast

Pittsburgh, Pennsylvania, *Press*, October 22, 1901

Meadville, Pa., October 22.—Harmonsburg, a hamlet eight miles west of here, is in a state of great excitement over the appearance of a strange animal which resembles a jaguar. Farmers have found partly devoured carcasses of their stock in the field and three report having seen the animal. Women and children are afraid to venture forth after dark and even the men fear the beast, as they go about armed. Calves and sheep are its favorite victims. A general hunt is being organized to slay the animal.

A Strange Beast on Pine Creek

Wellsboro, Pennsylvania, *Agitator*, November 26, 1902

Mr. James Kerr, who lives not far from Tiadaghton, reports an experience with a wild beast the other night which was thrilling—to him, at least. He believes that he was followed by a large panther or lynx.

Mr. Kerr has started for Wellsboro very early in the morning with a cow. It was long before daylight, but the mood was so bright he was able to distinguish objects plainly. While passing through a piece of woods not far from Tiadaghton he was startled by seeing a strange animal suddenly make its appearance and halt in a crouching attitude a few feet ahead of him. Mr. Kerr threw a stone of the beast and hallooed loudly, which appeared to frighten it somewhat, and it slunk out of sight. Mr. Kerr then proceeded on his way down the mountain, but had gone only a short distance when the animal again made its appearance, this time being perched above him about twelve feet, ready to make a spring. Mr. Kerr was now thoroughly frightened, having no weapon but a jackknife with which to defend himself. By resorting to the same tactics as before, hallooing loudly and throwing more stones, he succeeded in driving the beast off its perch.

The cow, which also became badly frightened by the capers of the wild beast, then started down the road at a good pace, with Mr. Kerr closely following holding the end of the rope.

Before leaving the woods the strange animal made its appearance for the third time, having followed Mr. Kerr for a distance of a mile or more. Mr. Kerr believes that had it not been for the presence of the cow the beast would have attacked him. He had a good view of the animal, it being bright moonlight. It was from four to five feet long, about the size of a big dog, and was spotted.

FABULOUS JOHN P. SWOPE,
LAST OF PENNSYLVANIA'S GREAT TRAPPERS, PART XIV.
Huntingdon, Pennsylvania, *Daily News*, March 16, 1957
By Albert M. Rung
(Excerpt from Swope's journal for November 1909.)

"A man told me that he and two other men and some dogs treed a large cat 9 feet long from the nose to the tip of the tail, and he said it was striped like a tiger and rings around its tail like a coon. I told him that I would have liked to have seen it as I never saw one like that before. It was as large as a panther, and that they had killed it with the first shot, that the ball went right through its head and it fell down dead, and the dogs would not go near as they were afraid of it."

WOLF-LIKE ANIMAL SHOT AND KILLED BY WILLIAM ROTH
IT HAD BEEN SEEN AND CHASED BY FARMERS IN VICINITY
OF SHAMBURG AND HAS BUSHY TAIL LIKE FOX.
ONE FRONT FOOT MISSING
Titusville, Pennsylvania, *Herald*, January 30, 1919

William Roth of Shamburg yesterday afternoon shot and killed the strange animal which has been seen recently and even chased on several occasions by farmers and hunters in that vicinity. Before capture it was believed by many that the animal was a wolf, while others thought it was a fox, and now after its capture those who have seen the pelt are at a loss to tell exactly just what it is.

On the morning of Jan. 18, the Herald published an account of a merry chase given to this animal on the Monday night previous by Ora Shields of Shamburg. At that time Mr. Shields was attracted to his barn late in the evening by the furious barking of his dog and saw the strange animal running toward the woods.

Had Been Chased.

Mr. Shields and dog gave chase and he was close enough to the animal several times to have shot it, but he did not have a gun. When near Lonesomehurst, the home of Edward Bellows, the dog was scared by the animal and would go no further.

Mr.. Shields then went to the home of George Schmidt, near Pleasantville, routed him out, and after resting for a few minutes the chase was continued. Messrs. Shields and Schmidt ran across the animal again in a haystack on the Emmett Smith farm, near Pithole, and a shot was taken at it, but the charge missed. Mr. Prall, on the Crawford farm also saw the animal the same morning and scared it away with a pitchfork.

One Front Foot Missing.

Mr., Roth shot the animal while hunting for fox today, it being killed near the Keech farm and Pithole road, about a mile and a half from the Rattlesnake school house. That it is the same animal which Messrs. Shields and Schmidt chased there is no doubt, because they were close enough to it in their chase to see that one of its front feet was gone and the dead animal also had a missing front foot.

Mr. Roth stated to the Herald last evening that the animal resembles a grey hound, excepting in color, it being a reddish dark brown, has long pointed nose, short ears, short hair and a long bushy tail. The pelt measured a little over six feet in length from the tip of the nose to the end of its tail. Mr. Roth skinned the animal where he killed it and brought only the pelt home with him.

WEIRD BEAST DEFIES HUNTERS
WILDCAT OR MOUNTAIN LION AS BIG AS A CALF
Frederick, Maryland, *Post*, January 25, 1921

Berlin, Pa., Jan. 24.—Excitement runs high in Brush Creek neighborhood about what is believed to be a huge wildcat or mountain lion that gave chase to several young men, who declare that the varmint is as big as a two-months-old calf, with a yell that curdles the blood in one's veins.

Hunting parties, armed to the teeth, had gone out after the animal, but as yet it still is at large, and residents of that section are keeping close indoors at nights, unless in groups or fully armed.

The most authentic story of the big cat comes from the three sons of Peter Saylor, who last week went to the home of a neighbor by the name of Brondie to spend the evening. On their way home they heard a bloodcurdling yell in the laurels a short distance from the road. The young men screwed up their courage and stopped to look around, and, as they did, out on the road stalked the big cat, its eyes shining like balls of fire.

The youngest of the three brothers, Paul, in telling the story, makes no secret of the fact that the trio broke all automobile speed regulations in getting out of that vicinity. With Alvey Marsh, world war veteran, who has a record of getting nineteen Boche soldiers single-handed, the Saylor boys organized a hunting party. Marsh fired two shots at the beast but it eluded death and still is at large.

Wild Cat Caught in Tinicum Twp.
Rare Animal, Supposed to Be Nearly Extinct,
Trapped by 16-Year-Old Lad
Sportsmen to See It Here

Bucks County, Pennsylvania, *News*, January 20, 1922

A genuine wild cat, a species of animal that is now said to be very rare in this country, was trapped and killed on Monday, January 16, by Tunis Brady, the 16-yearold son of Joseph Brady, on the Cook premises, on the edge of the State Auxiliary game preserve in the Tinicum Swamp, this county. With the possible exception of two similar animals killed about two years ago by a son of Daniel Trouts in the same locality, this is probably the first wild cat known to be taken in Bucks county for nearly three-quarters of a century. The last wild cat shot in central Bucks county was killed on Spruce Hill, about a mile below Chalfont fifteen or twenty years before the Civil War. This capture is said to be the only authenticated taking of a wild cat anywhere in the county until the capture last Monday—but the Spruce Hill cat may have been a bob cat.

The carcass of the Nockamixon cat is now in the possession of Game Protector Warren Fretz, who will exhibit it at the annual meeting and smoker of the Bucks County Fish, Game and Forestry Association next Tuesday night, when it will no doubt be an object of great curiosity to the sportsmen.

Terrified the Neighborhood

For three years residents in the vicinity of the Auxiliary Game Preserve have known of the presence of some strange wild animal in the vicinity, which uttered "unearthly" yells at night. So terrifying were the cries of this animal that timid women in that sparsely settled part of the county were afraid to go out of the house at night. Young Brady tracked two strange animals a few weeks ago to a den in some huge rocks near his home on the Cook place and set a trap with the idea of capturing them. Three times they sprung the traps without being caught.

Put Up a Big Fight

Last Monday the animals were less fortunate and one of them, the male, was securely caught with one hind foot and one front foot in the trap. When Brady went to look at the traps in the morning and found that he had captured some animal, he attempted to kill it with a club, but the animal put up such a terrific fight that the boy thought it would break the trap-chain and escape. He then procured a rifle and killed it with several shots in the head. It was the male cat, and its companion, believed to be a female, is still under the rocks.

Not a Domestic "Wild" Cat

The captured animal is not a domestic cat gone back to a wild state, but is, without much doubt a distinct species—a genuine wild cat. It is not a bob cat nor a Canada lynx, nor is it a cross between these two animals, as some people have pronounced it. It corresponds exactly with the description of the wild cat contained in Dr. J. G. Wood's "Natural History," and it occupied precisely the habitat which Dr. Wood says the wild cat frequents, "rocky and woody country, making its home in the deft of some rocks," a place just like that in which this wild cat was captured.

The Cat's Measurements

The animal captured in Nockamixon has strong and powerful claws and teeth, and its head is large in proportion to its body. It apparently has not an ounce of surplus flesh, being sinewy and wiry, and yet it weighs eight and a half pounds. Its length from the tip of

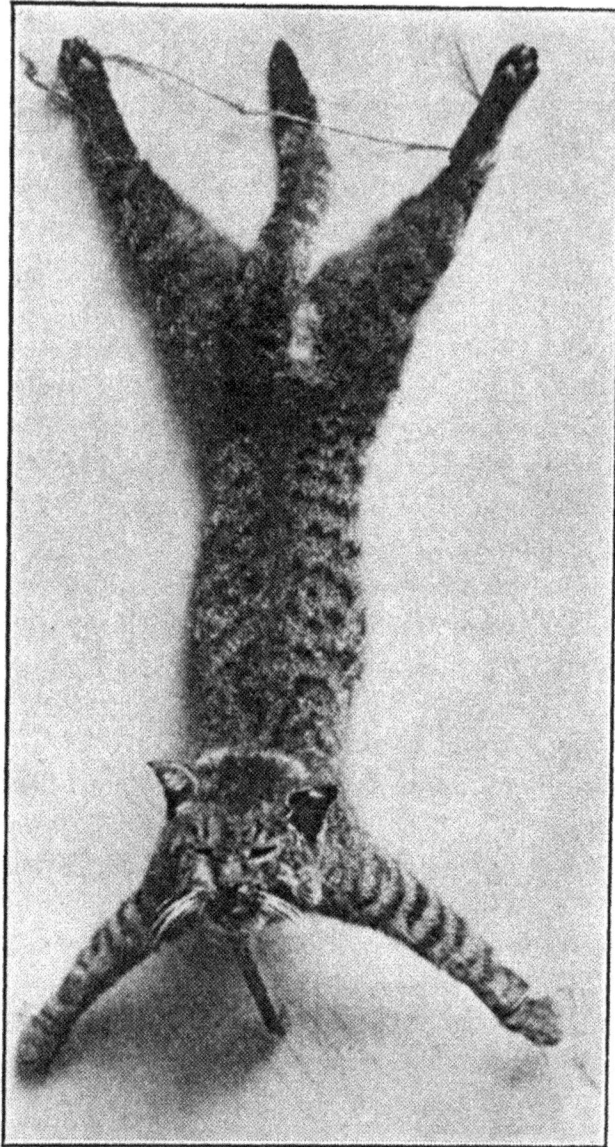

EUROPEAN WILD CAT, KILLED IN PENNSYLVANIA, 1922
(Front view)

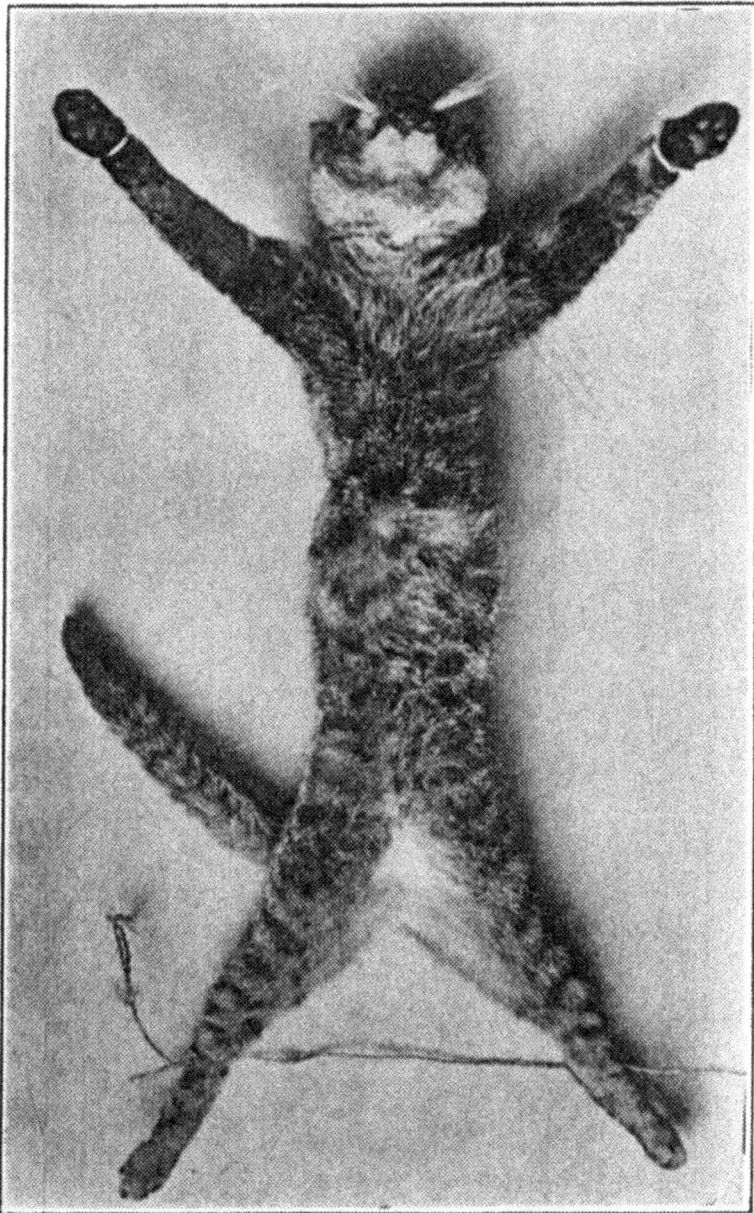

EUROPEAN WILD CAT KILLED IN PENNSYLVANIA
(Under view)

his nose to the tip of its tail is 30 inches. Its body length is 20 inches and it stands 13 inches high. Its front legs are 7 inches long and its back legs 13 inches. Its head measures 11 inches in circumference the broad way and 13 ½ inches the long way. It measures 3 inches between the ears and 7 inches across the ears. Its body is 12 inches around just behind its forelegs and 14 ½ inches around the centre. Its tail is scant 11 inches long, thick and inclined to be bushy, which distinguishes it from the domestic cat, which has a long tapering tail.

Its Peculiar Markings

Its markings are precisely those Dr. Wood describes for the wild cat—the fur being a sandy gray, with some yellow or huffy color, the buff color being particularly noticeable on the body under the hind legs. A dark line extends from the shoulders along the spine to the end of the tail. It has black "tiger" markings on the body, legs and tail. The black stripes extend entirely around the body and hind legs. On the front legs the black bands run into black patches on the under side.

The top of the head is black, and the face grey, marked with regular black lines. The "whiskers" or "feelers" are white, stiff and bristly and rather abundant. The chin and neck are lighter gray than the rest of the body and also huffy, with a median transverse double half-moon brown line. The throat has a conspicuous white spot.

The tail is of the pervailing gray color, ringed with bands of black, and has a long black tip, the whole resembling a raccoon's tail. The dark markings on the animal are not uniform in intensity, but they contrast with the grey with sufficient strength to make it a very striking-looking beast.

Distribution of the Wild Cat

The wild cat, though little known here, is widely spread, being found not only in this country, but in the British Isles, Continental Europe. Northern Asia and Nepaul, though it is not as common anywhere now as it once was. It has been claimed that it is a native of Ireland, though this is disputed. It is sometimes called the British tiger, though it now is seldom found in England.

It has been claimed that the domestic cat originated from the wild cat, but this theory has been rejected by most scientific investigators. Domestic cats sometimes revert to wild life, but no matter how many generations they continue wild, there are still certain distinctions between the wild domestic cat and the wild cat. Some naturalists claim that our race of domestic cats originated from the Egyptian cat, which more nearly resembles the domestic cat than the wild cat, but, while this theory is highly probable, it has never been satisfactorily proven.

The capture of the Nockamixon wild cat is likely to cause a stir in scientific circles, because of the rarity of the incident. It is believed that there are several more of his breed in the same territory. In fact the animal is now so little known that some recently published natural histories do not even mention it.

Tinicum Wild Cat Discusssion
Capture of European Species Raises Question of Its American Nativity
Great Interest in "Find"

Bucks County, Pennsylvania, *News*, January 21, 1922

The publication of the story yesterday in the Bucks County Daily News of the capture of an European wild cat in the Tinicum swamp by Tunis Brady, a sixteen-year-old resident on one of the tracts included in the Bucks Counts' Fish, Game and Forestry Association's Auxiliary Preserve, created great interest locally and among naturalists generally. It is also likely to raise a controversy as to whether or not the European wild cat is a native of this country. Dr. Wood, the author of the British work on natural history, quoted in yesterday's story, mentions this animal as being found in "this country," but he clearly means England and not America.

Should Be Investigated

Instead of becoming a matter of controversy, the question as to whether the European wild cat is indigenous to America ought to be a matter for serious and careful investigation. Present-day American writers on natural history do not mention the European wild cat as a member of the native fauna. While admitting that the animal captured in Tinicum may be the European wild cat (*Felis catus*), they say it is probably an escape from a menagerie and does not belong to this country's fauna.

There are others, however, who believe that the *Felis catus* is an American animal; that in the early settlement of this country it was somewhat common, hut, as it does not stand persisting hunting, had become practically extinct over a large area of the country. These claimants say that it has escaped notice in recent years because it so closely resembles the "tiger" type of domestic cat, and that hunters who have killed them in modern days have never brought them to the attention of naturalists because they thought them domestic cats, and thus they have escaped the attention of naturalists.

Naturalists May Have Overlooked It

Dr. Henry C. Mercer, of Doylestown, who saw the animal killed in Tinicum, not only says that the specimen is an European wild cat, but is inclined to the opinion that the animals are indigenous to this country. He cites a quotation from Oliver Goldsmith's "History of Animated Nature," in which Goldsmith states that Columbus was shown one of these wild cats by a hunter who killed it on this side of the Atlantic.

Well Established in Tinicum

It may be difficult to prove now that this animal is native to America. There is not much doubt, however, that it has become thoroughly established in the Tinicum Swamp. Residents in the vicinity have known of the existence of strange animals there for several years. Two similar animals were killed in that territory between two and three years ago, while the one trapped this week had a mate which is still in hiding in the rocky fastnesses of the swamp region.

The specimen captured on January 16 is still in the possession of Game Protector Fretz, who has called the attention of Dr. Woodruff, a noted naturalist of this State, to the capture. The animal will be exhibited at the annual meeting and smoker of the Bucks County Fish, Game and Forestry Association in the Sixth Regiment Armory next Tuesday night, after which it will be placed in the hands of a taxidermist for mounting.

Tinicum Cat True Wild Cat
Species Not So Near Extinct as Many Naturalists Think
Its Habits are Nocturnal

Bucks County, Pennsylvania, *News*, January 1922

By Miss Elizabeth C. Cox, of Holicong.

I have been very much interested in the Wild Cat, caught in the Tinicum swamp. I went to see it and was much surprised to read that the sportsmen generally pronounced it a "domestic cat."

Species Not Nearly Extinct

Instances that have come to my knowledge lead me to believe that it is a wild cat, and that the species is not so nearly extinct as is suggested. Years ago I read a book entitled "Early Settlers in Canada." It was a very old book, the story of an English family who settled in central Canada, while that country was a trackless wilderness. One night one of the boys of the family shot an animal in a tree near the house. When they examined it, they think it a domestic cat, but an old man, a hunter and trapper, who has spent practically all his life in the forest, tells them that it is a wild cat, that while it is smaller than either the Canadian lynx or the more common bobcat, it is much more fierce than either of them and cannot be tamed, and the reason they have not encountered it before is because it is entirely nocturnal in its habits. The description of the animal, in the book, as I remember it tallies perfectly with the specimen in Mr. Fretz's possession.

One in the Philadelphia Zoo

One time at the Philadelphia Zoo they had in a small cage an animal ticketed "Indian Devil." When I first saw it it was curled up asleep, and looked like a large house cat, but when aroused it jumped against the bars of the cage hissing and spitting, with ears laid close to its head, bit at the bars and seemed in a perfect frenzy to attack us.

Some time afterwards, when visiting the zoo, I asked an old man, who was working about the garden, if he remembered the "Indian Devil."

"Oh yes," he said "it was one of them wild cats out of the Blue Ridge Mountains. It soon died. You can't keep 'em in captivity. Why, that beast just spent every night while it lived buttin' its head agin the bars tryin' to git out. It eat enough, but it just killed itself tryin' to git away."

As I remember the cat at the zoo it was not quite as distinctly striped as Mr. Fretz's cat, and was, perhaps, a little larger, but it had the same large round head, flat ears and short ringed tail.

Found in Fayette County

A lady whose home was at one time on the summit of Chestnut Ridge, a spur of the Allegheny Mountains, in Fayette county, Pennsylvania, tells me that these long-tailed wild cats are by no means uncommon there; that often in driving at night they would see a wild cat cross the road in front of the car, taking his time and staring at the lights as he went. She says she never knew them to attack anyone, but that when they drove in a carriage they always carried a pistol as a protection against wildcats, and that they were always warned that if they had meat in the carriage and encountered a wild cat to throw the meat to it and get away as fast as possible.

(Miss Cox's contribution to the Wild Cat question is very interesting and valuable. Miss Cox is one of the most careful investigators and most conscientious naturalists in Bucks county, and her opinion and the facts she relates should have much weight. It is to be hoped that other naturalists will make similar inquiries.—Editor.)

KILLS MAINE TIGER CAT
Frederick, Maryland, *News*, February 9, 1924

Allentown, Pa., Feb. 9.—Game Protector James D. Geary, of this city, shot and killed on the Blue Mountain near here a rare animal known as the Maine tiger cat, said to be a cross between an Angora cat and a raccoon.

The animal was devouring a pheasant when the warden came upon it and shot it. The cat was still alive when Geary picked it up and put up a desperate struggle. The animal weighed eighteen pounds and is being mounted.

A WILD TAME CAT?

Pennsylvania Game News, June 1934

We were sitting around the large egg-stove in the hunting lodge of the Little Sanders Springs Camp. This hunting lodge is situated in the huge swale at the head of Foley's Draft, in the mountains about eight miles from Driftwood, a dilapidated town reminiscent of the great lumber days in this region. The stories were flying fast, some true, and some tainted with white lies hunters are apt to tell. Finally several of the boys made, break in the tales by getting up and going outside for a breath of fresh air and to limber up their legs, which were stiff from the day's hike. (We were hunting bear that day.)

On returning from outside they reported hearing a peculiar meowing. It sounded as if it was the noise of a kitten. Then, again and again, the same weak meow was heard in the lull between stories.

One of the automobiles returned to camp from a trip to the town to pick up a battery which had gone dead. Work progressed rapidly on replacing the charged battery, for some of the fellows expected to break camp and return home for Sunday. The hunter holding the lantern shouted, "There he is."

"There is what?" was the answer from the others.

"The two eyes that were making the meow we heard before," he replied.

Sure enough, out of the darkness shone two balls of fire. The other hunters in the camp rushed to the door.

"Get your gun, Frank," shouted Ray, "I don't want to go out in the snow in my slippers."

Frank reached for his 20 gauge and slipped in a few shells.

"Flash him again," Frank asked.

"There he is," replied Chick, "get him."

"Hold that flash on him."

Meow—then a long hard meow. It turned, stopped and turned around again. "Flash him again."

Bang!—spouted Frank's gun.

"You got him."

With the aid of a long rattlesnake club Chick removed the animal. First came the head, then the shoulders, the long body, the tail. Everyone was astonished at the size of the cat.

"It's a wildcat."

"No, it's a tame cat gone wild."

"Its tail is too long for a true wild cat," said Chick.

"Yes, but look at the tusks, and the broad face, and its eyes are too large, and its ears have the lashes of a wild cat," remonstrated Ray.

"Sure he's a wild cat," assured Frank.

Then the bets were on. What was this animal? It tested, the hunter's knowledge of Pennsylvania's wild animals.

After consulting several experts on wild life, and after all was said and done, the animal turned out to be a male cat, 37 inches long from the tip of his nose to the end of his tail and weighing 13 pounds. The head was large, the face was flat and blunt, with long whiskers, protruding tusks, and sharply pointed ears, which had long black tufts on the tips. The shoulders were broad, the body stout, the tail long. The color markings were light yellowish brown on his stomach, with dark spots. The back was dark brownish tan, with dark stripes. The tail was about eight inches long, dark brown with black rings.

The final conclusions were: cross between a wildcat and a common house (tiger) cat.

How much small game would an animal like this destroy during the nesting season in the spring? in a week? In a month? in a year? in comparison with a hunter, who only kills during hunting season? How large a territory would this animal cover? How would it get into the woods in the first place?—Ray Hornfeck and Frank Lang, McKeesport, Pa.

HOLD THAT TIGER!

JOBS CORNERS FOLK ORGANIZE POSSE TO TRAP STRANGE WILD BEAST

Wellsboro, Pennsylvania, *Gazette*, February 27, 1936

By W. A. Winner

Jobs Corners—As soon as the snow settles enough to permit traveling in the big woods east of the village, sportsmen and farmers will make up a posse to trap the wild animal that has terrorized man and beast in these regions for the last two years.

Fail to Capture Animal

The severe winter weather evidently drove the beast from his jungle living quarters to an open hillside last Tuesday, where he was discovered by Joseph Shuckwood, whose attention was attracted by the severe cries and screams of the beast as he crouched under a tree facing O. J. Furman's farm buildings. Mr. Furman took up his rifle and sent Joe around to drive the animal closer for a direct shot. When Joe became stuck in the snow about up to his waist, the animal just leaped back over the hill to the big [—] that is practically impenetrable. Mr. Furman describes the animal as being the size of a lion, black, with a long tail curled up in that fashion. The animal was in this vicinity again last Thursday night as his blood-curdling cries could be heard in the cold still night.

Lynn Kilgore, who lives near the north end of the village, said he has at times heard undescribably distressing screams coming out of the woods.

While hunting birds over a year ago, F. N. Garrison said his attention was attracted by a large animal close behind him, but by the time he had reloaded with heavier ammunition the beast had disappeared. During the past summer Myrt Scott, O. J. Furman, and Austin Prutsman lost cattle and sheep killed by some large unknown animal. Some of the carcasses were dragged to the wire fence which surrounds the big woods. Dogs were suspected of some of the killings.

Recalls Last Panther

While the farmers believe this noxious animal may be a specie of Puma or Mountain Lion, older settlers tell of the last one known to have been seen in this region.

The beast became lodged in a woodshed near Coryland where it had gone in to devour a fresh-killed porker. The door closed on the animal and it jumped from side to side, throwing its whole weight against the door, which was only held from breaking into the kitchen by a lone woman who held the door intact until men came home and shot the brute through a crack in the door.

Just a few years ago one of these animals terrorized a woman who was traveling through the woods south of Ralston. The animal would follow along the path, running ahead at times, climb trees, and wait for the traveler to pass through under.

CLAIM PANTHER AGAIN ROAMS IN NEARBY WOODS
Smethport, Pennsylvania, *McKean County Democrat*, April 21, 1938

For several years what the skeptics called a "phantom panther" has been seen roaming the forests of Tioga and Potter counties.

This so-called spirit from the "happy hunting ground" has recently taken on flesh and blood proportions as stories pour in from reliable sportsmen that they have seen this animal close enough to identify it as a large panther the size of a big police dog. It has a long tail and its body is almost black. It was last seen near Gaines by two men who authenticated the story to Game Protector Leslie Wood.

Wood states that at any gathering of sportsmen a mention of the Tioga-Potter county panther brings forth more proof of its reality. Recently a Potter county game protector told Wood of having seen the animal in this section.

Since panthers are extinct in this section and black panthers were never native to the Pine Creek forests, Mr. Wood gives it as his opinion that this animal was either imported for a pet when it was a cub and later released or that it escaped from a zoo or a circus.

MANY REPORTS OF WILD BEASTS 'ON THE LOOSE'
Gettysburg, Pennsylvania, *Star and Sentinel*, November 18, 1944

A good sized menagerie could be assembled if all of the wild animals reported running loose in this section of the country were captured—but local authorities said today that all of the rumors so far are "still a lot of hokum."

Stories from the lower end of the county this morning were that some kind of a large beast was running loose through the Mount Joy township section. The rumors began a week or more ago when a black panther supposedly had escaped from a circus train at Westminster and started a series of attacks on both cattle and humans in Carroll county.

Panther, Wildcat or Dog
The panther finally as worked into a wildcat and the stories of its exploits moved into the Union Mills section. Up until Tuesday the animal was supposed to be inhabiting the area around that Maryland town, but this morning the story was told that it had moved north for the winter and had invaded Mount Joy township.

Maryland State Game Warden E. Lee LeCompte said Tuesday that the attacker and destroyer of cattle in Carroll county is probably a pack of wild dogs. While discounting the panther tale, Maryland State police said there was foundation for reports of destruction by a wild animal.

With the panther theory discredited, the possibility of a wildcat arose, but LeCompte scoffed at such an idea. He said he never heard of a f wildcat attacking cattle. A wildcat will jump on the back of a deer and kill it, he said, but not a cow, and certainly not a bull.

Pennsylvania State Game Warden Lea Bushman reported Thursday that he had received no calls concerning a fierce cattle-attacking animal in the county—with the exception of a gorilla and a mountain lion on the loose near Fountaindale. However, he reported, he hadn't yet gotten around to hunting for either.

State police here also said they had not been asked to investigate the crimes of wild animals. If they wait for a few days perhaps the rumors will have the beast in Gettysburg and they won't have to waste any gasoline traveling south to find them, the officers observed.

Sunset Heights Receives Visit From Panther
Woman and Children See It,
and Armed Men and Boys Go On Hunt
Seen In Titusville

Titusville, Pennsylvania, *Herald*, August 23, 1945

Residents of Sunset Heights wiped the grease off their bear guns and went out yesterday in search of the "black panther" which had the temerity to invade the settlement shortly after noon.

Mrs. Judy Walter, 106 North Dillon Drive, was in her back yard husking sweet corn and near her were two little children three to four years old whose names she does not know. Hearing a slight noise in a clump of bushes less than 20 feet from her, she glanced up, then stiffened.

That dread spectacle, the black panther, to use her own words, went slinking through the grass along a path and into another patch of brush. She saw it walk along hurriedly for about 30 feet before it disappeared.

Was she scared? She screamed at the top of her voice. The little tots saw it too and exclaimed "Ooh! big cat."

Mr. McGaha, who has the garbage removal contract at the Heights, was next door, less than forty feet from Mrs. Walter, and when she screamed looked up from his work, but too late to see what she saw. He dropped what he was doing and ran to the woman, who seemed thoroughly frightened. After learning what had caused her fright, he ran into the bushes indicated, but found nothing. The screams had evidently frightened the animal away.

Had Beautiful Black Coat

Mrs. Walter described the body of the animal as being about 30 inches long, with a tail about the same length. She said that it was about knee high, but as it was close to the ground, it was probably taller.

"It was shiny and jet black," she said. "The sun was glistening on him and he had a pretty coat."

Mrs. Walter admitted her knees trembled for some time and she could not recall many other details about the big cat. She is sure that it did not look toward her, evidently being as scared as she was and concerned only in finding refuge. It was traveling east.

Within a few minutes, word of the incident got around the settlement and there was considerable excitement. Mr. McGaha borrowed a rifle at once and attempted to locate the animal without avail. Several other men and boys got out their rifles and took up the hunt in the afternoon and evening, but came back empty handed.

Matthew Miller, one of the Heights caretakers, joined in the hunt and stayed out for two hours. He was unable to find any trace of the animal around the big rocks where he figured that it probably would go.

Child Tells of "Black Bear"

Later in the day, he talked to a resident of the neighborhood who told of his child roaming among the big rocks Tuesday with his toy gun. When the boy came home the father asked, "Did you shoot anything?" The youthful hunter replied in pop-eyed seriousness, "I shot a big black bear and he went into a hole in the rocks." He convinced his father that he had seen something big and black, but the parent dismissed it in his own mind as being a dog, until yesterday's incident.

The Herald was further informed of two men who were working in a garden out over North Perry street hill one evening last week. They heard what they thought was a woman screaming in the nearby woods. They left their work and ran over into the woods to investigate the cause of the commotion. They found nothing, but now think it was the panther.

Seen Here by Two Persons

When the police were informed last evening of the panther scare on Sunset Heights. Officer Roy Loomis said, "That animal was seen up on West Elm street between First and Second streets just the other night. A fellow told me of seeing it cross Elm street and go back of Bob Smith's house toward Oak. He told me not to tell anyone because he didn't want to be guyed by his friends." The description of the animal seen yesterday was much the same as that given by the West End resident.

Officer Pete Brady added. "A fellow upon Union street was telling me that he saw a similar animal in the grass near his home last, night. He went into the house for a flashlight to get a better look at it, but the animal was gone when he came out with the light."

OUR BLACK PANTHER MAY BE OFFSPRING OF CARNIVAL ANIMALS
Titusville, Pennsylvania, *Herald*, August 25, 1945

Says the Meadville *Tribune-Republican*:

"The saga of the black panther which Titusville residents claim they have seen—is rapidly expanding.

"Latest theory is that said panther is the offspring of two other wandering black panthers—a unique situation, indeed.

"But to pick up loose ends of the story—according to reports of local residents familiar with the panther scare that originated at Corry, two black panthers escaped two or three years ago from a wrecked carnival in nearby New York state.

"Some time later one of the panthers was shot, the story goes, somewhere in the Corry vicinity. Not long afterwards, two panthers were sighted together in the same vicinity which led to the theory that the original two panthers had mated before one was shot.

"Proof that the Corry residents are taking the panther seriously is the report that a substantial bounty—quoted in the $1,500 bracket—has been raised by a group of citizens as a reward for the capture or killing of the animal."

PANTHER RETURNS TO FORMER HAUNTS
Titusville, Pennsylvania, *Herald*, September 1, 1945

Says the *Corry Evening Journal* of Friday:

"Reports to police headquarters early this morning would indicate that the black panther, last reported seen by several persons in the area near Titusville, may have returned to this city.

"For the first time, the big black cat is reported having been sighted on the south side of town. Arthur and Edward Giacoma, altar boys at St. Elizabeth's church, were leaving the sacristy door after the 6 o'clock mass this morning when they saw the big animal near the St. Elizabeth's Hall building in the rear of the church edifice. They declared that it was as large as a police dog and glossy black.

"Rev. Father Heidt was called by the boys, but the animal had disappeared and police officers on duty at that time were unable later to find a trace of it."

LATEST 'PANTHER' THEORY HAS IT A DOG
Titusville, Pennsylvania, *Herald*, October 4, 1945

The *Herald* was informed last evening that the latest theory regarding the mysterious creature known as "the black panther" is:

It is a black dog owned by Charles Foss of Pleasantville. The family formerly lived in Titusville near Swede Hill. It is theorized that the animal, which is much given to wandering, may merely have been roaming about his old haunts when the "panther" was seen in this city on the North side earlier in the fall.

He is a big fellow and presents a ferocious appearance, according to our informant, but. has never been known to harm anyone and actually has a friendly way with those he knows. The next time we hear of the panther's prowlings, we intend to check with Mr. Foss to find if his dog has been out visiting.

'Black Panther' Hunted in Jefferson County
Titusville, Pennsylvania, *Herald*, October 11, 1945

Ringgold, Pa., Oct. 10.—(AP)—Despite skepticism in official Harrisburg circles, residents near this Jefferson county, community declare a black panther is on the prowl and that with the first tracking snow they'll get the cat.

There is a definite wave of panther fear. Women stick close home, children are kept indoors after school. A band of hunters spent hours in the fields and woods after a report the animal had been sighted.

How do they know it's a panther? Dogs set on the trail start out enthusiastically but return shivering.

When "black panther" stories drifted in a few weeks ago, the State Game Commission scoffed, declared panthers have been extinct in the commonwealth for many years.

But, "you can kid all you want about it" declared one resident "To us, it's mighty real. When the snows come, we're going to track it down and kill it. We figure we'll have something to show that might surprise you."

Beware of "The Thing"
Stalk Mysterious Beast Near Pottstown, Pa.
Lowell, Massachusetts, *Sun*, November 11, 1945

Pottstown, Pa., Nov. 14 (AP)—The nocturnal tramp-tramp-tramp of a posse combing the woods of nearby Sheep's Hill for a mysterious beast that makes 20-foot leaps, cries and screams like a baby and steals chickens, is not good for the nerves.

Especially, says Mrs. Edward Creger of Sheep's Hill, "to mine."

She complained to police that hunting parties trespass across her property and take pot-shots at black shadows late at night.

"Maybe I'll start some shooting myself," she said, "if this nonsense doesn't stop."

Meantime "The Thing"—as residents in the area have tagged the mystery animal—is responsible for four casualties among those who have stalked "it" the past five days.

Police reported William J. Brandel, 18-year-old Pottstown youth, was struck in the thigh by a "pumpkin ball" bullet; Betty Hart, 17, of nearby Douglassville, was injured in the left arm by a "trigger-nervous" posse member, and a young couple—tracking "The Thing" by auto—were seriously injured when a frightful screech caused the driver to lose control of the car.

Variously it has been described as a panther, a puma, a wild chow dog, a bear and a black fox. But until the prowling posses nabs the shrieking beast it will be just "The Thing of Sheep's Hill."

BIRD, FISH OR ANIMAL?
Indiana, Pennsylvania, *Evening Gazette*, November 14, 1945

Pottstown. Pa., Nov, 14.—(AP)—The screech on Sheep's Hill had residents of this eastern Pennsylvania area stumped yesterday.

A mysterious animal has been feasting on chickens and turkeys— and a posse and 15 foxhounds joined the search last night.

Varying identifications of the animal with its wailing tons:

(A) Policeman John J. Boyle—Panther. "Last night I took along a fellow from upper New York state. He's an old hand with panthers. He told me the animal was bound to be a panther." It has a big tail and gleaming yellow eyes.

(B) Housewife Mrs. Mabel Herd—Puma. A pair of pumas escaped from a circus near here two years ago and an encyclopedia furnished adequate identification.

(C) Game Warden Peter J. Filkowski—Chow, "From what the Sheep's Hill people tell me about the animal, its face seems to resemble the face of a chow dog. Chow dogs are black; their eyes are yellow."

(D) Philadelphia Zoo officials—Bear. Bears' vocal chords often surprise with the diversity of their offerings.

No matter what it is, Boyle says, it's going to be a dead duck, or is it?

POSSES HUNT 'THE THING OF SHEEP'S HILL'
Danville, Virginia, *Bee*, November 14, 1945

Pottstown, Pa., Nov. 14.—(AP)—The nocturnal tramp-tramp-tramp of a posse combing the woods of nearby Sheep's Hill for a mysterious beast that makes 20-foot leaps, cries and screams like a baby and steals chickens, is not good for the nerves.

Especially, says Mrs. Edward Creger of Sheep's Hill, "to mine." She complained to police that hunting parties trespass across her property and take pot-shots s at black shadows late at night.

"Maybe I'll start some shooting myself," she said, "if this nonsense doesn't stop."

Meantime "the thing"—as residents in the area, have tagged the mystery animal—is responsible for four casualties among those who have stalked "it" the past five days.

Police reported William J. Brandel, 18-year-old Pottstown youth, was struck in the thigh by a "pumpkin ball" bullet; Betty Hart, 17, of nearby Douglassville, was injured in the left arm by a "trigger-nervous" posse member and a young couple—tracking "the thing" by auto—were seriously injured when a frightful screech caused the driver to lose control of the car.

John Hipple, a Montgomery county farmer, says he saw the animal.

"It was like a big cat," he recounted. "I shot at it and it leaped 20 feet into air and, screaming, disappeared."

Another—John Wojack of Pottstown—spotted the huge beast:

"It gave a shrill cry, then it bounded away in leaps of at least 10 feet in length each." Variously it has been described as a panther, a puma, a wild chow dog, a bear and a black fox. But until the prowling posse nabs the shrieking beast it will be just "the thing of Sheep's Hill."

TRIGGER-HAPPY HUNTERS ARE MORE DANGEROUS THAN MONSTER MONTIE
Lebanon, Pennsylvania, *Daily News*, November 15, 1945

Pottstown, Pa., Today.—(UP)—Residents of North Coventry Township were not worried about Montie the monster today.

They were more concerned with an invasion of trigger-happy hunters flocking to get a shot at the mysterious animal—panther, dog, puma or what have you—which has raided hen houses in the area for nearly two weeks.

One resident of Sheep's Hill, where the what-is-it reportedly was seen, described the situation as "worse than the battle of Gettysburg."

"They are not apt to shoot a wolf or whatever it is, but they will likely shoot me if I am a foot above the ground."

Another, Mrs. Edward Cregor, said she might "start some shooting of my own if this nonsense doesn't stop."

State police entered the picture as local inhabitants surveying roads clogged with automobiles jam-packed with hunters, threatened to go "up in arms."

Counting four casualties already as the result of searches organized by impromptu bands of youths, state troopers said they would assume the hunt, personally.

In addition, they fined two members of an unofficial posse $25 each for illegally carrying weapons on charges preferred by District Game Warden Peter J. Filkosky. Any other civilians caught bearing firearms in the area will be dealt with similarly, police said.

The mysterious Montie, which caused the furore, has been described variously as a panther, a chow dog, and a puma. People who swear they saw Montie say he is about three feet long and has a bushy tall "with a kink in it."

Some claim the animal has the wail of a banshee, others say he barks like a dog, while still others insist he laughs like a hyena.

SET TRAP FOR "THING" ON SHEEP HILL
Indiana, Pennsylvania, *Evening Gazette*, November 16, 1945

Pottstown, Pa., Nov 16.—(AP)—Police set bear traps around the community of Sheep Hill yesterday in an effort to capture a mysterious beast that stalks the countryside, giving vent to nocturnal, man-sounding screams.

County Game Protector Peter Filkowsky said residents have been warned by police to keep away from the area.

"The thing"—as residents refer to the animal— "must be getting hungry by now," Filkowsky said, "and if it is a panther or some similar creature he can't live on what he finds in the woods " Filkowsky, in fact, wasn't sure it was an animal.

"If it is," he remarked, "it's a freak" and pointed out that no two description of "the thing" have tallied. Some people say it may be a panther; others a puma, a bear a chow dog, a black fox and a huge cat.

And when it gives out with voice it really shrieks.

Lester Thompson of nearby Douglassville says, "it starts sort of low-pitched, gives a couple of short bursts first then it lets her rip."

To 15-year-old Claude Reinhart of Pottstown, "the thing" sounded like "a man screaming as loud a he could."

"The thing," has raided turkey runs and chicken farms in the vicinity, taking reported leaps of up to 20 feet in length to escape armed posses and hunting dogs.

Black Panther Seen Hear Corry
Dixon, Illinois, *Evening Telegraph*, July 30, 1946

Erie, PA., July 30—(AP)—A black panther was reported yesterday on Route 277, a mile from Corry. J. E. Fitch of County Line road said he hurled a rock at the animal and that it darted into bushes along the road.

Chester Fisher Sees White Panther In Yard
Lock Haven, Pennsylvania, *Express*, August 1, 1946

They swear that it's so—and not an hallucination.

Chester Fisher and his father, Harry Fisher of Allison Twp. saw a white panther early this
morning in the Fisher yard.

Chester, who had his attention attracted by the low growls of the family dog, sat up in bed, and saw the panther alight on the ground apparently from a leap from the steps of the house. Its color was so light that it stood out clearly in the darkness—about eight feet long, including its tail, he estimated.

The animal made two jumps, cleared a five foot bush, and went off toward the hills near the Fisher home.

Just what brought the prowler into civilization, the Fishers didn't know. But nearby the Lakners have a large number of chickens in a field, and this might have attracted the animal.

To prove that the visitation wasn't a dream in the middle of the night, the Fishers had tracks in their yard this morning.

MYSTERY OF 'PANTHER' SOLVED BY LITTLE GIRL
Titusville, Pennsylvania, *Herald*, August 10, 1946

Rumor that the "mysterious black panther," whose alleged prowlings have thrilled and chilled residents in the Bear Lake area, had met with a horrible but entirely civilized death sent about forty persons, including newspaper men and photographers from Warren and Jamestown to the state line on the Busti-Sugar Grove road Thursday where the "elusive" panther was reported to have been struck down by an automobile.

Inspecting the carcass, "panther experts" said the deceased beast was not the newsmaking panther, but a "long tailed wildcat." With the highway jammed with many excited observers, dozens of camera shutters clicked to immortalize the "long tailed wildcat" in photograph albums.

When the road became blocked by more and more of the curious, a little girl stepped up to the "panther and wildcat experts" and said, "Gee mister, that ain't no wildcat! That's my big tomcat. He was killed by a car yesterday."

BLACK PANTHER IS SEEN NEAR HEAD OF JEFFERSON AVENUE
Warren, Pennsylvania, *Times-Mirror*, October 3, 1946

Jamestown and Corry beware. Warren county now has a black panther of its own and the excitement at Corry and Jamestown may well die down and lay low while Warren county hunts and kills its slinky black fellow whose shrieks at night send cold chills up and down spines and curdle the blood and make goose pimples grow all round.

The panther of Warren county was reported to the police department last evening as having been sighted by Donald Heim, 217 Jefferson street. Young Heim who is a nature lover and also a hunter during the hunting season was out Tuesday evening in the woods above Prospect street when he saw the critter. It was near an old oil shanty on the hill and he describes it as being about three feet in length and 18 inches high and black in color. "As first I thought it was a big cat but on taking a second look I knew it was a panther."

Young Heim, who is 16, hurriedly went to his home and with a neighbor went back to the scene both carrying guns. Search in the vicinity failed to turn up the big black fellow. They have been hunting the animal since but have not gained sight of it. However Mrs. Lora Heim, mother of Donald, has heard the shrieks of the animal at night and states they are bloodcurdling and eerie.

It is also reported that Clifford Myers, Jefferson street, has sighted the animal, but states that he saw a pair of the big animals in the woods above Prospect street. Mr. Myers could not be contacted today.

Several persons have determined to hunt out the "painters" with dogs and get the animal or animals and residents in the vicinity have cautioned their children against playing in the woodlands of the vicinity.

A black panther was reportedly seen in old Irvinedale park some years ago by Dr. L. E. Chapman while he was making the rounds of his oil lease there. It was up on the limb of a tree and he had a good view of the animal and is certain it was a panther.

It's Panther Time Again; Cries Heard
Verified Reports of Screams Come from Newtontown Area
Dogs Are Scared
Titusville, Pennsylvania, *Herald*, October 4, 1946

Residents of the Newtontown neighborhood, about four miles northwest of Titusville, are keeping close to their homes these nights after being frightened by screams they think are those of a panther or other wild animal.

This calls to mind the reports of a year ago when an animal, described as a black panther, was said to be roaming through the region, spreading great apprehension.

In the interval since the animal was reported seen in the Titusville region, stories have appeared in other newspapers to the north of such an animal having been seen near Union City, North East and places east of those towns.

A resident of Cherrytree township told a Herald man about three weeks ago that the animal had been sighted in the Hamilton Corners vicinity in that township and a couple had been unsuccessful in running down the beast although their car was traveling 40 miles per hour before the big cat had darted into the brush.

Cries of a strange animal were heard later in the vicinity and although a search was made at night, the animal was not seen there afterward as far as could be learned.

A young man calling at the John Price home at Newtontown on Tuesday night was just leaving when he was startled by a blood-curdling scream high on the hill south of Oil Creek. He turned back and spoke to a member of the Price family who later heard it, but at a distance. He had previously heard an animal running in the nearby field.

Ruth and Patricia Price were doing chores at the barn on Wednesday night and when finished at the barn, decided to go to the rabbit pen, farther from the house. Just as they turned off the barn light, they heard a noise which they described as sounding like a cry made by one of their dogs when he was caught in a trap at one time.

They did not locate the distance or direction of the noise closely and paying little attention, went to the rabbit pen. They were suddenly terrorized by a piercing scream just a short distance ahead of them that momentarily froze them to the spot and then they fled for the shelter of the house, yelling at the top of their voices.

Mrs. Price said the older daughter, Ruth, was ill the rest of the evening. Patricia said she cleared a large dog house in a jump when running away. Their mother later heard a similar cry off in the direction of Graytown. The family's dogs would not leave the back porch for the rest of the night.

W. R. (Rusty) Orum, son-in-law of the Prices, told of having been aroused by a large animal prowling on his porch twice this week, but when he stirred in bed the animal dashed from the porch. The Orum cottage is about a quarter-mile from the Price home.

A pup owned by Orum was so badly scared that he crawled under the house and had to be forcibly dragged from his hiding place. Since that time, he howls at night until brought inside.

All this may not prove a thing except that one should not stand in the path of the Price girls when they are scared, as a matter of personal safety. Information has come from so many sources regarding the existence of a strange feline that such an animal may exist.

The only discrepancies in the panther theory are: (1) the animal sighted is said to be black. Black panthers are natives of Asia and their rare appearance here in this country is only behind the bars of strong cages, because they are untamable; (2) no livestock has been reported missing in the areas through which the panther is supposed to have wandered; (3) it is unlikely that an animal of this type would stay in a populous region, but would retreat instead to the remote fastness of the big woods.

HERE AND THERE IN SPORTS' LAND

New Castle, Pennsylvania, *News* (excerpt), October 7, 1946

Don Helm reports he saw a black panther on the outskirts of Warren, Pa. His mother also reported having heard the scream of a panther at night.

BLACK PANTHER IS SEEN NEAR WARREN

New Castle, Pennsylvania, *News*, October 16, 1946

Warren, Pa., Oct. 16.—(INS)—The report that a black panther has been on the loose in the vicinity of Warren gained impetus today when three women said they saw the animal sunning itself on the roof of a chicken coop.

Mrs. Elle Magini said she and two guests saw the panther leap from its resting place and escape in a wooded section.

During the past two weeks reports of the animal's presence has been received by game wardens and police.

PHANTOM PANTHER IS APPEARING IN CORRY

Dunkirk, New York, *Evening Observer*, February 1, 1947

Corry—Mrs. Merle Dodd of Columbus, who insists she knows a panther when she sees one, said that the phantom panther, which recently created an air of mystery throughout Chautauqua county, was now making matinee appearances in the Corry area.

Mrs. Dodd reported seeing the animal cross a road near Corry and gave the usual description as being dark in color and having a cat-like head. The panther, or mountain lion, appeared to come out of a swamp just before dusk and that she had ample time to observe its actions.

BLACK PANTHER BEING HUNTED
Lima, Ohio, *News*, July 10, 1947

Warren, July 10—(INS)—Searching parties were organized today in nearby Bloomfield to track down a reported "black panther" that was terrorizing residents of the area.

State highway patrolmen combed the wooded area in the Bloomfield sector yesterday but found nothing but sets of unidentified tracks.

Orville Earl told the police that the animal was a panther. He said he had "seen them before." Another resident of the area who reported spotted the "panther" was Alfred Carlson, who said he fired a shot at the "cat."

Many others have insisted they heard the animal scream in the night. Dogs were being pressed into service to aid in the organized hunt of the woods.

DEPUTIES HUNT 'BLACK PANTHER'
Massillon, Ohio, *Evening Independent*, September 12, 1947

Warren—Sheriff Ralph R. Millikin and his deputies Thursday were out after that "black panther" in the Parkman area of Trumbull country.

Mrs. Ida Bronson, who lives just off route 422, said she saw a "coal black beast jumping over a fence in a field near my home." The sheriff said that "the beast killed a calf Wednesday night on a neighboring farm and carried most of the carcass away."

THING BACK AGAIN? SOME THINK IT IS
Pottstown, Pennsylvania, *Mercury*, November 20, 1948

Frightened residents of North Coventry township, recalling the terror spread by the weird animal which roamed the Cedarville area three years ago, fear that the beast, or one of its offspring, has returned.

According to reports, Victor Wagner, 42 East Main street, South Pottstown, and Chief of Police Daniel Guldin are alleged to have spotted the strange beast roaming the property of a D. K. Bullens, Schuylkill road.

They have described it, it is said, as a black cat-like animal about the size of a bird-dog. It leaps across the fields with a bounding motion more than it runs, it was reported.

Three years ago, residents recall, "The Thing of Sheep Hill" was believed to be a mountain lion or wildcat which had ranged into the area. Large tracks and half-eaten carcasses of rabbits and birds were found in several places.

The inhuman screams emitted by the animal at that time were described as sounding like a "human being in mortal agony." For more than five weeks, the cries heard at night spread horror throughout North Coventry township.

At that time, posses were formed to hunt the beast, and while it was seen many times, no one could get a clear shot at it.

Now, residents fear it may have returned. If it is not the same animal, they said, it surely is one of its off-spring because it resembles it strongly.

Black Panther Again

Warren, Pennsylvania, *Times-Mirror*, October 13, 1948

After nearly a year without any reports of a black panther in this area, now comes the story that such an animal was seen last Friday by William, 11-year-old son of Ernest Dunnewold, who resides six miles north of Corry. Later in the day Mr. and Mrs. Dunnewold saw the animal in the pasture with the cattle, but before he could get a gun it had disappeared.

Black Panther 'Stalks' in Town

Syracuse, New York, *Post-Standard*, February 6, 1951

Clairton, Pa., Feb. 5. (AP)—This bustling steel mill town is enjoying the thrills and chills of a black panther scare.

Police began by scoffing at reports of excited residents who telephoned with news of seeing a black panther. Now police aren't so sure. Chief Peter Orsini and others have seen the animal.

The chief said the beast is about four feet long with a long tail. How the animal reached this industrial center near Pittsburgh remains a mystery.

In September 1951, Lynn Wycoff trapped a wildcat with a foot-long tail south of Wharton, Pennsylvania. The cat was kept alive, but could not be tamed. After keeping it for three months, Wycoff said, "It is a big cat now. Nobody can tell me that it is a bobcat. Their color lightens as they grow older but this animal's color has not changed" (Lyman 1973).

'BLACK PANTHER' IS REPORTED ROAMING MILE FROM THIS CITY
Titusville, Pennsylvania, *Herald*, September 5, 1951

A strange large animal described as a black panther was reported yesterday to be in this vicinity— in fact, about a mile south of Titusville.

Paul Olsen of Titusville, Route 3, said yesterday his son, Grant, had seen a large black animal in the woods about 100 yards from their home, which is on the left side of Route 8 going south about a mile from the top of South Franklin street hill.

Mr. Olson said his son first saw the animal about three weeks ago as it darted into the dense woods. It was bigger than a large dog and had a long black tail, he declared.

Two weeks ago the son saw tracks which were thought to have been made by the elusive animal. The paw prints, found near a spring run, were five inches long by about four and a half and resembled those of a cat. Mr. Olsen measured them.

Later more tracks were seen throughout the neighborhood and most recently were observed on Monday by the son and several other men from the neighborhood. The son said he thought he and the other men had aroused the animal since they found the spot where it had been stretched out, and there were fresh tracks.

Mr. Olsen said other persons have reported seeing the animal and one person reported hearing it give a terrifying cry as it sped into the woods. Another person said he saw it as it cleared the highway in one tremendous leap.

NOW COUNTY HAS 'THING'
Charleroi, Pennsylvania, *Mail*, January 27, 1951

"The Thing" popped up today in the Lone Pine section of Washington County, stealing wildcat thunder from Tidioute and the northern tier of Pennsylvania.

A varmint mountain lion, at least some kind of wild animal, has been reported in the local region. It is described as a large, sleek, black beast, about three feet long, weighing about 175 pounds, with a long tail.

Mrs. Anthony Chopp reported seeing the animal in the afternoon on the Henry Enstrom farm, four miles from Laboratory on Route 19. She was in the house at the time and called a neighborhood boy, Dick Enstrom, who came with his gun to chase "the thing." While Mrs. Chopp was getting a gun she could see the animal clearly from a second story window, fifty feet from her. They were unable to get closer.

"The Thing" was reported the same day along the Marianna-Lone Pine road.

R. B. Doetzbacher, Washington county game protector, stated last night no mountain lions had been reported in that area for many years. He also declared a mountain lion or cougar is generally light brown or tan. A wildcat has a short tail and averages about 25 pounds in weight, he declared. The possibility that it was a bear was discredited by the statements of Mrs. Chopp and the Enstrom boy, both of whom had a look at it and saw a long tail—to which was attached the rest of "The Thing," black and big.

Farmers in the region agreed that the animal may be a large dog, possibly a crossbreed of German shepherd and Doberman Pinscher, which would account for its size and black coat.

At any rate, it isn't only television that is keeping Lone Pine residents indoors these frosty winter nights.

The Thing Out Again

Charleroi, Pennsylvani, *Mail*, January 31, 1951

Marianna. Pa., Jan. 31—Amwell Township's strange animal, reported roaming over hill and through dale, was sighted last evening about 5:30 o'clock by Harold Cain, farmer and bird hunter, who resides on the ridge just above No. 1 dam of the Citizens Water Company.

"I don't know what kind of an animal it is but it doesn't look anything like any police dog I ever saw," commented Cain. "It is black and huge and has short ears.

"I had a young bird dog with me. It chased after the animal but couldn't begin to keep up with it and came back to me. It was moving rapidly. I was in a field making repairs to a line fence when I sighted the strange animal about 100 yards away.

"The animal is not a bobcat.

"I am willing to join others interested in staging a possible hunt," Cain concluded.

There have been no reports of any livestock having been killed by "the thing" which was reported previously seen along the Lone Pine-Marianna road.

Bradford, Pennsylvania, *Era*, September 9, 1952

Prowling Panther: A resident of West Branch informs RTS that he saw a black panther while he was driving home on a recent Saturday near midnight in the Bradford watershed in West Branch. When he got home the motorist, who said the beast measured four feet in length, warned neighbors to stay inside.

Hunt Panther At Meadville

Charleroi, Pennsylvania, *Mail*, April 7, 1952

Meadville. Pa., April 7—A "Black Panther" scare sent same 300 armed residents of the Cussewago Creek Valley out on a second day's determined hunt yesterday, despite freezing weather and continued snow that whitened the ground.

The group, under leadership of Morris Shellenberger, of Mosiertown, and Earnest Calhoun, Coons Corners General store operator, found two dead deer partly eaten, yesterday.

"The animal several persons have seen must be a panther, which lives on deer and other animals, while a tom cat does not," said Glen Burns, of the Fountain House area, who took out his prized dog, a 75-pound English big game hound, "Arizona Jim."

The dog, raised by the famed Lee Brothers in Arizona, who have hunted game for the government for years, followed a scent for some time. He was believed to have treed the animal several days ago, when no one was around.

The searchers, armed with rifles and shotguns, said they saw what looked like the tracks of their quarry around the deer, whose carcasses were found a mile and a half west of Saegertown. The deer had been killed beside a stream and dragged over a bank.

Tracks of a strange animal have been reported in several sections.

Wesley Sliter who lives a half mile south of Little Corners, brought in a set of tracks found in mud which were far larger than his 50-pound collie dog.

Mrs. Sliter saw a strange black animal "so black it shone" behind her home some time ago. It slinked away when a flashlight was brought out.

'Twas Long, Low, Catlike and Hunter Shot At It; We Won't Call It a P—r

Titusville, Pennsylvania, *Herald*, November 19, 1952

A Pleasantville hunter slammed two rounds of No. 4 shot into a long, low black animal near Pioneer in Cherrytree Township last Friday and knocked it down.

But the catlike beast sprang to its feet and leaped over a railroad embankment into a swamp thick with cattails. Two hunters patrolled the area with shotguns at the ready but failed to locate the animal.

The hunter and his 16-year-old companion preferred to remain anonymous because of all the joking about "black panthers." But to a reporter who interviewed them, the Pleasantville man, 23, told the following:

"First, I don't believe in panthers. But I don't know what sort of animal this was."

The two were driving along the abandoned railroad grade from Pioneer toward Miller Farm when the man thought he glimpsed something moving.

They stopped the car. But all their shells were in their hunting coats in the back seat and it took them some time to get one 12-gauge shotgun loaded.

The man then began running after the black animal.

"Then what amazed me," he declared, "was that it stopped and squared around to look at me.

"I shot the left barrel, the most powerful one. The animal was knocked down to one side. It quickly got on its feet. I gave it the other barrel.

"It jumped high in the air and sailed over the bank."

The two, rather reluctant to beat around in the thick undergrowth after the animal, walked up and down the bank and threw stones into the brush. However, they failed to flush it. It just disappeared.

The hunter was chagrined because the shotgun blasts did not kill the animal. The distance was estimated like that across two wide streets, about average shotgun distance.

Tufts of fur were torn out by the first shot. The results of the second could not be determined because the animal made a tremendous leap and was away.

The animal definitely was not a house cat, the hunter declared. Much too big. It was longer than his shotgun, by more than a foot, according to a sight measurement made when the animal turned broadside to the man. It appeared to be a little over half way to a man's knees in height.

The animal was black all over, the hunter said, and its fur seemed to be sleek. It trotted like a cat until shot at, when it took off in great leaps.

It made no outcry and no tracks could be found because the railroad grade was all cinders.

What the animal was is a puzzle to both hunters. They had never seen anything like it.

In the Miller Farm vicinity last April two oil lease workers saw a "long, low, glossy black thing" which resembled, they said, a half cow, half deer. It was chasing three deer but was not catching up. It had a rather large head and its belly was only a foot or so from the ground. In support of their cow-deer theory they offered as evidence the facts that Black Angus cows are kept in that vicinity, and that cows and deer are known to graze peacefully side by side.

PANTHER THOUGHT TO HAVE ESCAPED FROM CIRCUS
Williamsport, Pennsylvania, *Sun*, June 27, 1953

English Center—The opinion is being expressed in this section that the "black panther" that has been seen in the vicinity is actually an escaped wild animal.

Last night a well-known sportsman in the area said that many persons think the animal is of the cat family which escaped from a circus were traveling show. He said it could be a lion, tiger, leopard or panther.

There have been three incidents involving panthers in recent weeks.

Ernest Danley, his family and friends, reported seeing a "large, shiny, black cat" last Tuesday morning. It was described as 10 feet long.

Another was reported to have attacked three calves owned by George Danley, Trout Run RD 1, clawing one.

Gerald Kahler reported seeing one on Boak's Mountain.

BLACK, SCREAMING ANIMAL IS GLIMPSED
Titusville, Pennsylvania, *Herald*, July 24, 1953

Riceville, July 23.—A mysterious animal was seen here Sunday.

Duke Bradley and Harold Knight went down in the pasture on the Ash farm to hunt for a cow fresh with calf.

They found the two, and very near was a deer and a shiny black animal which ran and gave out a terrible scream. The men got only a fleeting glimpse of the black animal, but that glimpse was in the bright sunshine. They said it was larger than any dog and very shiny black.

Titusville, Pennsylvania, *Herald*, September 2, 1953

"Panther" Victim Practices Football—Phillip Reynolds, 14-year-old son of Mr. and Mrs. Howard Reynolds of Oil City, who was attacked by an unidentified animal last Thursday noon in Hasson Heights and was hospitalized several days for resulting injuries, is getting along well, although very nervous, his mother said yesterday. He was discharged from the Oil City Hospital Saturday and Monday went to football practice. Phil still sticks to his story that he was attacked by "a black, shiny animal with a long tail," she said, and a. number of men are continuing their search with guns in the Heights area where the boy was attacked.

BLACK, CAT-LIKE CREATURE STREAKS BY LOCAL ARCHER

Titusville, Pennsylvania, *Herald*, October 21, 1953

A local archer was caught with his arrow in his bow yesterday when a black cat-like creature streaked through the woods near him.

The critter was too speedy and he had no time to loose his arrow however.

The archer said the animal was "blacker'n heck" and had a long tail.

"It was too big for a house cat—at least two feet high—and panther was the first thought that came to my mind," the resident said.

The incident occurred on an old logging road between Shamburg and Jerusalem Corners at 3:30 p.m. as the hunter was out for deer He had walked part way down the road and stood at intervals, and was slowly making his way back to his car when he heard a rustle in the leaves beside the path

His view was blocked by a fallen tree. "There goes a deer," he thought.

The noise stopped. The man waited. He then took a step forward to see around the tree Crunch, went leaves under his foot. No other sound. Another step. Crunch. Still nothing. The archer then stepped hurriedly around the fallen tree.

Whish! Away went the creature. It shot straight away from the hunter and then angled up toward where he had parked his car. He got several good glimpses of the cat-like animal.

He doesn't know what it was but if you say "panther" to him the archer will be inclined to say there's one loose.

HOUSEWIFE SEES MOUNTAIN LION

Monessen, Pennsylvania, *Daily Independent*, November 12, 1953

Chambersburg, Pa. (UP)—When housewife Mrs. Corlena Simpson of Chambersburg saw a black, panther-like animal in a field near her home, she took the scientific approach.

Mrs. Simpson ran first for her binoculars and then an encyclopedia, and identified the animal as a black mountain lion. She said she saw the animal's green eyes clearly in the binoculars.

Mrs. Simpson said the animal, which eventually walked out of sight, was about 30 inches high, had a body three feet long plus a long tail. She said her collie dog, Tootsie, likes to bark at anything that moves in the field: but made an exception when she saw the big cat. Mrs. Simpson said the dog watched the animal while crouching quietly in the lawn.

WHAT'S THAT? PANTHER?
Titusville, Pennsylvania, *Herald*, September 14, 1953

The black panther is back. But he is a law-abidin' critter who stays on his side of the road and faces traffic when he goes for a stroll along the highway. And he is as nonchalant as they come.

A carload of local residents driving here along Route 8 just north of Centerville at 1:45 a.m. yesterday saw the black varmint as their car breezed near Breezeland.

By the time they did a second take and asked each other "What was that?" the thing was gone.

But in their one good look at the beastie they saw the following, they said:

A cold-black animal with back as high as the top strand of the guard rails, walking between the rails and the highway, facing traffic. It had greenish eyes, pointed ears and long body, much longer than a dog's. It had a catlike walk as it padded along unconcerned as could be.

"It looked like the panthers in the zoo," one resident said. "I think if it was a panther it was one which escaped from a zoo or circus and is used to people."

HUNDREDS EXPECTED TO JOIN HUNT FOR 'BLACK PANTHER'
Titusville, Pennsylvania, *Herald*, January 27, 1954

By John F. McNichol

City Editor, Williamsport *Gazette and Bulletin*

Williamsport, Pa., Jan. 26 (AP)—A king-sized dragnet will be spread over Bobst Mountain 20 miles northwest of here Saturday in a mass hunt for "The Critter of Cogan House Township."

An estimated 600-700 hunters are expected to register for the "black panther hunt," organized by conscientious hunters who actually believe such a beast is roaming the hills in the remote township.

President Judge Charles Scott Williams, of the Lycoming County Courts, is chairman of the hunt. Dozens of other county officials and prominent sportsmen in this section of the state are helping him.

It all started about 10 days ago when new reports reached Williamsport that tracks of a large cat had been found in the Cogan House region. Many sportsmen are of the opinion that some sort of "critter" is at large, attacking livestock and killing smaller game.

A cash prize of $300 for the carcass has been raised and there are other prizes for those bagging foxes or wildcats.

From headquarters which will be established at a turkey farm atop Steam Valley Mountain on Route 15, some 600-700 men from 16 years old up will seek the "critter."

Shotguns—with no load bigger than buckshot—will be the legal weapon. Rifles and sidearms are barred.

The only hitch to the plans came last week end when Jared W. Reaser of Williamsport, who's better known professionally as Prince El Kigordo, arrived home for a brief vacation between bookings.

Reaser, who has a wild animal act, was interviewed about the "critter." Said Reaser: "There are no black panthers, only black leopards." Then he conceded it could be a puma, or mountain lion, "which occasionally wander into this section of the country."

Reaser will be out of the city for a television appearance Saturday so his truck load of lions will be safe.

MASS HUNT BAGS ONLY ONE RED FOX
Titusville, Pennsylvania, *Herald*, February 1, 1954

Williamsport, Pa., Jan.— A lone red fox was the only "wild animal" caught in the dragnet of 360 hunters who combed Bobst Mountain yesterday in search of the "Critter of Cogan House Township."

The mass safari into the isolated timberland was beset by troubles from the start. First, snow and bitter cold weather cut down the number of hunters in the posse from an expected 800. Then darkness set in just as they thought they were on the trail at last.

'The "Critter" is an unknown variety of animal that has been frightening away game in the popular hunting area 20 miles northwest of here. It has been described variously as a black panther, mountain lion or oversized bob cat.

A $300 prize awaited the hunter who might have bagged the "Critter" with a 100-pound supply of dog food going to the dog first gaining the scent.

President Judge Charles Scott Williams of the Lycoming County Court, who was chairman of the expedition, said the hunt may be resumed at a later date.

Woman Hunter Sees Panther-Like Animal; Man's Rifle Shots Miss It

Titusville, Pennsylvania, *Herald*, November 1, 1954

The first black panther report in the Titusville area in a couple years—and the most authoritative—came in from a hunter on the opening day of small game season Saturday.

Mrs. K. Arden Bennett of Spartansburg saw an animal in a clearing off the Sanford Road which she identified positively as a panther.

Mrs. Bennett said she intended to go squirrel hunting and carried a 410-gauge shotgun. She got out of her car about three miles east of the Titusville-Spring Creek Road in the Sanford-Torpedo area.

Entering a clearing where she had seen squirrels last year, she looked up and saw, about 80 yards away, a black animal watching her movements.

"He was looking at me and I was looking at him," she said. "I was scared, but I put the gun on him and started to back out of the clearing. I backed out as far as I could and then got into the car. I didn't want to fire, because I was afraid I would wound him."

In the car, Mrs. Bennett said, she got a pair of binoculars and watched the animal, which she later found had been feeding on a rabbit. "The only thing to do," Mrs. Bennett continued, "was get a man with a high-powered rifle."

So she went to the nearby home of Leigh Smith, Torpedo, RD, and Mr. Smith went back with her to the clearing. They found the animal about 10 yards from where he had been when he was first sighted by Mrs. Bennett.

Mr. Smith shot at an 80-yard range. The shot was about a foot high, Mrs. Bennett calculated, and his second shot also missed the beast, which loped off at the first shot.

Mrs. Bennett said the animal was about four feet long, with a tail 2 ½ feet long, and standing it would come to her knee, she estimated. "It was just coal black, with not a mark on it, and it was built just like a leopard," she said.

Its movements were catlike, and the animal leaped and lunged, covering "probably 10 to 12 feet" at a bound, Mrs. Bennett said.

"I'm not kidding you, either," said Mrs. Bennett. "Lots of people probably take it for a hoax, but it isn't." Mrs. Bennett added that she was reluctant to tell the newspapers about it, because she was afraid people wouldn't believe it, but her neighbors said the people in the area where the animal is roaming should know about it.

Black Cat-Like Animal Sighted Slinking Over Caldwell Road
Lock Haven, Pennsylvania, *Express*, February 10, 1955

A strange cat-like animal, similar in appearance to a "black panther" sighted in Centre County, was seen in Gallagher Township, two miles east of Caldwell.

Dean Probst, State College businessman, and Mr. and Mrs. Howard Rippey, Lock Haven, RD, saw the animal on a recent afternoon.

They were driving out to Mr. Probst's camp, ten miles from Lock Haven.

Their sighting of the animal came about three weeks after a "black panther" was seen at Valley View, west of Bellefonte.

There have been repeated appearances during the last half year in Centre County.

Mr. Probst, a native of Lock Haven, said he and the Rippeys were traveling out to Gallagher Township in the late afternoon. The cat-like animal, dark in color, its belly close to the ground, crossed the highway in front of the car.

Familiar with Game
Quite familiar with animals found in the woods of Pennsylvania, Mr. Probst said that what he and the Rippeys saw was not a wildcat. He said he had seen a wildcat in the same region and that the two were not similar. He said the dark animal was as large or larger than a boxer dog and that it had pointed ears.

The animal crossed the road so quickly that there was only a moment or two to view it. Mr. Probst said he did not get enough time to see if the animal had a short or long tail.

During deer season Mr. Probst reported that he and others had heard a scream one night in the Gallagher Township woods. He said that first it was believed a wounded deer was making the cry but that a search with car headlights located a wildcat. The cry was described as bringing genuine shivers up and down the spines of those who heard it, Mr. Probst said.

Thought to Be Wounded

In Centre County the repeated reports of a black animal have caused considerable public concern. At Valley View a store owner saw the animal in a tree across the road from her store. Two boys chased it, armed with a rifle, and believed they winged the animal with a shot.

During the last six months sightings of the animal have formed a circle on the map, going from the south side of Centre County around to the west. Mr. Probst pointed out that if the animal he and the Rippeys saw is the same one, it would have had to cross the West Branch of the Susquehanna. The river has been frozen over in recent weeks.

STRANGE ANIMAL NEAR NEW CASTLE ELUDES HUNTERS
Monessen, Pennsylvania, *Daily Independent*, March 2, 1955

New Castle. Pa. (UP)—Game protectors said today they have determined nothing certain about a strange animal reported prowling in the area except that it sure is elusive.

The animal has been reported seen or heard in the vicinities of three nearby towns, Harbor, Pulaski and Frizzleburg.

Frank Steiner, 39, Ashtabula, Ohio, truck driver, told police Lt. Albert Russo that a "black panther" leaped at him early Tuesday as he checked his truck's lights along a lonely road near Barbor.

Mrs. Josephine Irwin, New Castle, checked in with a report that while visiting her parents' farm near Frizzleburg Feb. 25 she saw a "panther" in the yard. Residents of Pulaski said they heard weird howls coming from atop a nearby hill.

Game protector Calvin Hooper, his deputy, Merle Glitch, and state police searched the area without success. Local sportsmen believe the animal may have been a coyote, a wolf, a gray fox, or just a stray dog. A coyote was killed in the area about 20 years ago.

BIG "PANTHER" HUNT STIRRED
Uniontown, Pennsylvania, *Morning Herald*, March 3, 1955

New Castle, March 2 (AP)—A search began in Lawrence County for a strange, black animal which an Ohio truck driver says sprang at him early yesterday morning.

The animal was believed by some residents to be a black panther.

But Game Protector Calvin Hooper thought the animal might be a grey wolf or a lost, black coon dog.

Attacks Trucker

Frank Steiner, 39, a truck driver from Ashtabula, O., told New Castle police the animal attacked him along U. S. Route 422, near New Castle.

Steiner said he spent the night along the highway. He said the animal sprang at him this morning after he left the truck to inspect its headlights.

Steiner said he hopped into the truck and sped away. The animal loped away into the woods.

CAT PUTS WHOLE CITY INDOORS
Traverse City, Michigan, *Record-Eagle*, March 4, 1955

Pittsburgh, Mar. 4—(UP)—A reign of terror over Pittsburgh's Mt. Washington section ended abruptly last night when a "ferocious wild beast" being hunted by police turned out to be a bushy, overgrown house cat.

Tales of the beast, which kept most residents of rainswept Mt. Washington behind locked doors yesterday, grew from a prank by a trucking company employe who had locked an Angora cat in a tractor-trailer.

Police stalked the slippery slopes of Mt. Washington for eight hours, egged on by reports of a black panther, a wildcat, a mountain lion and a "thing four feet long with black hair."

Radio and television stations interrupted regular programs to warn parents to keep their children Indoors. Anxious parents called police, inquiring whether schools would be closed "as long as the lion is running around loose."

Another cat, a slow-footed tomcat, became a victim of the hunt when nervous police riddled it with bullets. A frightened rabbit flushed from a junkyard escaped unscratched but badly shaken.

The whole thing ended abruptly when a suspicious patrolman decided to check the report first issued by Richard Wateska, a helper at a trucking terminal. Wateska admitted he and fellow workers dreamed up a stunt to scare the day shift at the terminal.

Wateska said he and other employes were frightened momentarily Wednesday night when they discovered the overgrown Angora in the tractor-trailer, which had arrived from Indianapolis, Ind.

He said they decided to leave the cat in the trailer overnight to "scare the day shift."

The hoax worked better than Wateska and his co-conspirators had hoped. The day shift workers were so startled when the cat darted out of the trailer that their descriptions to investigating police were somewhat out of proportion.

They estimated its weight at 80 to 100 pounds and said, "It looked like a lion."

Police doubted that any charges could be lodged against the pranksters.

Wildcat Killed Near Here; Puts Up Fight

Titusville, Pennsylvania, *Herald*, October 13, 1955

Two coon hunters treed a wildcat last night near the Redfield School on the White City road southeast of Titusville in Venango County.

The squalling cat jumped two coon dogs. Jack, Earl and Bob Anthony of Pleasantville stood by while the animals fought for about 15 minutes before the wild cat was killed. The dogs were "chewed up pretty good," Earl said.

The animal was 32 inches long, had striped legs and a long tail, it was much larger than a domestic cat, which it resembled quite a bit in appearance.

And stink! Man, it gave off an aroma which could be detected 15 feet away.

The Anthony brothers were coon hunting when the dogs put up the wildcat. They trailed it for a ways before it treed. When the hunters came up, the cat jumped away. It was treed again, and the same thing happened.

The third time the hunters extinguished their gasoline lantern and put out their flashlights to approach in the dark. When the animal jumped, they put their flashlights on it and the battle was on.

"It put up an awful fight with the dogs," Jack Anthony said.

Others hunters looked over the animal and gave their verdict that it was a wildcat instead of a domesticated cat which had gone wild. One man said it must be about 15 years old. It was vicious in appearance even when dead.

PANTHER MYSTERY MAY BE RESOLVED

Pennsylvania State University *Daily Collegian*, April 19, 1956

By Terry Leach

A black panther reported sighted by several Centre County residents and similar in appearance to the stone Nittany Lion crouched in the woods near Recreation Hall, may have migrated to this area from Ohio.

Capt. Warren Aikens of the Air Force gave the following explanation for the many panther stories circulated in the area for the past four years.

A family of Mexican farm laborers came to Fostoria, Seneca County, Ohio, in 1948 to work in the sugar beet fields. They brought with them a pair of cougars which mated and bore a litter of three young. ...

Cougars Turned Loose

When the young cougars outgrew their cubhood, the beet-workers found the cats too much to handle and turned them loose.

Shortly afterward Seneca County residents reported sighting several panthers in woods and fields. An investigation by the county sheriff revealed the fact that the beet-workers had released the cougars.

Captain Aikens said his wife reported sighting one of the half-grown cubs in the yard of her parent's home in Tiffin, Ohio, in 1949.

A trapper was hired by the county to capture or kill the cougars in 1950. A search of two months failed to reveal any trace of the family of cats. Residents of the county no longer reported seeing the animals.

Migrated to Pennsylvania

Captain Aikens said he felt certain one or more of the cougars migrated to the hills of central Pennsylvania. Such a migration would be natural for a mountain bred cat seeking his natural habitat, he said.

Residents of Centre County reported seeing a black panther in the area for the first time in 1952. Several readers told of seeing the panther in articles published in the Centre Daily Times.

The mystery remained unsolved until Captain Aikens offered his solution in an article appearing in the Centre Daily Times early this month. ...

POLICEMAN SAYS HE SAW BLACK PANTHER HERE

Connellsville, Pennsylvania, *Daily Courier*, August 10, 1957

City Patrolman D. S. Mancuso today reported seeing a black panther on the hill side behind the Sandusky Lumber Yard in the North End.

According to the officer, he saw the animal early Friday morning while on a routine patrol in the police car.

Mancuso said, "I had stopped for a moment to talk to my brother-in-law, Guy DeLuca of 3-D North Manor, who was en route to work, when he pointed to the hill side behind Sandusky's and there, as big as you please, was a black panther."

"We both could hardly believe what we saw," continued the officer, "but there it was. It was a real beauty, walking there on the hill side. We watched it for at least a minute before it disappeared over the other side of the hill."

It was the first report received at the Courier office of any such animal in the vicinity.

CASE OF MISTAKEN IDENTITY, GAME COMMISSION SAYS
RUMORS OF BLOCK PANTHER IN GLEN RICHEY AREA REPORTED
Clearfield, Pennsylvania, *Progress*, December 3, 1959

By Betty Hamilton

Progress Staff Writer

The Pennsylvania Game Commission claims it's all a case of mistaken identity—but residents of the Glen Richey area aren't quite as quick to brush off rumors that a black panther stalks the woods near their homes.

For the past couple of months, families living along the Glen Richey road or the old Erie pike haven't been quite as concerned as some folks as to whether or not there's animal life on the planet Mars. They'd rather know if the large black animal seen in their area is a member of the species of leopard family that has been considered extinct in this section of Pennsylvania for 60 years.

Officially the Pennsylvania Game Commission pooh-poohs the idea. It claims the last panther in the Clearfield County section was killed in 1898. It would probably be likely to attribute any reported since that time to poor eyesight on the part of a terrified person.

But unofficially, some Game Commission employes aren't anxious to commit themselves.

"I certainly would not want to say there is a panther around here, because from everything I've read it isn't possible," Game Warden Victor Hollopeter of the Erie Pike says officially. But off the record he admits he is one of the persons who saw a large black animal and "it wasn't a bear, a dog or a bobcat."

Game Warden Hollopeter had his look at the mystery animal several months ago while he was searching for deer poachers in the Christmas tree nursery owned by James Long.

"I heard a noise and thought at first it was a bear," he recalls. "I stood still and then I saw this animal. It certainly looked like a panther—with low body, heavy muscular legs and a rope-like tail. Although I remained quiet it must have sensed something for it ran away fast."

This was not the first indication he had had that some unusual animal lurked in the vicinity. He, and other persons had previously seen unfamiliar tracks which he said "couldn't have been made by a dog or a bobcat."

The game warden is not the only person to have caught a glimpse of the black stranger. Several other persons also claim to have seen it. Each gave almost the same description and each was impressed most by the same feature—a heavy long black tail resembling a rope.

Eighteen-year-old Bob Bloom, son of Mr. and Mrs. Boyd Bloom of the Clover Hill section, saw the mystery animal the day before Thanksgiving.

Bob was driving a truck along one of the back roads to his home when an animal ran across the road in front of him. At first glance he thought it was a deer but a second look revealed it to be a large panther-like animal.

Bob went to his home and then with his father and Game Warden Hollopeter returned to the scene. A thorough search of the area failed to produce the creature, however.

The elusiveness of the beast seems to be one of its characteristics. From all reports it is as reluctant to have a face-to-face meeting with its human neighbors as they are.

About the only persons reported to have had much more than a fleeting look at it are members of the James Moriarity family. They saw it about a week ago near their home and then among the prize-winning Hereford cattle at nearby Browncrest Farms.

They contacted Robert Brown in the hope that he could kill the animal before it harmed his frightened cattle. But on reaching the herd, Mr. Brown found the animal to be gone, making him wonder if someone hadn't mistaken the large Brown family cat, "Zorro," for the panther.

But even Zorro who is reported to be as brave and adventuresome as his namesake, doesn't quite fit the description given of the animal. Although he is big and black, his 10 pounds could hardly make him appear as large as the animal which some persons have described as weighing "between 65 and 70 pounds."

(According to the Game Commission some panthers weigh as much as 120 pounds and are as long as nine feet).

So far, the animal—whatever it is—has been content with just worrying the Browncrest cattle. There have been no reports of any animals being killed there or on any of the other farms in the area.

"We haven't even had any reports of deer being killed by a wild animal," Game Warden Hollopeter says. "Usually if there is such an animal around, the hunters discover deer killed by it. We haven't heard anything like that this year."

Other persons advance the theory that probably the animal has killed enough small game that it hasn't had to resort to killing deer, cows or horses for food.

Reports of lone panthers in the Clearfield County area are nothing new. The game warden says they've been popping up ever since the 1930's.

Such rumors were going the rounds last year. Truckers said they saw a panther in the Curry Run section. A Grampian resident, out for a drive, claimed he saw one standing unafraid in the populated section known as Irishtown.

As for the residents of the Glen Richey area, most of them don't care whether or not they ever have an opportunity for a close look at the mystery animal. "I don't care what it is—all I hope is I never see it," is the way one housewife sums it up.

ONE TOUCH OF NATURE: A COLUMN FOR OUTDOORSMEN
Titusville, Pennsylvania, *Herald*, April 4, 1963
Steve Szalewicz
Imagination Unlimited

Ever wonder how many black panther stories got started? Well, we could have circulated a hot one on the first day of October, last.

We're hunting deer with a bow near Oleopolis. It's about 7:30 a.m. There is a slight fog drifting over the black top road which we are patrolling because deer cross here frequently. And because the fields are heavy with dew and we don't have waterproof gear, we stay on the road. Sharp hooves have broken down the bank along the road in several places indicating at least a half dozen well-worn deer trails. The trick is to be near the right trail when the deer come up out of the run after their morning drink and head for beds higher up on the hill.

Our thoughts this morning run to nothing more than the obvious. The acorns are both noisy and plentiful this year. They are rolling and collecting along the berm. What a pity they can't be saved for later in the winter. But maybe the leaves will preserve them.

Suddenly out of nowhere and out of hemlocks along the steep ravine to our right pops a black animal and almost crosses the road.

Somehow it senses that we are in the vicinity, about 100 yards away. The black thing looks at us and bounds back into the ravine, and appears no more.

The Thing Grows

We pay the oddity very little mind for a time. Later that morning en route home, we began to speculate. That was a funny looking animal. We know that it was black, it was too large for a house cat, yet it had a long tail, too long for a dog. It did run like a dog but seemed to lope along the road and bound.

We began to wonder just how we describe this creature without exposing ourselves to ridicule. The more we tried to recall how the black thing looked, the more it began to take on the physical characteristics of a ferocious panther.

The ears now looked erect and tufty, crowned apart on a wide skull. The mouth was opened and snarling, teeth menacing, whiskers bristling, tongue a live pink.

What once appeared too large for a domestic cat but not too large when compared to a dog, now had more definite measurements. The body lengthened, about three or four feet, suspended on muscular limbs that were ready to spring. The tail lengthened, and with it a bow in the middle, seemed to swing back and forth warning that the beast was in its worst mood.

Truthfully, we know there is no such thing like a black panther, least not around Oleopolis ... nor elsewhere. Certainly this was not a black bear ... nor a dog ... nor a cat ... but then what was it?

It only goes to prove that when you don't know for sure what you see the mind starts building up possibilities. We had speculated ourselves into a corner—and a panther was an only out. We dismissed what we saw. However, a few days later, a friend hunting in the same spot said, "I saw a funny looking black thing along the Oleopolis road...."

COOPERSTOWN PANTHER SCARE IS REVIVED

Titusville, Pennsylvania, *Herald*, June 4, 1963

Two reports of panther sightings in the Cooperstown area have revived the periodic panther scare in this area.

Bruce W. Krepp, whose property is on Beatty Run two miles west of Cooperstown, reported that he saw a black panther twice Sunday, the first time about 11 a.m., and the second time about 5:30 p.m. when it approached the driveway of his home.

Mr. Krepp reported that his son. Chuckie, 5, spotted the animal and called the family. David Young and Charles Young arrived with guns, and David Young shot once at it and missed. An attempt to track it in the woods failed, he said.

He described it as looking like a panther and moving with cat-like motions.

Sighting of a panther-like animal along Sugar Creek over the weekend was reported to the Franklin office of the Pennsylvania Game Commission.

Ed Mitchell of New Kensington said he saw the animal, the size of a collie, light tan in color and with a long tail, near his camp two miles north of Cooperstown.

A week ago, Harry Springer of Van shot a black wildcat, weighing six to eight pounds. The cat had four fangs three-quarters of an inch long.

A Game Commission spokesman said that reports of panther sightings have occurred in this area over the past three years, but no conclusive evidence of a panther's existence has been found.

PIG-KILLING ANIMAL MIGHT BE WOUNDED

Bedford, Pennsylvania, *Gazette*, December 21, 1963

The "black cat of the hills" near Hyndman reportedly has been wounded by a deer hunter.

The report came from John Clites who owns the mountain farm where "the cat" last month killed five pigs and drained their blood. The hunter, said to be a resident of Broad Top Area, saw the animal Monday on the Clites farm.

The element of surprise allowed but a fleeting shot that may have nicked the animal on a leg. The hunter followed a faint trail of blood several hundred yards but lost the track.

The hunter described the animal as "coal black" and five feet in length with a tail two to two and one-half feet long.

Clites and his sons visited the farm Friday morning and discovered one of three sheep set out as bait for "the cat" had been torn apart.

The condition of the sheep was in sharp contrast, Clites reported, to the five pigs. They showed little body damage except for a torn spot behind left front legs where the killer gained access to the heart.

The sheep was killed in an outside pen containing the three sheep and a sixth hog crippled in the earlier attack. Had the animals fled to an inside pen and the attacker followed, a capture might have resulted. The door to the inside pen was designed to fall shut.

Clites and his sons checked for tracks near the pen containing the sheep but drifting snow covered or erased them. He reported his sons saw large tracks Saturday in woods near Fairhope but failed to spot a cat-like animal.

MYSTERY SURROUNDS ATTACKING, EATING OF PIG BY SOMETHING
Connellsville, Pennsylvania, *Daily Courier*, January 22, 1969

A game warden received several reports of a large black feline, along with the account of a New Stanton pig being killed and partially eaten by an unknown animal (possibly a dog). A Youngwood woman told him she had seen the panther, "not a large dog or black kitty." Another man told him it had been prowling around the area for some time.

THE SPORTSMEN'S YEAR, BY FRANCIS KEMP
Huntingdon, Pennsylvania, *Daily News*, February 4, 1969

A "black mountain lion" and a pack of dogs were both reported killing deer in Henderson Township.

Sutherly (1976) noted that in 1973, "a mysterious creature was reported in the area of Upper Pottsgrove Township." He doesn't give details, though notes there were several articles in the Pottstown, PA, *Mercury*. I was unable to find corresponding articles, except for one account of a rabbit hutch smashed and one rabbit killed by an unknown animal in Hilltown Township. The article briefly notes that there were reports of a strange animal in Pottstown. (Noted in the Doylestown, PA, *Daily Intelligencer* of March 16, 1973.)

PITTSBURGH SAFARI SEEKS PANTHERS
Tyrone, Pennsylvania, *Daily Herald*, April 1, 1983

Several witnesses reported seeing a couple of black cat-like animals in the area. One, a mechanic, watched as a "60 pound" feline carrying something small in its mouth leaped over a 12-foot fence. Police were unable to find anything, and blamed it on large dogs roaming the neighborhood.

CHILDREN KEPT INSIDE AS POLICE HUNT FOR TIGER ON THE LOOSE IN STATE
Doylestown, Pennsylvania, *Daily Intelligencer*, July 28, 1986
A WILD BEAST ON THE LOOSE? OR JUST A PET?
Kokomo, Indiana, *Tribune*, July 30, 1986
SEARCH CALLED OFF FOR MYSTERIOUS TIGER
Doylestown, Pennsylvania, *Daily Intelligencer*, July 30, 1986

In late July 1986, a Nicholson resident reported seeing a big cat in the woods along the Susquehanna and Wyoming counties border. Police went in search, and spotted it as they flew over in a helicopter, claiming it was "a small tiger, about 200 to 300 pounds." It disappeared into the forest. Sightings over the next two days became confusing, as some said it was striped, others spotted, and others that it was a mountain lion. After a 2-day heavy search by police and wildlife officers, the hunt was called off.

CAT-LIKE ANIMAL SPOTTED
Doylestown, Pennsylvania, *Daily Intelligencer*, July 24, 1987

A large black cat was reported by two police officers crossing Route 611 near Nockamixon. Tracks found near the area were identified as canine by Philadelphia Zoo employees.

HOUSE CAT OR COUGAR?
Erie, Pennsylvania, *Times-News*, October 30, 2003

Ronald and Brenda Graham were fishing in Girard Borough Park when they saw and videotaped a black feline that they thought was a mountain lion. (Still images from the video clearly show a housecat profile.)

QUEBEC

What Was Strange Animal Which Once Roamed Eardley?

Ottawa, Ontario, *Citizen*, July 24, 1931

A strange nocturnal animal was reported in the area between Luskville and Beechgrove for almost four years. One witnessed described it as being a "creamy color and in appearance the shape of a young tiger," and "heavier in the front than the rear, and lower in the rear than at the front." It "never showed any desire to attack humans, nor yet did it show fear of them." It had a strange cry, a "whimpering, crying noise," as well as "a noise as if it was laughing." Some people thought it might be an escaped hyena.

Lowell, Massachusetts, *Sun*, March 14, 1959

Montreal—A panther hunt was organized today to trap a shadowy black animal reported prowling the woods on Montreal island only 20 miles from Canada's second largest city. A dozen persons have reported seeing it.

RHODE ISLAND

Does a 'Big Cat' Prowl South County?
Narragansett, Rhode Island, *Times*, September 17, 2003

In an article covering a spate of cougar-like sightings, the reporter interviewed a woman who saw a "black panther" crossing a road one evening, about a month prior. Another woman said she saw a "chocolate brown" panther in her Carolina, RI, backyard.

SOUTH CAROLINA

STRANGE ANIMAL AT LARGE.
IN CAROLINA CITIZENS ENCOUNTER A STRANGE QUADRUPED.
Atlanta, Georgia, *Constitution*, May 3, 1901

Columbia, S. C., May 2—(Special)—Two widely separated points in Aiken county report the presence of a strange wild animal. A large beast with the short end of a chain dangling from its neck, frightened a fisherman on upper Three Runs. He dropped his rod, abandoned his fish and fled. People in the vicinity of Ellenton expressed the belief that this animal is a tiger escaped from some circus. But no circus has been through the state recently nor has the escape of a tiger been reported.

A gentleman writing from Sally says a strange animal has been prowling around there, that its tracks resemble those of a dog, except the hind feet are larger than the forefeet.

STRANGE BEAST "TERRORIZES" SOUTH CAROLINA AREA.
Fitchburg, Massachusetts, *Sentinel*, July 1, 1935

Bluffton, S. C., July 1 (AP)—A strange beast that looks like a jaguar and kills cattle like a grizzly bear is doing much to stimulate stay-at-home habits after nightfall on the part of the fearful here about.

Forest Ranger Wyman Cook, one of the few persons to see the marauder, describes it as about eight feet long and dark brown or brindle in color with black spots on its body.

Showed a picture of a jaguar, Cook said it resembled the varmint he saw in the forest on the 40,000 acre estate of George W. Varn near here.

GENERALLY SPEAKING

Aiken, South Carolina, *Standard and Review*, June 15, 1955

by Ed Kenney

The "mystery animal" which is reportedly slaying stock in Martinez, Ga., sounds like the creature which harassed the western side of Aiken county during early May. We saw the animal one time and it appeared to be a large black panther. The animal crossed the superhighway just behind a Negro man on a bicycle who was unaware (and still is) that he had such a close brush with death....

SOUTH CAROLINA MONSTER HUNT IS SCHEDULED

Burlington, North Carolina, *Daily Times*, September 4, 1957

Yemassee, S.C. (AP)—residents of this lowland area are planning a monster hunt come cool weather.

The reported monster screams like a woman, is yellowish brown with a bushy tail, is about five feet long and a four feet tall.

Is it a lion? A panther? A myth?

Except for the bushy tail, it could be a panther. If the first was reported a year ago, but they hunt was fruitless.

The hot weather now has too many rattlesnakes out in this low lying area. When cool weather sends them into hiding, hunters in the area plan to take out after the monster.

WEIRD BEAST ROAMS JUNGLE

THE 'THING' HAUNTS VILLAGE AGAIN

Mansfield, Ohio, *News Journal*, September 5, 1957

Yemassee, S. C. (INS)— "The thing," a science fiction-type creature which first made its appearance in Yemassee, S. C., several years ago and then disappeared, is haunting the little community again.

Town officials met to plot a way of hunting down the weird animal which disappeared last year in the midst of a search.

Few people in the community have seen the creature which thrives in a neighboring forest, but all of them claim to have heard it.

It has been seen through field glasses, however, and is described as a lionish looking animal, four-legged, about five feet in length, four feet high with a bushy tail. It is yellow-brown colored. The creature's hideous bellowing is its main trademark.

A small armed search party went into the woodlands last year to track down "the thing," as the Yemassee people call it. During the search, the bellowing stopped and the town had a year of quiet until the beast's familiar growls were heard again this week.

Plans are again being made to form another hunting squad, but officials warned the search will have to wait for cooler weather until snakes go into hibernation.

GENERALLY SPEAKING: LOOKING AT LIONS

Aiken, South Carolina, *Standard and Review*, March 31, 1959

By Ed Kenney

In recent weeks the Sheriff's department has received reports of sightings of a large animal, described by one witness as a "lion."

In the past year there nave also been reports of domestic animals being killed and eaten by some type of ferocious animal.

Last December a black mountain lion was spotted in the Savannah River Plant area, and more recently panther tracks have been found in the plant area.

The plant refuses to make note of these reports, mainly because they do not feel these two instances are sufficient evidence for a strong report. However, neither do they deny the possibility of such an animal in this area.

There has been in the past few years a great increase in game of all types in the SRP area, mainly because it has become a sort of game reserve where shooting is not permitted.

A black mountain lion, or panther, could have reached this area from Florida as they are often seen in the Ocala National Forest and Okefenokee Swamp areas south of here.

Should this report be true all we can hope is that if it's a she, then there is no he around.

MYSTERY BEAST DISPLAYS A VERY NASTY TEMPER

Spartansburg, South Carolina, *Herald-Journal*, August 20, 1960

By John Black

Clover—Something with toe-nails two inches long, a brown hairy coat, a taste for meat and a generally bad temperament has aroused a mixture of alarm and curiosity in a wooded community midway between this York County town and the Cherokee County line.

An elusive creature that has killed chickens, chased cattle and torn up the woods in several sections, has been sighted briefly several times, shot at at least once and in each instance made a clean getaway.

Speculation has run a wide course, everything from a bear to a gorilla. It is most strongly believed that the unknown animal is some kind of a bear.

Charley Wells, a Bethany community farmer, shot at the animal with a 12 gauge shotgun at a distance of 100 yards. He says the animal is brown.

This confuses the idea that the creature might be a bear since most bears in this area are on the black side.

Wells quoted an Indian who saw the tracks as saying he believes the creature is a grizzly bear. This would explain the reported brownish color and footprints as long as a big man's shoe.

However, how a grizzly might come to be in this area is puzzling.

The father of Mr. Wells has lost about 65 chickens to the marauder. A brother, Johnny, apparently almost met the beast in his living room as claw marks on his trailer indicated the animal attempted to get in his window.

After Wells shot at the beast, he stated, it turned and growled at him then darted off into the woods. He said the animal left broken cornstalks in the fields, ate roots and pulled up stumps in the woods.

A farmer three miles from the Wells home reported recently that something had eaten a young hog and a calf.

The unknown animal's footprints were first seen last Sunday.

The beast at one time knocked all the chairs off the porch of one area house and left large splotches of mud on the front door where it tried to get in.

It has jumped for distances of 15 to 20 feet over fences, does not leave any feathers when it kills chickens and has caused many residents in the area between Clover and Kings Mountain to bolt their doors at night.

Bear and fox dogs have been set on the trail. Once the animal ran straight through the pack of dogs and on another occasion the dogs raced back frantically to their handlers with hairs bristling as if the beast had turned on them.

The elder Wells has been forced to place a barricade in his driveway leading from the main road to his house. A sign there reads, "The Bear Has Gone. Please Do Not Ask Questions."

His son explained this was done to turn away curious persons.

He said there were once so many cars in the area that five carloads of policemen were required to disperse the traffic. The younger Mr. Wells' one ambition is to see the animal, make its picture, and then shoot it.

HUNT WILD BEAST
Sumter, South Carolina, *Daily Item*, July 14, 1966

An animal nicknamed the "woffin nanny" was seen by witnesses in the Greensboro area, and was believed to have taken a piglet.

CRITTER: IT WAS BIG, BLACK AND SCARY
Rock Hill, South Carolina, *Herald*, June 29, 1977

A father and son, on separate occasions, reported seeing a strange long black creature they thought was a black panther. Robert Rockholt saw it jump a wall by the cemetery, "jumping too far, just big, long leaps" to be a dog. His son, Larry, saw it while driving at night, slowing down so as not to hit it. He said it stood higher than a dog, was black, and had a long tail. A Riverbanks Zoo zoologist suggested it could be a dog or an otter.

PELION AREA ON LOOK OUT FOR TIGER
Sumter, South Carolina, *Daily Item*, August 24, 1987
OFFICIALS SAY TIGER MIGHT HAVE LEFT AREA
Sumter, South Carolina, *Daily Item*, September 1, 1987

A Lexington County woman reported seeing a "big cat" behind her mobile home, and said her pit bull had tangled with something that left claw marks on it, though the dog wasn't seriously injured. Two exotic animal specialists examined paw prints for the sheriff's department, and said they were from a tiger, estimated at 10 months of age and weighing between 300 and 500 pounds. A massive search for the animal was unsuccessful, and eventually officials decided it had probably wandered out of the area. Captain Bob Ford of the sheriff's department said it wasn't a "ghost," that members of his department had also seen the animal.

PANTHER REPORTS PUT MAULDIN ON WATCH
Greenville, South Carolina, *News*, October 18, 2003

A large black cat was spotted several times around the town of Mauldin. An associate director with the Greenville Zoo noted that he had heard about several reports of a "black panther" in south Greenville County and in Laurens County. (Subsequent articles note unsuccessful attempts to trap the animal. One witness claimed it was four feet in body length with a three foot long tail.)

BIG CAT SIGHTED AGAIN?
Columbia, South Carolina, *WLTX.com*, March 3, 2006

A dog-sized feline, estimated by one witness to weigh up to 75 pounds, was spotted twice in the Irmo area. It was described as a "large brownish-orange exotic cat with black spots." Animal control suggested it might be a cougar.

BIG CATS MIGHT BE AROUND, BUT WHERE'S THE EVIDENCE?
Augusta, Georgia, *Chronicle*, November 26, 2006

Tim Glover was hunting in Edgefield County when he saw what looked like a "large, dark panther" 20 to 30 yards away. He tried to keep up with it in his scope, but it soon moved out of sight.

'LIFE CHANGING EVENT': PANTHER CHASES FORESTER
Columbia, South Carolina, *The State*, January 27, 2007

Terrance Fletcher, a technician with the US Forest Service, was chased into the Chatooga River by a 7-foot long black panther. "While taking a break near the river bank, Fletcher heard rustling in the woodsand looked in that direction. Staring back at him was what appeared to be ablack panther, crouched on the forest floor like a house cat stalking a bird, he said. When he stood up, the cat started running, prompting him to take the icy dip in the Chattooga. Soaked to the skin and freezing, he met up with his partner and walked through the woods to their Forest Service truck."

Wild Cats? Or Imaginations Revisited?
Anderson, South Carolina, *Independent Mail*, March 4, 2007

After soliciting opinions on big cat sightings in the area, the newspaper received several sighting reports. One man who lived near Pelzer stated: "I'm so relieved. I never told nobody and I've been living here 22 years. It was down by the Lee Steam Plant and that thing was the size of a German Shepherd. It was pitch black as if someone poured motor oil all over it." A woman wrote to say that she and her husband had seen a large black feline cross S.C. 29 South, and a few weeks later saw the same cat crossing a field.

Panther on the Loose at Clemson?
Columbia, South Carolina, *The State*, December 19, 2007

In separate incidents, a professor and a security guard at Clemson University claim they saw a large black feline, from 18 to 24 inches high at the shoulder. A university police officer later saw a black feral cat in the area.

Big Cat Sightings are Numerous, But Proof is Hard to Come By
Anderson, South Carolina, *Independent Mail,* August 3, 2008

This article discusses the latest efforts of the USFWS to document eastern cougar sightings, and notes some of the latest big cat sightings in the area. On July 30, a Lincoln County man saw an unusual feline: "It was around 6:40 this morning (July 30), I was leaving for work. At first I thought this was a black lab until I got a little closer. It was solid black with a long tail that sort of curled upwards more at the tail. Short pointy ears, all the classic cat looks. It was not a dog, bear, domestic cat, etc. I was maybe 50 yards from it!"

TENNESSEE

A Strange Beast In Tennessee
That Eats Nine Hogs in One Meal
Piqua, Ohio, *Daily Leader*, March 24, 1892

Clarksville, Tenn., March 24.—A strange and mysterious animal has invaded the Rossview neighborhood of Montgomery county, causing a sensation. J. McMurray states he can not imagine what it is that is disturbing the quietude of that vicinity, but after thorough investigation is satisfied that some strange, ferocious wild beast is roaming around destroying everything it can get its mouth on. It has visited the farm of Col. J. B. Killebram several times, committing depredations and killing hogs. It has eaten several large, fat hogs, leaving nothing but the feet and larger bones. One night a sow and eight pigs were devoured at a single meal. It travels over wheat fields, leaving tracks as large as a man's hand, with long claw prints. McMurray has fleeces of the animal's hair, which is three to four Inches long. The color next to the skin is white, while the outer ends are of a reddish or fox color. Neighbors go together with butcher-knives and double-barrel guns heavily charged. Men are going nights in squads of four hunting the beast. It has taken up its abode in a cave and no one will venture near the spot alone.

"Giant Kangaroo" in Tennessee Alarming; Mysterious Animal
Cape Girardeau, Missouri, *The Southeast Missourian*, January 16, 1934

South Pittsburg, Tenn., Jan. 16.—(AP)—A mysterious animal is "fast as lightning" and like a "giant kangaroo" is spreading terror through the Hamburg community.

The creature first appeared Saturday night. It killed and partially devoured several German police dogs. The next night it killed other dogs and some geese and ducks.

Farmers are carrying their shotguns to the fields for fear of the beast and others are going about their daily work armed with pistols.

Rev. W. J. Hancock, Negro minister, saw the animal.

He said "it was as fast as lightning and looked like a giant kangaroo running and leaping across the field." It had just killed a large police dog and had left nothing but the head and shoulders of the victim in the owner's yard. Frank Cobb also saw the thing. He said it was unlike anything he had ever seen but that it resembled a kangaroo.

A searching party tracked the animal up a mountainside to where the trail disappeared near a cave.

Lion-Like Beast Turns Out To Be Nothing But Dog

Kingsport, Tennessee, *News*, November 12, 1946

Carthage, Tenn.—AP—Week-end fears of a strange "lion-like" beast prowling the willow brakes around Carthage abated after Smith County Sheriff J. B. Davis announced the animal was dead.

The beast—a big red dog—was shot by a tenant on the farm of Congressman Albert Gore (D-Tenn.), Sheriff Davis said.

A group of some 50 men, women and children sloshed over a mile of muddy terrain to see the body of the beast, which several admitted "might have looked like a lion in the dark."

Sheriff Davis's announcement ended a systematic hunt planned after Frances Bryant said he sighted the animal which "looked at first like a calf," and had a large, shaggy head.

Bryant and a deputy sheriff Floyd Hunter found large tracks in the brakes near the Cumberland River and said they heard something nearby "roar like a lion."

Residents reported a number of dogs missing and others returning home bloodied and crippled, and fears mounted until Sheriff Davis reported the death of the "beast."

HUNT FOR WEIRD VARMINT TO BE HELD NEAR AIRPORT

Kingsport, Tennessee, *News*, October 28, 1948

Gunnings—(Spl.)—Plans for an organized "varmint" hunt with trained bear dogs were reported underway here Wednesday as jittery residents heard additional rumors of a mysterious prowling animal. Variously described as "large and black" to "brown-spotted," the animal has reportedly been sighted and heard several times in the Gunnings-Tri-City Airport area within the past month. Several hunters have attempted to corner the beast, which some believe to be a panther.

Bear dogs from Johnson City and North Carolina are to be used Thursday morning to track "it" down.

Tracks left by the animal in soft earth near waterholes are said to be double the width of a man's hand, and one theory is that it is a large cat which escaped from a wrecked circus truck some time ago.

Whatever it is—panther, lion, bear, or housecat—the tales of its weird screams have convinced some citizens it's best to stay indoors after dark.

Although no actual attacks by the varmint have been reported, one tenant farmer said it had a calf hemmed in a fence corner before daylight one morning, and he chased it away with a rock.

Grit, August 19, 1956

My neighbor recently killed a strange animal. It was snow white, weighed about 20 pounds, with a very ferocious appearance. It resembled a large cat, but was too large to be one. It had a short, flat tail, long claws, and long teeth, and preyed on chickens and pigs. No one ever heard of the likes around here.—Mrs. Edna Parker, Hartsville, Tenn.

Downing (1984) noted that a black panther shot near Cosby, Tennessee, was examined by a park ranger and determined to be a large house cat. It was mounted and displayed anyway.

Karl Shuker (1998a; 1998b) reported on a strange feline allegedly shot by hunters in 1996 near Jackson County, Tennessee. His contact (who saw a photograph) described it as "cheetah-like" with a red head and paws, a red dorsal stripe (shades of Rafinesque's *Felis dorsalis*...), golden-brown fur, and black stripes and spotting. Supposedly, the hunters were afraid it was something protected, and didn't report it to authorities. So, it was skinned and the pelt sold to a local university student. Shuker and others who looked into the case were unable to track down the specimen. (While I don't like to merely toss out theories that I can't prove one way or the other, it almost sounds like the hunters may have shot a wandering serval, with perhaps some memory confusion regarding the reddish coloration. Or, if it was a photo of the animal at the time of the kill, blood may stained the fur.)

LARGE BLACK CAT CAUGHT ON TAPE
PANTHER SEEN IN SMOKY MOUNTAINS
Memphis, Tennessee, *Commercial* Appeal, July 11, 1997

Joe Gattas videotaped what he believed was a black panther in late December, 1996, near Gatlinburg, Tennessee. He, his family, and a friend all watched from a chalet deck as both a black and a tawny feline walked off into the woods. The newspaper article's author, Larry Rea, viewed the tape and stated, "it's obvious this is no ordinary house cat or dog. It's big—150 to 200 pounds. And then, there's the way it walks—like an animal from Africa." The article does not mention if there are visual cues in the tape to the animals' size.

One individual told me (Arment 2000b) that there had been several sightings of a black feline by hunters near Indian Mound, Tennessee. She said that some had reported losing deer kills to it. The cat was reported to be about five feet long, larger than a Labrador retriever, but with a stub tail.

TEXAS

A Singular Sight.

Elyria, Ohio, *Independent Democrat*, March 17, 1869

A hunter in the wilds of Texas, who met with many startling adventures, was once witness to a singular battle on the bank of a lonely lagoon in the forest.

He had killed a black panther at this place—more in self defence than for game, for he was chasing wild cattle that day—and leaving the carcass, to return by and by for its skin, hurried forward on a trail which he expected would lead to the object of his hunt.

He came back before night with the trophy of a bull hide, and passed the lagoon where he had encountered and killed his dangerous assailant in the morning. Savage cries and sounds of brutal struggle informed him, before he came to the place, that some deadly battle was going on among the beasts of the forest.

He soon came in full view of the scene, and a sanguinary one it was.—Four black panthers were ferociously disputing the possession of the carcass of the dead panther with two enormous alligators. The object of the combatants on both sides appeared to be the same, viz., to eat the carcass; and for this both fought with painful tenacity, tugging at the bone of contention by way of seeing how much had been gained.

The panthers were superior in numbers, two to one; but the alligators had much the thickest armor, and could fight with their tails as well as with their heads, so that the battle was pretty nearly equal. One of the reptiles had a panther on his back, plying his flank vigorously with his hind claws, and another was holding him by the fore leg with jaws like a tiger.—When he succeeded in shaking off his savage assailants his foreleg was broken, and a slit was made in his side, nearly through the flesh into his entrails.

Meanwhile the other alligator was making frantic efforts to get the third panther into his mouth. He had nearly succeeded, when with a tremendous swing the huge tail of his fellow-saurian knocked out the panther, and wedged itself firmly between his jaws. The teeth snapped together like a pair of copper-mill shears, and one of the tail-thrashing combatants with minus his weapon.

The fourth panther, which had been viciously busy with teeth and claws at the eyes and throat of the curtailed reptile, now redoubled his attacks, with the aid of two others, in front and rear, and soon disabled him. The third panther, owing to his entrance and exit through his enemy's jaws, was *hors du combat* with a broken back.

The fight was now between a single alligator and the three remaining cat-savages. One of the three, however, was by this time badly damaged. Some terrific stroke or bite had completely scalped him, and the skin hung down the side of his neck, flapping as he fought. Another, apparently, had a rib or two broken, but did not seem to mind it. The odds in the battle were still not so very great.

Only the advantage of celerity was vastly on the side of the panther, and when the alligator with much difficulty succeeded in seizing one of them, he was so slow about crushing his prey, and made so much awkward mouthing of it, that he put himself almost at the mercy of his antagonists.

Still his powerful tail was free, and the cat-like creatures, in spring about to find his vulnerable points, were not so spry but they took some stunning cuffs from this caudal bludgeon. Besides, during all the combat, the amphibians had been working gradually toward the water, and now the survivor was almost at the edge.

Once in the lagoon, and his enemies would be powerless. The panthers seemed to be sensible of this, and by an artful movement both succeeded in getting at his throat, while his mouth was full. A few seconds of vigorous tugging and tearing at the tenderer flesh made the alligator's death-wounds, and he slid hopelessly into the water.

Two panthers survived to claim the victory. But they had scarce strength enough left to snuff about the bodies of their slain. Bother were much the worse for wear, and the hunter leveled his rifle, and easily brought them down with a single shot apiece, after which he took off their hides and the hides of their companions, and made his way back to his ranch.

In an eventful life of more than ten years subsequently to this in frontier countries, he never saw anything in the shape of a forest jungle fight that could compare with this combat between the panthers and alligators.

A Strange Varmint

Feasts on a Dog Every Night in the Week.

Galveston, Texas, *Daily News*, November 17, 1897

Henderson, Tex., Nov. 15.—A great deal of excitement has existed here lately on account of a beast that is making its living by killing and eating a dog almost every night, some of them being the most valuable dogs in the town. It has been described by some who claim to have seen it as an animal about two feet in length, a foot high, white legs and head, with a brindle body. So far it has killed about ten dogs and devoured all except the head of at least half of them. Its first appearance in the city was about two weeks ago.

Strange Wild Animal

Victoria, Texas, *Advocate*, February 4, 1911

Westfield, Tex., Jan. 24.—During the past week considerable excitement has been caused by a strange looking, wild animal in this settlement. Last week it came to the county road camp, probably looking for food, but the road superintendent began shooting at it before determining why it came. He describes it as being about two feet high and that it carries its head low like a hyena, and runs life a wolf. Several other parties saw the same animal, and claim that it is no wolf. It has the color of a panther and feeds on dead animals, as last Saturday it was found near a dead cow, upon which signs of its claws and teeth were found. The hunters ran it with well-trained wolf hounds, but it beat them to the timber, and got away in Cypress Creek bottom.

This animal has been seen here many times, and all the wolf and cat hunters are very anxious to kill it, so as to find out what it is.

It serves a very useful purpose in keeping the school children from being tardy at school in the morning and from coming home late in the evening.

DOZENS OF HUNTERS SEEK BLACK PANTHER

Coshocton, Ohio, *Tribune*, April 2, 1931

San Antonio, Tex. April 2—Somewhere in the brush country between Three Rivers and George West, a big black panther, rarely found in this country, is matching his wits against no less than a dozen hunters who for weeks have been stalking him

An invasion of that part of the state by panthers from Mexico this year has made panther hunting one of the chief diversions for ranchmen in that section. Three have been killed within the last month, but the ambition of every rancher is to bring in the big cat that has evaded hunters and steadily ravaged cattle herds all over the section.

Several persons have seen the huge cat, usually in the early morning as he stole quietly across an open plot near a corral or ranch house.

He has always managed to stay out of gun range, however, and efforts to track him have led his pursuers weary miles along a trail spotted with the carcasses of half-devoured colts and yearlings.

BLACK PANTHER HUNT REPORTED IN HOUSTON

Abilene, Texas, *Reporter-News*, September 13, 1946

Houston, Sept. 13.—(UP)—A snarling black panther was hunted today in Texas' biggest city.

The beast snapped the ear off a "cat" dog last night in the Cloverleaf section of Houston as he escaped 15 such recruited dogs.

Officials had scoffed politely at reports of the big jungle cat in Houston streets. But deputies saw for themselves, while cruising the neighborhood to appease the womenfolk.

With the tracking dogs they had obtained through a public appeal, the officers spotted the panther in a back yard. They gave chase and the dogs did too.

A brief tangle ensued, one dog lost an ear, and the whole panther party lost the quarry.

Today with the posse larger the hunt went on.

BEXAR DEPUTIES FOLLOW TRAIL OF SAVAGE, SNARLING ANIMAL

San Antonio, Texas, *Express*, September 15, 1951

A snarling black leopard or panther is apparently loose on the east edge of town, deputy sheriffs agreed.

Deputies Tony Wiatt and Carl Fritsch and a posse of eight or 10 neighboring families tracked the animal—believed to be wild—for half a mile down Salado Creek before the large claw-like tracks disappeared in the brush Friday.

The animal hunt started after the Rev. R. A. Mickles, 614 G St., called the sheriff's office that has 13-year-old son, Tommy, had seen a big black animal that "looked like a tiger." Tommy's big German police dog was badly cut in a fight with the animal.

Two more calls about the animal were received later.

"The tracks looked like it was a leopard or a panther," said Deputy Wiatt. "Whatever it was it sure did whip the daylights out of that dog.

"That was a big dog, and he was badly cut up and bleeding.

"The tracks were fresh, and we tracked it down the creek through the mud. We followed it about half a mile.

"Then it started zigzagging through the brush and we lost its trail.

"We had about 18 or 20 people from that neighborhood out helping us look up and down the creek.

"I'd say it was a wild animal, and not something that escaped from a circus. Those animals can travel a long way. These tracks were about the size of a man's hand pushed down in the mud."

Deputy Wiatt said the sheriff's office would wait until they got another call about the panther before organizing a full-scale posse. He said they would get the famous Walker coon hounds to track the animal.

They had no dogs on Friday's hunt.

Last Tuesday the police department received the first call about a panther being loose in the city. A gutted chicken was the only trace of the big cat.

Accompanying the deputies on the search was Game Warden Arthur Hitzfelder. He said the tracks looked like those of a panther to him.

A Frightened Boy Flew Home on Wings of Fear

"When I saw that big old tiger a'fighting my dog, I didn't wait to look no more."

Thirteen-year-old Tommy Mickles, son of the Rev. and Mrs. R. A. Mickles, 614 G Street, was coming home from doing an errand for his mother with his German Police dog, Poochie, Friday.

"I was crossing the bridge on Yucca when I looked over in the ditch.

"There was a big, black animal with a long black tail drinking water. Looked like a tiger. Poochie ran over to him, and that animal raired up and swiped at him. They started fighting.

"I didn't think nothin'. I just ran. I was that scairt. Poochie, he caught up with me. We both hit home running about the same time. He shore is a bad cut up dog."

Panther Hunt Ends With a Handshake

Amarillo, Texas, *Daily News*, September 20, 1951

San Antonio, Sept 19 (AP)—The panther hunt has been called off—it was a great dane all the time.

San Antonio's "panther" story got started last week with reports of a "big black cat" and "big black panther."

The "phantom panther" is Gus, a friendly dog who had been missing 24 hours.

Gus, sleek and black and 7 feet long from tip of his nose to the end of his tail, is owned by George Husos.

One look at Gus will convince you that—if you were expecting to see a panther—Gus would look like a panther.

Deputy Sheriff Tony Wiatt, who's been in the thick of things since the panther reports started coming in, saw Gus Monday night and swore he was a panther, until he shook hands with him.

'BLACK PANTHER' SCARE DISCOUNTED
Abilene, Texas, *Reporter-News*, December 14, 1952

Denison, Dec. 13 (AP)—Denison's recent "black panther" is discounted by Luther Johnson, Grayson County's state trapper.

"There is no wildlife native to the United States that is black," says Johnson, who has trapped and killed about 200 wolves in the county in the past two and a half years.

"It couldn't have been a black panther unless one had escaped from a circus. It wasn't a wolf or the dogs in the neighborhood where it was reported would have raised more of a howl.

"It could have been a bobcat, but everyone who saw it described it as having a long tail, which a bobcat doesn't have. I think it was a dog and as far as I'm concerned it's a closed incident "

Johnson gets a $5 bounty for each wolf pelt he turns in to the county auditor. He uses a cyanide shell which he paints with a broth made of a ground deer and javelina and places just beneath the surface of the ground. The mechanism of the shell is exploded by the wolf as he digs the non-existent morsel which he smells.

'BLACK PANTHER' ON SCENE AGAIN
San Antonio, Texas, *Express*, March 26, 1953

The famous "black panther" may have re-entered San Antonio Wednesday.

Augustin Cincarni, 140 Wilma Jean St., reported to police he had seen the panther in the rear of his house alongside a drainage ditch about 11 a.m. Wednesday.

Police rushed to the scene, remembering that a little more than a year ago a "black panther" scare had San Antonians in every section of the city "seeing" the animal.

Wednesday, in an official report, Patrolman J. G. Caroll said: "Upon investigation I scared up three dogs, two cows, and one horse, but no panther."

'MONSTER' SEEN IN BIG THICKET
Galveston, Texas, *Daily News*, July 28, 1955

Hardin, Tex., July 27—UP—It's monster season again in the Big Thicket.

Harvey Mecom, a rancher who lives near Herdin, reported Wednesday that his neighbors flushed a "big creature as large as a circus lion" from the brush.

One neighbor, he said, was knocking down and piling stumps with a bulldozer, when the animal appeared, leaped over a bulldozer, just cleared the man's head, ran a few feet, slowed down and walked casually away. The neighbor was not identified by Mecom.

He said the animal was as tall as a small pony, with long hair on his fore parts and short hair on the back part of his body, with a long tail.

Some persons might identify a small, wild horse from this description, but the implication was that it wasn't a horse.

Mecom said other farmers had seen it dashing through the thicket but their dogs looked frightened when ordered to trail it and refused.

OUTDOORS: FROM BLACK PANTHERS TO TIGERS
San Antonio, Texas, *Express*, February 28, 1963

By Dan Klepper

Great shades of the phantom black panther of the brush! Now we have none other than the regal Bengal tiger roaming the thorn-infested lower country along the Rio Grande.

I have received two reports of the majestic striped cat. It has been sighted four times recently in the San Ygnacio area above Falcon Lake by two different men.

Perhaps our elusive black panthers will cross with the Bengal tiger and we can start tracking down reports of either black tigers or Bengal panthers.

However, I sincerely hope the tiger doesn't come in contact with the blue-gray cat with charcoal eyes. I don't think I could stand the combination!

I have never heard an explanation of how black panthers got started in Texas. The Bengal tiger is even more preposterous than the panthers, of course, but at least the tiger has a reason for being here . . . or so the story goes.

It's like this: A man attempted to bring several large cats and a giraffe across the border from Mexico to Texas but was turned back at the bridge.

The man was so irritated he turned the cats loose! Now, how does that strike your funny bone?

RUMORS GROW
VARMINT SIGHTED, BUT SHOT MISSES
Abilene, Texas, *Reporter-News*, June 19, 1963

Haskell (RNS)—Two definite sightings of strange animals several miles apart in Haskell County during the past two days have been recorded by Haskell County authorities as having some background in fact among the many fictitious ones gathering momentum here.

Law enforcement officers in Haskell County have issued a plea for cooperation requesting practical jokers to desist in prank playing about varmint sightings in the territory because it definitely hampers handling of calls that may aid in catching the carnivorous culprit, they said here Tuesday.

A caller who identified himself only as "Carter," a Midland used car dealer, reported an incident 10:30 a.m. Tuesday to Ina May Allen, dispatcher at the sheriff's office. He said he saw a "black panther" in a pasture on the west side of the Haskell-Stamford highway and that it was running a small herd of cattle there at the time.

Carter told the Sheriff's office he shot at the animal, but missed and it ran away.

Thirty-five minutes later, a call from the Stamford Police Dept. reported "Carter" had told authorities there the same story and had driven on toward Midland.

Chief Haskell County Deputy Sheriff Pete Mercer called in veteran hunter Pete Callaway from his work on his farm and together the pair took dogs to the reported scene. No track was found nor trail picked up.

An earlier report, better substantiated, was that Mr. and Mrs. Bill Arend of the Mattson Community, near where the varmint has been reported seen, the last time two weeks ago yesterday, saw a panther-like animal cross the road in front of their car as they drove toward Haskell. This was on State Highway 24, the Throckmorton Highway, about eight miles from Haskell from Bill Brannon's Service Station.

The Arends said the large cat crossed from south to north into a pasture of the Jack Chapman property and disappeared. Arend came to Haskell to secure help in hunting the creature, but it left no tracks, reported Chief Deputy Mercer.

TEXAS TRAILS, BY HART STILWELL
San Antonio, Texas, *Light*, December 6, 1968

The columnist noted the sighting of two black panthers by a woman driving on Bootlegger's Lane in Yoakum.

POLICE TRYING TO TRAP LARGE CAT SEEN HERE
Corpus Christi, Texas, *Times*, August 10, 1972

Police were setting up live animal traps on the far west end of Corpus Christi to try and catch a large black feline. It was said to be "five feet long excluding his tail." Witnesses, including two police officers, said it looked like a black panther; a local wildlife expert suspected it was a jaguarundi.

PEOPLE & PLACES, BY ED BRYSON
Paris, Texas, *News*, March 8, 1976

The columnist noted that Raymond B. Criss of Dimple (Red River County) has been hunting a "black catamount" or cougar for some time. He hadn't yet had a good opportunity to shoot it. It had already killed two calves and was black, "not buff or gray."

PLUM CREEK MONSTER STALKS AREA
Lockhart, Texas, *Post-Register*, May 29, 1980
MONSTER THEORIES ABOUND
Lockhart, Texas, *Post-Register*, June 5, 1980

One or two strange animals (perhaps a pair) were blamed for killing livestock on ranches along Plum Creek. In one case, five calves were gutted, "their stomachs ripped open and their intestines tore out." Witnesses described the animal as "built exactly like a hyena," about 120 pounds, having the long snout of a Doberman, big in the shoulders, low hind quarters, pointed ears, a bobbed tail, and a white V marking on its snouts. The larger one apparently had dark brown shaggy fur, while the smaller one (still twice as big as a "cow dog") had lighter brown-gray fur. A veterinarian suggested the calves could have died from natural causes and then scavenged by coyotes. There were also rumors that an exotic game farm had lost a few hyenas several years earlier.

BIG CAT OF THE SAND HILLS? SOME HAVE SEEN IT
Seguin, Texas, *Gazette Enterprise*, December 31, 1980

Sightings of a large black feline have been reported from this region for years. While deer hunting in November, 1980, Edward Woehler, Sr., watched an animal through binoculars that he first thought was one of his sister's Dobermans. He stated: "It was beautiful. It would run, then crouch down, then get up and run a little farther, then crouch. It appeared to be a cub, not fully grown, but bluish black and sleek with that long slick tail." He estimated it was about 2 ½ feet tall. Going back later with his brother-in-law, they found tracks 4 inches in diameter. Another couple watched a solid black cat, about 20 inches at the shoulder, attempt to catch a duck in their yard. Barbara Mueller saw a cat similar to a black jaguar, and about the size of a German shepherd crossing a road in Hickory Forest.

HENRY'S JOURNAL: BY HENRY WOLFF, JR.
Victoria, Texas, *Advocate*, January 13, 1993

The columnist noted several mountain lion stories, but also mentioned the sighting several years prior of a black panther near the Guadalupe River just south of Cuero. A hunter and his wife saw it, thinking at first it was a big black dog. Upon examination through field-glasses, they saw it was feline.

OFFICERS TRY TO TRAP PANTHER IN GRAPEVINE AREA
Fort Worth, Texas, *Star-Telegram*, August 21, 1998
TARRANT COUNTY'S ELUSIVE PANTHER IS ONE ALOOF CAT
Fort Worth, Texas, *Star-Telegram*, August 22, 1998
PANTHER SEEN, BUT STILL LOOSE
Fort Worth, Texas, *Star-Telegram*, August 23, 1998
THE BIG CAT STAYS OUT OF THE BAG
Fort Worth, Texas, *Star-Telegram*, August 25, 1998

A black panther was reported from the woods along the Colleyville-Grapevine border, and animal control officers thought it was probably the same animal reported near Haltom City several weeks before. It was estimated to weigh 60 pounds. Multiple sightings triggered a hunt (and baited traps) by animal control, but they were unsuccessful. Police said that a tracker found tracks of a cub along with the larger feline.

Where the Wild Things Aren't
Dallas, Texas, *Morning News*, August 30, 2005

While discussing rumored animals in Texas, the article notes that "black panthers are an East Texas tradition." State biologist Gary Calkins, stationed in Jasper County, noted that they get one report a week, and that most turn out to be misidentifications (large black dogs, feral pigs, otters, etc.).

Residents Claim Beast Roams Cherokee County
Tyler, Texas, *KLTV.com*, March 23, 2007

Residents in the Dialville area were terrified of a large black panther that was apparently roaming the woods there. One man saw a panther with three cubs, while others reported strange screams and animal kills.

Black Panther Sightings in Upshur County
Tyler, Texas, *KLTV.com*, March 28, 2007

Residents in a wooded region called Raintree Lake, near Rhonesboro, reported seeing a large black panther. Dogs were missing in the area, and some livestock had been found killed.

Jameson (2007) notes a number of sightings of a black panther in the Guadalupe Mountains since the 1940s. In one early instance, a rancher watched it kill a lamb and carry it up Manzanita Ridge.

VERMONT

Troy, New York, *Record*, October 23, 1945
James Johnson of East Pownal reports seeing a large black animal which he believes was a panther standing near his chicken house when he went out in the early; morning. The animal went away toward the mountains.

BLACK 'PANTHER' SEEN IN VERMONT
Portsmouth, New Hampshire, *Herald*, April 11, 1950
Stowe, Vt., April 11 (AP)—That big black cat—believed by some to be a panther—is in again.

E. C. Hillman reported last night that he and members of his family saw the animal sneaking through their yard in the direction of Mount Mansfield during the weekend.

Hillman said it is as "big as a dog, with a long tail." For the past several years Vermonters have reported seeing the big cat. But no one got close enough to get a fatal shot at it.

VERMONT PANTHER IS REPORTED SEEN ON BREAD LOAF MT.
North Adams, Massachusetts, *Transcript*, August 28, 1951
Middlebury, Vt. (AP)—Doubting Vermonters who disbelieve the "seen-again" tales of a panther, usually spotted in the southern part of the state, can take the word of a seasoned World war 2 ski trooper that the animal does get around occasionally.

The ski trooper, Ralph Myhre of Middlebury, says he saw a black panther on Bread Loaf mountain in Ripton between the home of

Saturday Evening Post Writer William Hazlett Upson and Middlebury college's Bread Loaf schools.

According to Myhre, he stopped his car, took out a woodsman's ax and got to within 30 feet of the animal. He turned tail and ran after the panther had growled at him in defiance.

Later, Myhre and a companion, both armed with rifles, combed the woods for two hours but failed to catch up with the elusive panther.

Myhre said the animal was about three feet long and was probably a cub.

'BLACK PANTHER' SEEN IN WOODS IN POWNAL
North Adams, Massachusetts, *Transcript*, November 13, 1953

Vermont's panther has changed its color... now it's black.

A large, shiny black animal has been seen on two occasions in woods off Route 7 in Pownal Center, and was labeled by the observers as a black panther.

It was first seen by Mrs. Franklin Hoag and her two sons as the animal was drinking water at a pond near her home. Mrs. Hoag described it as very long, about two feet high and with a huge head, small ears, and large glowing eyes.

On another occasion, the "black panther" was seen in the same wooded area, by Davie King, 12-year-old son of Rev. Isabelle King, pastor of the Pownal Center Community church. He said the animal had a long tail and made a sound like an angry cat.

BLACK PANTHER STORY FROM PETERSBURG WAS NO HOAX
Troy, New York, *Times Record*, October 8, 1955

By George Schelling

The black panther which terrorized the Petersburg area three summers ago and an area outside of Greenwich about five years ago was no hoax. These black "cats" came from the Stratton National Forest about 9 miles east of Arlington, Vt.

No less an authority for this is James Torpey, well known hunter and sportsman of Hoosick Falls, who came upon one of these ferocious-looking animals sunning itself on the trail leading through the mountain wilderness of the park. With him were Mr. and Mrs. Edward Fanning of Washington, D. C., who vacation in the area. The party made the trip several weeks ago.

Torpey and his guests were enroute to the site in the park where Daniel Webster made a speech at a Whigs outing in 1840. They had just driven around a bend in the trail when they saw the panther.

Saw Animal.

"I leaped from the car and was within 30 feet of the beast when it jumped on a boulder, bared its fangs, spit and then slunk into woods," Torpey said. "A strong east wind and the noise from the roaring Branch River probably prevented the cat from hearing our approach," he added. Torpey and another hunting companion, James Rourk, formerly of Hoosick Falls and now of Woburn, Mass., made a second trip into the park not long ago with guns and Torpey's Labrador retriever, "Mike." They didn't see the panther but saw evidence that there is probably more than one black panther or catamount.

An interesting sidelight of the second safari was that the mystery of the missing burros was solved. A few years ago, Torpey related, a crew from the Fishery Division of the National Park's Service used the burros to carry cans of fingerlings to Bourne and Branch ponds through the rocky terrain. He explained that while the fish planting was going on the burros got away and were never found.

Found Burros.

Rourk and Torpey came upon the burros in the park. They had increased in number. More than that, one of the burros which stood by as her young one showed signs of having battled one of the big cats. "Its ear was torn in such a manner as to indicate that a powerful animal had attacked it," Torpey said.

Torpey described the panther as being about four feet long with a "chuckled" head familiar to the cat family. He said that its powerful forelegs were bowed beneath a deep chest and by its lope-like, walk indicated that it was definitely a breed of panther or like the cougar familiar to the far west.

A study of wild life reports of that area in the early 19th century indicated that the catamount were plentiful, and posed a problem to sheep herders and farmers. To date there have been no complaints that the big cats have been attacking cattle,

Torpey declared that the terrain of this section of the Green Mountains is ideal and a natural habitat of the catamount.

Dog Gone Chasing Panther
North Adams, Massachusetts, *Transcript*, March 24, 1959

Groton—For years, every time things are lagging around Vermont, the famed "black panther" has been making his appearance—like the "Loch Ness monster" of Scotland.

Yesterday two milkmen, Kenneth Williams and Lyman Blanchard, reported they saw the black panther walking slowly across a field during the weekend.

The men say they borrowed a hound dog and the dog disappeared while on the trail of the panther. Neither the dog nor the panther has been seen since, they said.

Game Warden Says Big Cat a Loose Pet
Burlington, Vermont, *Free Press*, November 6, 1997
The Cat That Roared
Burlington, Vermont, *Free Press*, November 8, 1997

Residents in South Burlington were unnerved by sightings of a black feline, estimated at "3 feet long, 18 inches high and weighs more than 50 pounds." One woman photographed it, and a game warden who visited the site and examined the photos thought it might be a South American exotic, as it didn't look like a North American native species. The game officer finally tracked down the feline, discovering it was Porgy, a large 14-pound domestic tomcat that lived in the area.

VIRGINIA

A Tiger Too Fond of Hogs

New York, New York, *Times*, April 13, 1885

Portsmouth, Va., April 9.—Last Fall a black tiger escaped from a circus in this city and was diligently hunted and could not be found. On Monday last J. F. Williams, of Greenville County, discovered two of his hogs dead and could not account for their death. The next day two more were found mysteriously dead. An investigation showed the tracks of some large animal like a dog, and they were followed to a swamp about a mile from where the hogs had been killed, and a large black tiger was discovered asleep under a tree. Mr. Williams returned home for his gun, his dogs, and his negroes. When he got back the tiger had been alarmed by the dogs and had gone up the tree. Mr. Williams fired, the tiger fell out and a fight ensued with the dogs. While this sport was going on one of the Negroes picked the opportunity and killed the tiger with an axe. He measured eight feet from tip to tip. The skin was carried home by Mr. Williams, and was stuffed as a trophy. The dead animal is believed to have been the same one that escaped from the circus.

Scared by Strange Beast.
Virginia Farmers Alarmed by a Monster Which Has Its Refuge in the Dismal Swamp.

Belleville, Illinois, *News Democrat*, March 29, 1902

Another monster has come forth from the dismal swamp jungles, and is spreading alarm among farmers around Driver, 12 miles from Suffolk, Va. Last week in unclassified animal not indigenous to high lands terrorized residents of Pleasant Hill so that they were afraid to leave home at night. After being chased by numerous armed posses this beast was killed by Harrison Walker, a colored hunter.

Now Edward Smith, a farmer of Driver, says that a strange being visited his premises and killed seven dogs, two of which were eaten, while the other five were mutilated. Another dog took refuge under a barn and Smith, hearing the dog's yells, went out with a pistol. The monster sprang upon him. Smith fell, and the beast tore his clothing to shreds. He finally beat it off, but the revolver, being corroded, would not fire.

Whit Walker, of the same neighborhood, says the animal visited his home, too. From Smith's description the animal is larger than a wolf, with shaggy, yellow hair, long head and sunken eyes. It is gaunt and vicious.

People are wondering what will next come from the unexplored swamp made famous by Tom Moore's verses. The more superstitious regard the visitation as supernatural, and families are having much trouble with their servants.

Strange Beast.
Possessed of Remarkable Strength, Terrorizes Whole County.
Pittsburgh, Pennsylvania, *Press*, December 29, 1902

Remarkable stories of a strange beast come form Prince Edward county, and a large portion of that section is in terror on account of its depredations. These stories vary to some extent, but they come from many sources, and are strongly vouched for by many persons.

The animal has been seen often at night, and sometimes in the day, and is described as being as large as a dog, though much longer and very tall in proportion. Dogs will scent the animal, but show the greatest signs of fear as soon as they come in sight of the beast.

On one occasion the animal came into a clearing where several children were at play, and started to attack them. Two dogs defended the children, and were both killed. When grown persons arrived, the beast retreated into the bushes. It has killed and eaten a number of dogs. All attempts to catch it with steel traps have failed.

It has been seen a number of times at Jones' Mill, at Prospect and at widely separated parts of the county within the last three weeks.

Samuel Gibson, residing near Prospect, tells an interesting story regarding the strange visitor. He says he caught it in a trap one night last week. The trap was chained to a tree, and the strength of the animal was so great that after being caught it carried the trap to parts unknown. The next night, however, the visitor brought the trap back. In attempting to get through the fence around his yard, the trap got hung in the palings and the animal pulled out of it, leaving in the trap a lot of hair. Mr. Gibson says it is impossible to catch the beast, for the reason that nearly every person in that section is afraid to participate in a hunting expedition, and, as already stated, all dogs are afraid of it.—Richmond (Va.) Cor. Baltimore American.

Taxidermist Jim Hunter Has Never Before Seen or Heard of Such Animal as Was Shot in the Dismal Swamps of Virginia

Burlington, North Carolina, *Daily Times-News*, October 5, 1943

The strangest animal known to have come from the dismal swamps in Virginia has been mounted by Jim Hunter, taxidermist of Graham, for J. A. Spivey, of Suffolk, Va.

During the many years since Hunter has worked at his hobby during spare hours from his textile work, mounting animals, fish and birds, he has never seen anything like it. He isn't sure that anybody else ever has.

Mr. Spivey, who killed the animal with buckshot while hunting in the swamps, could not identify it. He had not found a person who could when he took it from cold storage and shipped it to Graham.

Some have called it a Coati, a cross between a Honey bear and an Ant eater; some have named it a Wolverine. In Jim Hunter's opinion it is neither. It is to him "just a mistake of some sort in the animal kingdom."

Over in Virginia there is speculation that the animal originated somewhere in South America and "shipped" to the United States where it escaped to roam the swamps. There is no verification.

It has facial resemblance to the weasel family. It has the feet of a raccoon and its forelegs are bowed to suggest that it was a climber. The hind legs and tail are heavy in comparison to the fore part of the body.

"Your guess is as good as mine," Jim says, "but I sure would like to know just what the thing is to satisfy my own curiosity."

COUNTY WOMAN REPORTES BLACK PANTHER IN YARD

Charlottesville, Virginia, *Daily Progress*, March 10, 1988

EXOTIC CAT ON THE PROWL RAISING FUR IN ALBEMARLE

Richmond, Virginia, *Times-Dispatch*, March 14, 1988

Police and animal control officers were looking for a large black feline, "about the same size as a mountain lion," reported by Rebecca Tate in Blenheim (Albemarle County). She described it as six to seven feet in length, about 2 ½ feet at the shoulder, coal-black, and having green eyes and a long tail. Another Blenheim resident who saw it called the animal "beautiful but frightening."

'BLACK PANTHER' ELUSIVE

Richmond, Virginia, *Times-Dispatch*, March 31, 2005

The previous December, a deer-hunting couple driving along a Sussex County dirt road went round a bend and saw a large coal-black cat about sixty yards away, which bounded off into the woods.

WASHINGTON

HUNT SPOTTED LION

Placerville, California, *Mountain Democrat*, February 16, 1924

Ashford, Wash—A mountain lion curiously marked with black and gray blotches on his tawny hide has been seen by three woodsmen near here at various times this winter. The

spotted cat is being hunted to death in the big drive government hunters are making in Mount Rainier National park. It is believed the cougar was caught in a forest fire and severely burned in spots, and the hair is returning black and gray where the tawny skin was wounded.

'PANTHER' HUNT ENDS

Pacific Stars and Stripes, May 12, 1952

Everett, Wash. (UP)—An alarm to Snohomish County, Wash., residents to be on the lookout for a black panther was ended Saturday.

The animal, believed to have been a panther and reported by John Ellis to the sheriff's office, actually was a cocker spaniel with big feet.

Ellis told the sheriff's office Monday that a panther was on the loose and the sheriff's office began an extensive search with bloodhounds. The alarm ended when the dog's paws matched the "panther tracks" found near a ravine between here and Marysville.

Tacoma Police Saw Lion, or Did They

Pasco, Kennewick, and Richland, Washington, *Tri-City Herald*, July 6, 1976

Dog Makes Mutts of Lion Hunters

Daytona Beach, Florida, *Morning Journal*, July 7, 1976

Two police officers in south Tacoma shot at a "100-pound African lion," with a black mane, but it escaped into the woods. The local humane officer suggested that a 100-pound African lion would be too young to have a mane.

The next day, a large dog with a "leonine haircut" was captured.

Local Safari Fails to Find Lion King

Spokesman, Washington, *Spokesman-Review*, March 22, 1995

Animal control officers and sheriff's deputies searched for a reported "African lion" without success. The woman who saw the animal was adamant it was not a dog.

Spokane Lion Still on the Lam

Spokane, Washington, *Spokesman-Review*, June 5, 1996

Dog's Citation Raises Questions About Elusive Lion

Spokane, Washington, *Spokesman-Review*, February 19, 1997

A "lion" (witnesses were unsure if it was a mountain lion or an African lion) was described as "500 pounds" and "scruffy." A few reports turned out to be dogs. A lamb was killed on a local farm by something "with major jaw power," but the culprit wasn't captured. Eight months later, a local Great Pyrenees was cited for biting joggers, and suspicions were raised that it might be the animal responsible for the "lion" sightings.

Sightings of Big Black Cats in Area Reported

Davenport, Washington, *Times*, September 10, 1998

Rural Officials Hope to Cross Big Black Cats' Path

Spokane, Washington, *Spokesman-Review*, September 19, 1998

Mysterious Black Cats in Quad Cities

Oelwein, Iowa, *Daily Register*, September 21, 1998

If It's Black as a Panther and Looks Like a Cougar, It Could be a River Otter

Spokane, Washington, *Spokesman-Review*, October 2, 1998

A pair of black cougar-sized cats (with "about three-foot tails") were reported seen over a half dozen times near Davenport, Washington. One witness described them as three feet tall and "lightning fast." Wildlife officers later suggested that river otters were responsible. Another witness, an avid hunter who spotted a large black feline cross the road in front of him "about 13 miles north of Cusick in Pend Oreille County" in "bright sunlight" and from a distance of "probably less than 100 yards" said it "was just too much of a cat to be anything else."

WEST VIRGINIA

"A person of the name of Draper had gone in the year 1770, to hunt on the Kanhawa. He had turned his horse loose with a bell on, and had not yet got out of hearing when his attention was recalled by the rapid ringing of the bell. Suspecting that Indians might be attempting to take off his horse, he immediately returned to him, but before he arrived he was half eaten up. His dog scenting the trace of a wild beast, he followed him on it, and soon came in sight of an animal of such enormous size, that though one of our most daring hunters and best marksmen, he withdrew instantly, and as silently as possible, checking and bring off his dog. He could recollect no more of the animal than his terrific bulk, and that his general outlines were those of the cat kind. He was familiar with our animal miscalled the panther, with our wolves and wild beasts generally, and would not have mistaken nor shrunk from them." (Jefferson 1799)

Newton (2005) noted the 1901 appearance of a maned feline with a tufted tail in Pocahontas County.

STRANGE ANIMAL REPORTED FROM THE FIRECO SECTION
Raleigh, West Virginia, *Register*, May 11, 1934
From Fireco come reports of a strange creature that, appearing in that vicinity three weeks ago, has slain hogs and dogs, carried away chickens, and frightened people by violent antics.

Henry Beard, eighth grade Sloco student, and the Reverend Artie Ward, colored minister, profess to have seen the creature, and their combined meagre description indicate only that it is "about two feet fall," has "yellowish gray hair," and "runs like the wind."

627

Three dogs owned by Floyd Terry, of Fireco, chased the strange varmint into an old drift-mouth. Presently two dogs came forth running as though possessed; the third dog remained behind, his head snatched off.

Upon its first appearance three weeks ago, the animal, if animal it be, attacked three hogs, tore off their heads, ate their brains and then decamped, leaving the carcasses behind. Shortly thereafter it was reported elsewhere in the community, and this time two hogs were left separated from heads and brains.

The creature, according to information, is fond of hog brains.

Last Wednesday night the Reverend Artie Ward heard a commotion among his chickens. He arrived at the chicken-yard fence barely in time to see a strange, yellowish-gray-haired animal clearer the six-foot fence at a leap and disappear, leaving some of the hair clinging to the fence.

As a result of these depredations, residents of the community are frightened and their dogs are so terrified that they remain in their owners' yards and howl mournfully.

Thus far nobody has been found who knows exactly what the dog-killing, hog-head-snatching creature is, nor is there any accurate description of him. Residents of the Fireco vicinity speak speculatively of wolves, panthers, and it has been suggested that some unnatural hybrid has resulted from an unusual association of animal life.

BLACK PANTHER SEEN IN HINTON SECTION

Beckley, West Virginia, *Post-Herald*, June 17, 1949

Hinton, June 16—An animal, to have been a black panther was seen on the Chestnut Mountain road on Elk Knob about 12 miles from Hinton, about midnight Wednesday night, by Constable Jesse Romanello.

The constable said that as he was driving along the road about 11:45 he saw eyes reflected in the road ahead and flipped up his headlights, revealing a black animal from 15 to 18 inches high and about four feel long exclusive of the long slim tail, which was carried with an up-curl at the end. The animal remained motionless for a few seconds, then turning at right angles to the car, entered the dense woods through which the road ran.

Romanello, certain the animal could not have been a dog, said it might have escapee from a road show somewhere near, but no reports of escaping animals have been received here, so far as is known.

State Joins Varmint Hunt; Isn't Panther, Experts Say
Charleston, West Virginia, *Daily Mail*, February 8, 1951

Conservation commission experts Thursday discounted stories that a "panther" is running Coal Branch Heights section.

J. H. Branham, district conservationist, said the last known appearance of either a panther or mountain lion in West Virginia dates back to 1916. "We're just not in the section of the country for panthers," the officer added.

He told the Daily Mail he had been informed of the varmint's appearance and plans to make an investigation of the occurrence Thursday afternoon. Meanwhile, stories of the strange animal continued to flow out of the hills above the city's incinerator plant.

One resident who claimed he had seen the "varmint" said the animal was "black and shaggy and stood about waist high."

Other residents of the area said the throats of a half a dozen dogs had been slashed during the last three weeks.

A conservation officer said the animal could conceivably be a stray bear. He, too, said the habitat of the nearest panthers was in the southern tip of Florida.

Branham said that almost every year similar cases are reported to conservationists. "The animals, we find, are either vicious dogs or some small game strayed away from its home."

Mrs. Velma Pagley, who first reported the animal's appearance, said she had received an offer from E. H. Sams of 541 Bream St. to set out traps in an effort to clear up the mystery.

Conservation officers said the theft of a pork carcass could have been the work of a "two-legged hog."

'Posse' Hunts Beast in Vain
People Terrorized, Says 'Heights' Man
Charleston, West Virginia, *Daily Mail*, February 10, 1951

The "Black Devil" of Coal Branch Heights may be a tomcat to the conservation commission but it is a dangerous demon to residents of the neighborhood and an active campaign to trap or kill it continues nightly.

A brush-beating band of men before dawn Saturday awakened the neighborhood just northwest of the city as they tramped through a hollow believed to be the hiding-place of the black creature with the blazing eyes.

"They fired at least seven shots and yelled and whooped," one woman reported, "but they didn't kill the animal." Other hunters prowled the brushy hills until midnight without success.

Mrs. Ernest Elkins of the community said she heard the marauding beast scream from the wooded valley back of her home and that eerie, chilling cry was heard distinctly by her neighbors.

Conservation commission experts have attributed the screams in the night to tomcats.

"No tomcat ever yelled that loud," one woman said.

R. L. Harrah, in a letter to the Daily Mail, said that a "calamity" had hit Coal Branch Heights.

He said it was all due to the refuse dumped in the vicinity by the city and that "this dump is luring wild animals that are killing the citizens' best friends, their dogs."

Mr. Harrah declared that the people are "so terrorized they are barring their doors and sitting with whatever weapons they have on their laps, ready for the worst to happen."

He said refuse should be hauled to more distant spots "where the wild beasts won't frighten so many people."

A bear hunter called the neighborhood and said he would bring four bear dogs into the community if all the people would lock up their dogs.

"My dogs," he warned, "will kill another dog just as quick as they will attack a bear."

PANTHER HUNT BEGUN

Beckley, West Virginia, *Raleigh Register*, July 26, 1960

Hepzibah, W. Va. (AP)—A hunt was started in this area last night for a panther whose screams have been heard hereabouts for more than a week.

That it is a panther rests on an eyewitness account. Mrs. Holland Vernon of Hepzibah, who reported the screams being heard, said her uncle saw it at a water barrel. His description was that the animal was either dark brown or black.

Harrison County deputy sheriffs and game wardens, hearing of this, started the hunt.

Some Residents on Alert for Dog-Scaring 'Thing'
Charleston, West Virginia, *Gazette*, June 29, 1965

Several residents of North Charleston claim to have seen an animal similar to a black panther. One resident "said it was pretty big, had a long tail and a face like a cat." Dogs in the neighborhood seemed scared of the animal.

A Jungle Out There? Hunter Says Lion in Woods
Beckley, West Virginia, *Register-Herald*, October 23, 2007
Search on for Lion in West Virginia Hills
Charleston, West Virginia, *Daily Mail*, October 26, 2007

Officials in Greenbrier County were searching for a maned lion after a 72-year old bowhunter said it approached him. He was able to watch it for more than 40 minutes. He thought it weighed between 250 and 300 pounds. "It had a mane, so I could tell it was a male. And I'm sure it wasn't a bear. Bears are all over Cold Knob. I see six to eight of them every time I go hunting, and I can tell the difference. Bears don't shake me up at all. This lion made me pretty nervous." When he reported it to wildlife officials, he was told he was the second person to report the animal.

WISCONSIN

Avon Farmers in Arms

Hunting for a Large, Strange Animal in the Woods

Janesville, Wisconsin, *Gazette*, July 28, 1891

The Beloit *Free Press* is responsible for the statement that a large, strange animal was seen in the woods on the farm of Mr. George Cox, of Avon, two weeks ago, and again last Friday evening. This fact, and a description of the animal as given by the parties who saw it (John Nelson, Arthur Stoneburner and others) has caused much excitement in that vicinity. Sunday thirty men, armed with rifles and revolvers, searched through forty acres of woods in which it is thought the animal now is, and found only its tracks in the soft ground. These tracks were examined by Mr. Yagla, an old hunter and trapper, who says they are like those made by a large bear, but the animal is described as being gray with black spots or stripes on its body. Plans are being made for a searching party of about one hundred men, who will soon spend a night in the woods, when an important capture will probably be made.

Milwaukee Police Can't Find "Tiger"

Jefferson City, Missouri, *Post-Tribune*, May 4, 1928

Milwaukee, May 4—(AP)—A thorough search by police this morning failed to reveal any trace of the "tiger" that had been reported roaming the west side residential area of Milwaukee.

The search was continued, although officers expressed belief that the "tiger" probably was a large dog.

Residents of the west side still are doubtful, however. Hundreds of phone calls poured into the police station this morning asking if it would be safe to send the children to school and demanding a search at once. No animals are loose from the city zoo.

Several people reported seeing the animal and all thought it was a tiger.

What Next? New Mystery Beast Reported on Prowl
Wisconsin Rapids, Wisconsin, *Daily Tribune*, July 7, 1949

Now it's Waushara County residents who are beginning to wear that harried look which comes from too many stories about strange critters on the prowl.

The data given by eyewitnesses adds up to an animal which resembles a baby polar bear with a few black spots.

For some time now persons occupying cottages around Long lake east of Plainfield have complained that an animal with outsized footprints had been knocking over their garbage cans.

Wednesday morning Herald Eastling, Plainfield, was tending his minnow stock at the lake shore when he heard a noise. Looking up he saw a creature gallop past him pursued by a hound dog.

He declared emphatically that the furry apparition was not another dog nor a wolf. He estimated its size at three feet high, weight at a hundred pounds and described its body as a long one.

Those who've seen the footprints say they are huge.

Wood County residents who have been pestered with a "lion" think the Waushara folks are being right neighborly now.

Hunters Find No Trace of Vicious Beast
Oshkosh, Wisconsin, *Daily Northwestern*, January 30, 1961

Poy Sippi—Hunters have their shooting irons at the ready here and a few of them are looking for the mysterious beast which has killed one dog, injured at least two others and probably mauled a heifer.

Even the veteran outdoorsmen have not been able to identify the animal positively, but most of them are skeptical that it is a wolverine. The Waushara County game warden was due here later today and hopes to identify the animal by its tracks.

Meanwhile, life was quite normal here and at Bloomfield, where the belligerent animal was first reported. Speculation varied. Some thought it might be a timber wolf, others believed it might be one or more large dogs. Wallace Nitzke, whose heifer was badly bitten earlier this week, thought a pair of wolves might have attacked his farm animal.

Nothing unusual was uncovered by the hunters, according to Conservation Department game warden Gilbert Voss. The hunters had been seeking tracks of the animal in the snow.

At least five persons are reported to have seen the animal. Norman Papp, who saw it fighting with his dog, said it was a dark yellow animal. Besides injury Papp's dog, the animal killed Ervin Schoenick's Irish setter, skinned it, and partly buried it.

Bits of hair from the animal has been sent to the conservation department in Madison for identification.

The wolverine, says game manager Ralph Hopkins of Wautoma, is, pound for pound, "the most wicked animal alive."

WAUSHARA'S 'BEAST' MAY BE WILD DOG

Madison, Wisconsin, *State Journal*, February 3, 1961

By William Stokes

It would be a dog of a thing to happen, but the mystery beast of the Waushara County, may well turn out to be a wild dog.

The dog theory is offered by Prof. Robert McCabe, University of Wisconsin wildlife expert. It comes on the heels of speculation from Waushara County residents in which the beast was called a lynx, bobcat, wolverine, panther, cougar, mountain lion, puma, etc.

The death of a large Irish setter is blamed on the beast as is other mysterious nocturnal activity in the area of West Bloomfield, along the Waushara Counties northern line. Several residents have reported sighting the animal, among them a farmer who said it looked like a dog.

Left Behind

McCabe said the animal may be a feral dog—one which is untamed or undomesticated and has no home association.

"It happens occasionally," McCabe said, "that when a dog is left behind when its owner moves or the animal is dumped out in the country, it becomes wild and begins to live off the land."

The animal has been sighted on the outskirts of the village of West Bloomfield, and McCabe said this would follow, since a wild dog would likely nose around garbage cans at night looking for food.

Could Be Bobcat

The incidence of a domestic animal turning wild is much more prevalent with cats, McCabe said, since many people leave them behind when moving or simply dump them in the country to get rid of them.

"It is my opinion." McCabe said, "that it is much more likely that the animal is a dog rather than a bobcat, although the bobcat theory is entirely possible."

A cluster of hair, allegedly from the animal, was collected in the vicinity of the dead Irish setter, and has been sent to University of Wisconsin biologists for identification. No report was yet available Thursday afternoon.

Offer Reward

A reward of $100 for capture of a mystery animal has been offered by the Grand Marsh Wildlife Ranch at Grand Marsh. The beast must be captured alive in order to collect the reward, according to the notice from the wildlife ranch.

The animal is wanted alive "so that the ranch can ascertain certain scientific facts as to the abundance of these animals in the state."

West Bloomfield residents, meanwhile, are anxiously awaiting tracking and hunting conditions to try and solve their backyard, backwoods mystery.

On the House: Charlie's Search Proves Futile; Barroom Talk About 'It' Goes On

Appleton, Wisconsin, *Post-Crescent*, February 7, 1961

By Charles House

West Bloomfield—Under nearly perfect tracking conditions I prowled the woods and all the fields, the thickets, ditches and road

sides of many miles of this section of Waushara County, but my plod-
ding hunt for the trail of a mystery critter has come to naught.

Wolverine, cougar, werewolf, lynx, bobcat, badger—whatever the
spectre may be—it left no track where I could find it.

I hunted alone and in parties for three days, hungering for nothing more than a clearly decipherable track in the soft, moist snow of
the environs.

I found the trails of fox, deer, rabbits, porcupine, squirrel and
innumerable birds. But the mystery prowler which has so excited
the people of this and nearby communities has eluded me or has
moved in areas where I did not.

I hunted that woods where a 60 pound dog was killed. I scouted
the ditches and roadsides where a strange animal had been sighted.
I plodded circular and zigzag courses over Pine Hill where a family
was routed by a mystery beast. I studied the trails of a hundred animals, including that of man, but my quest was futile.

I saw to the peculiar wolf-like trail of the wolverine, gulo luscus,
whose five-toad mark should have been easy to read in the softening
snow. I searched for the larger, wider foretrack than the hindtrack
and I studied the drifts for the furrow of its clumsy body.

Three-Legged Deer

I hunted in vain for the sign of a kill, a hint of a woodland burial,
the bloodied snow from a fight in the woods, the carcass or the remains of a bird or beast, a gnawed bone, but these signs alluded me,
too.

I searched also for a trail of a lynx or a bobcat but discovered
only the little stories of lesser creatures and one of a three-legged
deer. But my hunt came to nothing but the fun of being in the woods
and the tests of reading little stories in the trails of woodland creatures.

But no matter. Somewhere in the woods of the area a mystery
critter moves slyly and stirs the entire region in quick, lunging hundreds, into hours-long discussion and conjecture.

"It'll turn up to be a big, yellow dog," said someone.

"I'm positive it's nothing more than a great, big badger," said
another.

"Big bobcat!"

"Cougar!"

"Lynx!"

"A panther escaped from a circus!"

"A pack of dogs!"

Animal Called 'It'

All those guesses are just that. Nobody knows what animal, strange or common, moves through these parts to so excite the people.

Conservation men, lacking good prints or accurate sightings, are unable to identify the critter, the "it" as the locals are calling the animal under so much discussion.

But they have learned not to negate the possibility of the presence of almost any animal scarce in these parts. Only 20 years ago a moose was killed in Florence County. Bobcats and lynx have turned up places where they do not ordinarily frequent.

A wolverine is not an impossibility all though it is far from its normal range in Canada and along its borders. Skeptics use the word "impossible" in connection with the unwelcome wolverine, but it is a wary, wily animal which ought to be ruled out because of common sense.

Cat-Like Habits

There have been no defilement of trap lines, no damaging vandalism, no horrible odor which is one of the unpleasant proofs of a wolverine's presence. Whatever signs there have been suggest an animal with cat-like habits—but there has been no spoor found, no evidence to prove up the theory.

I found an almost unbelievable number of dog prints in the woods, evidence that the mystery animal should surely be well-pursued. I found the trails of many a hunter with those of their questing dogs. I found three places in different parts of the woods where several dogs were running deer or trailing rabbits.

The hunters hereabout are sensitive to hints and suggestion. Example: in one sorty, I jumped a herd of four deer which fled towards a nearby road.

An hour later when I returned to West Bloomfield for "new reports," I learned that some deer had been seen running across a road and that a group of hunters had backtracked them in search of the animal which frightened them. They will have found my trail.

Answers Must Come

Some day—and recent suggests that it ought to be soon—someone must get a good look at a track or at the creature itself. And then the identity of the strange critter may be known.

Meanwhile, such capable hunters as Clayton Looker and Dale Ludtke of Fremont will plod through the woods with the hundreds of huntsmen and farmers of the regions in search of the spoor of the animal.

And off in the taverns, like Wendts and Harry's in West Bloomfield, people will talk and rumors will linger. And Norman Schoenick's grocery store in that hamlet will be a central place for conversation and conjecture.

Off in Fremont, the customers at the Fremont Hotel or in Morack's Restaurant, will have discussion and conjecture with their coffee, and all of them will terminate their discussions with ... "Well, we ought to find out pretty soon."

HAIR SAMPLE NO HELP IDENTIFYING
WAUSHARA COUNTY MYSTERY BEAST

Madison, Wisconsin, *State Journal*, February 10, 1961

By William Stokes

There is some hair in the mystery animal stew, but it isn't enough to aid identification, according to biologists.

A tuft of hair from the unknown animal reportedly roaming the West Bloomfield area of Waushara County has been studied by biologists, but identification of the beast will remain a mystery until more evidence is found.

The hair sample, found near where the animal killed and partially skinned a large Irish setter, was made up of belly hair which is a type that defies definite analysis, the biologists report.

Examined by Socha

Only one hair in the sample was a guard-hair or the longest hair from a furbearing animal, which is the best for analytical purposes.

The rest was made up of under-fur, which is short and difficult to work with in obtaining identification.

George Knudsen, Conservation Department research biologist, received the hair in Madison and sent it to the University of Wisconsin department of zoology. The analysis work at the department was done by George Socha.

Knudsen said that Socha spent a long time working on the hair in an attempt to pin down identification. Microscopic viewing was done and the hair was compared with a large collection of hair samples. No conclusion can be reached as to the source of the Waushara County hair.

The mystery beast has not been sighted in the West Bloomfield area for the past two weeks, according to Harry Hudziak, a tavern-keeper in the small village.

Hunters Ready

"Hunters are ready to go around here the minute we get some sign," Hudziak said, "but there haven't been any reports for the past couple of weeks."

Hudziak is one of a number of West Bloomfield area people who has reported sighting the animal. He said he saw it one night when his dog made an unusual fuss. The beast was about three feet tall, and had a long tail, Hudziak said.

According to a report from Green Bay, a restauranteur there photographed the animal near Fremont last summer. Viewers of the movie film report that the beast most resembles a wolverine.

Knudsen said that he and other biologists are most anxious to receive further information on the animal. Knudsen said he had not seen the Green Bay film.

MAN REPORTS BLACK COUGAR, BUT DNR OFFICIAL DUBIOUS
Milwaukee, Wisconsin, *Journal*
undated clipping, 1990s, sent to researcher Gary Mangiacopra

A Friendship, Wisconsin, man reported that he had seen and photographed a black cougar. He stated, "I have six Polaroids of it, but the photos were taken at a distance where the cougar won't show up in a newspaper reproduction." John

Haugh, DNR wildlife manager, said it might be a black bear, but noted that the DNR had received other cougar sightings, including black cougar reports.

Godfrey (2003; 2006) noted a number of cases of a wolf-like animal, sometimes erect, particularly in the vicinity of Elkhorn, but also elsewhere in the state.

WYOMING

Strange Animal Shot Near Merna Monday

Pinedale, Wyoming, *Roundup*, November 14, 1963

Dick Roberts, Daniel rancher, shot a rather strange animal in the Merna area Monday. The animal is a coyote, apparently crossed with some other animal, according to Jim Straley, state game biologist for this area.

The animal, a female, is the same size as a coyote, but colored considerably darker. The skull will be sent to the Zoology Department at the University of Wyoming for positive identification.

The hide will be mounted in life size by Faler Bros. Taxidermists.

YUKON TERRITORY

TRAPPER GETS PELT OF STRANGE NEW ANIMAL
New Castle, Pennsylvania, *News*, June 27, 1923

Tacoma, Wash., June 27.—Robert McConnell, old trapper and hunter of the Upper Yukon region, brought to Fort Yukon a fine bale of various furs, among them a freak skin which has been exciting comment among fur traders.

Some think the animal from which the pelt was taken an offshoot of the tiger family, and its markings resemble the royal outlaw. The strange skin is 38 inches long, the average width eight inches. It is tan color, the line markings on breast and ribs like a large gray striped house cat. Some think a cat strayed into the forest and allied with a wild creature in bringing out a new species. Others declare the skin too large for anything of that sort and that it is an entirely new variety of fur bearer. The animal when trapped, according to McConnell, weighing about 50 pounds.

EVIDENTIARY REQUIREMENTS

We have hundreds of published accounts of strange carnivore-like animals here in North America. (There are likely many, many more yet to uncover.) Quite a few of them, if accurate and reliable, suggest the presence of unknown species. But there is the problem: *if accurate and reliable.* The sighting accounts, the historical records, even such evidence as images and track print casts, are circumstantial. They can point to a possibility, but they cannot confirm. For scientific confirmation (and the "holy grail" of end results: the scientific description of a new species), we need physical evidence that can be examined, tested, and verified. While there is no doubt that personal beliefs within science (i.e., interpretations of scientific data differing due to assumptions, experience, and personal agendas) commonly create acrimonious debate, such controversies are founded on recognized data. The interpretations may differ, but the data remain the same. There is no legitimate excuse for a cryptozoological investigator to gainsay the necessity for physical evidence, or to criticize mainstream scientists for not showing interest when the evidence presented is strictly circumstantial. Personally, I don't consider any group to be scientific that obligates or touts a "no-kill" policy, particularly for pseudo-conservationist reasons. I don't think every researcher needs to be carrying a rifle in the field (and there are many cases where specimens have already been taken, they just need to be tracked down and tested), but it is irresponsible to claim that a single specimen would jeopardize a population, when we see so many contributing factors to extinction events in North America today: habitat destruction, climate change, spreading wildlife diseases, etc. We can't save a species if we don't know it exists.

So why are sighting reports so untrustworthy? It comes down to the fact that perception itself is malleable (Balcetis and Dunning 2006). What we actually see undergoes unconscious interpretation, and there are numerous biases that can distort the reality of what is being seen. If someone has heard strange rumors, an ambiguous strange animal sighting may be interpreted as a black panther. And, if someone thinks they've seen a black panther, the perception of size and shape may be involuntarily molded to fit what the witness "knows" to be the right size and shape of a black panther. This is not an argument that every sighting report must be mistaken, but it is a well-deserved caution for investigators.

Confirmation bias is a particular problem among both cryptozoology enthusiasts and skeptics. We see this especially in the perception of alleged photographs of mystery animals where vague (sometimes imaginative) outlines are "identified" as Bigfoot or some other cryptid. And, it certainly plays a part in the cursory investigations some wildlife officers make as they break out familiar mantras and try to pacify public agitation instead of seriously examining the evidence.

Occam's razor offers little help, though there are many attempts to use (or misuse) this concept. (In effect, Occam's razor argues that when multiple explanations are possible for a given observation, the simplest explanation is the best.) Here, one problem comes down to a definition of simplicity: what is truly more simple, a correct identification or a misidentification? Isn't it more likely that an enthusiast or a skeptic will consider "more simple" (read, "more correct"), the explanation that he or she already believes most likely? (And, of course, there are many studies suggesting that parsimony in science may not only mislead researchers, but may create a bias towards incorrect explanations. We can't hide from complexity in the real world just because it isn't elegant.) A more rational approach dictates caution in forcing any explanation, and a closer look at contextual parameters of sightings.

And then, of course, we are met with "extraordinary claims require extraordinary evidence," that infamous skeptical creed. It is, of course, nonsense. The evidentiary requirements for the description of an "extraordinary" new species (say, Bigfoot) are the same as are needed for the description of a new species of field mouse. We need a physical body in either case. There's nothing extraordinary about such evidence. That's the problem with so many skeptical propositions: there's a tendency towards subjectivity. An objective principle would be, "All claims require sufficient evidence."

Given the propensity for error in eyewitness reports, the cryptozoological researcher must carefully evaluate evidence, discerning that while circumstantial evidence can initiate or direct an investigation (and may show evidence for misidentification or error), it cannot successfully conclude with the description of a new species. For that, we need physical confirmative evidence. This isn't possible to acquire in every case, particularly when dealing with historical accounts, but as investigators, we have a responsibility to strive for the best evidence possible, and not force available evidence to say more than it does.

Appendix A
Predator Kill Patterns

These are general characteristics of predator kills, but individuals may vary in their behavior for any number of reasons (i.e., type of prey, habitat, age of predator, presence of scavengers.). Other characters (dentition traits, surrounding tracks and traces, scat, etc.), not denoted here, will also be useful in determining the responsible predator.

Foxes

Domestic prey includes poultry, cats, lambs, and young pigs.

Foxes often attack the throats of young livestock, but may also bite the head and neck repeatedly. Foxes don't have the strength of other canids, so often multiple bites are required to bring its prey down.

Smaller prey may be carried off to the den, or pieces may be buried.

Foxes start eating behind the ribs, preferring the viscera.

Sources: Phillips and Schmidt (1994); Mapston (n.d.)

Coyotes

Coyotes will hunt sheep, calves, pigs, poultry, and domestic pets.

Coyotes usually go for the throat with sheep, trying to suffocate them, but attack calves at the anus or abdomen. There will usually be multiple sets of bite marks. Coyotes are adaptable in behavior, however, and in new situations may develop different predatory tactics.

Coyotes usually start feeding on the rump, or the flank behind the ribs, then eating the organs and entrails. The prey's nose is often consumed, particularly with younger prey. Ribs are often chewed and splintered.

Coyotes also scavenge on dead animals, so feeding is not necessarily indicative of predation. If there is bleeding under the skin at the bite marks, this indicates that the animal was alive when bitten.

Surplus killing is possible, though a coyote will usually only feed on one animal killed.

Coyotes don't attempt to cover and hide large carcasses, but they will cache smaller prey or pieces of a larger animal.

Sources: Green et al (1994); Main (2001); Mapston (n.d.); Cougar Network (n.d.)

WOLVES

Some wolves will take cattle, sheep, poultry, and domestic pets. Where livestock kills occur, calves are particularly targeted.

Wolves chase their prey, biting at the animal's sides and hindquarters. Larger prey will show numerous bites and ragged wounds. When it is weakened to the point that it falls, the wolves tear open the belly. Wolves will also target the animal's nose and neck during an attack. Smaller prey will be bitten on the head, neck, throat, back, and legs.

Wolves usually start feeding on the viscera and hindquarters first. A large carcass is often torn apart by a pack, bones are chewed and broken, and any remaining edible parts may be carried off to a den or buried.

Surplus killing is known, especially with sheep and poultry. (In some cases, up to 200 turkeys have been killed in a single night.)

Smaller prey usually only have a single set of large puncture wounds from the wolf's teeth. Coyotes often readjust their grip, leaving multiple sets.

Wolves will scavenge dead animals they find, so signs of a struggle and bleeding under the skin near bite marks should be used to distinguish between predation and scavenging.

Sources: Paul and Gipson (1994); Buskirk and Gipson (1978)

Dogs

Both wandering and feral dogs will chase and kill livestock. The carcass is often mutilated with indiscriminate bites, and usually isn't eaten unless the dogs are truly feral.

Dogs don't always have a specific manner of taking down prey, but could be confused with young inexperienced coyotes. Truly feral dogs may learn to hunt in a similar manner to coyotes.

Sources: Green and Gipson (1994)

Bobcats

Bobcats may take prey as large as calves and deer.

Larger prey will often show claw marks on their backs and shoulders. (Bobcats jump on the back to gain position.) When attacking the throat, the bobcat holds on with a single strong bite over the larynx, suffocating the victim. Smaller prey is taken down by bites to the neck, head, or throat.

Feeding often starts behind the ribs, into the visceral organs. Sometimes, however, bobcats start feeding on the meat in the neck, shoulders, or hindquarters first.

Bobcats (and mountain lions) leave neat edges when tearing into skin and tissue, while coyotes and wolves leave ragged edges. Hair may be plucked.

Larger prey may be covered for later consumption, leaving scratch marks up to 15 inches in length in the surrounding soil. Moderate-sized prey may be dragged to better cover.

Sources: Virchow and Hogeland (1994); Main (2001); Mapston (n.d.)

Mountain Lion

Prey includes deer, cattle, goats, sheep, horses, dogs, cats, pigs, and poultry.

Prey is stalked, then grabbed after a sudden rush. Sheep, goats, deer, and calves may be killed by a bite to the top of the skull or back of the neck, often breaking the neck.

Sometimes, the mountain lion will bite the throat, leaving marks similar to a coyote kill. These are differentiated by the cougar's larger canine teeth, which are also farther apart.

Claw puncture marks will often be seen on larger prey, where the mountain lion is holding on, but not necessarily slashes.

Mountain lions use their teeth to shear open clean-edged access behind the ribs. They often remove (by shearing or plucking) hair from the carcass. The entrails are eviscerated and moved aside. Several organs (lungs, hearts, and liver) are eaten, along with larger leg muscles. The intestines and stomach usually aren't eaten. Large bones are usually left unbroken.

Sometimes prey will be dragged to better cover for feeding. Mountain lions often try to cover remaining prey for later consumption. This may leave claw marks more than three feet in length in the soil around the prey as leaves and grass are scratched onto the carcass.

Kills are known to occur within a few yards of occupied homes.

Surplus killing occurs.

Sources: Knight (1994); Sitton et al (1978); Cougar Network (n.d.)

JAGUARS

Jaguars attack the neck or base of the skull in large prey, often breaking the neck. It is very rare for a jaguar to attack the throat.

Jaguars often leave large tooth marks in dense bones like the femur, the base of the cranium, and the neck vertebrae, breaking and rupturing them.

Jaguars usually start feeding in the front part of larger prey, especially the throat, neck, shoulders, and chest. They often leave the forelimbs and posterior sections untouched.

Jaguars rarely try to cover prey, but rather drag the carcass to more suitable hidden cover.

Source: Silveira et al (2008)

LEOPARDS

Leopards will take a wide range of prey, though have a preference for small to medium-sized mammals.

Leopards usually kill larger prey with a bite on the throat, suffocating the animal, but may kill smaller prey with a bite to the head or neck.

Prey is often dragged to better cover, and is sometimes cached in trees, under ground cover, or in caves. (Leopards are strong, and prey as large as giraffe calves have been found cached in trees.) Some may cover prey with surrounding plants and debris, but this doesn't appear to be a common practice. Tree caching appears to be a response to disturbance while feeding, or to a lack of appropriate ground cover.

Hair is plucked neatly from many kills, and intestines are removed from the carcass uneaten. Leopards start from the vital organs, moving forward to the upper abdomen, then back to the hindquarters. Leopards eventually finish everything except the head, legs, hooves, and skin.

Leopards scavenge opportunistically.

Sources: Schaller (1972); De Ruiter and Berger (2001); Maheshwari (2006); Swanepoel (2008)

Lions

When a lion lunges at prey, it uses its forepaws to force down the hindquarters, then moves forward to bite the neck (sometimes the throat).

Smaller prey may be dragged to cover for feeding.

Lions will start feeding at the hindquarters, then move forward through the abdomen and chest, to the neck and head. Of course, with a pride of lions, the carcass is torn to pieces, and each lion eats whatever it can grab first.

In some regions, lions will discard the intestines and stomach, but Schaller noted that lions in the Serengeti often consume it quite readily.

Sources: Schaller (1972); Maheshwari (2006)

Tigers

A tiger prefers a throat attack on prey, suffocating it, sometimes using its paws to first drag down the animal, or swat it down. Sometimes a bite to the throat (or to the prey's neck) will sever or crush the spinal cord.

Tigers drag their prey to appropriate cover, then begin feeding, starting from the hindquarters. They may cover prey with grass if it is too large to finish in a single meal.

Sources: Schaller (1972); Seidensticker and McDougal (1993)

Black Bears

Few black bears will prey on livestock, but once the habit starts, it is unlikely to be broken of it.

Black bears will give a crushing bite to the back of the neck, leaving deep tooth marks behind the prey's ears. There may also be claw slashes on the shoulders and sides.

The black bear usually starts by opening the body cavity and removing and eating the organs, especially the liver. Intestines are often removed and left intact. Udders on lactating female prey will also be eaten, often before anything else. The hide is often left "skinned out."

Larger prey may be dragged to cover and buried in soil, leaves, and debris, for later consumption.

Sources: Hygnstrom (1994); Main (2001)

Grizzly Bears

The grizzly will chase down prey at high speed, slashing with their claws at the hindquarters, and pulling the animal down. They use their weight to hold down the prey while biting its neck and head and aiming powerful blows. There will also often be crushing bites to the back.

It starts to feed by first ripping open the belly. The prey's hide may be peeled off, or may be eaten. Any remains will often be covered up with debris, and the bear will stay nearby.

Surplus killing, especially with sheep, may occur, but sometimes is a result of panic by the sheep rather than direct attack by the bear.

Bears will usually leave plenty of other evidence (tracks, hair, scat).

Sources: Jonkel (1994); Buskirk and Gipson (1978)

Appendix B
Basic Profiles and Tracks

(CC) Claudio Matsuoka

(CC) Craig ONeal

A good side view often offers clues to the animal's identity. Bobcats, for example, have 6-7 inch long tails, a short muzzle, and a ruff of fur around the face. Remember it is easy to overestimate size when there are no visual cues nearby.

©Twildlife

©Roberto

One key character distinguishing the cougar is the proportionally smaller head. A cougar's ears are also smaller in proportion to its head than is seen in most domestic cats. I have seen other traits being used to try and distinguish cougar, but some of these appear speculative, or supported by poor sampling efforts.

©Jennifer Richards

©Larry Allen

Wolf Track vs Cougar Track

TRACKS

There are a few rules of thumb for distinguishing a feline track from a canine track. (There are a few anomalous track-makers, like the cheetah which does not have retractible claws, but for tracks found in North America, these are unlikely exceptions.) Make certain you examine more than one track if there is a chance (especially with mud) that a print could have been distorted.

A) A canine print has enough space between the toes and heel pad to allow an X to be drawn across the ridge formed between the heel pad and the outer toes. You will not be able to do this with a feline track.

B) It is very rare to find claw marks on a feline track. Claw marks are usually apparent on canine tracks (unless the claws are very worn or heavily trimmed).

C) The front of the heel pad on a canine is pointed (single-lobed). The front of the heel pad on a feline is blunt (double-lobed).

D) The rear of the heel pad on a canine is double-lobed. The rear of the heel pad on a feline is three-lobed.

When you have identified a feline track, size is then a good indicator for a given species. If it is about the size of a quarter, it is probably a domestic cat. If it is golf ball-sized, it is probably a bobcat. If it is baseball-sized, it could be a cougar.

Other characters which should be considered can be found in guides like the Puma Identification Guide, downloadable from the Cougar Network online.

Bibliography

Adams, J. R., J. A. Leonard, and L. P. Waits. 2003. Widespread occurrence of a domestic dog mitochondrial DNA haplotype in southeastern US coyotes. *Molecular Ecology* 12(2): 541-546.

Adcox, Seanna. 2002. Experts not sure why bear killed baby. Milwaukee, Wisconsin, *Journal Sentinel*. (August 21).

Adkins Giese, Collette L. 2006. The big bad wolf hybrid: how molecular genetics research may undermine protection for gray wolves under the Endangered Species Act. *Minnesota Journal of Law, Science & Technology* 6(2): 865-872.

Alexander, Autumn. 1987. Lava bears to Indian lairs: legends pepper central Oregon's high desert. *Bend, Oregon, Bulletin* (March 1): 9.

Allendorf, Fred W., et al. 2001. The problems with hybrids: setting conservation guidelines. *TRENDS in Ecology & Evolution* 16(11): 613-622.

Allison, Harold. 2009. Do you know about the wampus cat? Bloomington, Indiana, *Herald Times* (March 8): F3.

American Bear Association (ABA). 2009-2010. Size of the black bear. www.american-bear.org

Anderson, Tovi M., et al. 2009. Molecular and evolutionary history of melanism in North American gray wolves. *Science* 323(March 6): 1339-1343.

Anderson, Traci, and Stacy Langley. 2005. It's back!! Wolverine sighted a few weeks ago. *Huron, Michigan, Daily Tribune* (April 15) www.michigansthumb.com

Anonymous. 1820. American Leopard. *The Kaleidoscope; or, Literary and Scientific Mirror*. (No. 22—November 28): 175.

Anonymous. 1927. Hunt's Circus loses its gentle hyena. Kingston, New York, *Daily Freeman* (April 13): 13.

Anonymous. 1936. Escaped leopard shot in mountain north of Bedford. Connellsville, Pennsylvania, *Daily Courier* (October 10): 1.

Anonymous. 1950. Puzzled gunner kills 'kinkajou.' Hagerstown, Maryland, *Morning Herald* (June 19): 16.

Anonymous. 1953. Territorial News Items: Kodiak. Fairbanks, Alaska, *Daily News-Miner* (May 1).

Anonymous. 1955. Escaped leopard keeps 'em at home. Galveston, Texas, *Daily News* (February 16): 7.

Anonymous. 1960. Monster of Alaska ice proof of hunter's skill. Lakeland, Florida, *Ledger* (May 15).

Anonymous 1963. Gospel-singing marksman kills escaped black panther. Auburn, New York, *Citizen-Advertiser* (March 2): 1.

Anonymous. 1995. Tiger tales. US Army Corps of Engineers: *Tulsa District Record* (March): 9.

Anonymous. 1997. Texas woman, would-be rescuer killed by bear as children watch. *Victoria (Australia) Advocate.* (August 17): 15A.

Anonymous. 2001. Wrong bear shot after teen mauled to death. North Bay, Ontario, *Nugget* (June 4): A7.

Anonymous. 2003. Bear kills campaigner in Alaska. BBC News. (October 8). news.bbc.co.uk

Anonymous. 2003b. U.S. Forest Service to study wolverines. Ann Arbor, Michigan, *Mlive.com* (October 24)

Anonymous. 2004. Army calls off search for Bengal tiger spotted outside base. WFTV News (September 9) www.wftv.com

Anonymous. 2006. Chilhowee bear attack kills child. Chattanoga, Tennessee, *The Chattanoogan.* (April 13.) www.chattanoogan.com.

Anonymous. 2006b. Grizzly kills man staking claim. Fort Frances, Ontario, *Times Online* (May 2). www.fftimes.com

Anonymous. 2006c. Authorities search for tiger over weekend. Valley Center, California, *Roadrunner* (February 15) www.valleycenter.com

Anonymous. 2007. Student might be first North American killed by wolves. Victoria, British Columbia, *Times Colonist* (October 31). www.canada.com

Anonymous. 2008. Wolves in New England? Sudbury, Massachusetts, *Town Crier* (July 1) www.wickedlocal.com

Anonymous. 2010. State suspends hunt for wolves that killed village teacher. Anchorage, Alaska, *Daily News.* (March 18). www.adn.com

Arment, Chad. 1999. Eyewitness account: A mystery cat in Missouri. *North American BioFortean Review* 1(1): 16.

Arment, Chad. 2000. Black panthers in North America: Examining the published explanations. *North American BioFortean Review* 2(1): 38-56.

Arment, Chad. 2000b. More odd "wildcat" reports. *North American BioFortean Review* 2(2): 41.

Arment, Chad, and Brad LaGrange. 2000. Crypto-Varmints. *North American Bio-Fortean Review* 2(3):18-20.

Associated Press. 1935. Zoo Keepers Astonished by Birth of Hybrid Bear. Galveston, Texas, *Daily News* (February 20).

Associated Press. 1950. Doped leopard is recaptured. Syracuse, New York, *Herald Journal* (February 28): 1.

Associated Press. 1951. Escaped leopard killed in Arkansas. AP wirephoto. Moverly, Missouri, *Monitor-Index* (November 5): 1.

Associated Press. 1956. Hey, What About Me? (wirephoto) Syracuse, New York, *Herald Journal* (February 9).

Associated Press. 1961. Some polar bears come a bit big. (wirephoto) Modesto, California, *Bee* (April 6): B8.

Associated Press. 1973. Florida's Become a Haven for Exotic Animals. *Sarasota, FL, Herald-Tribune* (February 25): 3B.

Associated Press. 1976. Jaguar captured. Statesville, North Carolina, *Record and Landmark* (July 21): 16A.

Associated Press. 1976b. Last escaped lioness killed. Oakland, California, *Tribune* (October 27): 12F.

Associated Press. 1981. Child killed by coyote. Lewiston, Maine, *Journal* (August 27).

Associated Press. 2001. Rare fatal bear attack jars New Mexico. Gainesville, Florida, *Sun* (September 1): 4A.

Associated Press. 2004. Wolverine pays state rare visit. Lansing, Michigan, *State Journal* (February 26) www.lsj.com

Associated Press. 2005. Grizzly bear kills couple in tent. Sydney, Australia, *Morning Herald Online* (June 27). www.smh.com.au

Associated Press. 2005b. Grizzly bear kills woman in Canada. Fox News (June 6). FoxNews.com.

Associated Press. 2005c. Hunter finds a rare white mountain lion. KTVB News. (August 26) www.ktvb.com

Associated Press. 2007. Pa. hunter bags rare albino black bear. *NewsVine.* (November 20). Newsvine.com.

Associated Press. 2007b. Alaska biologist certain wolves killed Canadian. Anchorage, Alaska, *Daily Press* (November 21). www.adn.com

Associated Press. 2008. New Mexico man torn apart by mountain lion. Fox News (June 25) www.foxnews.com

Aubry, Keith A., Kevin S. McKelvey, and Jeffrey P. Copeland. 2007. Distribution and broadscale habitat relations of the wolverine in the contiguous United States. *Journal of Wildlife Management* 71(7): 2147-2158.

Aughey, Samuel. 1884. Curious Companionship of the Coyote and Badger. *American Naturalist* 18(6): 644-645.

Bachay, George S. 1957. Photographing Kodiak bear is a thrilling experience. Janesville, Wisconsin, *Daily Gazette* (August 29).

Bailey, Vernon. 1905. Biological survey of Texas. *North American Fauna*. No. 25. Washington, DC: USDA Biological Survey.

Balcetis, Emily, and David Dunning. 2006. See what you want to see: motivational influences on visual perception. *Journal of Personality and Social Psychology* 91(4): 612-625.

Bangs, O. 1898. The land mammals of peninsular Florida and the coast region of Georgia. *Proceedings of the Boston Society of Natural History* 28: 157-235.

Barsh, Gregory S., et al. 2009. Response: How the gray wolf got its color. *Science* 325(July 3): 34.

Bass, Rick. 1997. *The Lost Grizzlies: A Search for Survivors in the Wilderness of Colorado*. New York: Houghton Mifflin Harcourt.

Beale, Donald M., and Arthur D. Smith. 1973. Mortality of pronghorn antelope fawns in western Utah. *Journal of Wildlife Management* 37(3): 343-352.

Beck, Karen. 2005. *Epidemiology of Coyote Introgression into the Red Wolf Genome*. PhD thesis. North Carolina State University.

Beier, Paul, et al. 2006. Evaluating scientific inferences about the Florida panther. *Journal of Wildlife Management* 70(1): 236-245.

Bell, Dawson. 2010. Only known wolverine in the Michigan wild dies. *Lansing, Michigan, Free Press* (March 15) www.freep.com

Black-footed Ferret Recovery Implementation Team. 2009. *Black-Footed Ferret Recovery Program*. www.blackfootedferret.org

Bogoras, Waldemar. 1902. The folklore of northeastern Asia, as compared with that of northwestern America. *American Anthropologist* 4(4): 577-683.

Bogue, Gary. 2009. Big, black cat really may be roaming hills near Pleasanton. Contra Costa, California, *Times* (August 21) www.insidebayarea.com

Bohn, Dean. 2006. DNA samples show Michigan wolverine out of its league. Saginaw, Michigan, *News* (April 13) www.mlive.com

Bolgiano, Chris, and Jerry Roberts, eds. 2005. *The Eastern Cougar: Historic Accounts, Scientific Investigations, and New Evidence*. Mechanicsburg, PA: Stackpole.

Boshoff, Andre F., and Graham I. H. Kerley. 2010. Historical mammal distribution data: how reliable are written records? *South African Journal of Science* 106(1/2): 26-33.

Boughner-Blair, Sheila. 2006. 'Naked' skunk found under Oil City porch. Oil City, Pennsylvania, *Derrick* (February 23) www.thederrick.com

Bowers, Annalea K., et al. 2001. Early history of the wolf, black bear, and mountain lion in Arkansas. *Journal of the Arkansas Academy of Science* 55: 22-27.

Boydston, Erin E., and Carlos A. López González. 2005. Sexual differentiation in the distribution potential of northern jaguars (*Panthera onca*). *USDA Forest Service Proceedings RMRS-P*-36. pp. 51-56.

Branham, Lowell. 1995. The mystery of the midnight possum. Defiance, Ohio, *Crescent-News* (May 19)

Brezosky, Lynn. 2004. Biologists set photo traps for endangered wildcat. Fort Worth, Texas, *Star-Telegram* (November 28) www.dfw.com

Brickell, John. 1737. *The Natural History of North-Carolina*. Dublin.

Bright, Jill L., and John J. Hervert. 2005. Adult and fawn mortality of Sonoran pronghorn. *Wildlife Society Bulletin* 33(1):43-50.

Britt, Tony. 2004. Jaguarundi lurking on Florida soil? Lake City, Florida, *Reporter* (August 29) www.lakecityreporter.com

Brown, David E. 1996. *The Grizzly in the Southwest: Documentary of an Extinction*. Norman, OK: University of Oklahoma Press.

Brown, David E., and Carlos A. López González. 1999. Jaguarundi (*Herpailurus yagouaroundi* Geoffroy 1803) not in Arizona or Sonora. *Journal of the Arizona-Nevada Academy of Sciences* 32(2): 155-157.

Brown, David E., and Carlos A. López González. 2001. *Borderland Jaguars: Tigres de la Frontera*. 2nd edition. Salt Lake City, UT: University of Utah Press.

Brown, Gary. 1996. *The Great Bear Almanac*. Guilford, CT: Lyons Press.

Burnett, Andrew W. 2008. Fisher: Return of a native. *New Jersey Fish & Wildlife Digest* (August): 70.

Busch, Robert H. 2004. *The Cougar Almanac*. Guilford, CT: Lyons Press.

Buskirk, Steven W., and Philip S. Gipson. 1978. Characteristics of wolf attacks on moose in Mount McKinley National Park, Alaska. *Arctic* 31(4): 499-502.

Butz, Bob. 2005. *Beast of Never, Cat of God: The Search for the Eastern Puma*. Guilford, CT: Lyons Press.

Callahan, Jody. 2007. Grizzly kills ex-Memphian who was hunting in Canada. *Memphis, Tennessee, Commercial Appeal* (December 7). www.commercialappeal.com

Campbell, Malcolm R. 1981. Records of albinotic opossum from central California. *Southwestern Naturalist* 25(4): 560.

Canwest News Service. 2008. Couple says it spotted albino black bear in B. C. Dose.ca (May 23). www.dose.ca.

Capparella, III, Angelo. 1977. The Santer: North Carolina's own mystery cat? *Shadows* (4): 1-3, (5): 1-3.

Carmony, Neil B. 1995. *Onza! The Hunt for a Legendary Cat*. Silver City, NM: High-Lonesome Books.

Carr, Steven M., and Shawn A. Hicks. 1997. Are there two species of marten in North America? in Proulx, Gilbert, et al., eds. Genetic and evolutionary relationships within *Martes*. *Martes: Taxonomy, Ecology, Techniques, and Management: Proceedings of the Second International Martes Symposium*. Edmonton, AB: University of Alberta.

CBC News. 2000. Bear killed Mary Beth Miller. CBC News (July 6) www.cbc.ca

CBC News. 2005. Ontario man believed killed by wolves in Saskatchewan. CBC News (November 10). www.cbc.ca

CBC News. 2006. Strange bear was grizzly-polar hybrid, tests show. CBC News (May 10) www.cbc.ca

CBC News 2006b. Mystery beast blamed for killing 3 dogs. CBC News (August 10) www.cbc.ca

CBC News. 2007. Calgary cyclist killed in possible bear attack. CBC News. (July 24) www.cbc.ca.

CBC News. 2010. Bear shot in N.W.T. was grizzly-polar hybrid. CBC News. (April 30) www.cbc.ca

Chapman, Frank Michler. 1894. Remarks on certain land mammals from Florida, with a list of the species known to occur in the state. *Bulletin of the American Museum of Natural History* 6(14): 333-346.

Chasnoff, Brian. 2005. Tiger tales swish through Atascosa. San Antonio, Texas, *Express-News* (July 14) www.mysanantonio.com

Chew, Jeff. 2009. 'This guy just likes killing and leaving them'—is this cougar a serial killer? *Peninsula, Washington, Daily News* (August 29) www.peninsula-dailynews.com

Chorvinsky, Mark, and Mark Opsasnick. 1988. Notes on the Dwayyo. *Strange Magazine* No. 2: 28-29.

Chorvinsky, Mark, and Mark Opsasnick. 1989. A field guide to the monsters and mystery animals of Maryland. *Strange Magazine* 5: 41+.

Christiansen, Per. 1999. What size were *Arctodus simus* and *Ursus spelaeus* (Carnivora: Ursidae)? *Annals Zool. Fennici* 36: 93-102.

Clapp, Henry. 1868. Notes of a fur hunter. *American Naturalist* 1(12): 652-666.

Clark, Brychan M., et al. 2008. Case-control study of armadillo contact and Hansen's disease. *American Journal of Tropical Medicine and Hygiene*. 78(6): 962-967.

Clark, David W., et al. 2002. A survey of recent accounts of the mountain lion (*Puma concolor*) in Arkansas. *Southeastern Naturalist* 1(3): 269-278.

Clark, Jerome. 1993. *Unexplained!* Detroit: Visible Ink.

Clark, Tim W. 1978. Current status of the black-footed ferret in Wyoming. *Journal of Wildlife Management* 42(1): 128-134.

Clark, Tim W. 1987. *Martes americana. Mammalian Species* (289): 1-8.

Coleman, Loren. 1979. Black 'mountain lions' in California? *Pursuit* 12(2): 61-62.

Coleman, Loren. 2001. Wampus cats, mystery felids, and the Santer. *North American BioFortean Review* 3(1): 33-35.

Coleman, Loren. 2007. *Mysterious America*. New York: Paraview.

Coleman, Loren, and Jerome Clark. 1999. *Cryptozoology A to Z*. New York: Fireside.

Colorado Division of Wildlife. 2002. Record mountain lion is evidence of Colorado's healthy mountain lion population. Press release. (September 6).

Cosgrove-Mather, Bootie. 2003. NJ 'tiger lady' loses her cats. CBS News (November 11) www.cbsnews.com

Cougar Network. n.d. *Puma Identification Guide*. CougarNet.org

Cox, Isaac Joslin, ed. 1905. *The Journeys of Réné Robert Cavelier, Sieur de La Salle.* Vol. 1. New York: A. S. Barnes & Co.

Cross, E. C. 1941. Colour phases of the red fox (*Vulpes fulva*) in Ontario. *Journal of Mammalogy* 22(1): 25-39.

Crutchfield, Gail. 2006. Tiger spotted in Berlin. Cullman, Alabama, *Times* (March 24) www.cullmantimes.com

CTV.ca News Staff. 2005. Manitoba man killed in black bear attack. CTV (August 27.) www.ctvbc.ctv.ca.

CTVCalgary.ca. 2008. Killer bear shot by wildlife officials. CTV Calgary. (Oct. 11). calgary.ctv.ca

Culver, M., et al. 2000. Genomic ancestry of the American puma (*Puma concolor*). *Journal of Heredity* 91(3): 186-197.

Cusick, David. 1848. *David Cusick's Sketches of Ancient History of the Six Nations...* Lockport, NY: Turner & McCollum.

Cuyler, W. Kenneth. 1934. Cinnamon and albino opossums found at Austin, Texas. *Journal of Mammalogy* 5(1): 130.

Daggett, Pierre M, and Dale R. Henning. 1974. The jaguar in North America. *American Antiquity* 39(3): 465-69

Davis, Mark. 2009. Mountain lions on the loose? Atlanta, Georgia, *Journal-Constitution* (August 9) www.ajc.com

Davis, Tony. 2006. Jaguar sighted this year during N.M. hunt trip. Tucson, Arizona, *Daily Star* (March 1) www.azstarnet.com

De Oliveira, Tadeu G. 1998. *Leopardus wiedii. Mammalian Species* (579): 1-6.

De Ruiter, Darryl J., and Lee R. Berger. 2001. Leopard (*Panthera pardus* Linnaeus) cave caching related to anti-theft behavior in the John Nash Nature Reserve, South Africa. *African Journal of Ecology* 39: 396-398.

Di Massa, Michael. 2007. Loose tiger rumors prove false. Nye County, Nevada, *Mirror* (October 18): 5.

Dinets, Vladimir, and Paul J. Polechla, Jr. 2005. First documentation of melanism in the jaguar (*Panthera onca*) from northern Mexico. Online publication. dinets.travel.ru

Doupé, J. P., et al. 2007. Most northerly observation of a grizzly bear (*Ursus arctos*) in Canada: photographic and DNA evidence from Melville Island, Northwest Territories. *Arctic* 60(3): 271-276.

Doutt, J. Kenneth. 1969. Mountain lions in Pennsylvania? *American Midland Naturalist* 82(1): 281-285.

Downing, Robert L. 1984. The search for cougars in the eastern United States. *Cryptozoology* 3: 31-49.

Dratch, Peter A., et al. 1993-1996. Molecular genetic identification of a Mexican Onza specimen as a puma (*Puma concolor*). *Cryptozoology* 12: 42-49.

Drew, R. E., et al. 2003. Conservation genetics of the fisher (*Martes pennanti*) based on mitochondrial DNA sequencing. *Molecular Ecology* 12: 51-62.

Dunbar, William. 1805. Letter to editor, dated Natchez, March 1st, 1801. *Philadelphia Medical and Physical Journal* 2(Part 1—November 27)

Eizirik, Eduardo, et al. 2001. Phylogeography, population history and conservation genetics of jaguars (*Panthera onca*, Mammalia, Felidae). *Molecular Ecology* 10: 65-79.

Eizirik, Eduardo, et al. 2003. Molecular genetics and evolution of melanism in the cat family. *Current Biology* 13(March 4): 448-453.

Ellis, Richard. 2009. *On Thin Ice: The Changing World of the Polar Bear*. New York: Random House.

Elton, C. S. 1954. Further evidence about the barren-ground grizzly bear in northeast Labrador and Quebec. *Journal of Mammalogy* 35(3): 345-357.

Etling, Kathy. 2003. *Hunting Bears*. South New Berlin, NY: Woods n' Water.

Findlay, Prentiss. 2006. Odd animal believed to be a rare type of fox. *Charleston, South Carolina, Post and Courier* (August 7) www.charleston.net

Floyd, Timothy. 1999. Bear-inflicted human injury and fatality. *Wilderness and Environmental Medicine* 10(2): 75-87.

Freeman, Patricia W., and Hugh H. Genoways. 1998. Recent northern records of the nine-banded armadillo (Dasypodidae) in Nebraska. *Southwestern Naturalist* 43(4): 491-504.

Frey, Jennifer K. 2006. Inferring species distributions in the absence of occurrence records: an example considering wolverine (*Gulo gulo*) and Canada lynx (*Lynx canadensis*) in New Mexico. *Biological Conservation* 130(2006): 16-24.

Fritzell, Erik K., and Kurt J. Haroldson. 1982. *Urocyon cinereoargenteus*. *Mammalian Species* (No. 189): 1-8.

Fuller, Angela K. 2004. Canada lynx predation on white-tailed deer. *Northeastern Naturalist* 11(4): 395-398.

Gallo-Reynoso, Juan-Pablo, et al. 2008. Probable occurrence of a brown bear (*Urctos arctos*) in Sonora, Mexico, in 1976. *Southwestern Naturalist* 53(2): 256-260.

Gammons, Daniel J., Michael T. Mengak, and L. Mike Conner. 2009. Translocation of nine-banded armadillos. *Human-Wildlife Conflicts* 3(1): 64-71.

Gashwilder, Jay S., W. Leslie Robinette, and Owen W. Morris. 1961. Breeding habits of bobcats in Utah. *Journal of Mammalogy* 83(1): 110-124.

Gazette News Services. 2009. Albino black bear moved to Glacier park. Billings, Montana, *Gazette* (November 1). www.billingsgazette.com.

Geluso, Keith. 2009. Distributional records for seven species of mammals in southern New Mexico. *Occasional Papers of the Museum of Texas Tech University* (287): 1-7.

Gippoliti, Spartaco, and Erik Meijaard. 2007. Taxonomic uniqueness of the Javan leopard; an opportunity for zoos to save it. *Contributions to Zoology* 76(1): 55-58.

Gipson, Philip S. 1976. Melanistic *Canis* in Arkansas. *Southwestern Naturalist* 21(1): 124-126.

Godfrey, Linda S. 2003. *The Beast of Bray Road: Tailing Wisconsin's Werewolf*. Madison, WI: Prairie Oak Press.

Godfrey, Linda S. 2006. *Hunting the American Werewolf*. Madison, WI: Trails Books.

Godman, John D. 1826. *American Natural History*. Vol. I, Part I. Philadelphia, PA: H. C. Carey & I. Lea.

Goldman, E. A. 1943. The races of the ocelot and margay in Middle America. *Journal of Mammalogy* 24(3): 372-385.

Gomez, Gloria. 2010. Otter attacks elderly man. Fox Tampa Bay (March 5) www.myfox-tampabay.com

Gompper, Matthew E. 2002. The ecology of northeast coyotes. *Wildlife Conservation Society Working Paper No. 17*.

Goodpaster, Woodrow W., and Donald F. Hoffmeister. 1968. Notes on Ohioan mammals. *Ohio Journal of Science* 68(2): 116-117.

Goodwin, George C. 1946. Inopinatus—The Unexpected. *Natural History* 55(November): 404-6.

Green, Jeffrey S., and Philip S. Gipson. 1994. Feral dogs. in *Prevention and Control of Wildlife Damage*. Hygnstrom, Scott E., Robert M. Timm, and Gary E. Larson, eds. Lincoln, NE: University of Nebraska-Lincoln.

Green, Jeffrey S., F. Robert Henderson, and Mark D. Collinge. 1994. Coyotes. in *Prevention and Control of Wildlife Damage*. Hygnstrom, Scott E., Robert M. Timm, and Gary E. Larson, eds. Lincoln, NE: University of Nebraska-Lincoln.

Greenwood, Tom. 2010. Biologists: Michigan wolverine died of natural causes. Detroit, Michigan, *News* (April 6) www.detnews.com

Gregg, Leanne. 2007. 11-year-old boy killed by a bear attack. Jacksonville, Florida, *First Coast News* (June 20). www.firstcoastnews.com

Grewal, Sonya K., et al. 2004. A genetic assessment of the eastern wolf (*Canis lycaon*) in Algonquin Provincial Park. *Journal of Mammalogy* 85:625–632.

Grigione, Melissa, Alison Scoville, Gerald Scoville, and Kevin Crooks. 2007. Neotropical cats in southeast Arizona and surrounding areas: past and present status of jaguars, ocelots and jaguarundis. *Mastozoologia Neotropical* 14(2): 189-199.

Grinnell, George Bird. 1920. As to the wolverine. *Journal of Mammalogy* 1(4):182-184.

Grondahl, Paul. 2004. Adirondack sightings stir a 4-footed debate. Albany, New York, *Times Union* (July 31) www.timesunion.com

Groves, Craig R. 1988. Distribution of the wolverine in Idaho as determined by mail questionnaire. *Northwest Science* 62(4): 181-185.

Hailer, Frank, and Jennifer A. Leonard. 2008. Hybridization among three native North American *Canis* species in a region of natural sympatry. *PLoS One* 3(10) www.plosone.org

Halpin, James. 2010. Wolves may have killed teacher in Alaskan village. Salt Lake City, Utah, *Deseret News*. (March 10). www.deseretnews.com

Handwerk, Brian. 2003. Did Carolina dogs arrive with ancient Americans? National Geographic News (March 11) news.nationalgeographic.com

Harper, Francis, and Delma E. Presley. 1981. *Okefinokee Album*. Athens, GA: University of Georgia Press.

Hartman, Carl. 1922. A brown mutation in the opossum (*Didelphis virginiana*) with remarks upon the gray and the black phases in this species. *Journal of Mammalogy* 3(3): 146-149.

Harwood, Jennifer Rich. 2009. Crying wolf? Experts skeptical of "red wolf" claims, but warn of coyotes. Destin, Florida, *Log* (September 8) www.thedestinlog.com

Haugen, Arnold O. 1961. Wolverine in Iowa. *Journal of Mammalogy* 42(4): 546-547.

Heckewelder, John. 1797. A letter from Mr. John Heckewelder to Benjamin Smith Barton, M.D. containing an account of an animal called the Big Naked Bear. *Transactions of the American Philosophical Society* 4(31): 260-262.

Heckewelder, John. 1805. Indian account of a remarkably strong and ferocious beast, which (they say) existed in the northern parts of the state of New-York, about two hundred years ago. *Philadelphia Medical and Physical Journal* 1(2): 161-165.

Helgen, Kristofer M., and Don E. Wilson. 2003. Taxonomic status and conservation relevance of the raccoons (*Procyon* spp.) of the West Indies. *Journal of Zoology, London.* 259: 69-76.

Hernandez, Marjorie. 2005. Big cat has crossed Hwy. 23. Ventura County, California, *Star* (February 20) www.venturacountystar.com

Hersteinsson, Pall, et al. 2007. The naked fox: hypotrichosis in arctic foxes (*Alopex lagopus*). *Polar Biology* 30(8): 1047-1058.

Hewitt, Paige. 2009. Reports of tiger still have county on alert. Houston, Texas, *Chronicle* (December 18) www.chron.com

Hillman, Conrad N., and Tim W. Clark. *Mustela nigripes. Mammalian Species* (126): 1-3.

Hittell, Theodore H. 1861. *The Adventures of James Capen Adams, Mountaineer and Grizzly Bear Hunter, of California*. Boston: Crosby, Nichols, Lee and Co.

Hock, Raymond J. 1955. Southwestern exotic felids. *American Midland Naturalist* 53(2): 324-328.

Hoffman, Craig. 2004. Pet snow leopard captured in Bullitt County. WAVE 3 (August 4) www.wave3.com

Hofmann, Joyce E. 2005. A survey for the nine-banded armadillo (*Dasypus novemcinctus* in Illinois. *Center for Biodiversity Technical Report* 2005 (16).

Homyack, Jessica A., et al. 2008. Canada lynx-bobcat (*Lynx canadensis* x *L. rufus*) hybrids at the southern periphery of lynx range in Maine, Minnesota and New Brunswick. *American Midland Naturalist* 159(2): 504-508.

Hornaday, William T. 1905. A new white bear, from British Columbia. *Annual Report of the New York Zoological Society* 9:81–86.

Hornfeck, Ray, and Frank Lang. 1934. A wild tame cat? *Pennsylvania Game News* 3(6): 10-12.

Huchon, Dorothée, et al. 1999. Armadillos exhibit less genetic polymorphism in North America than South America: nuclear and mitochondrial data confirm a founder effect in *Dasypus novemcinctus* (Xenarthra). *Molecular Ecology* 8: 1743-1748.

Humphrey, Stephen R. 1974. Zoogeography of the nine-banded armadillo (*Dasypus novemcinctus*) in the United States. *BioScience* 24(8): 457-462.

Hurley, Timothy. 2003a. Big cat traps to be dismantled. Honolulu, Hawaii, *Advertiser*. (July 8) www.honoluluadvertiser.com

Hurley, Timothy. 2003b. Officials look into possible attack by Maui mystery cat. Honolulu, Hawaii, *Advertiser*. (July 15) www.honoluluadvertiser.com

Hurley, Timothy. 2003c. Maui's big-cat mystery may draw in expert. Honolulu, Hawaii, *Advertiser*. (August 1) www.honoluluadvertiser.com

Hurley, Timothy. 2003d. Big-cat search to begin. Honolulu, Hawaii, *Advertiser*. (August 8) www.honoluluadvertiser.com

Hurley, Timothy. 2003e. High-tech gear vs. Maui mystery cat. Honolulu, Hawaii, *Advertiser*. (August 12) www.honoluluadvertiser.com

Hurley, Timothy. 2003f. Weather drowns out tape to taunt big cat on Maui. Honolulu, Hawaii, *Advertiser*. (August 12) www.honoluluadvertiser.com

Hurley, Timothy. 2003g. Big-cat search turns up marks high up on tree. Honolulu, Hawaii, *Advertiser*. (August 13) www.honoluluadvertiser.com

Hurley, Timothy. 2003h. Expert says he's sure big cat is out there. Honolulu, Hawaii, *Advertiser*. (August 14) www.honoluluadvertiser.com

Hurley, Timothy. 2003i. Another big-cat sighting reported on Maui. Honolulu, Hawaii, *Advertiser*. (August 26) www.honoluluadvertiser.com

Hurley, Timothy. 2003j. Mystery cat on Maui suspected in dog attack. Honolulu, Hawaii, *Advertiser*. (September 10) www.honoluluadvertiser.com

Hurley, Timothy. 2003k. Trapping expert's plan is to snare big cat. Honolulu, Hawaii, *Advertiser*. (October 23) www.honoluluadvertiser.com

Hurley, Timothy. 2003l. Second expert says Maui big cat is "out there." Honolulu, Hawaii, *Advertiser*. (October 29) www.honoluluadvertiser.com

Hurley, Timothy. 2003m. State suspends hunt for Maui cat. Honolulu, Hawaii, *Advertiser*. (November 22) www.honoluluadvertiser.com

Hurley, Timothy. 2003n. For Maui, it was year of the cat. Honolulu, Hawaii, *Advertiser*. (November 30) www.honoluluadvertiser.com

Hurley, Timothy. 2004. Doubt cast on new big-cat sightings. Honolulu, Hawaii, *Advertiser*. (January 14) www.honoluluadvertiser.com

Hutchins, Ross E. 1977. *Trails to Nature's Mysteries*. New York: Dodd, Mead & Co.

Hutchinson, Jeffrey T., and Thea Hutchinson. 2000. Observation of a melanistic bobcat in the Ocala National Forest. *Florida Field Naturalist* 28(1): 25-26.

Hygnstrom, Scott E. 1994. Black bears. in *Prevention and Control of Wildlife Damage*. Hygnstrom, Scott E., Robert M. Timm, and Gary E. Larson, eds. Lincoln, NE: University of Nebraska-Lincoln.

International News Service. 1926. Still seek escaped leopard. Hammond, Indiana, *Times* (August 9): 29.

Jackson, Hartley H. T. 1954. Wolverine (*Gulo luscus*) specimens from Wisconsin. *Journal of Mammalogy* 35(2): 254.

Jameson, W. C. 2007. *Legend and Lore of the Guadalupe Mountains*. Albuquerque, NM: University of New Mexico Press.

Jaques, Francis Lee. 1946. *Outdoor Life's Gallery of North American Game*. New York: Outdoor Life.

Jefferson, Thomas. 1799. A memoir on the discovery of certain bones of a quadruped of the claw kind in western parts of Virginia. *Transactions of the American Philosophical Society* 4: 246-260.

Jenness, Diamond. 1953. Stray notes on the Eskimo of Arctic Alaska. *Anthropological Papers of the University of Alaska*. 1(May): 5-13.

Johnson, Johnny. 2004. Expert: It's a "mange coyote." Lufkin, Texas, *Daily News* (October 15) www.lufkindailynews.com

Johnson, Terry B., William E. Van Pelt, and James N. Stewart. 2009. *Jaguar Conservation Assessment for Arizona, New Mexico, and Northern Mexico.* AZ-NM Jaguar Conservation Team.

Jones, Fred L. 1955. Records of southern wolverine, *Gulo luscus luteus*, in California. *Journal of Mammalogy* 36(4): 569.

Jones, Robert. (Fall) 1979. Black panther sighted near headquarters. *Vestigia Newsletter* 3(2): 1-2.

Jones, Sarah V. H. 1923. Color variations in wild animals. *Journal of Mammalogy* 4(3): 172-177.

Jonkel, Charles. 1994. Grizzly/brown bears. in *Prevention and Control of Wildlife Damage.* Hygnstrom, Scott E., Robert M. Timm, and Gary E. Larson, eds. Lincoln, NE: University of Nebraska-Lincoln.

Kamler, Jan F., Warren Ballard, and Philip S. Gipson. 2003. Occurrence of feral dogs (*Canis lupus familiaris*) in northwest Texas: an observation. *Texas Journal of Agriculture and Natural Resources* 16: 75-77.

Kaufmann, John H., Dirk V. Lanning, and Sarah E. Poole. 1976. Current status and distribution of the coati in the United States. *Journal of Mammalogy* 57(4): 621-637.

Kays, Roland, Abigail Curtis, and Jeremy J. Kirchman. 2009. Rapid adaptive evolution of northeastern coyotes via hybridization with wolves. *Biology Letters* 2009 Sept. 23 [epub ahead of print]

Kennedy, Michael L., et al. 2002. Geographic variation in the black bear (*Ursus americanus*) in the eastern United States and Canada. *Southwestern Naturalist* 47(2): 257-266.

Kenney, Edward L. 2008. Cougar or kitty? Wilmington, Delaware, *News Journal* (May 8) www.delawareonline.com

Kentucky Fish and Wildlife Commission. 2007. Hairless raccoon causes a stir. *Kentucky Fish & Wildlife Commisioner's Newsletter* 2(11): 7.

Kidd, A. G., et al. 2009. Hybridization between escaped domestic and wild American mink (*Neovison vison*). *Molecular Ecology* 18: 1175-1186.

King, Carolyn M. 1983. *Mustela erminea. Mammalian Species* (195):1-8.

Kiraly, Andrew. 2001. Something is out there. Las Vegas, Nevada, *Mercury* (March 30) www.lasvegasmercury.com

Klinka, Dan R., and Thomas E. Reimchen. 2009. Adaptive coat colour polymorphism in the Kermode bear of coastal British Columbia. *Biological Journal of the Linnean Society* 98: 479-488.

Knight, James E. 1994. Mountain lions. in *Prevention and Control of Wildlife Damage*. Hygnstrom, Scott E., Robert M. Timm, and Gary E. Larson, eds. Lincoln, NE: University of Nebraska-Lincoln.

Koblmüller, Stephan, et al. 2009. Origin and status of the Great Lakes wolf. *Molecular Ecology* 18: 2313-2326.

Kontos, Charles C., and Paul A. X. Bologna. 2008. Extirpation and reappearance of the fisher (*Martes pennanti*) in New Jersey. *Bulletin of the New Jersey Academy of Science* 53(3): 1-4.

Kramer, Kari. 2004. Odd creature, killed in East Texas, identified. Sulphur Springs, Texas, *Country World* (October 28) www.countryworldnews.com

Kubota, Gary T. 2003. State hopes perfume will attract mystery cat. Honolulu, Hawaii, *Star Bulletin*. (August 1) www.starbulletin.com

Kubota, Gary T. 2003a. Visiting expert on Maui to place call to wild cat. Honolulu, Hawaii, *Star Bulletin*. (August 12) www.starbulletin.com

Kubota, Gary T. 2003b. Expert says fur not best identifier of Maui creature. Honolulu, Hawaii, *Star Bulletin*. (October 2) www.starbulletin.com

Kubota, Gary T. 2003c. Another expert to track Maui cat. Honolulu, Hawaii, *Star Bulletin*. (October 9) www.starbulletin.com

Kubota, Gary T. 2004. Residents say big cat is in South Maui area. Honolulu, Hawaii, *Star Bulletin*. (January 13) www.starbulletin.com

Kurtén, Björn, and Elaine Anderson. 1980. *Pleistocene Mammals of North America*. New York: Columbia University Press.

Kyle, C. J., et al. 2006. Genetic nature of eastern wolves: past, present and future. *Conservation Genetics* 7: 272-283.

Kyle, C. J., et al. 2008. The conspecific nature of eastern and red wolves: conservation and management implications. *Conservation Genetics* 9:699-701.

Laacke, Robert J., et al. 2006. Erythrism in the North American badger, *Taxidea taxus*. *Southwestern Naturalist* 51(2): 289-291.

Labisky, Ronald F., and Margaret C. Boulay. 1998. Behaviors of bobcats preying on white-tailed deer in the Everglades. *American Midland Naturalist* 139(2): 275-281.

LaGrange, Brad. 2000. An old black panther report. *North American BioFortean Review* 2(2): 40.

LaGrange, Brad. 2000b. Black panthers in Perry County, Indiana. *North American BioFortean Review* 2(3): 4.

LaGrange, Brad. 2001. Black panther sighting. *North American BioFortean Review* 3(1): 27-28.

LaGrange, Brad. 2001b. Giant armadillos in Florida? *North American BioFortean Review* 3(2): 26-27.

Lane, Joshua E., et al. 2006. Borderline tuberculoid leprosy in a woman from the state of Georgia with armadillo exposure. *Journal of the American Academy of Dermatology* 55(4): 714-716.

Lanning, Dirk V. 1976. Density and movements of the coati in Arizona. *Journal of Mammalogy* 57(3): 609-611.

Lapinski, Mike. 2006. *Wilderness Predators of the Rockies: The Bond Between Predator and Prey*. Guilford, CT: Globe Pequot Press.

Lankalis, J. A. 2006. Are there leopards in America? in H. J. McGinnis, J. W. Tischendorf, and S. J. Ropski, eds. *Proceedings of the Eastern Cougar Conference 2004*. Morgantown, West Virginia: ECF/AERIE.

Larivière, Serge. 1999. *Mustela vison. Mammalian Species* (608): 1-9.

Leberg, Paul L., et al. 2004. Recent record of a cougar (*Puma concolor*) in Louisiana, with notes on diet, based on analysis of fecal materials. *Southeastern Naturalist* 3(4): 653-658.

Leggett, Mike. 2008. Is this a coydog eat dog world? East Texas abounds with reports of monstrous coyotes. *Austin, Texas, American-Statesman* (August 7) www.statesman.com

Leonard, Jennifer A., and Robert K. Wayne. 2008. Native Great Lakes wolves were not restored. *Biology Letters* 4: 95-98.

Leonard, Jennifer A., and Robert K. Wayne. 2009. Wishful thinking: imagining that the current Great Lakes wolf is the same entity that existed historically. *Biology Letters* 5: 67-68.

Long, Charles A. 2008. *The Wild Mammals of Wisconsin*. Museum of Natural History Publication No. 56. Stevens Point, Wisconsin: University of Wisconsin-Stevens Point.

Long, Charles A., and H. Bradford House. 1961. *Bassariscus astutus* in Wyoming. *Journal of Mammalogy* 42(2): 274-275.

Loring, J. Alden. 1899. Occurrence of the Virginia opossum in southern central New York. *Science* 9(January 13): 71.

Loring, Stephen, and Arthur Spiess. 2007. Further documentation supporting the former existence of grizzly bears (*Ursus arctos*) in northern Quecec-Labrador. *Arctic* 60(1): 7-16.

Loughry, W. J., et al. 2009. Is leprosy spreading among nine-banded armadillos in the southeastern United States? *Journal of Wildlife Diseases* 45(1): 144-152.

Lush, Tamara. 2001. Unusual otter attack kills dog. St. Petersburg, Florida, *Times* (May 4) www.sptimes.com

Lyman, R. R., Sr. 1973. *Amazing Indeed—Strange Events in the Black Forest*. Vol. II. Coudersport, PA: Leader Publishing.

Maheshwari, Aishwarya. 2006. Food habits and prey abundance of leopard (*Panthera pardus fusca*) in Gir National Park and Wildlife Sanctuary. M. Sc. Dissertation. Department of Wildlife Sciences, Aligarh Muslim University.

Main, Martin B. 2001. Interpreting the physical evidence of predation on domestic livestock. Document WEC141. Department of Wildlife Ecology and Conservation, Florida Cooperative Extension Service, Institute of Food and Agricultural Sciences, University of Florida.

Majerus, Michael E. N., and Nicholas I. Mundy. 2003. Mammalian melanism: natural selection in black and white. *TRENDS in Genetics* 19(11): 585-588.

Mangiacopra, Gary S., and Dwight G. Smith. 1995-6. Connecticut's mystery felines: The Glastonbury Glawackus, 1939-1967. *The Anomalist* 3: 90-123.

Manville, Richard H. 1961. Notes on behavior of marten. *Journal of Mammalogy* 42(1): 112.

Manville, Richard H. 1961b. Angora cottontail from Georgia. *Journal of Mammalogy* 42(2): 255.

Mapston, Mark E. n.d. *Predation . . . Who done it?* Uvalde, TX: Texas Cooperative Extension, Wildlife Services.

Marshall, H. D., and K. Ritland. 2002. Genetic diversity and differentiation of Kermode bear populations. *Molecular Ecology* 11: 685-697.

Marshall, Robert E. 1961. *The Onza*. New York: Exposition Press.

Matheus, Paul E. 1995. Diet and co-ecology of Pleistocene short-faced bears and brown bears in Eastern Beringia. *Quaternary Research* 44: 447-453.

Matlock, Staci. 2008. Otters make return to New Mexico waters. *Santa Fe, New Mexico, New Mexican.* (October 14) www.santafenewmexican.com

May, George. 1965a. Mysterious 'Dwayyo' on loose in county. Frederick, Maryland, *News* (November 29): 1.

May, George. 1965b. 'Dwayyo' monster is still running loose. Frederick, Maryland, *News* (December 1): 1, 5.

May, George. 1965c. Elusive 'Dwayyo' still uncaptured. Frederick, Maryland, *News* (December 2): 1.

May, George. 1965d. 'Dwayyo' could be a modern Snallygaster. Frederick, Maryland, *News* (December 3): 1.

May, George. 1965e. Dwayyo hunt planned. Frederick, Maryland, *News* (December 6): 1.

May, George. 1965f. Dwayyo hunt tonight. Frederick, Maryland, *News* (December 8): 1, 5.

May, George. 1965g. Dwayyo hunt flops. Frederick, Maryland, *News* (December 9): 1.

May, George. 1965h. Dwayyo for Christmas? Frederick, Maryland, *News* (December15): 1, 5.

McBee, Karen, and Robert J. Baker. 1982. *Dasypus novemcinctus. Mammalian Species* (162): 1-9.

McCain, Emil B., and Jack L. Childs. 2008. Evidence of resident jaguars (*Panthera onca*) in the southwestern United States and the implications for conservation. *Journal of Mammalogy* 89(1): 1-10.

McCrady, Edward, H. T. Kirby-Smith, and Harvey Templeton. 1951. New finds of Pleistocene jaguar skeletons from Tennessee caves. *Proceedings of the United States National Museum* 101(3287): 497-511.

McKelvey, Kevin S., Keith B. Aubry, and Michael K. Schwartz. 2008. Using anecdotal occurrence data for rare or elusive species: the illusion of reality and a call for evidentiary standards. *BioScience* 58(6): 549-555.

McKibbon, Sean. 1999. Polar bear kills one, injures two others. Nunatsiaq, Nunavut, *News* (July 16) www.nunatsiaqonline.com

McManus, John J. 1974. *Didelphis virginiana. Mammalian Species* (40): 1-6.

McNay, Mark E. 2002. A case history of wolf-human encounters in Alaska and Canada. *Alaska Department of Fish and Game Wildlife Technical Bulletin* 13.

Meaney, Carron A., Steven J. Bissell, and Jennifer S. Slater. 1987. A nine-banded armadillo, *Dasypus novemcinctus* (Dasypodidae), in Colorado. *Southwestern Naturalist* 32(4): 507-508.

Mech, L. David. 2009. Crying wolf: concluding that wolves were not restored. *Biology Letters* 5: 65-66.

Meisner, D. H. 1983. Psychedelic opossums: fluorescence of the skin and fur of *Didelphis virginiana* Kerr. *Ohio Journal of Science* 83: 4.

Merriam, C. Hart. 1896. Preliminary synopsis of the American bears. *Proceedings of the Biological Society of Washington* 10: 65-83.

Merriam, C. Hart. 1918. Review of the grizzly and brown bears of North America (genus *Ursus*), with description of a new genus, *Vetularctos. North American Fauna, No.* 41. Washington, D.C.: USDA Bureau of Biological Survey.

Merriam, C. Hart. 1919. Is the jaguar entitled to a place in the California fauna? *Journal of Mammalogy* 1(1): 38-40.

Merriam, D. F. 2002. The armadillo [*Dasypus novemcinctus* (Linnaeus)] invasion of Kansas. *Transactions of the Kansas Academy of Science* 105(1-2): 44-50.

Michael, Edwin D. 1968. Unusual pelage and skeleton of a raccoon. *Southwestern Naturalist* 13(1): 105.

Miller, Bill. 2010. Wise County 'chupacabra' a hairless raccoon, biologist says. Fort Worth, Texas, *Star-Telegram* (January 20) www.star-telegram.com

Miller, Katie. 2009. Unusual cat breed spotted in Orange Beach. Gulf Shores, Alabama, *BaldwinCountyNow.com* (February 21)

Mlot, Christine. 1997. Stalking the ancient dog. *Science News* 151(26): 400.

Monson, Valerie. 2003. Oh where, oh where did the big cat go; oh where, oh where could it be? Wailuku, Maui, *The Maui News.* (December 30) www.mauinews.com

Moore, Jr., Chester. 2002. Are US 'black panthers' actually jaguarundi? Online publication. www.anomalist.com

Moore, Kelly. 2004. No tiger yet as search continues. Leesville, Louisiana, *Daily Leader* (September 3) www.leesvilledailyleader.com

Moore, Willard B. 1977. The written and oral narratives of Sara Cowan. *Indiana Folklore* 10(1): 7-91.

Moriarty, Katie M., et al. 2009. Wolverine confirmation in California after nearly a century: native or long-distance immigrant? *Northwest Science* 83(2): 154-162.

Morrison, Shawna. 2006. Possible rabid otter attack creates creature discomfort. Roanoke, Virginia, *Times* (August 29) www.roanoke.com

Mott, Maryann. 2006. Carolina mystery beast is a rare abnormal fox, experts say. *National Geographic News* (March 21) news.nationalgeographic.com

Muñoz-Fuentes, Violeta, et al. 2009. Ecological factors drive differentiation in wolves from British Columbia. *Journal of Biogeography* 36(8): 1516-1531.

Murdoch, John. 1886. A few legendary fragments from the Point Barrow Eskimos. *American Naturalist* 20(7): 593-599.

Murdoch, John. 1892. Ethnological Results of the Point Barrow Expedition. *Ninth Annual Report of the Bureau of Ethnology to the Secretary of the Smithsonian Institution* 1887-'88. Washington D.C.: GPO.

NABC (North American Bear Center). 2010. Website. www.bear.org

Naish, Darren. 2007. Australia's new feral mega-cats. *Tetrapod Zoology* (March 4) scienceblogs.com/tetrapodzoology/

Nead, David M., James C. Halfpenny, and Steve Bissell. 1985. The status of wolverines in Colorado. *Northwest Science* 8(4): 286-289.

Nellans, Joanna Dodder. 2009. Prescott's lion finally loses to urban perils. Prescott, Arizona, *Daily Courier* (July 6) www.dcourier.com

Nellis, David W., et al. 1978. Mongoose in Florida. *Wildlife Society Bulletin* 6(4): 249-250.

Nelson, Edward William. 1899. The Eskimo About Bering Strait. *Eighteenth Annual Report of the Bureau of American Ethnology to the Secretary of the Smithsonian Institution* 1896-97. Part I. Washington D.C.: GPO.

Newton, Michael. 2005. *Encyclopedia of Cryptozoology.* Jefferson, NC: McFarland.

N. Y. Times News Service. 1999. Cat hunt leads police to tiger preserve. Syracuse, New York, *Post-Standard* (January 31): A2.

Nichols, Mike. 2005. Wildlife sightings sounding less wild. *Milwaukee, Wisconsin, Journal* (April 30) www3.jsonline.com

North Carolina Wildlife Resources Commission (NCWRC). 2008. Black Bear. *North Carolina Wildlife Profiles.* www.ncwildlife.org

Nowak, Ronald M. 1973. A possible occurrence of the jaguar in Louisiana. *Southwestern Naturalist* 17(4): 430-432.

Nowak, Ronald M. 1992. The red wolf is not a hybrid. *Conservation Biology* 6(4): 593-595.

Nowak, Ronald M. 2005. *Walker's Carnivores of the World.* Baltimore, MD: Johns Hopkins University Press.

Oakes, Larry. 2006. That's not a hyena; it's a wolf with mange. Duluth, Minnesota, *Star Tribune* (August 10) www.startribune.com

Paige, Christopher F., Daniel T. Scholl, and Richard W. Truman. 2002. Prevalence and incidence density of *Mycobacterium leprae* and *Trypanosoma cruzi* infections within a population of wild nine-banded armadillos. *American Journal of Tropical Medicine and Hygiene* 67(5): 528-532.

Palmeira, Francesca, et al. 2008. Cattle depredation by puma (*Puma concolor*) and jaguar (*Panthera onca*) in central-western Brazil. *Biological Conservation* 141(2008): 118-125.

Pankratz, Howard. 2009. Young wolverine travels to Colorado after 500 mile hike. Vail, Colorado, *Daily News* (June 18) www.vaildaily.com

Park, Ed. 1968. Impressive adventure film. Bend, Oregon, *Bulletin* (March 23)

Patton, Naomi R. 2006. Big, black cat is caught on videotape in White Lake. Detroit, Michigan, *Observer & Eccentric* (September 21) www.hometownlife.com

Paul, William J., and Philip S. Gipson. 1994. Wolves. in *Prevention and Control of Wildlife Damage.* Hygnstrom, Scott E., Robert M. Timm, and Gary E. Larson, eds. Lincoln, NE: University of Nebraska-Lincoln.

Pavey, Rob. 2009. Panther taken was wild, not a pet. Augusta, Georgia, *Chronicle* (August 9) chronicle.augusta.com

Pavlik, Steve. 2003. Rohonas and spotted lions: the historical and cultural occurrence of the jaguar, *Panthera onca*, among the native tribes of the American southwest. *Wicazo Sa Review* 18(1): 157-175.

Paxton, C. G. M. 2009. The plural of "anecdote" can be "data": statistical analysis of viewing distances in reports of unidentified giant marine animals 1758 – 2000. *Journal of Zoology* (279): 381–387.

Pearson, A. M. 1976. Population characteristics of the Arctic mountain grizzly bear. *Bears: Their Biology and Management.* Vol. 3. IUCN Publications New Series no. 40(1976). pp. 247-260.

Petersen, David. 2009. *Ghost Grizzlies: Does the Great Bear Still Haunt Colorado?* 3rd ed. Durango, CO: Ravens Eye Press.

Phillips, Michael K., V. Gary Henry, and Brian T. Kelly. 2003. Restoration of the red wolf. pp. 272-288 in Mech, David L., and Luigi Boitani, eds. *Wolves: Behavior, Ecology, and Conservation.* Chicago: University of Chicago Press.

Phillips, Robert L., and Robert H. Schmidt. 1994. Foxes. in *Prevention and Control of Wildlife Damage.* Hygnstrom, Scott E., Robert M. Timm, and Gary E. Larson, eds. Lincoln, NE: University of Nebraska-Lincoln.

Pils, Douglas. 2002. Arkansas town on edge after 4 lions killed in woods. Syracuse, New York, *Post-Standard* (September 24): A3.

Platt, Steven G., et al. 2009. Distribution records and comments on mammals in western South Dakota. *Western North American Naturalist* 69(3): 329-334.

Poglayen-Neuwall, Ivo, and Dale E. Toweill. 1988. *Bassariscus astutus. Mammalian Species* (327): 1-8.

Polechla, Jr., Paul J., et al. 2005. First physical evidence of the nearctic river otter (*Lontra canadensis*) collected in New Mexico, USA, since 1953. *IUCN Otter Specialist Group Bulletin* 21(2): 70-75.

Pooley, J. H. 1884. The American Badger in Ohio. *American Naturalist* 18(12): 1276.

Poszig, D., C. D. Apps, and A. Dibb. 2004. Predation on two mule deer, *Odocoileus hemionus*, by a Canada lynx, *Lynx canadensis*, in the southern Canadian Rocky Mountains. *Canadian Field-Naturalist* 118(2): 191-194.

Powell, Roger A. 1981. *Martes pennanti. Mammalian Species* (156): 1-6.

Prevo, Robert. 2001. Arkansas' "black panthers." *North American BioFortean Review* 3(2): 51-53.

Puckett, Karl. 2008. Young grizzlies accused of killing 71 sheep. Great Falls, Montana, *Tribune* (July 17). www.greatfallstribune.com

R., F. 1894. Notes on Trapping and Wood-Craft, in Newhouse, S. *Trapper's Guide and Manual of Instructions for Capturing All Kinds of Fur-Bearing Animals...* Ninth (Revised) Edition. New York: Forest and Stream Publishing Co.

Rabinowitz, Alan. 2010. Opinion: Jaguars don't live here anymore. New York, New York, *Times* (January 24) www.nytimes.com

Raesly, Elaine J. 2001. Progress and status of river otter reintroduction projects in the United States. *Wildlife Society Bulletin* 29(3): 856-862.

Rafinesque, C. S. 1832. On the large wandering tygers or jaguars of the United States. *Atlantic Journal* 1(1): 18-19.

Raisor, Michelle J. 2004. *Determining the Antiquity of Dog Origins: Canine Domestication as a Model for the Consilience between Molecular Genetics and Archaeology.* PhD Thesis. Texas A&M University.

Rausch, Robert. 1953. On the status of some Arctic mammals. *Arctic* 6: 91-148.

Regan, Timothy W., and David S. Maehr. 1990. Melanistic bobcats in Florida. *Florida Field Naturalist* 18(4): 84-87.

Reich, D. E., R. K. Wayne, and D. B. Goldstein. 1999. Genetic evidence for a recent origin by hybridization of red wolves. *Molecular Ecology* 8: 139-144.

Ritland, Kermit, Craig Newton, and H. Dawn Marshall. 2001. Inheritance and population structure of the white-phased "Kermode" black bear. *Current Biology* 11: 1468-1472.

Robbins, John Charles. 2004. Local historian makes case for native wolverines. Holland, Michigan, *Sentinel* (June 14) www.thehollandsentinel.net

Robinson, Reuel. 1907. *History of Camden and Rockport, Maine*. Camden, Maine: Camden Publishing.

Robinson, Walker. 2004. Chupacabra? Strange Animal Found in Elmendorf. WOAI News (July 28) www.woai.com

Roest, Aryan I. 1961. Partially albino badger from California. *Journal of Mammalogy* 42(2): 275-276.

Rounds, Richard C. 1987. Distribution and analysis of colourmorphs of the black bear (*Ursus americanus*). *Journal of Biogeography* 14(6): 521-538.

Roy, Michael S., et al. 1994. Patterns of differentiation and hybridization in North American wolflike canids, revealed by analysis of microsatellite loci. *Molecular Biology and Evolution* 11(4): 553-570.

Rozell, Ned. 2006. Hybrid grizzly-polar bear a curiosity. Ketchikan, Alaska, *Stories in the News*. (May 19) www.sitnews.us

Rue, Leonard Lee. 1968. *Sportsman's Guide to Game Animals*. New York: Outdoor Life.

Rutledge, Linda Y., et al. 2009. How the gray wolf got its color. *Science* 325(July 3): 33-34.

Rutledge, L. Y., et al. 2010. Genetic differentiation of eastern wolves in Algonquin Park despite bridging gene flow between coyotes and grey wolves. *Heredity* (2010): 1-12.

Ryan, James M. 2008. *Adirondack Wildlife: A Field Guide*. Lebanon, NH: University of New Hampshire Press.

Saile, Bob. 1983. 'Phantom' grizzly the real thing, guide can tell you. Tuscaloosa, Alabama, *News* (September 14).

Sanderson, Ivan T. 1973. More new cats? *Pursuit* 6(April): 35-36.

Sanderson, Ivan T. 1974. The dire wolf. *Pursuit* 7(October): 91-94.

Schaller, George B. 1972. *The Serengeti Lion: A Study of Predator-Prey Relations*. Chicago: University of Chicago Press.

Schemnitz, Sanford D. 1972. Populations of bear, panther, alligator, and deer in the Florida Everglades. *South Florida Environmental Project: Ecological Report No. DI-SFEP-74-33*. Tallahassee, FL: Florida State Game and Fresh Water Fish Commission.

Schmutz, Sheila M., et al. 2007. Agouti sequence polymorphisms in coyotes, wolves and dogs suggest hybridization. *Journal of Heredity* 98(4): 351-355.

Schwartz, Michael K. 2007. Ancient DNA confirms native Rocky Mountain fisher (*Martes pennanti*) avoided early 20th century extinction. *Journal of Mammalogy* 88(4): 921-925.

Schwartz, Michael K., and John A. Vucetich. 2009. Molecules and beyond: assessing the distinctness of the Great Lakes wolf. Molecular Ecology 18: 2307-2309.

Schwartz, Michael K., et al. 2004. Hybridization between Canada lynx and bobcats: genetic results and management implications. *Conservation Genetics* 5: 349-355.

Schwartz, Michael K., et al. 2007. Inferring geographic isolation of wolverines in California using historical DNA. *Journal of Wildlife Management* 71(7): 2170-2179.

Sealfon, Rebecca A. 2007. Dental divergence supports species status of the extinct sea mink (Carniovora: Mustelidae: *Neovison macrodon*). *Journal of Mammalogy* 88(2): 371-383.

Seidensticker, John., and Charles McDougal. 1993. Tiger predatory behavior, ecology and conservation. *Symposia of the Zoological Society of London* 65: 105-125.

Seton, Ernest Thompson. 1920. The jaguar in Colorado. *Journal of Mammalogy* 1(5): 241

Shaw, Liz. 2010. Deckerville man's quest brought him up close to wolverine. Flint, Michigan, *Journal* (April 2) mlive.com

Sheffield, Steven R., and Carolyn M. King. 1994. *Mustela nivalis. Mammalian Species* (454): 1-10.

Shoemaker, Henry W. 1912. *More Pennsylvania Mountain Stories*. Reading, PA: Bright Printing Co.

Shoemaker, Henry W. 1917. A game paradise. *Forest and Stream* 87(April): 122+.

Shoemaker, Henry W. 1919. *Extinct Pennsylvania Animals, Part II*. Altoona, PA: Altoona Tribune Co.

Shoemaker, Henry W. 1922. *Felis Catus in Pennsylvania?* Altoona, PA: Times Tribune.

Shuker, Karl P. N. 1989. *Mystery Cats of the World*. London: Robert Hale.

Shuker, Karl P. N. 1990. The Kellas cat: reviewing an enigma. *Cryptozoology* 9: 26-40.

Shuker, Karl P. N. 1998a. Menagerie of mystery. *Strange Magazine* 19:20+.

Shuker, Karl P. N. 1998b. Menagerie of mystery. *Strange Magazine* 20:36+.

Siebert, Jr., F. T. 1937. Mammoth or "Stiff-Legged Bear." *American Anthropologist* 39: 721-725.

Silveira, Leandro, et al. 2008. Management of domestic livestock predation by jaguars in Brazil. *CAT News* 4: 26-30.

Simberloff, Daniel, Don C. Schmitz, and Tom C. Brown, eds. 1997. *Strangers in Paradise: Impact and Management of Nonindigenous Species in Florida.* Washington, DC: Island Press.

Simmons, Morgan. 2000. Black bear mauling rare, out of character. Spartanburg, South Carolina, *Herald-Journal.* (May 29): A9.

Simms, Jimmy. 2006. Still no sign of animal. Cullman, Alabama, *Times* (March 31) www.cullmantimes.com

Simpson, George Gaylord. 1941. Discovery of jaguar bones and footprints in a cave in Tennessee. *American Museum Novitates* (1131): 1-12.

Sinclair, Steve. 2008. The last jaguar. Brownsville, Texas, *Herald* (June 23) www.brownsvilleherald.com

Sinclair, Steve. 2009. Wild cat sightings creating a stir. Brownsville, Texas, *Herald* (February 8) www.brownsvilleherald.com

Sitton, Larry W., Susan Sitton, and Dick Weaver. 1978. Mountain lion predation on livestock in California. *Cal-Neva Wildlife: Transactions* 1:174-186.

Small, Maureen P., Karen D. Stone, and Joseph A. Cook. 2003. American marten (*Martes americana*) in the Pacific Northwest: population differentiation across a landscape fragmented in time and space. *Molecular Ecology* 12: 89-103.

Spiess, Arthur. 1976. Labrador grizzly (*Ursus arctos* L.): First skeletal evidence. *Journal of Mammalogy* 57(4): 787-790.

Spiess, Arthur, and Steven Cox. 1977. Discovery of the skull of a grizzly bear in Labrador. *Arctic* 29(4): 194-200.

Sprinkle, Bruce Allen. 1994. Black cats in Indiana. *INFO Journal* 70: 29.

Staats, Eric. 2007. Crying wolf? Animal seen in North Naples likely a wolf-dog or coyote. Naples, Florida, *News* (December 15) www.naplesnews.com

Star-Bulletin staff. September 28, 2003. Maui big cat sightings reported in Olinda area. Honolulu, Hawaii, *Star Bulletin*. www.starbulletin.com

Stefánsson, Vilhjálmur. 1914. The Stefánsson-Anderson Arctic Expedition of the American Museum: Preliminary Ethnological Report. *Anthropological Papers of the American Museum of Natural History*. XIV(I): 1-396.

Stephenson, R. O., D. V. Grangaard, and J. Burch. 1991. Lynx, *Felis lynx*, predation on red foxes, *Vulpes vulpes*, caribou, *Rangifer tarandus*, and Dall sheep, *Ovis dalli*, in Alaska. *Canadian Field-Naturalist* 105(2): 255-262.

Stewart, Paul A., and Norman C. Negus. 1961. Recent record of wolf in Ohio. *Journal of Mammalogy* 42(3): 420-421.

Storer, Tracy I., and Lloyd P. Tevis. 1996. *California Grizzly*. Berkeley, CA: University of California Press.

Strong, William Duncan. 1926. Indian records of California carnivores. *Journal of Mammalogy* 7(1): 59-60.

Sucik, Nick. 2002. Unidentified canids in North America. *North American BioFortean Review* 4(2):25-29.

Suckley, Geo., and Geo. Gibbs. 1860. Chapter III. in No. 2, Report upon the Mammals Collected on the Survey. House of Representatives, 36th Congress, 1st Session. *Reports of Explorations and Surveys to Ascertain the Most Practicable and Economical Route for a Railroad from the Mississippi River to the Pacific Ocean.* Vol. XII. Book II.

Sutherly, Curt. 1976. The "thing" of Sheep's Hill. *Pursuit* 9(1): 9-10.

Swanepoel, Lourens Hendrik. 2008. Ecology and conservation of leopard, *Panthera pardus*, on selected game ranches in the Waterberg region, Limpopo, South Africa. M. Sc. Dissertation. Natural and Agricultural Sciences, University of Pretoria.

Taber, F. Wallace. 1940. Range of the coati in the United States. *Journal of Mammalogy* 21(1): 11-14.

Taulman, James F., and Lynn W. Robbins. 1996. Recent range expansion and distributional limits of the nine-banded armadillo (*Dasypus novemcinctus*) in the United States. *Journal of Biogeography* 23: 635-648.

Tompkins, Shannon. 2010. Wild about these cats: state of ocelot population in Texas dire. Houston, Texas, *Chronicle* (February 18) www.chron.com

Traylor, Waverley. 2010. *The Great Dismal Swamp in Myth and Legend*. Pittsburgh, Pennsylvania: RoseDog Books.

Truman, Richard. 2005. Leprosy in wild armadillos. *Leprosy Review* 76: 198-208.

Tyler, Jack D., and Wesley D. Webb. 1992. Occurrence of the ringtail (*Bassariscus astutus*) in Oklahoma. *Southwestern Naturalist* 37(2): 202-205.

Ulmer, Fred A., Jr. 1941. Melanism in the Felidae, with special reference to the genus *Lynx*. *Journal of Mammalogy* 22(3): 285-288.

United States Fish & Wildlife Service. 2010. *Red Wolf Recovery Program: 1st Quarter Report, October-December 2009.* www.fws.gov/redwolf

Uphyrkina, Olga, et al. 2001. Phylogenetics, genome diversity and origin of modern leopard, *Panthera pardus*. *Molecular Ecology* 10: 2617-2633.

United Press. 1926. Escaped leopard shot by farmer. Marshall, Michigan, *Evening Chronicle* (October 15): 1.

United Press. 1950. Sleeping pills kill leopard; body will be stuffed for zoo. Oakland, California, *Tribune* (March 1): 8B.

United Press. 1974. Jaguar captured. Hayward, California, *Daily Review* (July 2): 20.

United Press. 1976. Escaped lion is hunted. Tucson, Arizona, *Daily Citizen* (October 26): 1.

United Press. 1979. Charging bear killed by hunter. *Bend, Oregon, Bulletin* (September 26).

United Press. 2009. Bear kills Colo. woman on her property. UPI News (August 11). www.upi.com

United Press. 2009b. Otters attack woman at Wisconsin lake. UPI News (August 12) www.upi.com

United Press. 2010. Rescuers capture hairless raccoon. UPI News (March 2) www.upi.com

Vallis, Mary. 2009. Toronto singer dies in Cape Breton coyote attack. Toronto, Ontario, *National Post* (October 28). www.nationalpost.com

Van Deelen, Timothy R., Judy D. Parrish, and Edward J. Heske. 2002. A nine-banded armadillo (*Dasypus novemcinctus*) from central Illinois. *Southwestern Naturalist* 47(3): 489-491.

Vanderklippe, Nathan. 2006. Possible 'pizzly' stumps experts. CanWest News Service (May 3) www.canada.com

Van Gelder, Richard G. 1979. Mongooses on mainland North America. *Wildlife Society Bulletin* 7(3): 197-198.

Vermont Agency of Natural Resources. 2007. *Large Canids in Vermont Have Puzzling Ancestry*. Press release. (October 9) www.vtfishandwildlife.com

Vinkey, Ray S., et al. 2006. When reintroductions are augmentations: the genetic legacy of fishers (*Martes pennanti*) in Montana. *Journal of Mammalogy* 87(2): 265-271.

Virchow, Dallas, and Denny Hogeland. 1994. Bobcats. in *Prevention and Control of Wildlife Damage*. Hygnstrom, Scott E., Robert M. Timm, and Gary E. Larson, eds. Lincoln, NE: University of Nebraska-Lincoln.

Virginia Department of Game & Inland Fisheries. 2002-2010. Life history of black bears. www.dgif.virginia.gov

Wallmo, O. C., and Steve Gallizioli. 1954. Status of the coati in Arizona. *Journal of Mammalogy* 35(1): 48-54.

Ward, Charles C. 1882. The black bear. *The Century* 23: 718-725.

Watercutter, Angela. 2004. One killed, another injured in mountain lion attacks. McCook, Nebraska, *Daily Gazette* (January 8): 9.

WBAL News. 2004. Mystery creature lurks in Baltimore County. WBAL (Baltimore, MD) News (July 19) www.thewbalchannel.com

Weinstein, Jack. 2009. Resident spots return of the hairless fox. Steamboat Springs, Colorado, *Pilot* (August 4) www.steamboatpilot.com

Weise, Elizabeth. 2010. Calif. wolverine looks in vain for a mate, nearest is 800 miles away. *USA Today*. (March 10) www.usatoday.com

Wheeldon, Tyler, and Bradley N. White. 2009. Genetic analysis of historic western Great Lakes region wolf samples reveals early *Canis lupus/lycaon* hybridization. *Biology Letters* 5: 101-104.

Whitehurst, Patrick. 2009. A mountain lion becomes a legend. Grand Canyon, Arizona, *News* (July 21) www.grandcanyonnews.com

Wides, Laura. 2005. Tiger shot, killed near Reagan library. Grand Forks, North Dakota, *Herald* (February 23) www.grandforks.com

Wilkins, Korie. 2006. 'Big cat' on tape was just fat cat. Oakland, Michigan, *Press* (September 20) www.theoaklandpress.com

Williams, Walt. 2007. Mystery monster returns home after 121 years. Bozeman, Montana, *Daily Chronicle* (November 15)

Williams, Bronwyn W., Jonathan H. Gilbert, and Patrick A. Zollner. 2006. Historical perspective on the reintroduction of the fisher and American marten in Wisconsin and Michigan. *General Technical Report NRS-5*. USDA-Forest Service-Northern Research Station.

Wilson, Paul J., et al. 2000. DNA profiles of the eastern Canadian wolf and the red wolf provide evidence for a common evolutionary history independent of the gray wolf. *Canadian Journal of Science* 78: 2156-2166.

Wilson, Paul J., et al. 2003. Mitochondrial DNA extracted from eastern North American wolves killed in the 1800s is not of gray wolf origin. *Canadian Journal of Zoology* 81(5): 936-940.

Wilson, Paul J., Walter J. Jakubas, and Shevenell Mullen. 2004. Genetic status and morphological characteristics of Maine coyotes as related to neighboring coyote and wolf populations. *Final report to the Maine Outdoor Heritage Fund Board*, Grant #011-3-7. Bangor, ME: Maine Department of Inland Fisheries and Wildlife.

Wilson, Paul J., et al. 2009. Genetic characterization of hybrid wolves across Ontario. *Journal of Heredity* 100(Supplement 1): S80-S89.

Winthrop, Lynn. 2004. Another Texas Chupacabra? WOAI News (October 13) www.woai.com

Wojcik, Lisa. n.d. *The Florida Jaguarundi*. Online publication. t4studios.com

Woodford, Riley. 2007. White black bears and blonde grizzlies: Alaska bears wear coat of many colors. *Alaska Fish & Wildlife News* (September 2007). Online. http://www.wc.adfg.state.ak.us/index.cfm

Wright, Bruce S. 1959. *The Ghost of North America: The Story of the Eastern Panther*. New York: Vantage Press.

Wright, Bruce S. 1972. *The Eastern Panther: A Question of Survival*. Toronto: Clarke, Irwin & Co. Ltd.

WSB News. 2008. Killer coyote captured in DeKalb subdivision. WSB Atlanta (March 26) www.wsbtv.com

Wydeven, Adrian P., and Jane E. Wiedenhoeft. n.d. *Rare Mammal Observations 2003*. Online publication. dnr.wi.gov

Yanchunas, Dom. 1999. Loose tiger's origins in question. Doylestown, Pennsylvania, *Intelligencer* (January 29): A11.

Young, Stanley P. 1958. *The Bobcat of North America*. Harrisburg, PA: Stackpole.

Zeveloff, Samuel I. 2002. *Raccoons: A Natural History*. Vancouver: UBC Press

Zotter, Jr., Frank. 1998. Judicial follies: The black cat. Ukiah, California, *Daily Journal*. October 25.

Coachwhip Publications
CoachwhipBooks.com

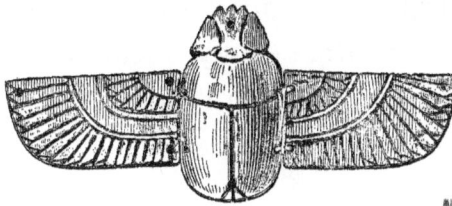

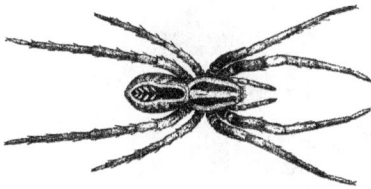

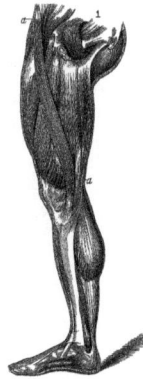

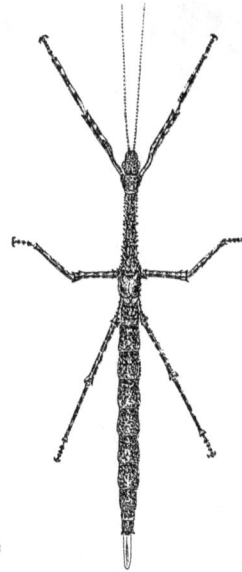

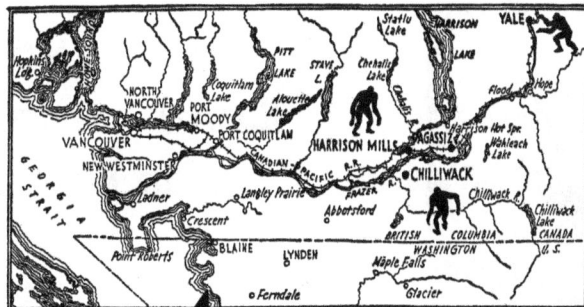

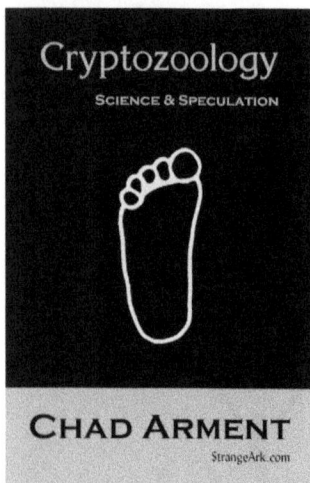

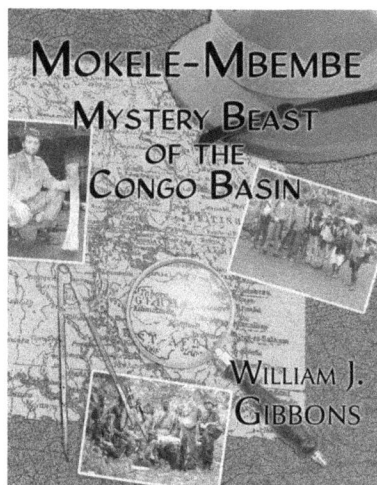

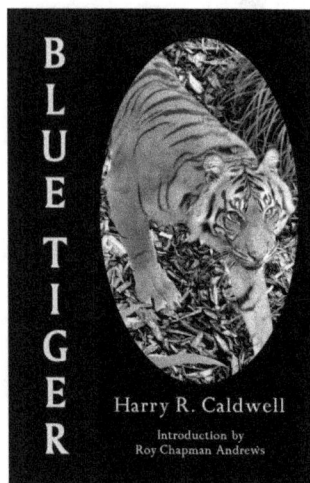

COACHWHIP PUBLICATIONS
CRYPTOFICTION

Coachwhip Publications offers a wide range of short story anthologies and novels that integrate cryptozoological (and cryptobotanical) themes in classic science fiction, fantasy, and horror. This is the most comprehensive collection of a long-overlooked theme in speculative fiction.

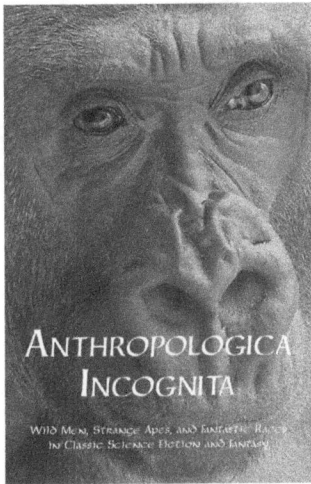

ANTHROPOLOGICA INCOGNITA
Wild Men, Strange Apes, and Fantastic Races in Classic Science Fiction and Fantasy

Bestiarium Cryptozoologicum
Mystery Animals and Unknown Species in Classic Science Fiction and Fantasy

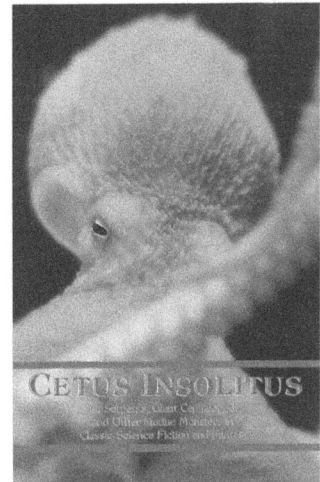

CETUS INSOLITUS
Serpents, Giant Cephalopods, and Other Marine Monsters in Classic Science Fiction and Fantasy

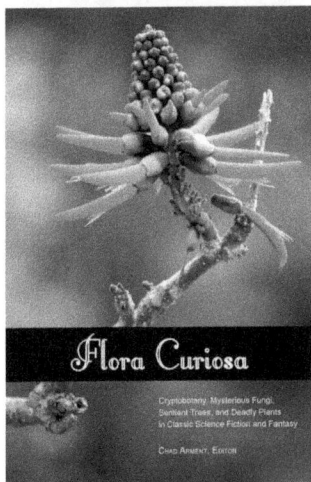

Flora Curiosa
Cryptobotany, Mysterious Fungi, Sentient Trees, and Deadly Plants in Classic Science Fiction and Fantasy
CHAD ARMENT, EDITOR

INVERTEBRATA ENIGMATICA
Giant Spiders, Dangerous Insects, and Other Strange Invertebrates in Classic Science Fiction and Fantasy

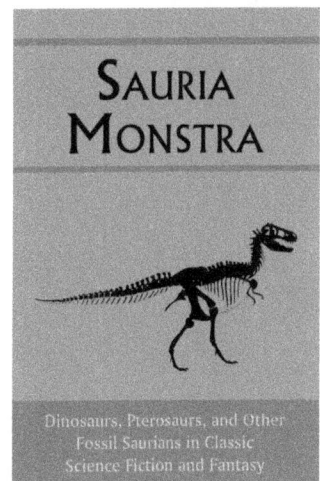

SAURIA MONSTRA
Dinosaurs, Pterosaurs, and Other Fossil Saurians in Classic Science Fiction and Fantasy

www.ingramcontent.com/pod-product-compliance
Lightning Source LLC
Chambersburg PA
CBHW080407270326
41929CB00018B/2924